Hardware and Computer Organization

Hardware and Computer Organization

The Software Perspective

By

Arnold S. Berger

AMSTERDAM • BOSTON • HEIDELBERG • LONDON
NEW YORK • OXFORD • PARIS • SAN DIEGO
SAN FRANCISCO • SINGAPORE • SYDNEY • TOKYO

Newnes is an imprint of Elsevier

ELSEVIER

Newnes

Newnes is an imprint of Elsevier
30 Corporate Drive, Suite 400, Burlington, MA 01803, USA
Linacre House, Jordan Hill, Oxford OX2 8DP, UK

Recognizing the importance of preserving what has been written,
Elsevier prints its books on acid-free paper whenever possible.

Library of Congress Cataloging-in-Publication Data

Berger, Arnold S.
  Hardware and computer organization : a guide for software professionals / by Arnold S. Berger.
    p. cm.
  ISBN 0-7506-7886-0
    1. Computer organization. 2. Computer engineering. 3. Computer interfaces. I. Title.

  QA76.9.C643B47 2005
  004.2'2--dc22

                                                         2005040553

British Library Cataloguing-in-Publication Data
A catalogue record for this book is available from the British Library.

For information on all Newnes publications
visit our Web site at www.books.elsevier.com

04 05 06 07 08 09   10 9 8 7 6 5 4 3 2 1

Printed in the United States of America

For Vivian and Andrea

Contents

Contents

Preface

Thank you for buying my book. I know that may ring hollow if you are a poor student and your instructor made it the required text for your course, but I thank you nevertheless. I hope that you find it informative and easy to read. At least that was one of my goals when I set out to write this book.

This text is an outgrowth of a course that I've been teaching in the Computing and Software Systems Department of the University of Washington-Bothell. The course, CSS 422, *Hardware and Computer Organization,* is one of the required core courses for our undergraduate students. Also, it is the only required architecture course in our curriculum. While our students learn about algorithms and data structures, comparative languages, numeric methods and operating systems, this is their only exposure to "what's under the hood." Since the University of Washington is on the quarter system, I'm faced with the uphill battle to teach as much about the architecture of computers as I can in about 10 weeks.

The material that forms the core of this book was developed over a period of 5 years in the form of about 500 Microsoft PowerPoint® slides. Later, I converted the material in the slides to HTML so that I could also teach the course via a distance learning (DL) format. Since first teaching this course in the fall of 1999, I've taught it 3 or 4 times each academic year. I've also taught it 3 times via DL, with excellent results. In fact, the DL students as a whole have done equally well as the students attending lectures in class. So, if you think that attending class is a highly overrated part of the college experience, then this book is for you.

The text is appropriate for a first course in computer architecture at the sophomore through senior level. It is reasonably self-contained so that it should be able to serve as the only hardware course that CS students need to take in order to understand the implications of the code that they are writing. At the University of Washington-Bothell (UWB), this course is predominantly taught to seniors. As a faculty, we've found that the level of sophistication achieved through learning programming concepts in other classes makes for an easier transition to low-level programming. If the book is to be used with lower division students, then additional time should be allotted for gaining fluency with assembly language programming concepts. For example, in introducing certain assembly language branching and loop constructs, an advanced student will easily grasp the similarity to WHILE, DO-WHILE, FOR and IF-THEN-ELSE constructs. A less sophisticated student may need more concrete examples in order to see the similarities.

Why write a book on Computer Architecture? I'm glad you asked. In the 5+ years that I taught the course, I changed the textbook 4 times. At the end of the quarter, when I held an informal course debriefing with the students, they universally panned every book that I used. The "gold standard" textbooks, the texts that almost every Computer Science student uses in their architecture class, were just not relevant to their needs. For the majority of these students, they were not going to go on and study architecture in graduate schools, or design computers for Intel or AMD. They needed to understand the architecture of a computer and its supporting hardware in order to write efficient and defect-free code to run on the machines. Recently, I did find a text that at least approached the subject matter in the same way that I thought it should be done, but I also found that text lacking in several key areas. On the plus side, switching to the new text eliminated the complaints from my students and it also reinforced my opinion that I wasn't alone in seeing a need for a text with a different perspective. Unfortunately, this text, even though it was a great improvement, still did not cover several areas that I considered to be very important, so I resolved to write one that did, without losing the essence of what I think the new perspective correctly accomplished.

It's not surprising that, given the UWB campus is less than 10 miles from Microsoft's main campus in Redmond, WA, we are strongly influenced by the Microsoft culture. The vast majority of my students have only written software for Windows and the Intel architecture. The designers of this architecture would have you believe that these computers are infinitely fast machines with unlimited resources. How do you counter this view of the world?

Often, my students will cry out in frustration, "Why are you making me learn this (deleted)?" (Actually it is more of a whimper.) This usually happens right around the mid-term examination. Since our campus is also approximately equidistant from where Boeing builds the 737 and 757 aircraft in Renton, WA and the wide body 767 and 777 aircraft in Everett, WA, analogies to the aircraft industry are usually very effective. I simply answer their question this way, "Would you fly on an airplane that was designed by someone who is clueless about what keeps an airplane in the air?" Sometimes it works.

The book is divided into four major topic areas:
1. Introduction to hardware and asynchronous logic.
2. Synchronous logic, state machines and memory organization.
3. Modern computer architectures and assembly language programming.
4. I/O, computer performance, the hierarchy of memory and future directions of computer organization.

There is no sharp line of demarcation between the subject areas, and the subject matter builds upon the knowledge base from prior chapters. However, I've tried to limit the interdependencies so later chapters may be skipped, depending upon the available time and desired syllabus.

Each chapter ends with some exercises. The solutions to the odd-numbered problems are located in Appendix A, and the solutions to the even-numbered problems are available through the instructor's resource website at http://www.elsevier.com/0750678860.

The text approach that we'll take is to describe the hardware from the ground up. Just as a Geneticist can describe the most complex of organic beings in terms of a DNA molecule that contains only four nucleotides, adenine, cytosine, guanine, and thymine, abbreviated, A, C, G and T, we can describe the most complex computer or memory system in terms of four logical building blocks,

AND, OR, NOT and TRI-STATE. Strictly speaking, TRI-STATE isn't a logical building block like AND, it is more like the "glue" that enables us to interconnect the elements of a computer in such a way that the complexity doesn't overwhelm us. Also, I really like the DNA analogy, so we'll need to have 4 electronic building blocks to keep up with the A, C, G and T idea.

I once gave a talk to a group of middle school teachers who were trying to earn some in-service credits during their summer break. I was a volunteer with the Air Academy School District in Colorado Springs, Colorado while I worked for the Logic Systems Division of Hewlett-Packard. None of the teachers were computer literate and I had two hours to give them some appreciation of the technology. I decided to start with Aristotle and concept of the logical operators as a branch of philosophy and then proceeded with the DNA analogy up through the concept of registers. I seemed to be getting through to them, but they may have been stroking my ego so I would sign-off on their attendance sheets. Anyway, I think there is value in demonstrating that even the most complex computer functionality can be described in terms of the logic primitives that we study in the first part of the text.

We will take the DNA or building-block approach through most of the first half of the text. We will start with the simplest of gates and build compound gates. From these compound gates we'll progress to the posing and solution of asynchronous logical equations. We'll learn the methods of truth table design and simplification using Boolean algebra and then Karnaugh Map (K-map) methodology. The exercises and examples will stress the statement of the problem as a set of specifications which are then translated into a truth table, and from there to K-maps and finally to the gate design. At this point the student is encouraged to actually "build" the circuit in simulation using the Digital Works® software simulator (see the following) included on the DVD-ROM that accompanies the text. I have found this combination of the abstract design and the actual simulation to be an extremely powerful teaching and learning combination.

One of the benefits of taking this approach is that the students become accustomed to dealing with variables at the bit level. While most students are familiar with the C/C++ Boolean constructs, the concept of a single wire carrying the state of a variable seems to be quite new.

Once the idea of creating arbitrarily complex, asynchronous algebraic functions is under control, we add the dimension of the clock and of synchronous logic. Synchronous logic takes us to flip-flops, counters, shifters, registers and state machines. We actually spend a lot of effort in this area, and the concepts are reintroduced several times as we look at micro-code and instruction decomposition later on.

The middle part of the book focuses on the architecture of a computer system. In particular, the memory to CPU interface will get a great deal of attention. We'll design simple memory systems and decoding circuits using our knowledge gained in the preceding chapters. We'll also take a brief look at memory timing in order to better understand some of the more global issues of system design.

We'll then make the transition to looking at the architecture of the 68K, ARM and X86 processor families. This will be our introduction to assembly language programming.

Each of the processor architectures will be handled separately so that it may be skipped without creating too much discontinuity in the text.

This text does emphasize assembly language programming in the three architectures. The reason for this is twofold: First, assembly language may or may not be taught as a part of a CS student's curriculum, and it may be their only exposure to programming at the machine level. Even though you as a CS student may never have to write an assembly language program, there's a high probability that you'll have to debug some parts of your C++ program at the assembly language level, so this is as good a time to learn it as any. Also, by looking at three very different instruction sets we will actually reinforce the concept that once you understand the architecture of a processor, you can program it in assembly language. This leads to the second reason to study assembly language. Assembly language is a metaphor for studying computer architecture from the software developer's point of view.

I'm a big fan of "Dr. Science." He's often on National Public Radio and does tours of college campuses. His famous tag line is, "I have a Master's degree...in science." Anyway, I was at a Dr. Science lecture when he said, "I like to scan columns of random numbers, looking for patterns." I remembered that line and I often use it in my lectures to describe how you can begin to see the architecture of the computer emerging through the seeming randomness of the machine language instruction set. I could just see in my mind's eye a bunch of Motorola CPU architects and engineers sitting around a table in a restaurant, pizza trays scattered hither and yon, trying to figure out the correct bit patterns for the last few instructions so that they don't have a bloated and inefficient microcode ROM table. If you are a student reading this and it doesn't make any sense to you now, don't worry...yet.

The last parts of the text steps back and looks at general issues of computer architecture. We'll look at CISC versus RISC, modern techniques, such as pipelines and caches, virtual memory and memory management. However, the overriding theme will be computer performance. We will keep returning to the issues associated with the software-to-hardware interface and the implications of coding methods on the hardware and of the hardware on the coding methods.

One unique aspect of the text is the material included on the accompanying DVD-ROM. I've included the following programs to use with the material in the text:

- Digital Works (freeware): A hardware design and simulation tool
- Easy68K: A freeware assembler/simulator/debugger package for the Motorola (Now Freescale) 68,000 architecture.
- X86emul: A shareware assembler/simulator/debugger package for the X86 architecture.
- GNU ARM Tools: The ARM developers toolset with Instruction Set Simulator from the Free Software Foundation.

The ARM company has an excellent tool suite that you can obtain directly from ARM. It comes with a free 45-day evaluation license. This should be long enough to use in your course. Unfortunately, I was unable to negotiate a license agreement with ARM that would enable me to include the ARM tools on the DVD-ROM that accompanies this text. This tool suite is excellent and easy to use. If you want to spend some additional time examining the world's most popular RISC architecture, then contact ARM directly and ask them nicely for a copy of the ARM tools suite. Tell them Arnie sent you.

I have also used the Easy68K assembler/simulator extensively in my CSS 422 class. It works well and has extensive debugging capabilities associated with it. Also, since it is freeware, the logistical problems of licenses and evaluation periods need not be dealt with. However, we will be making some references to the other tools in the text, so it is probably a good idea to install them just before you intend to use them, rather than at the beginning of your class.

Also included on the DVD-ROM are 11 short lectures on various topics of interest in this text by experts in the field of computer architecture. These videos were made under a grant by the Worthington Technology Endowment for 2004 at UWB. Each lecture is an informal 15 to 30 minute "chalk talk." I hope you take the time to view them and integrate them into your learning experience for this subject matter.

Even though the editors, my students and I have read through this material several times, Murphy's Law predicts that there is still a high probability of errors in the text. After all, it's software. So if you come across any "bugs" in the text, please let me know about it. Send your corrections to aberger@u.washington.edu. I'll see to it that the corrections will gets posted on my website at UW (http://faculty.uwb.edu/aberger).

The last point I wanted to make is that textbooks can just go so far. Whether you are a student or an instructor reading this, please try to seek out experts and original sources if you can. Professor James Patterson, writing in the July, 2004 issue of *Physics Today,* writes,

> *When we want to know something, there is a tendency to seek a quick answer in a textbook. This often works, but we need to get in the habit of looking at original papers. Textbooks are often abbreviated second-or third-hand distortions of the facts...*[1]

Let's get started.

Arnold S. Berger
Sammamish, Washington

---

[1] James D. Patterson, *An Open Letter to the Next Generation, Physics Today,* Vol. 57, No. 7, July, 2004, p. 56.

Acknowledgments

First, and foremost, I would like to acknowledge the sponsorship and support of Professor William Erdly, former Director of the Department of Computing and Software Systems at the University of Washington-Bothell. Professor Erdly first hired me as an Adjunct Professor in the fall of 1999 and asked me to teach a course called Hardware and Computer Organization, even though the course I wanted to teach was on Embedded System Design.

Professor Erdly then provided me with financial support to convert my Hardware and Computer Organization course from a series of lectures on PowerPoint slides to a series of online lessons in HTML. These lessons became the core of the material that led to this book.

Professors Charles Jackels and Frank Cioch, in their capacities as Acting Directors, both supported my work in perfecting the on-line material and bringing multimedia into the distance learning experience. Their support helped me to see the educational value of technology in the classroom.

I would also like to thank the Richard P. and Lois M. Worthington Foundation for their 2004 Technology Grant Award which enabled me to travel around the United States and videotape short talks in Computer Architecture. Also, I would like to thank the 11 speakers who gave of their time to participate in this project.

Carol Lewis, my Acquisitions Editor at Newnes Books saw the value of my approach and I thank her for it. I'm sorry that she had to leave Newnes before she could see this book finally completed. Tiffany Gasbarrini of Elsevier Press and Kelly Johnson of Borrego Publishing brought the book to life. Thank you both.

In large measure, this book was designed by the students who have taken my CSS 422 course. Their end-of-Quarter evaluations and feedback were invaluable in helping me to see how this book should be structured.

Finally, and most important, I want to thank my wife Vivian for her support and understanding. I could not have written 500+ pages without it.

What's on the DVD-ROM?

One unique aspect of the text is the material included on the accompanying DVD-ROM. I've included the following programs to use with the material in the text:

- Digital Works (freeware): A hardware design and simulation tool.
- Easy68K: A freeware assembler/simulator/debugger package for the Motorola (now Freescale) 68,000 architecture.
- X86emul: A shareware assembler/simulator/debugger package for the X86 architecture.
- GNU ARM Tools: The ARM developers toolset with Instruction Set Simulator from the Free Software Foundation.
- Eleven industry expert video lectures on significant hardware design and development topics.

CHAPTER 1

Introduction and Overview of Hardware Architecture

Learning Objectives

When you've finished this lesson, you will be able to:
▶ *Describe the evolution of computing devices and the way most computer-based devices are organized;*
▶ *Make simple conversions between the binary, octal and hexadecimal number systems, and explain the importance of these systems to computing devices;*
▶ *Demonstrate the way that the atomic elements of computer hardware and logic gates are used, and detail the rules that govern their operation.*

Introduction

Today, we often take for granted the impressive array of computing machinery that surrounds us and helps us manage our daily lives. Because you are studying computer architecture and digital hardware, you no doubt have a good understanding of these machines, and you've probably written countless programs on your PCs and workstations. However, it is very easy to become jaded and forget the evolution of the technology that has led us to the point where every Nintendo Game Boy® has 100 times the computing power of the computer systems on the first Mercury space missions.

A Brief History of Computing

Computing machines have been around for a long time, hundreds of years. The Chinese abacus, the calculators with gears and wheels and the first analog computers are all examples of computing machinery; in some cases quite complex, that predates the introduction of digital computing systems. The computing machines that we're interested in came about in the 1940s because World War II artillery needed a more accurate way to calculate the trajectories of the shells fired from battleships.

Today, the primary reason that computers have become so pervasive is the advances made in integrated circuit manufacturing technology. What was once primarily orange groves in California, north of San Jose and south of Palo Alto, is today the region known as Silicon Valley. Silicon Valley is the home to many of the companies that are the locomotives of this technology. Intel, AMD, Cypress, Cirrus Logic and so on are household names (if you live in a geek-speak household) anywhere in the world.

About 30 years ago, Gordon Moore, one of the founders of Intel, observed that the density of transistors being placed on individual silicon chips was doubling about every eighteen months. This observation has been remarkably accurate since Moore first stated it, and it has since become known as Moore's Law. Moore's Law has been remarkably accurate since Gordon Moore first articulated it. Memory capacity, more then anything else, has been an excellent example of the accuracy of Moore's Law. Figure 1.1 contains a semi-logarithmic graph of memory capacity versus time. Many circuit designers and device physicists are arguing about the continued viability of Moore's Law. Transistors cannot continue to shrink indefinitely, nor can manufacturers easily afford the cost of the manufacturing equipment required to produce silicon wafers of such minute dimensions. At some point, the laws of quantum physics will begin to alter the behavior of these tiny transistors in very profound ways.

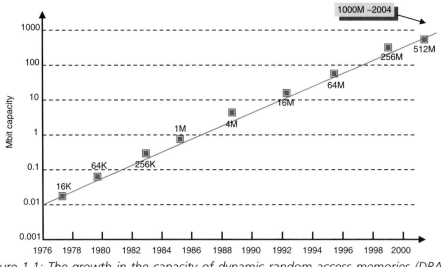

Figure 1.1: The growth in the capacity of dynamic random access memories (DRAM) with time. Note the semi-logarithmic behavior, characteristic of Moore's Law.

Today, we are capable of placing hundreds of millions of transistors (the "active" switching device from which we create logic gates) onto a single perfect piece of silicon, perhaps 2 cm on a side. From the point of view of designing computer chips, the big breakthrough came about when Mead and Conway[1] described a method of creating hardware designs by writing software. Called *silicon compilation*, it has led to the creation of *hardware description language*, or HDL. Hardware description languages such as Verilog® and VHDL® enable hardware designers to write a program that looks remarkably like the C programming language, and then to compile that program to a recipe that a semiconductor manufacturer can use to build a chip.

Back to the beginning; the first generation of computing engines was comprised of the mechanical devices. The abacus, the adding machine, the punch card reader for textile machines fit into this category. The next generation spanned the period from 1940–1960. Here electronic devices— vacuum tubes—were used as the active device or switching element. Even a miniature vacuum tube is millions of times larger then the transistor on a silicon wafer. It consumes millions of times the power of the transistor, and its useful lifetime is hundreds or thousands of times less then a transistor. Although the vacuum tube computers were much faster then the mechanical computers of the preceding generation, they are thousands of times slower then the computers of today. If you are a fan of the grade B science fiction movies of the 1950's, these computers were the ones that filled the room with lights flashing and meters bouncing.

The third generation covered roughly the period of time from 1960 to 1968. Here the transistor replaced the vacuum tube, and suddenly the computers began to be able to do real work. Companies such as IBM®, Burroughs® and Univac® built large mainframe computers. The IBM 360 family is a representative example of the mainframe computer of the day. Also at this time, Xerox® was carrying out some pioneering work on the human/computer interface at their Palo Alto Research Center, Xerox PARC. Here they studied what later would become computer networks, Windows® operating system and the ubiquitous mouse. Programmers stopped programming in machine language and assembly language and began to use FORTRAN, COBOL and BASIC.

The fourth generation, roughly 1969–1977 was the age of the minicomputer. The minicomputer was the computer of the masses. It wasn't quite the PC, but it moved the computer out of the sterile environment of the "computer room," protected by technicians in white coats, to a computer in your lab. The minicomputer also represented the replacement of individual electronic parts, such as transistors and resistors, mounted on printed circuit boards (called discrete devices), with integrated circuits, or collections of logic functions in a single package. Here was the introduction of the small and medium scale integrated circuits. Companies such as Digital Equipment Company (DEC), Data General and Hewlett-Packard all built this generation of minicomputer.[2] Also within this timeframe, simple integrated-circuit microprocessors were introduced and commercially produced by companies like Intel, Texas Instruments, Motorola, MOS Technology and Zilog. Early microcomputer devices that best represent this generation are the 4004, 8008 and 8080 from Intel, the 9900 from Texas Instruments and the 6800 from Motorola. The computer languages of the fourth generation were: assembly, C, Pascal, Modula, Smalltalk and Microsoft BASIC.

We are currently in the fifth generation, although it could be argued that the fifth generation ended with the Intel® 80486 microprocessor and the introduction of the Pentium® represents the sixth generation. We'll ignore that distinction until it is more widely accepted. The advances made in semiconductor manufacturing technology best characterize the fifth generation of computers. Today's semiconductor processes typify what is referred to as Very Large Scale Integration, or VLSI technology. The next step, Ultra Large Scale Integration, or ULSI is either here today or right around the corner. Dr. Daniel Mann[3], an AMD Fellow, recently told me that a modern AMD Athlon XP processor contains approximately 60 million transistors.

The fifth generation also saw the growth of the personal computer and the operating system as the primary focus of the machine. Standard hardware platforms controlled by standard operating systems enabled thousands of developers to create programs for these systems. In terms of software, the dominant languages became ADA, C++, JAVA, HTML and XML. In addition, graphical design language, based upon the universal modeling language (UML), began to appear.

Two Views of Today's Computer

The modern computer has become faster and more powerful but the basic architecture of a computing machine has essentially stayed the same for many years. Today we can take two equivalent views of the machine: The hardware view and the software view. The hardware view, not surprisingly, focuses on the machine and does allow for the fact that the software has something to do with its reason to exist. From 50,000 feet, our computer looks like Figure 1.2.

In this course, we'll focus primarily on the CPU and memory systems, with some consideration of the software that drives this hardware. We'll touch briefly on I/O, since a computer isn't much good without it.

The software developer's view is roughly equivalent, but the perspective does change somewhat. Figure 1.3 shows the computer from the software designer's point of view. Note that the view of the system shown in this figure is somewhat problematic because it isn't always clear that the user interface communicates directly with the application program. In many cases, the user interface first communicates with the operating system. However, let's look at this diagram somewhat loosely and consider it to be the flow of information, rather than the flow of control.

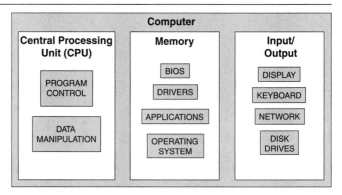

Figure 1.2: Abstract view of a computer. The three main elements are the control and data processor, the input and output, or I/O devices, and the program that is executing on the machine.

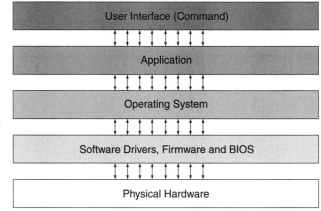

Figure 1.3: Representing the computer in terms of the abstraction levels of the system.

Abstraction Levels

One of the more modern concepts of computer design is the idea of abstraction levels. Each level provides an abstraction of the level below it. At the lowest level is the hardware. In order to control the hardware it is necessary to create small programs, called *drivers*, which actually manipulate the individual control bits of the hardware.

Sitting above the drivers is the operating system and other system programs. The operating system, or *OS,* communicates with the drivers through a standard set of *application programming interfaces*, or *APIs*. The APIs provide a structure by which the next level of up in the abstraction level can communicate with the layer below it. Thus, in order to read a character from the keyboard of your computer, there is a low-level driver program that becomes active when a key is struck. The operating system communicates with this driver through its API.

At the next level, the application software communicates with the OS through system API's that once again, abstract the lower levels so that the individual differences in behavior of the hardware and the drivers may be ignored. However, we need to be a bit careful about taking the viewpoint of Figure 1.3 too literally. We could also argue that the Application and Operating System layers should be reversed because the user interacts with the application through the Operating System layer as well. Thus, mouse and keyboard inputs are really passed to the application through the Operating System and do not go directly from the User to the Application. Anyway, you get the picture.

The computer hardware, as represented by a desktop PC, can be thought of as being comprised of four basic parts:

1. Input devices can include components such as the mouse, keyboard, microphone, disks, modem and the network.
2. Output devices are components such as the display, disk, modem, sound card and speakers, and the network.
3. The memory system is comprised of internal and external caches, main memory, video memory and disk.
4. The central processing unit, or CPU, is comprised of the arithmetic and logic unit (ALU), control system and busses.

Busses

The busses are the nervous system of the computer. They connect the various functional block of the computer both internally and externally. Within a computer, a bus is a grouping of similar signals. Thus, your Pentium processor has a 32-bit address bus and a 32-bit data bus. In terms of the bus structure, this means that there are two bundles of 32 wires with each bundle containing 32 individual signals, but with a common function. We'll discuss busses much more thoroughly in a later lesson.

The typical computer has three busses: One for memory addresses, one for data and one for status (housekeeping and control). There are also industry standard busses such as PCI, ISA, AGP, PC-105, VXI and so forth. Since the signal definitions and timing requirements for these industry-standard busses are carefully controlled by the standards association that maintains them, hardware devices from different manufacturers can generally be expected to work properly and interchangeably. Some busses are quite simple—only one wire—but the signals sent down that wire are quite complex and require special hardware and standard protocols to understand it. Examples of these types of busses are the universal serial bus (USB), the small computer system interface bus (SCSI), Ethernet and Firewire.

Memory

From the point of view of a software developer, the memory system is the most visible part of the computer. If we didn't have memory, we'd never have a problem with an errant pointer. But that's another story. The computer memory is the place where program code (instructions) and variables (data) are stored. We can make a simple analogy about instructions and data. Consider a recipe to bake a cake. The recipe itself is the collection of instructions that tell us how to create the cake. The data represents the ingredients we need that the instructions manipulate. It doesn't make much sense to sift the flour if you don't have flour to sift.

We may also describe memory as a hierarchy, based upon speed. In this case, speed means how fast the data can be retrieved from the memory when the computer requests it. The fastest memory is also the most expensive so as the memory access times become slower, the cost per bit decreases, so we can have more of it. The fastest memory is also the memory that's closest to the CPU. Thus, our CPU might have a small number of on-chip data registers, or storage locations, several thousand locations of off-chip cache memory, several million locations of main memory and several billion locations of disk storage. The ratio of the access time of the fastest on-chip memory to the slowest memory, the hard disk, is about 10,000 to one. The ratio of the cost of the two memories is somewhat more difficult to calculate because the fastest semiconductor memory is the on-chip cache memory, and you cannot buy that separately from the microprocessor itself. However, if we estimate the ratio of the cost per gigabyte of the main memory in your PC to the cost per gigabyte of hard disk storage (and taking into account the mail-in rebates) then we find that the faster semiconductor storage with an average access time of 20–40 nanoseconds is 300 times more costly then hard disk storage, with an average access time of 1 millisecond.

Today, because of the economy of scale provided by the PC industry, memory is incredibly inexpensive. A standard memory module (SIMM) with a capacity of 512 million storage locations costs about $60. PC memory is dominated by a memory technology called *dynamic random access memory*, or *DRAM*. There are several variations of DRAM, and we'll cover them in greater depth later on. DRAM is characterized by the fact that it must be constantly accessed or it will lose its stored data. This forces us to create highly specialized and complex support hardware to interface the memory systems to the CPU. These devices are contained in support chipsets that have become as important to the modern PC as the CPU. Why use these complex memories? DRAM's are inherently very dense and can hold upwards of 512 million bits of information. In order to achieve these densities, the complexity of accessing and controlling them was moved to the chipset.

Static RAM (SRAM)

The memory that we'll focus on is called *static random access memory* or *SRAM*. Each memory cell of an SRAM device is more complicated then the DRAM, but the overall operation of the device is easier to understand, so we'll focus on this type of memory in our discussions. The term, static random access memory, or SRAM, refers to the fact that:

1. we may read from the chip or write data to it, and
2. any memory cell in the chip may be accessed at any time, once the appropriate address of the cell is presented to the chip.

3. As long as power is applied to the memory, we only a required to provide an address to the SRAM cell, together with a READ or a WRITE signal, in order to access, or modify, the data in the cell. This is quite a bit different then the effort required to maintain data integrity in DRAM cells. We'll discuss this point in greater detail in a later chapter.

With a RAM memory there is no need to search through all the preceding memory cells in order to get to the one you want to read. In other words, data can be read from the last cell of the RAM as quickly as it could be read from the first cell. In contrast, a tape backup device must stream through the entire tape before the last piece of data can be retrieved. That's why the term "random access" is used. Also note that when are discussing SRAM or DRAM in the general sense, as listed in items 1 and 2, we'll just use the term RAM, without the SRAM or DRAM distinction.

Memory Hierarchy

Figure 1.4 shows the memory hierarchy in real terms, where we see can see how the various memory elements grow exponentially in size and decrease exponentially in access time. Closest memory to the CPU is the smallest, typically in the range of 1 Kbyte of data (1,000 8-bit characters) to 1 Mbyte of data (1,000,000 characters). These on-chip cache memories can have access times as short as ½ to 1 nanosecond or one-billionth of a second. As a benchmark, light can travel about one foot through the air in one nanosecond. We call this cache the *Level 1*, or *L1* cache.

Below the L1 cache is the *Level 2*, or *L2* cache. In today's Pentium-class processors, the L2 cache is usually on the processor chip itself. In fact, if you could lift the lid of a Pentium or Athlon processor and look at the silicon die itself under a microscope you might be surprised to find that the biggest percentage of chip area was taken up by the cache memories.

The secondary cache sits your computer's main memory. These are the "memory sticks" that you can purchase in a computer shop to increase your computer's performance. This memory is considerably slower than on-chip cache memory, but you usually have a much larger main memory then the on-chip memory. You might wonder why adding more memory will give you better performance. To see this, consider the next level down from main memory, the hard disk drive. Your hard drive has lots of capacity, but it comes at a price, speed. The hard disk is an electro-mechanical device. The data is stored on a rotating platter and, even at a rotational speed of 7200 revolutions per minute, it takes precious time for the correct data to come under the disk read heads. Also, the data is organized as individual tracks on each side of multiple platters. The heads have to move from track-to-track in order to access the correct data. This takes time as well.

Now, your operating system, whether its MAC O/S, Linux or one of the flavors of Windows, uses the hard

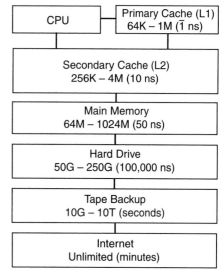

Figure 1.4: The memory hierarchy. Notice the inverse relationship between the size of the memory and the access time.

disk as a handy place to swap out programs or data that can't fit in main memory at a particular point in time. So, if you have several windows open on your computer, and only 64 megabytes of main memory, you might see the hourglass form of the cursor appearing quite often because the operating system is constantly swapping the different applications in and out of main memory. From Figure 1.4, we see that the ratio of access times between the hard disk and main memory can be 10,000 to one, so any time that we go to the hard disk, we will have to wait. The moral here is that the best performance boost you can give to your computer is to add as much memory as it can hold.

Finally, there are several symbols used in Figure 1.4 that might be strange to you. We'll discuss them in detail in due course, but in case you were wondering what it meant, here's a little preview:

Symbol	Name	Meaning
ns	nanosecond	billionth of a second
K	kilobytes	2^{10} or 1024 8-bit characters (bytes)
M	megabytes	2^{20} or 1,048,576 bytes
G	gigabytes	2^{30} or 1,073,741,824 bytes
T	terabytes	2^{40} or 1,099,511,627,776 bytes

Hopefully, these numbers will soon become quite familiar to you because they are the dimensions of modern computer technology. However, one note of caution: The terms *kilo, mega, giga and tera* are often overloaded. Sometimes they are used in the strictly scientific sense of a shorthand notation for a multiplication factor of 10^3, 10^6, 10^9 and 10^{12}, respectively. So, how do you know the computer-speak versions, 2^{10}, 2^{20}, 2^{30}, 2^{40} are being used, or the traditional science and engineering meaning is being used? That's a good question. Sometimes it isn't so obvious and mistakes could be made. For example, whenever we are dealing with memory size and memory issues, we are almost always using the base 2 sense of the terms. However, this isn't always the case. Hard disk drives, even though they are storage devices, use the terms in the engineering sense. Therefore, an old 1 gigabyte hard drive does not hold the same amount of data as 1 gigabyte of memory because the "giga" term is being used in two different senses. In any case, you are generally safe to assume the base 2 versions of the term in this text, unless I specifically state otherwise. In the real world, *caveat emptor.*

Hard Disk Drive

Let's look a bit further into the dynamics of the hard disk drive. Disk drives are a marvel of engineering technology. For years, electronic industry analysts were predicting the demise of the hard drive. Economical semiconductor memories that were as cost effective as a hard disk were always "a few years away." However, the disk drive manufacturers ignored the pundits and just continued to increase the capacity and performance, improve the reliability and reduce the cost of the disk drives. Today, an average disk drive costs about 60 cents per gigabyte of storage capacity.

Consider a modern, high-performance disk drive. Specifically, let's look at the Model ST3146807LC from Seagate Technology®[4]. Following are the relevant specifications for this drive:

- Rotational Speed: 10,000 rpm
- Interface: Ultra320 SCSI
- Discs/Heads: 4/8
- Formatted Capacity (512 bytes/sector) Gbytes: 146.8
- Cylinders: 49,855
- Sectors per drive: 286,749,488
- External transfer rate: 320 Mbytes per second
- Track-to-track Seek Read/Write (msec): 0.35/0.55
- Average Seek Read/Write (msec): 4.7/5.3
- Average Latency (msec): 2.99

What does all this mean? Consider Figure 1.5. Here we see a highly simplified schematic diagram of the Seagate *Cheetah*® disk drive discussed above. The hard disk is comprised of 4 aluminum platters. Each side of the platter is coated with magnetic material that records the stored data. Above each platter is a tiny magnetic pick-up, or *head*, that floats above the disk platter (disc) on a cushion of air. In a very real sense, the head is flying over the surface of the platter, a distance much less than the thickness of a human hair. Thus, when you get a disk crash, the results are quite analogous to when an airplane crashes. In either case, the airplane or the read/write head loses lift and hits the ground or the surface of the disk. When that happens, the magnetic material is scrapped away and the disk is no longer usable.

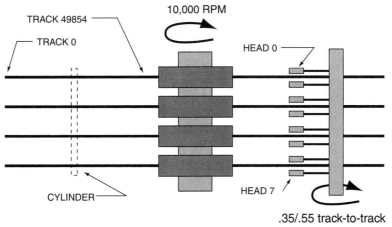

Figure 1.5: Schematic diagram of the Seagate Cheetah® hard disk drive.

Each surface contains 49855 concentric tracks. Each track is discreet. It is not connected to the adjacent track, as a spiral. Thus, in order to move from one track to an adjacent track, the head must physically move a slight amount. All heads are connected to a common shaft that can quickly and accurately rotate the heads to a new position. Since the tracks on all the surfaces are in vertical alignment, we call this a *cylinder*. Thus, cylinder 0 contains 8 tracks from the 4 platters.

Now, how do we get a 146.8 Gbyte drive? First, each platter contains two surfaces and we have four platters, so that the total capacity of the drive will be 8 times the capacity of one surface. Each sector holds 512 bytes of data. By dividing the total number of sectors by 8, we see that there are 35,843,686 sectors per surface. Dividing again by 49855, we see that there are approximately 719 sectors per track. It is interesting that the actual number is 718.96 sectors per track. Why isn't this value a whole number? In other words, how can we have a fractional number of sectors per track?

There are a number of possibilities and we need not dwell on it. However, one possibility is that the number of sectors per track is not uniform across the disk surface because the sector spacing changes as we move from the center of the disk to the outside. In a CD-ROM or DVD drive this is corrected by changing the rotational speed of the drive as the laser moves in and out. But, since the hard disk rotates at a fixed rate, changing the recording density as we move from inner to outer tracks makes the most sense.

Anyway, back to our calculation. If each track holds 719 sectors and each sector is 512 bytes, then each track holds 368,128 bytes of data. Since there are 8 tracks per cylinder, each cylinder holds 2,945,024 bytes. Now, here's where the big numbers come in. Since we have 49855 cylinders, our total capacity is 146,824,171,520 bytes, or 146.8 Gbytes.

Before we leave the subject of disk drives, let's consider one more issue. Considering the hard drive specifications, above, we see that access times are measured in units of milliseconds (msec), or thousandths of a second. Thus, if your data sectors are spread out over the disk, then accessing each block of 512 bytes can easily take seconds of time. Comparing this to the time required to access data stored in main memory, it is easy to see why the hard drive is 10,000 times slower than main memory.

Complex Instruction Set Architecture, and RISC or Reduced Instruction Set Computer

Today we have two dominant computer architectures, complex instruction set architecture or *CISC* and reduced instruction set computer, or *RISC*. CISC is typified by what is referred to as the von Neumann Architecture, invented by John von Neumann of Princeton University. In the von Neumann architecture, both the instruction memory and the data memory share the same physical memory space. This can lead to a condition called the *von Neumann bottleneck*, where the same external address and data busses must serve double duty, transferring instructions from memory to the processor for program execution, and moving data to and from memory for the storage and retrieval of program variables.

When the processor is moving data, it can't fetch the next instruction, and vice versa. The solution to this dilemma, as we shall see later, is the introduction of separate on-chip instruction and data caches. Motorola's 68000 processor and its successors, and Intel's 8086 processor and its successors are all characteristic of CISC processors. We'll take a more in-depth look at the differences between CISC and RISC processors again later when we study pipelining.

Note: You will often see the processor represented as 80X86, or 680X0. The "X" is used as a placeholder to represent a member of a family of devices. Thus, the 80X86 (often written as X86) represents the 8086, 80186, 80286, 80386, 80486 and 80586 (the designation of the first Pentium processor). The Motorola processors are the 68000, 68010, 68020, 68030, 68040 and 68060.

Howard Aiken of Harvard University designed an alternate computer architecture that we commonly associate today with the Reduced Instruction Set Computer, or RISC architecture. The classic Harvard architecture computer has two entirely separate memory spaces, one for instructions and one for data. A processor with these memories could operate much more efficiently because data and instructions could be fetched from the computer's memory when needed, not just when the busses were available. The first popular microprocessor that used the Harvard Architecture was the Am29000 from AMD. This device achieved reasonable popularity in the early Hewlett-Packard laser printers, but later fell from favor because designing a computer with two separate memories was just not cost effective. The obvious performance gain from the two memory spaces was the driving force to inspire the CPU designers to move the memory spaces onto the chips, in the form of the instruction and data caches that are common in today's high-performance microprocessors.

The gains in CPU power are quite evident if we look at advances in workstation performance over the past few years. It is interesting that just four years later, a high-end PC, such as a 2.4 GHz Pentium 4 from Intel could easily outperform the DEC Alpha 21264/600 workstation.

Figure 1.6[5] is a graph of workstation performance over time.

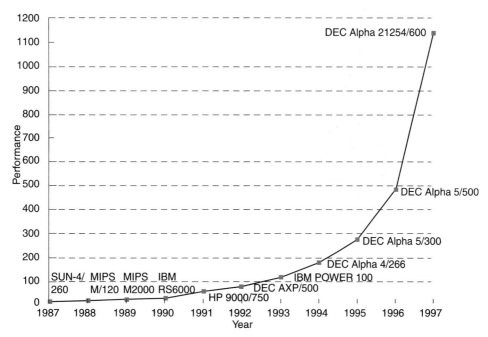

Figure 1.6: Improvement in workstation performance over time. (From Patterson and Hennessy.)

The relative performance of these computers was measured using a standard set of programs called *benchmarks*. In this case, the industry standard SPECbase_int92 benchmark was used. It is difficult to compare performance based on these numbers with more modern performance measurements because the benchmark changed along the way. Today, the benchmark of choice is the SPECint95, which does not directly correlate with the earlier SPECint92 benchmark[6]. However, according to Mann, a conversion factor of about 38 allows for a rough comparison. Thus, according to the published results[7], a 1.0 GHz AMD Athlon processor achieved a SPECint95 benchmark result of 42.9, which roughly compares to a SPECint92 result of 1630. The Digital Equipment Corporation (DEC) AlphaStation 5/300 is one workstation that has published results for both benchmark tests. It measures about 280 in the graph of Figure 1.6 and 7.33 according to the SPECint95 benchmark. Multiplying by 38, we get 278.5, which is in reasonable agreement with the earlier result. We'll return to the issue of performance measurements in a later chapter.

Number Systems

How do you represent a number in computer? How do you send that number, whatever it may be, a *char*, an *int*, a *float* or perhaps a *double* between the processor and memory, or within the microprocessor itself? This is a fair question to ask and the answer leads us naturally to an understanding of why modern digital computers are based on the binary (base 2) number system. In order to investigate this, consider Figure 1.7.

In Figure 1.7 we'll do a simple-minded experiment. Let's pretend that we can place an electrical voltage on the wire that represents the number we would like to transmit between two functional elements of the computer. The method might work for simple numbers, but I wouldn't want to touch the wire if I was sending 2000.456! In fact, this method would be extremely slow, expensive and would only work for a narrow range of values.

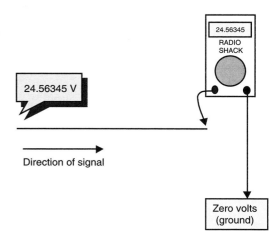

However, that doesn't imply that this method isn't used at all. In fact, one of the first families of electronic computers was the *analog computer*. The analog computer is based upon *linear amplifiers*, or the kind of electronic circuitry that you might find in your stereo receiver at home. The key point is that variables (in this case the voltages on wires)

Figure 1.7: Representing the value of a number by the voltage on a wire.

can assume an infinite range of values between some limits imposed by the nature of the circuitry. In many of the early analog computers this range might be between –25 volts and +25 volts. Thus, any quantity that could be represented as a steady, or time varying voltage within this range could be used as a variable within an analog computer.

The analog computer takes advantage of the fact that there are electronic circuits that can do the following mathematical operations:

- Add / subtract
- Log / anti-log
- Multiply / divide
- Differentiate / integrate

By combining this circuits one after another with intermediate amplification and scaling, real-time systems could be easily modeled and the solution to complex linear differential equations could be obtained as the system was operating.

However, the analog computer suffers from the same limitations as does your stereo system. That is, its amplification accuracy is not infinitely perfect, so the best accuracy that could be hoped for is about 0.01%, or about 1 part in 10,000. Figure 1.8 shows an analog computer of the type used by the United States submarines during World War II. The Torpedo Data Computer, or TDC, would take as its inputs the compass heading and speed of the target ship, the heading and speed of the submarine, the desired firing distance. The correct speed and heading was then sent to the torpedoes and they would track the course, speed and depth transmitted to them by the TDC.

Figure 1.8: An analog computer from a WWII submarine. Photo courtesy of www. fleetsubmarine.com.

Thus, within the limitations imposed by the electronic circuitry of the 1940's, an entire family of computers based upon the idea of inputs and outputs based upon continuous variables. In that sense, your stereo amplifier is an analog computer. An amplifier *amplifies*, or boosts an electrical signal. An amplifier with a *gain* of 10, has an output voltage that is, at every instant of time, 10 times greater than the input voltage. Thus, $V_{out} = 10$ V_{in}. Here we have an analog computing block that happens to be a multiplication block with a constant multiplier.

Anyway, let's get back to discussing to number systems. We might be able to improve on this method by breaking the number into more manageable parts and send a more limited signal range over several wires at the same time (in parallel). Thus, each wire would only need to transmit a narrow range of values. Figure 1.9 shows how this might work.

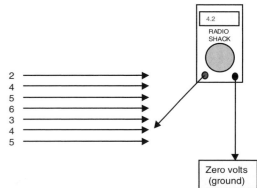

Figure 1.9: Using a parallel bundle of wires to transmit a numeric value in a computer. The wire's position in the bundle determines its digital weight. Each wire carries between 0 volts and 9 volts.

In this case, each wire in the bundle represents a decimal decade and each number that we send would be represented by the corresponding voltages on the wires. Thus, instead of needing to transmit potentially lethal voltages, such as 12,567 volts, the voltage in each wire would never become greater than that of a 9-volt battery. Let's stop for a moment because this approach looks promising. How accurate would the voltage on the wire have to be in so that the circuitry interprets the number as 4, and not 3 or 5? In Figure 1.7, our voltmeter shows that the second wire from the bottom measures 4.2 V, not 4 volts. Is that good enough? Should it really be 4.000 ± .0005 volts? In all probability, this system might work just fine if each voltage increment has a "slop" of about 0.3 volts. So we would only need to send 4 volts ± 0.3 volts (3.7–4.3 volts) in order to guarantee that the circuitry received the correct number. What if the circuit erred and sent 4.5 volts instead? This is too large to be a 4 but too small to be a 5. The answer is that we don't know what will happen. The value represented by 4.5 volts is *undefined*. Hopefully, our computer works properly and this is not a problem.

The method proposed in Figure 1.9 is actually very close to reality, but it isn't quite what we need. With the speed of modern computers, it is still far too difficult to design circuitry that is both fast enough and accurate enough to switch the voltage on a wire between 10 different values. However, this idea is being looked at for the next generation of computer memory cells. More on that later, stay tuned!

Modern transistors are excellent switches. They can switch a voltage or a current on or off in trillionths of a second (picoseconds). Can we make use of this fact? Let's see. Suppose we extend the concept of a bundle of wires but let's restrict even further the values that can exist on any individual wire. Since each wire is controlled by a switch, we'll switch between nothing (0 volts) and something (~3 volts). This implies that just two numbers may be carried on each wire, 0 or something (not 0). Will it work? Let's look at Figure 1.10.

Figure 1.10: Sending numbers as binary values. Each arrow represents a wire with the arrowhead representing the direction of signal transmission. The position of each wire in the bundle represents its numerical weight. Each row represents an increasing power of 2.

In this scenario, the amount of information that we can carry on a single wire is limited to nothing, 0, or something (let's call something "1" or "on"), so we'll need a lot of wires in order to transmit anything of significance. Figure 1.10 shows 16 wires, and as you'll soon see, this limits us to numbers between 0 and 65,535 if we are dealing with unsigned numbers, or the signed range of −32,768 to +32,767. Here the decimal number 0 would be represented by the binary number 00000000000000 and the decimal number 65,535 would be represented by the binary number 1111111111111111.

Note: For many years, most standard digital circuits used 5 volts for a 1. However, as the integrated circuits became smaller and denser, the logical voltage levels also had to be reduced. Today, the core of a modern Pentium or Athlon processor runs at a voltage of around 1.7–1.8 volts, not very different from a standard AA battery.

Now we're finally there. We'll take advantage of the fact that electronic switching elements, or transistors, can rapidly switch the voltage on a wire between two values. The most common form of this is between almost 0 volts and something (about 3 volts). If our system is working properly, then what we define as "nothing", or 0, might never exceed about ½ of a volt. So, we can define the number 0 to be any voltage less than ½ volt (actually, it is usually 0.4 volts). Similarly, if the number that we define as a 1, would never be less than 2.5 volts, then we have all the information we need to define our number system. Here, the number 0 is never greater than 0.4 volts and the number 1 is never less than 2.5 volts. Anything between these two ranges is considered to be undefined and is not allowed.

It should be mentioned that we've been referring to "the voltage on a wire." Just where are the wires in our computer? Strictly speaking, we should call the wires "electrical conductors." They can be real wires, such as the wires in the cable that you connect from your printer to the parallel port on the back of your computer. They can also be thin conducting paths on printed circuit boards within your computer. Finally, they can be tiny aluminum conductors on the processor chip itself. Figure 1.11 shows a portion of a printed circuit board from a computer designed by the author.

Notice that some of the integrated circuit's (ICs) pins appear have wires connecting them to another device while others seem to be uncon-

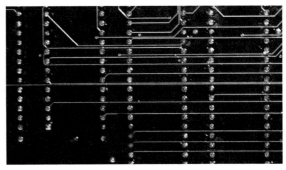

Figure 1.11: Printed wires on a computer circuit board. Each wire is actual a copper trace approximately 0.08 mm wide. Traces can be as close as 0.08 mm apart from each other. The large spots are the soldered pins of the integrated circuits coming through from the other side of the board.

nected. The reason for this is that this printed circuit board is actually a sandwich made up of five thinner layers with wires printed on either side, giving a total of ten layers. The eight inner layers also have a thin insulating layer between then to prevent electric short circuits. During the manufacturing process, the five conducting layers and the four insulating layers are carefully aligned and bonded together. The resultant, ten-layer printed circuit board is approximately 2.5 mm thick.

Without this multilayer manufacturing technique, it would be impossible to build complex computer systems because it would not be possible to connect the wires between components without having to cross a separate wire with a different purpose.

[NOTE: A color version of the following figure is included on the DVD-ROM.] Figure 1.12 shows us just what's going on with the inner layers. Here is an X-ray view of another computer system hardware circuit. This is about the same level of complexity that you might find on the motherboard of your PC. The view is looking through the layers of the board and the conductive traces on each layer are shown in a different color.

While this may appear quite imposing, most of the layout was done using computer-aided design (CAD) software. It would take altogether too much time for even a skilled designer to complete the layout of this board. Figure 1.13 is a magnification of a smaller portion of Figure 1.12. Here you can clearly see the various traces on the different layers. Each printed wire is approximately 0.03 mm wide.

[NOTE: A color version of the following figure is included on the DVD-ROM.] If you look carefully at Figure 1.13, you'll notice that certain colored wires touch a black dot and then seem to go off in another direction as a wire of a different color. The black dots are called *vias*, and they represent places in the circuit where a wire leaves its layer and traverses to another layer. Vias are vertical conductors that allow signals to cross between layers. Without vias, wires couldn't cross each other on the board without short circuiting to each other. Thus, when you see a green wire (for purposes of the grayscale image on this page, the green wire appears as a dotted line) crossing a red wire, the two wires are not in physical contact with other, but are passing over each other on different layers of the board. This is an important concept to keep in mind because we'll soon be looking at, and drawing our own electronic circuit diagrams, called *schematic diagrams*, and we'll need to keep in mind how to represent wires that appear to cross each other without being physically connected, and those wires that are connected to each other.

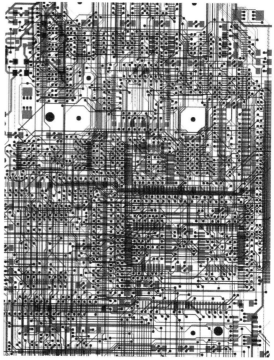

Figure 1.12: An X-ray view of a portion of a computer systems board.

Let's review what we've just discussed. Modern digital computers use the binary (base 2) number system. They do so because a number system that has only two digits in its natural sequence of numbers lends itself to a hardware system which utilizes switches to indicate if a circuit is in a "1" state (on) or a "0" state (off). Also, the fundamental circuit elements that are used to create complex digital networks are also based on these principles as logical expressions. Thus, just as we might say, logically, that an expression is TRUE or FALSE, we can just as easily describe it as a "1" (TRUE) or "0" (FALSE). As you'll soon see, the association of 1 with TRUE and

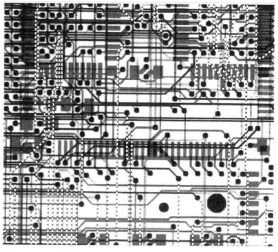

Figure 1.13: A magnified view of a portion of the board shown in Figure 1.12.

0 with FALSE is completely arbitrary, and we may reverse the designations with little or no ill effects. However, for now, let's adopt the convention that a binary 1 represents a TRUE or ON condition, and a binary 0 represents a FALSE or OFF condition. We can summarize this in the following table:

Binary Value	Electrical Circuit Value	Logical Value
0	*OFF*	*FALSE*
1	*ON*	*TRUE*

A Simple Binary Example

Since you have probably never been exposed to electrical circuit diagrams, let's dive right in. Figure 1.14 is a simple schematic diagram of a circuit containing a battery, two switches, labeled A and B, and a light bulb, C. The positive terminal on the battery is labeled with the plus (**+**) sign and the negative battery terminal is labeled with the minus (**−**) sign. Think of a typical AA battery that you might use in your portable MP3 player. The little bump on the end is the positive terminal and the flat portion on the opposite end is the negative terminal. Referring to Figure 1.11, it might seem curious that the positive terminal is drawn as a wide line, and the negative terminal is drawn as a narrow line. There's a reason for it, but we won't discuss that here. Electrical Engineering students are taught the reason for this during their initiation ceremony, but I'm sworn to secrecy.

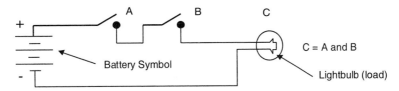

Figure 1.14: A simple circuit using two switches in series to represent the AND function.

The light bulb, C, will illuminate when enough current flows through it to heat the filament. We assume that in electrical circuits such as this one, that current flows from positive to negative. Thus, current exits the battery at the **+** terminal and flows through the closed switches (A and B), then through the lamp, and finally to the **−** terminal of the battery. Now, you might wonder about this because, as we all know from our high school science classes, that electrical current is actually made up of electrons and electrons, being negatively charged, actually flow from the negative terminal of the battery to the positive terminal; the reverse direction.

The answer to this apparent paradox is historical precedent. As long as we think of the current as being positively charged, then everything works out just fine.

Anyway, in order for current to flow through the filament, two things must happen: switch A must be closed (ON) *and* switch B must be closed (ON). When this condition is met, the output variable, C, will be ON (illuminated). Thus, we can talk about our first example of a logical equation:

$$C = A \text{ AND } B$$

This is a very interesting result. We've seen two apparently very different consequences of using switches to build computer systems. The first is that we are lead to having to deal with numbers as binary (base 2) values and the second is that these switches also allow us to create logical equations. For now, let's keep item two as an interesting consequence. We'll deal with it more thoroughly in the next chapter. Before we leave Figure 1.14, we should point out that the switches, A and B, are actuated mechanically. Someone flips the switch to turn it on or off. In general, a switch is a *three-terminal device*. There is a control input that determines the signal propagation between the other two terminals.

Bases

Let's return to our discussion of the binary number system. We are accustomed to using the decimal (base 10) number system because we had ten fingers before we had an iMAC®. The base (or *radix*) of a number system is just the number of distinct digits in that number system. Consider the following table:

Base 2	0,1	Binary
Base 8	0,1,2,3,4,5,6,7	Octal
Base 10	0,1,2,3,4,5,6,7,8,9	Decimal
Base 16	0,1,2,3,4,5,6,7,8,9,A,B,C,D,E,F	Hexadecimal

Look at the hexadecimal numbers in the table above. There are 16, distinct digits, 0 through 9 and A through F, representing the decimal numbers 0 through 15, but expressing them in the hexadecimal system.

Now, if you've ever had your PC lockup with the "blue screen of death," you might recall seeing some funny looking letters and numbers on the screen. That cryptic message was trying to show you an address value in hexadecimal where something bad has just happened. It may be of little solace to you that from now on the message on the blue screen will not only tell you that a bad thing has happened and you've just lost four hours of work, but with your new-found insight, you will know where in your PC's memory the illegal event took place.

When we write a number in binary, octal decimal or hexadecimal, we are representing the number in exactly the same way, although the number will look quite different to us, depending upon the base we're using. Let's consider the decimal number 65,536. This happens to be 2^{16}. Later, we'll see that this has special significance, but for now, it's just a number. Figure 1.15, shows how each digit of the number, 65,536, represents the column value multiplied by the numerical weight of the column. The leftmost, or *most significant digit,* is the number 6. The rightmost, or *least significant digit*, also happens to be 6. The column weight of the most significant digit is 10,000 (10^4) so the value in that column is 6 x 10,000, or 60,000. If we multiply out each column value

10^4	10^3	10^2	10^1	10^0			
6	5	5	3	6			
					$6 \times 10^0 =$	6	
• Notice how each column is weighted by					$3 \times 10^1 =$	30	
the value of the base raised to the power					$5 \times 10^2 =$	500	
					$5 \times 10^3 =$	5000	
					$+ \quad 6 \times 10^4 =$	60000	
					$=$	65536	

Figure 1.15: Representing a number in base 10. Going from right to left, each digit multiplies the value of the base, raised to a power. The number is just the sum of these multiples.

and then arrange them as a list of numbers to be added together, as we've done on the right side of Figure 1.15, we can add them together and get the same number as we started with. OK, perhaps we're overstating the obvious here, but stay tuned, because it does get better. This little example should be obvious to you because you're accustomed to manipulating decimal numbers. The key point is that the column value happens to be the base value raised to a power that starts at 0 in the rightmost column and increases by 1 as we move to the left. Since decimal is base 10, the column weights moving leftward are 1, 10, 100, 1000, 10000 and so on.

If we can generalize this method of representing number, then it follows that we would use the same method to represent numbers in any other base.

Translating Numbers Between Bases

Let's repeat the above exercise, but this time we'll use a binary number. Let's consider the 8-bit binary number 10101100. Because this number has 8 binary numbers, or bits associated with it, we call it an 8-bit number. It is customary to call an 8-bit binary number a *byte* (in C or C++ this is a *char*). It should now be obvious to you why binary numbers are all 1's and 0's. Aside from the fact that these happen to be the two states of our switching circuits (transistors) they are the only numbers available in a base 2 number system.

The byte is perhaps most notable because we measure storage capacity in byte-size chunks (sorry). The memory in your PC is probably at least 256 Mbytes (256 million bytes) and your hard disk has a capacity of 40 Gbytes (40 billion bytes), or more.

Consider Figure 1.16. We use the same method as we used in the decimal example of Figure 1.15. However, this time the column weight is a multiple of base 2, not base 10. The column weights go from 2^7, or 128, the most significant digit, down to 2^0, or 1. Each column is smaller by a power of 2. To see what this binary number is in decimal, we use the same process as we did before; we multiply the number in the column by the weight of the column.

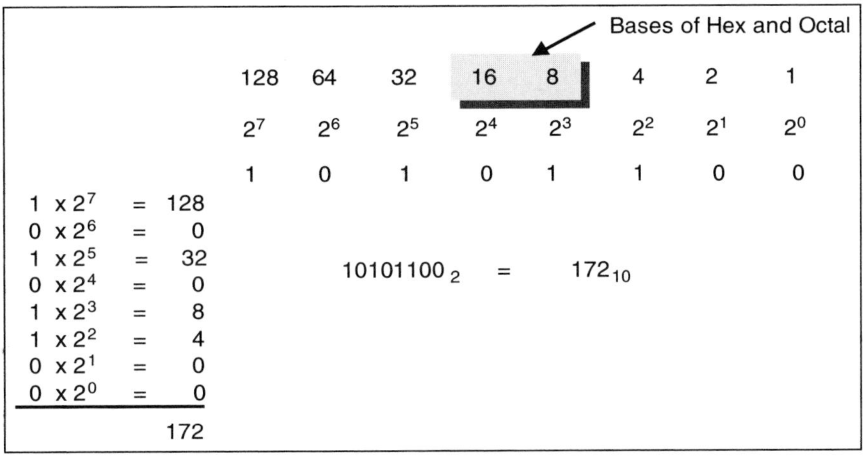

				Bases of Hex and Octal			
128	64	32	16	8	4	2	1
2^7	2^6	2^5	2^4	2^3	2^2	2^1	2^0
1	0	1	0	1	1	0	0

$$
\begin{aligned}
1 \times 2^7 &= 128 \\
0 \times 2^6 &= 0 \\
1 \times 2^5 &= 32 \\
0 \times 2^4 &= 0 \\
1 \times 2^3 &= 8 \\
1 \times 2^2 &= 4 \\
0 \times 2^1 &= 0 \\
0 \times 2^0 &= 0 \\
\hline
&\ 172
\end{aligned}
$$

$$10101100_2 = 172_{10}$$

Figure 1.16: *Representing a binary number in terms of the powers of the base 2. Notice that the bases of the octal (8) and hexadecimal (16) number systems are also powers of 2.*

Thus, we can conclude that the decimal number 172 is equal to the binary number 10101100. It is also noteworthy that the bases of the hexadecimal (Hex) number system, 16 and the octal number system, 8, are also 2^4 and 2^3, respectively. This might give you a hint as to why we commonly use the hexadecimal representation and the less common octal representation instead of binary when we are dealing with our computer system. Quite simply, writing binary numbers gets extremely tedious very quickly and is highly prone to human errors.

To see this in all of its stark reality, consider the binary equivalent of the decimal value:

$$2,098,236,812$$

In binary, this number would be written as:

$$1111101000100001000110110001100$$

Now, binary numbers are particularly easy to convert to decimal by this process because the number is either 1 or 0. This makes the multiplication easy for those of us who can't remember the times tables because our PDA's have allowed significant portions of our cerebral cortex to atrophy.

Since there seems to be a connection between the bases 2, 8 and 16, then it is reasonable to assume that converting numbers between the three bases would be easier than converting to decimal, since base 10 is not a natural power of base 2. To see how we convert from binary to octal consider Figure 1.17.

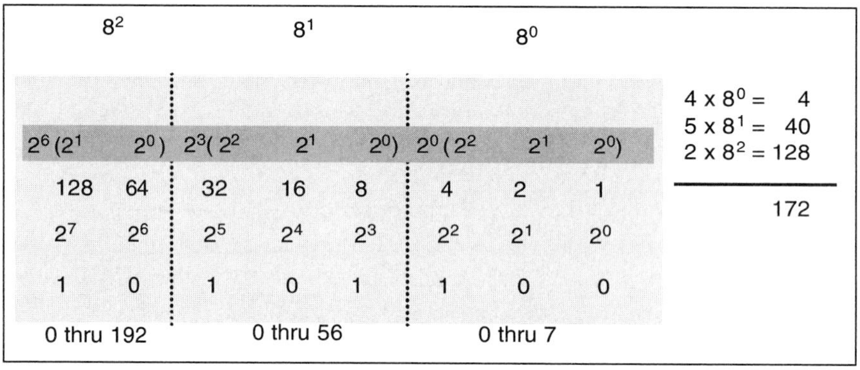

Figure 1.17: Translating a binary number into an octal number. By factoring out the value of the base, we can combine the binary number into groups of three and write down the octal number by inspection.

Figure 1.17 takes the example of Figure 1.16 one step further. Figure 1.17 starts with the same binary number, 10101100, or 172 in base 10. However, simple arithmetic shows us that we can factor out various powers of 2 that happen to also be powers of 8. Consider the dark gray highlighted stripe in Figure 1.17. We can make the following simplifications.

Since any number to the 0 power = 1,

$$(2^2\, 2^1\, 2^0) = 2^0 \times (2^2\, 2^1\, 2^0)$$

Thus,

$$2^0 = 8^0 = 1$$

We can perform the same simplification with the next group of three binary numbers:

$$(2^5\ 2^4\ 2^3\ ) = 2^3 \times (2^2\ 2^1\ 2^0\ )$$

since 2^3 is a common factor of the group.

However, $2^3 = 8^1$, which is the column weight of the next column in the octal number system.

If we repeat this exercise one more time with the final group of two numbers, we see that:

$$(2^7\ 2^6\ ) = 2^6 \times (2^1\ 2^0\ )$$

since 2^6 is a common factor of the group. Again, $2^6 = 8^2$, which is just the column weight of the next column in the octal number system.

Since there is this natural relationship between base 8 and base 2, it is very easy to translate numbers between the bases. Each group of three binary numbers, starting from the right side (least significant digit) can be translated to an octal digit from 0 to 7 by simply looking at the binary value and writing down the equivalent octal value. In Figure 1.14, the rightmost group of three binary numbers is 100. Referring to the column weights this is just 1 * (1 * 4 + 0 * 2 + 0 * 1), or 4. The middle group of three gives us 8 * (1 * 4 + 0 * 2 + 1 * 1), or 8 * 5 (40). The two remaining numbers gives us 64 * (1 * 2 + 0 * 1) or 128. Thus, 4 + 40 + 128 = 172, our binary number from Figure 1.8. But where's the octal number? Simple, each group of three binary numbers gave us the column value of the octal digit, so our binary number is 254. Therefore, 10101100 in binary is equal to 254 in octal, which equals 172 in decimal.

Neat! Thus, we can convert between binary and octal as follows:

- If the number is in octal, write each octal digit in terms of three binary digits. For example:
  $$256773 = 10\ 101\ 110\ 111\ 111\ 011$$
- If the number is in binary, then gather the binary digits into groups of three, starting from the least significant digit and write down the octal (0 through 7) equivalent. For example:
  $$110001010100110111_2 = 110\ 001\ 010\ 100\ 110\ 111 = 612467_8$$
- If the most significant grouping of binary digits has only 1 or 2 digits remaining, just pad the group with 0's to complete a group of three for the most significant octal digit.

Today, octal is not as commonly used as it once was, but you will still see it used occasionally. For example, in the UNIX (Linux) command **chmod 777**, the number 777 is the octal representation of the individual bits that define file status. The command changes the file permissions for the users who may then have access to the file.

We can now extend our discussion of the relationship between binary numbers and octal numbers to consider the relationship between binary and hexadecimal. Hexadecimal (hex) numbers are converted to and from binary in exactly the same way as we did with octal numbers, except that now we use 2^4, as the common factor rather than 2^3. Referring to Figure 1.18, we see the same process for hex numbers as we used for octal.

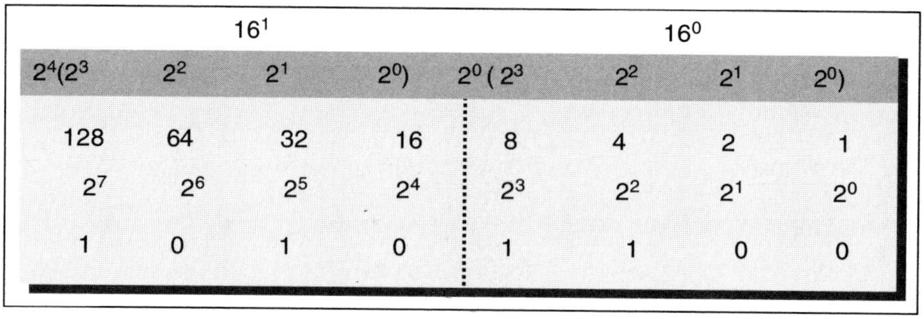

16^1				16^0			
$2^4(2^3$	2^2	2^1	$2^0)$	$2^0(2^3$	2^2	2^1	$2^0)$
128	64	32	16	8	4	2	1
2^7	2^6	2^5	2^4	2^3	2^2	2^1	2^0
1	0	1	0	1	1	0	0

Figure 1.18: Converting a binary number to base 16 (hexadecimal). Binary numbers are grouped by four, starting from the least significant digit, and the hexadecimal equivalent is written down by inspection.

In this case, we factor out the common power of the base, 2^4, and we're left with a repeating group of binary numbers with column values represented as

$$2^3 \; 2^2 \; 2^1 \; 2^0$$

It is simple to see that the binary number $1111 = 15_{10}$, so groups of four binary digits may be used to represent a number between 0 and 15 in decimal, or 0 through F in hex. Referring back to Figure 1.16, now that we know how to do the conversion, what's the hex equivalent of 10101100? Referring to the leftmost group of 4 binary digits, 1010, this is just $8 + 0 + 2 + 0$, or A. The rightmost group, 1100 equals $8 + 4 + 0 + 0$, or C. Therefore, our number in hex is AC.

Let's do a 16-bit binary number conversion example.

Binary number:	0101111111010111
Octal:	0 101 111 111 010 111 = 057727 (grouped by threes)
Hex:	0101 1111 1101 0111 = 5FD7 (grouped by fours)
Decimal:	To convert to decimal, see below:

Octal to Decimal	Hex to Decimal
$7 \times 8^0 = 7$	$7 \times 16^0 = 7$
$2 \times 8^1 = 16$	$13 \times 16^1 = 208$
$7 \times 8^2 = 448$	$15 \times 16^2 = 3840$
$7 \times 8^3 = 3584$	$5 \times 16^3 = 20480$
$5 \times 8^4 = 20480$	
24,535	24,535

Definitions

Before we go full circle and consider the reverse process, converting from decimal to hex, octal and binary, we should define some terms. These terms are particular to computer numbers and give us a shorthand way to represent the size of the numbers that we'll be working with later on. By size, we don't mean the magnitude of the number, we actually mean the number of binary bits that the number refers to. You are already familiar with this concept because most compilers require that you declare the type of a variable before you can use it. Declaring the type really means two things:

1. How much storage space will this variable occupy, and
2. What type of assembly language algorithms must be generated to manipulate this number?

The following table summarizes the various groupings of binary bits and defines them.

bit	*The simplest binary number is 1 digit long*
nibble	*A number comprised of four binary bits. A NIBBLE is also one hexadecimal digit*
byte	*Eight binary bits taken together form a byte. A byte is the fundamental unit of measuring storage capacity in computer memories and disks. The byte is also equal to a* char *in C and C++.*
word	*A word is 16 binary bits in length. It is also 4 hex digits in length or 2 bytes in length. This will become more important when we discuss memory organization. In C or C++, a word is sometimes called a* short.
long word	*Also called a LONG, the long word is 32 binary bits or 8 hex digits. Today, this is an* int *in C or C++.*
double word	*Also called DOUBLE, the double word is 64 binary bits in length, or 16 hex digits.*

From the table you may get a clue as to why the octal number representation has been mostly supplanted by hex numbers. Since octal is formed by groups of three, we are usually left with those pesky remainders to deal with. We always seem to have an extra 1, 2, or 3 as the most significant octal digit. If the computer designers had settled on 15 and 33 bits for the bus widths instead of 16 and 32 bits, perhaps octal would still be alive and kicking. Also, hexadecimal representation is a far more compact way of representing number, so it has become today's standard. Figure 1.19 summarizes the various sizes of data elements as they stand today and as they'll probably be in the near future.

Today we already have computers that can manipulate 64-bit numbers in one operation. The Athlon64® from Advanced Micro Devices Corporation is one such processor. Another example is the processor in the Nintendo N64 Game Cube®. Also, if you consider yourself a *PC Gamer*, and you like to play fast action video games on your PC, then you likely have a high-performance video card in your game machine. It is

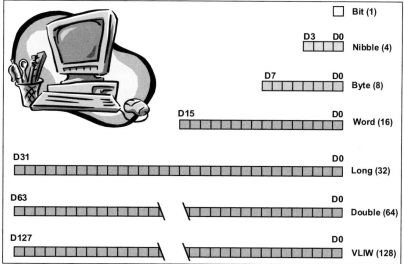

Figure 1.19: Size of the various data elements in a computer system.

likely that your video card has a video processing computer chip on it that can process 128 bits at a time. Is a 256-bit processor far behind?

Fractional Numbers

We deal with fractions in the same manner as we deal with whole numbers. For example, consider Figure 1.20.

10^2	10^1	10^0		10^{-1}	10^{-2}	10^{-3}	10^{-4}
5	6	7	.	4	3	2	1

Figure 1.20: Representing a fractional number in base 10.

We see that for a decimal number, the columns to the right of the decimal point go in increasing negative powers of ten. We would apply the same methods that we just learned for converting between bases to fractional numbers. However, having said that, it should be mentioned that fractional numbers are not usually represented this way in a computer. Any fractional number is immediately converted to a floating point number, or *float*. The floating-point numbers have their own representation, typically as a 64-bit value consisting of a mantissa and an exponent. We will discuss floating point numbers in a later chapter.

Binary-Coded Decimal

There's one last form of binary number representation that we should mention in passing, mostly for reasons of completeness. In the early days of computers when there was a transition taking place from instrumentation based on digital logic, but not truly computer-based, as they are today, it was convenient to represent numbers in a form called *binary coded decimal*, or BCD. A BCD number was represented as 4 binary digits, just like a hex number, except the highest number in the sequence is 9, rather than F. Devices like counters and meters used BCD because it was a convenient way to connect a digital value to some sort of a display device, like a 7-segment display. Figure 1.21 shows the digits of a seven-segment display.

The seven-segment display consists of 7 bars and usually also contains a decimal point and each of the elements is illuminated by a light emitting diode (LED). Figure 1.21 shows how the 7-segment display can be used to show the numbers 0 through 9. In fact, with a little creativity, it can also show the hexadecimal numbers A through F.

BCD was an easy way to convert digital counters and voltmeters to an easy to read display. Imagine what your reaction would be if the Radio Shack® voltmeter read 7A volts, instead of 122 volts. Figure 1.21 shows what happens when we count in BCD. The numbers that are displayed make sense to us because they look like decimal digits. When the count reaches 1001 (9), the next increment causes the display to roll around to 0 and carry a 1, instead of displaying A. Many microprocessors today still contain vestiges of the transition period from BCD to hex numbers by containing special instructions, such as *decimal add adjust*, that are used to create algorithms for implementing BCD arithmetic.

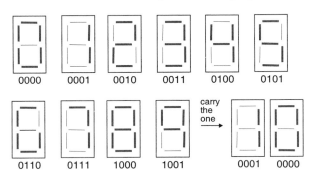

Figure 1.21: Binary coded decimal (BCD) number representation. When the number exceeds the count of 9, a carry operation takes place to the next most significant decade and the binary digits roll around to zero.

Converting Decimals to Bases

Converting a decimal number to binary, octal or hex is a bit more involved because base 10 does not have a natural relationship to any of the other bases. However, it is a rather straightforward process to describe an algorithm to use to translate a number in decimal to another base. For example, let's convert the decimal number, 38,070 to hexadecimal.

1. Find the largest value of the base (in this case 16), raised to an integer power that is still less than the number that you are trying to convert. In order to do the conversion, we'll need to refer to the table of powers of 16, shown below. From the table we see that the number 38,070 is greater than 4,096 but less than 65,536. Therefore, we know that the largest column value for our conversion is 16^3.

$16^0 = 1$	$16^1 = 16$	$16^2 = 256$	$16^3 = 4096$
$16^4 = 65,536$	$16^5 = 1,048,576$	$16^6 = 16,777,216$	$16^7 = 268,435,456$

2. Perform an integer division on the number to convert:
   a. 38,070 DIV 4096 = 9
   b. 38,070 MOD 4096 = 1206
3. The most significant hex digit is 9. Repeat step 1 with the MOD (remainder) from step 1. 256 is less than 1206 and 4096 is greater than 1206.
   a. 1206 DIV 256 = 4
   b. 1206 MOD 256 = 182
4. The next most significant digit is 4. Repeat step 2 with the MOD from step 2. 182 is greater than 16 but less than 256.
   a. 182 DIV 16 = 11 (B)
   b. 4b 182 MOD 16 = 6
5. The third most significant digit is B. We can stop here because the least significant digit is, by inspection,
6. Therefore: $38,070_{10} = 94B6_{16}$

Before we move on to the next topic, logic gates, it is worthwhile to summarize why we did what we did. It wouldn't take you very long, writing 32-bit numbers down in binary, to realize that there has to be a better way. There is a better way, hex and octal. Hexadecimal and octal numbers, because their bases, 16 and 8 respectively, have a natural relationship to binary numbers, base 2. We can simplify our number manipulations by gathering the binary numbers into groups of three or four. Thus, we can write a 32-bit binary value such as 10101010111101011110000010110110 in hex as AAF5E0B6.

However, remember that we are still dealing with binary values. Only the way we choose to represent these numbers is different. As you'll see shortly, this natural relationship between binary and hex also extends to arithmetic operations as well. To prove this to yourself, perform the following hexadecimal addition:

0B + 1A = 25 (Remember, that's 25 in hex, not decimal. 25 in hex is 37 in decimal.)

Now convert 0B and 1A to binary and perform the same addition. Remember that in binary 1 + 1 = 0, with a carry of 1.

Since it is very easy to mistake numbers in hex, binary and octal, assemblers and compilers allow you to easily specify the base of a number. In C and C++ we represent a hex number, such as AA55, by 0xAA55. In assembly language, we use the dollar sign. So the same number in assembly language is $AA55. However, assembly language does not have a standard, such as ANSI C, so different assemblers may use different notation. Another common method is to precede the hex number with a zero, if the most significant digit is A through F, and append an "H" to the number. Thus $AA55 could be represented as 0AA55H by a different vendor's assembler.

Engineering Notation

While most students have learned the basics of using scientific notation to represent numbers that are either very small or very large, not everyone has also learned to extend scientific notation somewhat to simplify the expression of many of the common quantities that we deal with in digital systems. Therefore, let's take a very brief detour and cover this topic so that we'll have a common starting point. For those of you who already know this, you may take your bio break about 10 minutes earlier.

Engineering notation is just a shorthand way of representing very large or very small numbers in a format that lends itself to communication simplicity among engineers. Let's start with an example that I remember from a nature program about bats that I saw on TV a number of years. Bats locate insects in absolute blackness by using the echoes from ultrasonic sound wave that they emit. An insect reflects the sound bursts and the bat is able to locate dinner. What I remember is the narrator saying that the nervous system of the bat is so good at echo location that the bat can discern sound pulses that arrive less than a few millionths of a second apart. Wow!

Anyway, what does a "few millionths" mean? Let's say that a few millionths is 5 millionths. That's 0.000005 seconds. In scientific notation 0.000005 seconds would be written as 5×10^{-6} seconds. In engineering notation it would be written as 5 μs, and pronounced 5 microseconds. We use the Greek symbol, μ (mu) to represent the micro portion of microseconds. What are the symbols that we might commonly encounter? The table below lists the common values:

TERA = 10^{12} (T)	**PICO = 10^{-12} (p)**
GIGA = 10^{9} (G)	**NANO = 10^{-9} (n)**
MEGA = 10^{6} (M)	**MICRO = 10^{-6} (μ)**
KILO = 10^{3} (K)	**MILLI = 10^{-3} (m)**
	FEMTO = 10^{-15} (f)

So, how do we change a number in scientific notation to an equivalent one in engineering notation? Here's the recipe:

1. Adjust the mantissa and the exponent so that the exponent is divisible by 3 and the mantissa is not a fraction. Thus, 3.05×10^{4} bytes becomes 30.5×10^{3} and not 0.03×10^{6}.
2. Replace the exponent terms with the appropriate designation. Thus, 30.5×10^{3} bytes becomes 30.5 Kbytes, or 30.5 Kilobytes.

About 99.99% of the time, the unit will be within the exponent range of ±12. However, as computers get ever faster, we will be measuring times in the fractions of a picosecond, so it's appropriate to include the femtosecond on our table. As an exercise, try to calculate how far light will travel in one femtosecond, given that light travels at a rate of about 15 cm per nanosecond on a printed circuit board.

Although we discussed this earlier, it should be mentioned again in this context that we have to be careful when we use the engineering terms for *kilo, mega* and *giga*. That's a problem that computer folk have created by misappropriating standard engineering notations for their own use. Since $2^{10} = 1024$, computer "Geekspeakers" decided that it was just too close to 1000 to let it go, so the overloaded the K, M and G symbols to mean 1024, 1048576 and 1073741824, respectively, rather than 1000, 1000000 or 1000000000, respectively.

Fortunately, we rarely get mixed up because the computer definition is usually confined to measurements involving memory size, or byte capacities. Any time that we measure anything else, such as clock speed or time, we use the conventional meaning of the units.

Summary of Chapter 1

- The growth modern digital computer progressed rapidly, driven by the improvements made in the manufacturing of microprocessors made from integrated circuits.
- The speed of computer memory has an inverse relationship to its capacity. The faster a memory, the closer it is to the computer core.
- Modern computers are based upon two basic designs, CISC and RISC.
- Since an electronic circuit can switch on and off very quickly, we can use the binary numbers system, or base 2, to as the natural number system for our computer.
- Binary, octal and hexadecimal are the natural number bases of computers and there are simple ways to convert between them and decimal.

Chapter 1: *Endnotes*

[1] Carver Mead, Lynn Conway, *Introduction to VLSI Sysyems*, ISBN 0-2010-4358-0, Addison-Wesley, Reading, MA, 1980.

[2] For an excellent and highly readable description of the creation of a new super minicomputer, see *The Soul of a New Machine*, by Tracy Kidder, ISBN 0-3164-9170-5, Little, Brown and Company, 1981.

[3] Daniel Mann, Private Communication.

[4] http://www.seagate.com/cda/products/discsales/enterprise/family/0,1086,530,00.html.

[5] David A. Patterson and John L. Hennessy, *Computer Organization and Design, Second Edition*, ISBN 1-5586-0428-6, Morgan Kaufmann, San Francisco, CA, p. 30.

[6] Daniel Mann, Private Communication.

[7] http://www.specbench.org/cgi-bin/osgresults?conf=cint95.

Exercises for Chapter 1

1. Define *Moore's Law*. What is the implication of Moore's Law in understanding trends in computer performance? Limit your answer to no more than two paragraphs.

2. Suppose that in January, 2004, AMD announces a new microprocessor with 100 million transistors. According to Moore's Law, when will AMD introduce a microprocessor with 200 million transistors?

3. Describe an advantage and a disadvantage of the organization of a computer around abstraction layers.

4. What are the industry standard busses in a typical PC?

5. Suppose that the average memory access time is 35 nanoseconds (ns) and the average access time for a hard disk drive is 12 milliseconds (ms). How much faster is semiconductor memory than the memory on the hard drive?

6. What is the decimal number 357 in base 9?

7. Convert the following hexadecimal numbers to decimal:

    (a) 0xFE57
    (b) 0xA3011
    (c) 0xDE01
    (d) 0x3AB2

8. Convert the following decimal numbers to binary:

    (e) 510
    (f) 64,200
    (g) 4,001
    (h) 255

9. Suppose that you were traveling at 14 furlongs per fortnight. How fast are you going in feet per second? Express your answer in engineering notation.

CHAPTER 2

Introduction to Digital Logic

Objectives

▶ *Learn the electronic circuit basis for digital logic gates;*
▶ *Understand how modern CMOS logic works;*
▶ *Become familiar with the basics of logic gates.*

• •

Remember the simple battery and flashlight circuit we saw in Figure 1.14? The two on/off switches, wired as they were in series, implemented the logical *AND* function. We can express that example as, "If switch A is closed *AND* switch B is closed, *THEN* lamp *C* will be illuminated." Admittedly, this is pretty far removed from your PC, but the logical function implemented by the two switches in this circuit is one of the four key elements of a modern computer.

It may surprise you to know that all of the primary digital elements of a modern computer, the central processing unit (CPU), memory and I/O can be constructed from four primary logical functions: *AND, OR, NOT* and *Tri-State (TS)*. Now TS is not a logical function, it is actually closer to an electronic circuit implementation tool. However, without tri-state logic, modern computers would be impossible to build.

As we'll soon see, tri-state logic introduces a third logical condition called *Hi-Z*. "Z" is the electronic symbol for impedance, a measure of the easy by which electrical current can flow in a circuit. So, Hi-Z, seems to imply a lot of resistance to current flow. As you'll soon see, this is critical for building our system.

Having said that we could build a computer from the ground up using the four fundamental logic gates: AND, OR, NOT and TS, it doesn't necessarily follow that we would build it that way. This is because engineers will usually take implementation shortcuts in designing more complex functions and these design efficiencies will tend to blur the distinctions between the fundamental logic elements. However, it does not diminish the conceptual importance of these four fundamental logical functions.

In writing this, I was struck by the similarity between the DNA molecule and its four nucleotides: adenine, cytosine, guanine, and thymine; abbreviated, A, C, G and T, and the fact that a computer can also be described by four "electronic nucleotides." We shouldn't look too deeply into this coincidence because the differences certainly far outweigh the similarities. Anyway, it's fun to imagine an "electronic DNA molecule" of the future that can be used as a blueprint for replicating itself. Could there be a science-fiction novel in this somewhere? Anyway, Figure 2.1

Enough of the DNA analogy! Let's move on. Let me make one last point before we do leave. The key point of making the analogy is that everything that we will be doing from now on in the realm of digital hardware is based upon these four fundamental logic functions. That may not make it any easier for you, but that's where we're headed.

Figure 2.2 is a schematic diagram of a digital logic gate, which can execute the logical "AND" function. The symbol, represented by the label F(A,B) is the standard symbol that you would use to represent an AND gate in the schematic diagram for a digital circuit design. The output, C, is a function of the two binary input variables A and B. A and B are binary variables, but in this circuit they are represented by values of voltage. In the case of *positive logic*, where the more positive (higher) voltage is a "1" and the less positive (lower) voltage is a "0", this circuit element is most commonly used in circuits where a logic "0" is represented by a voltage in the range of 0 to 0.8 volts and a "1" is a voltage between approximately 3.0 and 5.0 volts.* From 0.8 volts to 3.0 volts is "no man's land."

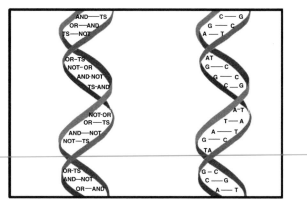

Figure 2.1: Thinking about a computer built from "logical DNA" on the left and schematic picture of a portion of a real DNA molecule on the right. (DNA molecule picture from the Dolan Learning Center, Cold Springs Harbor Laboratory[1].)

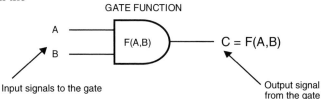

Figure 2.2: Logical AND gate. The gate is an electronic switching circuit that implements the logical AND function for the two input values A and B.

Signals travel through this region very quickly on their way up or down, but don't dwell there. If a logic value were to be measured in this region, there would be an electrical fault in the circuit.

The logical function of Figure 2.2 can be described in several equivalent ways:

- *IF* A is TRUE *AND* B is TRUE *THEN* C is TRUE
- *IF* A is HIGH *AND* B is HIGH *THEN* C is HIGH
- *IF* A is 1 *AND* B is 1 *THEN* C is 1
- *IF* A is ON *AND* B is ON *THEN* C is ON
- *IF* A is 5 volts *AND* B is 5 volts *THEN* C is 5 volts

The last bullet is a little iffy because we allow the values to exist in ranges, rather than absolute numbers. It's just easier to say "5 volts" instead of "a range of 3 to 5 volts."

As we've discussed in the last chapter, in a real circuit, A and B are signals on individual wires. These wires may be actual wires that are used to form a circuit, or they may be extremely fine copper paths (traces) on a printed circuit (PC) board, as in Figure 1.11, or they may be microscopic paths on an integrated circuit. In all cases, it is a single wire conducting a digital signal that may take on two digital values: "0" or "1".

---

* *Assuming that we are using the older 5-volt logic families, rather than the newer 3.3-volt families.*

The next point that we want to consider is why we call this circuit element a "gate." You can imagine a real gate in front of your house. Someone has to open the gate in order to gain entrance. The AND gate can be thought of in exactly the same way. It is a gate for the passage of the logical signal. The previous bulleted statements describe the AND gate as a logic statement. However, we can also describe it this way:

- *IF* A *is* 1 *THEN* C *EQUALS* B
- *IF* A *is* 0 *THEN* C *EQUALS* 0

Of course, we could exchange A and B because these inputs are equivalent. We can see this graphically in Figure 2.3. Notice the strange line at inputs B and output C. This is a way to represent a digital signal that is varying with time. This is called a *waveform* representation, or a *timing diagram*. The idea is that the waveform is on some kind of graph paper, such as a strip chart recorder, and time is changing on the horizontal axis. The vertical axis represents the logical value a point in the circuit, in this case points B and C.

When A = 1, the same signal appears at output C that is input at B. The change that occurs at C as a result of the change at B is not instantaneous because the speed of light is finite, and circuits are not infinitely fast. However, in terms of a human scale, it's pretty fast. In a typical circuit, if input B went from a 0 value to a 1, the same change would occur at output C, delayed by about 5 billionths of a second (5 nanoseconds).

You've actually seen these timing diagrams before in the context of an EKG (electrocardiogram) when a physician checks your heart. Each of the signals on the chart represents the electrical voltage at various parts of your heart muscles over time. Time is traveling along the long axis of the chart and the voltage at any point in time is represented by the vertical displacement of the ink. Since typical digital signals change much more rapidly than we could see on a strip chart recorder, we need specialized equipment, such as oscilloscopes and logic analyzers to record the waveforms and display them for us in a way that we can comprehend. We'll discuss waveforms and timing diagrams in the upcoming chapters.

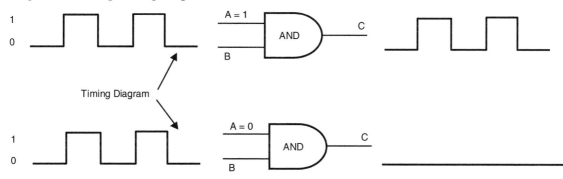

Figure 2.3: The logical AND circuit represented as a gating device. When the input A = 1, output C follows input B (the gate is open). When input A = 0, output C = 0, independent of how input B varies with time (the gate is closed).

Thus, Figure 2.3 shows us an input waveform at B. If we could get really small, and we had a very fast stopwatch and a fast voltmeter, we could imagine that we are sitting on the wire connected to the gate at point B. As the value of the voltage at B changes, we check our watch and plot the voltage versus time on the graph. That's the waveform that is shown in Figure 2.3.

Also notice that the vertical line that represents the transition from the logic level 0 to the logic level 1. This is called the *rising edge*. Likewise, the vertical line that represents the transition from logic level 1 to logic level 0 is called the *falling edge*. We typically want the time durations of the rising edge and falling edge to be very small, a few billionths of a second (nanoseconds) because in digital systems, the space between 0 and 1 is not defined, and we want to get through there as quickly as possible. That's not to say that these edges are useless to us. Quite the contrary, they are pretty important. In fact, you'll see the value of these edges when we study system clocks later on.

If, in Figure 2.3, A = 1 (upper figure), the output at C follows the input at A with a slight time delay, called the *propagation delay*, because it represents the time required for the input change to work its way, or propagate, through the circuit element. When the *control input* at A = 0, output C will always equal 0, independent of whatever input B does. Thus, input B is *gated* by input A. For now, we'll leave the timing diagram in the wonderful world of hardware design and return to our studies of logic gates.

Earlier, we touched on the fact that the AND gate is one of three logic gates. We'll keep the tri-state gate separate for now and study it in more depth when we discuss bus organization in a later chapter. In terms of "atomic," or unique elements, there are actually three types of logical gates: AND, OR and NOT. These are the fundamental building blocks of all the complex digital logic circuits to follow. Consider Figure 2.4.

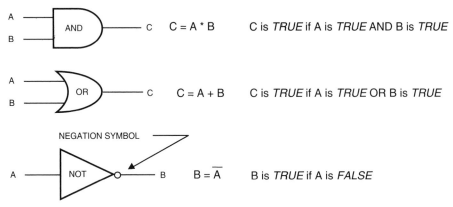

Figure 2.4: The three "atomic" logic gates: AND, OR and NOT.

The symbol for the AND function is the same as the multiplication symbol that we use in algebra. As well see later on, the AND operation is similar to multiplying two binary numbers, since $1 \times 1 = 1$ and $1 \times 0 = 0$. The asterisk is a convenient symbol to use for "ANDing" two variables.

The symbol for the OR function is the plus sign, and it is "sort of" like addition, because $0 + 0 = 0$, $1 + 0 = 1$. The analogy fails with $1 + 1$, because if we're adding, then $1 + 1 = 0$ (carry the 1) but if we are OR'ing, then $1 + 1 = 1$. OK, so they overloaded the + sign. Don't get angry with me, I'm only the messenger.

The negation symbol takes many forms, mostly because it is difficult to draw a line over a variable using only ASCII text characters. In Figure 2.4, we use the bar over the variable A to indicate that the output B is the negation, or opposite of the input A. If A = 1, then B = 0. If A = 0, then B = 1. Using only ASCII text characters, you might see the NOT symbol written as, B = ~A, or B = /A with the forward slash or the tilde representing negation. Negation is also called the *complement.*

The NOT gate also uses a small open circle on the output to indicate negation. A gate with a single input and a single output, but without the negation symbol is called a *buffer.* The output waveform of a buffer always follows the input waveform, minus the propagation delay. Logically there is no obvious need for a buffer gate, but electrically (those pesky hardware engineers again!) the buffer is an important circuit element.

We'll actually return to the concept of the buffer gate when we study analog to digital conversion in a future lesson. Unlike the AND gate and the OR gate, the NOT gate always has a single input and a single output. Furthermore, for all of its simplicity, there is no obvious way to show the NOT gate in terms of a simple flashlight bulb circuit. So we'll turn our attention to the OR gate.

We can look at the OR gate in the same way we first examined the AND gate. We'll use our simple flashlight circuit from Figure 1.14, but this time we'll rearrange the switches to create the logical OR function. Figure 2.5 is our flashlight circuit.

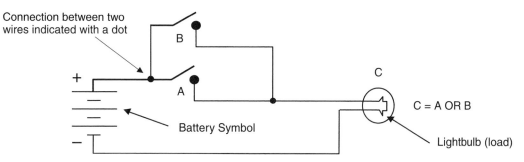

Figure 2.5: The logical OR function implemented as two switches in parallel. Closing either switch A or B turns on the lightbulb, C.

The circuit in Figure 2.5 shows the two switches, A and B wired in parallel. Closing either switch will allow the current from the battery to flow through the switch into the bulb and turn it on. Closing both switches doesn't change the fact that the lamp is illuminated. The lamp won't shine any brighter with both switches closed. The only way to turn it off is to open both switches and interrupt the flow of current to the bulb. Finally, there is a small dot in Figure 2.5 that indicate that two wires come together and actually make electrical contact. As we've discussed earlier, since our schematic diagram is only two dimensional, and printed circuit boards often have ten layers of electrical conductors insulated from each other, it is sometimes confusing to see two wires cross each other and not be physically connected together. Usually when two wires cross each other, we get lots of sparks and the house goes dark. In our schematic diagrams, we'll represent two wires that actually touch each other with a small dot. Any other wires that cross each other we'll consider to be insulated from each other.

We still haven't looked at the tri-state logic gate. Perhaps we should, even though the reason for the gate's existence won't be obvious to you now. So, at the risk of letting the cat out of the bag, let's look at the fourth member of our atomic group of logic elements, the tri-state logic gate shown schematically in Figure 2.6.

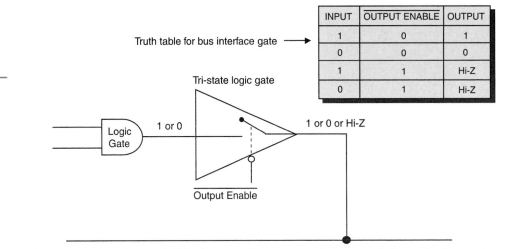

INPUT	OUTPUT ENABLE	OUTPUT
1	0	1
0	0	0
1	1	Hi-Z
0	1	Hi-Z

Truth table for bus interface gate

Tri-state logic gate

Logic Gate

1 or 0

1 or 0 or Hi-Z

Output Enable

Figure 2.6: The tri-state (TS) logic gate. The output of the gate follows the input along as the Output Enable (OE) input is low. When the OE input goes high, the output of the gate enters the Hi-Z state.

There are several important concepts here so we should spend some time on this figure. First, the schematic diagram for the gate appears to be the same as an inverter, except there is no *negation bubble* on the output. That means that the logic sense of the output follows the logic sense of the input, after the propagation delay. That makes this gate a *buffer gate*. This doesn't imply that we couldn't have a tri-state gate with a built-in NOT gate, but that would not be atomic. That would be a *compound gate*.

The TS gate has a third input, labeled *Output Enable*. The input on the gate also has a negation bubble, indicating that the gate is *active low*. We've introduced a new term here. What does "active low" mean? In the previous chapter, we made a passing reference to the fact that logic levels were somewhat arbitrary. For convenience, we were going make the more positive voltage a 1, or TRUE, and the less positive voltage a 0, or FALSE. We call this convention *positive logic*. There's nothing special about positive logic, it is simply the convention that we've adopted.

However, there are times when we will want to assign the "TRUENESS" of a logic state to be low, or 0. Also, there is a more fundamental issue here. Even though we are dealing with a logical condition, the meaning of the signal is not so clear-cut. There are many instances where the signal is used as a controlling, or enabling device. Under these conditions, TRUE and FALSE don't really apply in the same way as they would if the signal was part of a complex logical equation. This is the situation we are faced with in the case of a tri-state buffer.

The Output Enable (OE) input to the tri-state buffer is active when the signal is in its low state. In the case of the TS buffer, when the OE input is low, the buffer is active, and the output will follow the input. Now, at this point you might be getting ready to say, "Whoa there Bucko, that's an AND gate. Isn't it?" and you'd almost be correct. In Figure 2.3, we introduced the idea of the AND logic function as a gate and when the *A* input was 0, the output of the gate was also 0, independent of the logic state of the *B* input. The TS buffer is different in a critical way. When $\overline{OE}$ is high, from an electrical circuit point of view, the output of the gate ceases to exist. It is just as if the gate wasn't there. This is the Hi-Z logic state. So, in Figure 2.3 we have the unique situation that the TS buffer acts behaves like a closed switch when $\overline{OE}$ is low, and it acts like an open switch when $\overline{OE}$ is high. In other words, Hi-Z is not a 1 or 0 logic state, it is a unique state of it own, and has less to do with digital logic then with the electronic realities of building computers. We'll return to tri-state buffers in a later chapter, stay tuned!

There's one last new concept that we've introduced in Figure 2.6. Notice the *truth table* in the upper right of the figure. A truth table is a shorthand way of describing all of the possible states of a logical system. In this case, the input to the TS buffer can have two states and the $\overline{OE}$ control input can have two states, so we have a total of four possible combinations for this gate. When $\overline{OE}$ is low, the output agrees with the input, when $\overline{OE}$ is high, the output is in the Hi-Z logic state and the input cannot be seen. Thus, we've described in a tidy little chart all of the possible operational states of this device.

We now have in our vocabulary of logical elements AND, OR, NOT and TS. Just like the building block life in DNA, these are the building blocks of digital systems. In actuality, these three gates are most often combined to form slightly different gates called NAND, NOR and XOR. The NAND, NOR and XOR gates are *compound gates*, because they are constructed by combining the AND, OR and NOT gates. Electrically, these compound circuits are just as fast as the primary circuits because the compound function is easily implemented by itself. It is only from a logical perspective do we draw a distinction between them.

Figure 2.7 shows the compound gates NAND and NOR. The NAND gate is an AND gate followed by a NOT gate. The logical function of the NAND gate may be stated as:

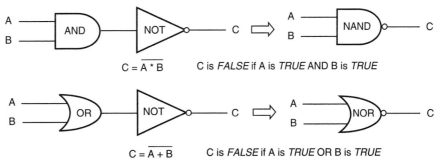

$$C = \overline{A * B} \qquad \text{C is } \textit{FALSE} \text{ if A is } \textit{TRUE} \text{ AND B is } \textit{TRUE}$$

$$C = \overline{A + B} \qquad \text{C is } \textit{FALSE} \text{ if A is } \textit{TRUE} \text{ OR B is } \textit{TRUE}$$

Figure 2.7: A schematic representation of the NAND and NOR gates as a combination of the AND gate with the NOT gate, and the OR gate with the NOT gate, respectively.

- OUTPUT C goes LOW if and only if input A is HIGH AND input B is HIGH.

The logical function of the NOR gate may be stated as:

- OUTPUT C goes LOW if input A is HIGH, or input B is HIGH, or if both inputs A and B are HIGH.

Finally, we want to study one more compound gate construct, the XOR gate. XOR is a shorthand notation for the *exclusive OR gate* (pronounced as "ex or"). The XOR gate is almost like an OR gate except that the condition when both inputs A and B equals 1 will cause the output C to be 0, rather than 1.

Figure 2.8 illustrates the circuit diagram for the XOR compound gate. Since this is a lot more complex than anything we've seen so far, let's take our time and walk through it. The XOR gate has two inputs, A and B, and a single output, C. Input A goes to AND gate #3 and to NOT gate #1, where it is inverted. Likewise, input B goes to AND gate #4 and its *complement* (negation) goes to AND gate #3. Thus, each of the AND gates has as its input one of the variables A or B, and the complement, or negation of the other variable, $\overline{A}$ or $\overline{B}$, respectively. As an aside, you should now appreciate the value of the black dot on the schematic diagram. Without it, we would not be able to discern wires that are connect to each other from wires that are simply crossing over each other.

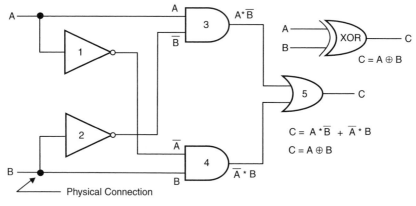

C is *TRUE* if A is *TRUE* OR B is *TRUE, but not if* A is *TRUE* AND B is *TRUE*

Figure 2.8: Schematic circuit diagram for an exclusive OR (XOR) gate.

Thus, the output of AND gate #3 can be represented as the logical expression $A * \overline{B}$ and similarly, the output of AND gate #4 is $B * \overline{A}$. Finally, OR gate #5 is used to combine the two expressions and allow us to express the output variable C, as a function of the two input variables, A and B as, $C = A * \overline{B} + B * \overline{A}$. The symbol for the compound XOR gate is shown in Figure 2.8 as the OR gate with an added line on the input. The XOR symbol is the plus sign with a circle around it.

Let's walk through the circuit to verify that it does, indeed, do what we think it does. Suppose that A and B are both 0. This means that the two AND gates see one input as a 0, so their outputs must be zero as well. Gate #5, the OR gate, has both inputs equal to 0, so its output is also 0. If A and B are both 1, then the two NOT gates, #1 and #2, negate the value, and we have the same situation as before, each AND gate has one input equal to 0.

In the third situation, either A is 0 and B is 1, or vice versa. In either case, one of the AND gates will have both of its inputs equal to 1 so the output of the gate will also be 1. This means that at least one input to the OR gate will be 1, so the output of the OR gate will also be 1. Whew! Another way to describe the XOR gate is to say that the output is TRUE if either input A is TRUE OR input B is TRUE, but not if both inputs A and B are TRUE.

You might be asking yourself, "What's all the fuss about?" As you'll soon see, the XOR gate forms one of the key circuit elements of a computer, the addition circuit. In order to understand this, let's suppose that we are adding two single-bit binary numbers together. We can have the following possibilities for A and B:

- A = 0 and B = 0: A + B = 0
- A = 1 and B = 0: A + B = 1
- A = 0 and B = 1: A + B = 1
- A = 1 and B = 1: A + B = 0, carry the 1

These conditions look suspiciously like A XOR B, provided that we can somehow figure out what to do about the "carry" situation.

Also, notice in Figure 2.8 that we've shown that the output of the XOR gate, C, can be expressed as a logical equation. In this case,

$$C = \overline{A} * B + \overline{B} * A$$

This is another reason why we call the XOR gate a compound gate. We can describe it in terms of a logical equation of the more atomic gates that we've previously discussed.

Now suppose that we modify the XOR gate slightly and add a NOT gate to it after the output. In effect, we'll construct an XNOR gate. In this situation the output will be 1, if both inputs are the same, and the output will be 0 if the inputs are different. Thus, we've just constructed a circuit element that detects equality. With 32 XNOR gates, we can immediately tell (after the appropriate propagation delay) if two 32-bit numbers are equal to each other or are different from each other.

Just as we might use 32 XNOR gates to compare the values of two 32-bit numbers, let's see how that might also work if we wanted to do a logical AND operation on two 32-bit numbers. You already know from your other programming courses that you can do a logical AND operation on variables that evaluate to TRUE or FALSE (Booleans), but what does ANDing two 32-bit numbers mean? In this case, the AND operation is called a *bitwise AND*, because it ANDs together the corresponding bit pairs of the two numbers. Refer to Figure 2.9. For the sake of simplicity, we'll show the bitwise AND operation on two 8-bit

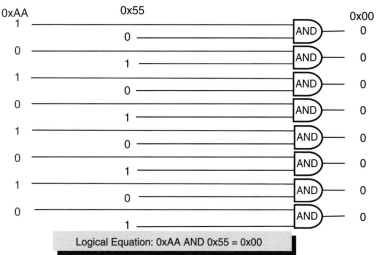

Figure 2.9: Bitwise AND operation on two 8-bit (byte) values. The operation performed is called a bitwise AND because it operates on the corresponding bit pairs of the two numbers. In the C and C++ language, a hexadecimal number is represented with the prefix 0x.

37

numbers, rather than two 32-bit numbers. The only difference in the two situations is the number of AND gates used in the operation.

In Figure 2.9, we are performing the bitwise AND operation on the two byte values 0xAA and 0x55. Since each AND gate has one input equal to 1 and the other equal to 0, every AND gate has a 0 for its output. In C/C++, the ampersand, &, is the bitwise AND operator, so we can write the code snippet:

Code Example

```
char    inA = 0xAA;

char    inB = 0x55;

char    outC = inA & inB;

cout << "The bitwise ANDing of 0x55 and 0xAA = ," outC << endl;
```

If we wrote a real program and ran it, the result would be 0. As a preview of 68,000 assembly language, this same equation would be written:

Code Example

```
MOVE.B    #$AA,D0

ANDI.B    #$55,D0
```

The first assembly language instruction copies the byte, 0xAA into an internal storage register, D0. The second instruction does a bitwise AND of the byte 0x55 with the contents of D0, 0xAA. The "ANDI.*B*" is interpreted as, "Do a logical AND operation on the byte value portion of register D0 with the immediate (literal) value $55." We could have written "ANDI.W" or "ANDI.L" to indicate 16-bit or 32-bit operations, respectively. The result, 0, is now stored in the register D0. If this doesn't make any sense to you now, don't worry. It might not make any sense to you later (but it probably will).

Up to now the AND, OR and their derivative gates were all represented with two inputs and one output. However, the AND gate and the OR gate may have an arbitrary number of inputs. To see this, let's consider Figure 2.10.

Case 1 shows three, 2-input AND gates connected such that output C will be true if and only if inputs A, B, G and D are all true. This is simplified in Case 2. In fact, we can

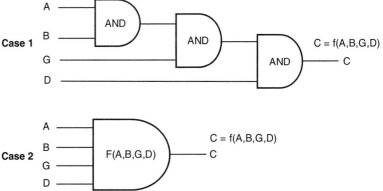

Figure 2.10: Logical equivalence of AND gates with more than 2 inputs. Although Case 1 and Case 2 perform identical logical functions, they are not identical when logic speed (propagation delay) is taken into account.

design the electrical circuit of case two to be almost identical to the circuit of one of the gates in Case 1, the exception being the number of inputs being AND'ed together. However, if we assume that all the gates in Figure 2.10 have exactly the same propagation delay, then the 4-input gate in Case 2 would have a propagation delay approximately 1/3 that of Case 1. You can apply the same analysis to the OR gate. However, this fails with the XOR gate because the mathematical function of exclusive OR is always applied to only 2-input variables at a time.

Electronic Gate Description

Today, most gates are electronic devices made from integrated circuits (ICs). You might be wondering what is inside of a gate. That's a fair question. It's not the purpose of this book to also teach transistor theory or IC design, even though it might be fun to try. Let's let the Electrical Engineers have their mysteries, after all, you've seen what a mess they make when they to write software. Let's take a quick peek at such an IC. Figure 2.11 is a picture of an industry standard part, the 74LS00. The number '74' refers to the logic series of parts and the temperature range over which these parts are designed to work. '74' series parts are considered to be commercial parts, while '54' series are designated as "military specified parts and will operate over a wider

range of temperatures. The letters 'LS' are an abbreviation for *low-power Schottky*, one of several standard IC manufacturing technologies. The last designation, '00' is the designation for an IC package containing four NAND gates with each NAND gate having two inputs and a single output. The part is contained in a plastic package with 14 pins, 7 on each side. This is called a *DIP* package, an abbreviation for dual-inline plastic. ICs like this are often called "bugs," for obvious reasons.

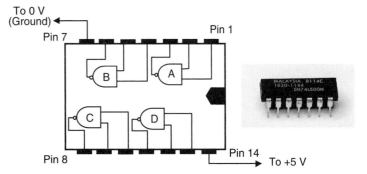

Figure 2.11: A quad, two-input NAND gate (Industry Standard Designation 74LS00). The plastic package contains four independent NAND gates. The integrated circuit is encapsulated in a plastic package (shown on the right). The input pins for gate A are 1 and 2, and the output pin is pin 3.

The integrated circuit, or IC, shown in Figure 2.11 contains four independent NAND gates. Each gate has two inputs and one output. Thus, the package requires twelve I/O pins for the gates themselves, plus one pin for power, in the case of the LS logic family +5 volts, and one pin for ground. This package costs about 10 cents to buy and any one of the gates has a propagation delay from input change to output response of about 5 nanoseconds.

We still haven't answered the question of what's actually inside a gate. I did assert that transistors make good switches and switches make good logic circuits, so let's see how it works. Rather than discuss the LS logic family, which is built on an older technology, let's perform our analysis with a more modern technology called *CMOS*, which is pronounced "sea moss." CMOS is an abbreviation for *complementary metal-oxide silicon*. CMOS is the dominant integrated circuit technology today and will be into the foreseeable future. Almost all modern microprocessors are built in CMOS technology. Also, you may have heard about this technology in the context of the image sensors for digital cameras. Since this is such an important technology, we should spend a few moments and try to understand it, at least in a cursory way.

In order to understand the basic CMOS construct, let's return to something we do understand, mechanical switches. See Figure 2.12.

Try not to pay too much attention to the quality of the artwork. Even with my stick figures, the operation of the circuit should be clear. Image that we

Figure 2.12: Mechanical switch representation of a CMOS switching circuit.

have two mechanical switches connected in series (one after the other) between our logic level 1 and logic level 0. In a real circuit, logic 1 would be the power supply and logic 0 would be the circuit ground, or 0 volts. Now, it should be obvious to you that you would never want to have both switches closed at the same time, otherwise bad things could happen. This would be similar to what happened when I was 5 years old and unbent a paper clip to see what would happen if I stuck it into the light socket on the wall. Although my experiment was a bit more spectacular, and perhaps more of a learning experience then Figure 2.12 might provide, conceptually we have the same result.

In order to output a logic level 0, we close the switch to the logic 0 line and open the switch to the logic 1 line. An observer looking into the output of the circuit would see a logic 0, because that switch is connecting us to that logic level. The alternative situation gives us a logic 1 condition. When we close the upper switch and open the lower switch, we "see" logic level 1.

Hopefully, one important fact should be coming clear now. When we first discussed the concept of rising and falling edges for logic signals, I stated that these edges must be very rapid transitions between the logic states. Figure 2.12 shows us why. We never want to have a situation where both of these switches are closed at the same time. So, if the logic transition is somehow controlling the opening and closing of these switches, then any overlap that might occur between one switch opening and the other switch closing should be as brief as possible.

Figure 2.13 is an actual circuit configuration for a CMOS logic gate. It happens to be a NOT gate. Let me first say that I'm showing you this figure with great trepidation. For one, I am concerned that I am letting too much sacred information out of the bag and the brotherhood and sisterhood of EE's will seek retribution. Also, I'm not sure that in the greater subject area of Computer Architecture that this is a must-know kind of a discussion. Anyway, let's forge ahead. You can be the judge.

The two circuit devices represented by the symbols in the gray circles are called *MOSFETs*. MOSFET is an abbreviation for Metal Oxide Silicon Field-Effect Transistor. Whew! MOSFETs come in two flavors: n-type or n-channel, and p-type or p-channel. The symbol for the n-channel device has the arrowhead pointing towards the device and the arrowhead in the p-channel device has the arrowhead pointing away from the device. Whether you have an n-type or p-type depends upon how the device is manufactured. In particular, the resulting device depends upon what type of impurities are added to the silicon in order to impart the desired electrical characteristics.

CMOS devices use the two MOSFETs in pairs. An n-type and a p-type are paired together. Aside from their manufacturing differences, the main distinguishing characteristic is that the n-channel device is used in the sense of a positive voltage, and the p-channel device is used in a negative voltage sense. If this doesn't mean anything to you now, keep reading. I promise that it gets better. Anyway, aside from the fact that one transistor is sort of a positive device and the other is sort of a negative device, both transistors behave pretty nearly identically, so it is customary to call them *complementary*. Thus, by pairing an n-type MOSFET with a p-type MOSFET of similar characteristics, we have a complementary pair, or a CMOS gate.

Each of the devices has three terminals of interest. The terminal labeled with a 'G' is the *gate*, the terminal with the 'D' is the *drain*, and the terminal with the 'S' is called the *source*. There is sometimes a fourth terminal, the *substrate, or body (B)*. This terminal is sometimes brought out as separate control, but in our circuit configuration it is connected to the source and we need not be concerned about it. In order to see how the CMOS gate behaves the way it does, we should take a quick peek at a simple graph of the behavior of a MOSFET device. Figure 2.14 shows schematically the electrical resistance from the drain to the source (R_{DS}) of a MOSFET device as a function of voltage on the gate, relative to the source, or V_{GS}. Now, it's true that we really haven't defined resistance yet, so I don't expect that it should mean much to you at this point, but conceptually, it is the same as the electrical impedance that we discussed earlier in the context of the tri-state gate.

Consider the curve for the n-channel device in Figure 2.14. We see that the resistance of the device drops dramatically from a value of several tens of millions of *ohms* to a

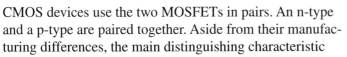

Figure 2.13: CMOS circuit configuration.

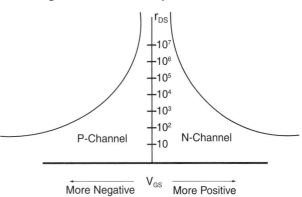

Figure 2.14: Electrical behavior of an n-channel and p-channel MOSFET device. The electric resistance across the device is plotted on a logarithmic scale as a function of the voltage on the gate of the device. From Watson[2].

value around 10 ohms as the voltage on the gate, relative to the source increases. In real terms, the amount that the voltage on the gate has to vary to cause this dramatic change is exactly the voltage swing from logic 0 to logic 1. As far as our electrical circuit analogy of Figure 2.12 is concerned, this is equivalent to closing the switch as we raise the voltage on the gate. As you can see, the p-channel device behaves in a similar manner as the voltage on the gate becomes more negative then the voltage of the source. In other words, it exhibits complementary behavior. Now, we can finally put this all together and understand CMOS gates in general and Figure 2.13 in particular.

Refer back to Figure 2.13. Recall that this is a NOT gate. Let's see why. Assume that the voltage on the gate is a logic 0. The n-type transistor is essentially an open switch. The resistance is extremely high (because V_{GS} is nearly zero). However, the complementary device, the p-type device has a very low resistance because the voltage on the gate is much lower then the voltage of the source ($-V_{GS}$ is large). This is the closed switch of Figure 2.12.

Thus, at the output of the gate, we see a low resistance (closed switch) to logic 1 and a high resistance (open switch to logic 0) and the output of the gate is logic 1. So, when the input to the gate is logic 0, the output from the gate is logic 1. You should be able to analyze the complementary situation for an input of logic level 1. So, we can summarize the behavior of the CMOS NOT gate as follows. When the input is at logic 0, the output "sees" the power supply voltage, or logic 1. When the input is logic 1, the output sees the ground reference voltage, or logic 0.

Figure 2.14 illustrates the electrical behavior of MOS transistors. Consider the N-channel device on the right-hand side of the graph. As the voltage on the gate, measured relative to the source (Vgs) becomes higher, or more positive, the electrical resistance between the drain and the source decreases exponentially. Conversely, when Vgs approaches zero, or becomes negative, the resistance between the drain and the source approaches infinity. Essentially, we have an open switch when Vgs is zero or negative, and almost a closed switch when Vgs is a few volts. The behavior of the P-channel device is identical to the N-channel device, except that the relative polarity of the voltages are reversed.

Before we leave this topic of the electrical behavior of a CMOS gate, you might be asking yourself, "What about the situation where the voltage on the gate, relative to the source, isn't all one way or the other, but in the middle?" In other words, what happens when the rising edge or the falling edge of the logic input to the gate is transitioning between the two logic states? According to the graph in Figure 2.14, if the resistance is too high, and it isn't too low, it is kind of like the temperature of the porridge in "Goldilocks and the Three Bears." At that point, there is a path for the current to flow from the power supply to ground, so we do waste some energy. Fortunately, the transition between states is fast, so we don't waste a lot of energy, but nevertheless, there is some power dissipation occurring.

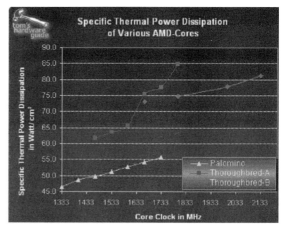

Figure 2.15: Power dissipation versus clock rate for various AMD processor families. From www.tomshardware.com[3].

Well how bad is it. Recall that a modern processor has tens of millions of CMOS gates and a large fraction of these gates are switching a billion or more times per second. Figure 2.15 should give you some idea of how bad it could be.

Notice two things. First, these modern processors really run hot. In fact, they are dissipating as much heat as an ordinary 60–75 watt lightbulb. Also, in each case, as we turn up the clock and make the processors run faster, we see the power dissipation going up as well.

You might also be asking another question, "Suppose we turn the clock down until it is really slow, or even stop it, will that have the reverse effect?" Absolutely, by slowing the clock, or shutting down portions of the chip, we can decrease the power requirements accordingly. This is a very important strategy in laptop computers, which have to make their batteries last for the duration of an airplane flight from New York to Los Angeles. This is also the strategy of how many other microprocessors, the kind that are used in embedded applications, can be attached to a collar around the neck of a moose and track the animal's movements for over two years on no more power than a AAA battery. No wonder CMOS is so popular.

OK, we've seen a NOT gate. That's a pretty simple gate. How about something a little more complicated? Let's do a NAND gate in CMOS.

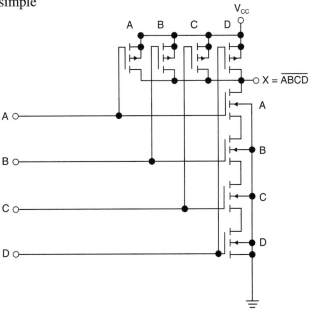

Recall that for a NAND gate, the output is 0 if all the inputs are logic 1. In Figure 2.16, we see that if all of the inputs are at logic 1, then all of the n-channel devices are turned on to the low resistance state. All of the p-channel devices are turned off, so, looking back into the device from the output, we see a low resistance path to the ground reference point (logic 0). Now, if any one of the four inputs A, B, C or D is in logic state 0, then its n-channel MOSFET is in the high-resistance state and its p-channel device is in the low-resistance state. This will effectively block the ground reference point from the output and open a low resistance path to the power supply through the corresponding p-channel device.

Figure 2.16: Schematic diagram of a CMOS, 4-input NAND gate. From Fairchild Semiconductor.[4]

Hopefully, at this point you are getting to be a little more comfortable with your burgeoning hardware skills and insights. Let's analyze a circuit configuration. Consider Figure 2.17. This is

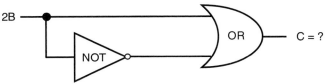

Figure 2.17: The Shakespeare circuit.

often called the "Shakespeare Circuit." You may feel free to ponder the deeper significance of the design. Note that I included this example just to prove that all computer designers are NOT humorless geeks.

Truth Tables

The last concept that we'll discuss in this lesson is the idea of the *truth table*. You had a brief introduction to the truth table when we discussed the behavior of the tri-state logic gate. However, it is appropriate at this point in our analysis to discuss it more formally. The truth table is, as its name implies a tabular form that represents the TRUE/FALSE conditions of a

AND				OR				NAND		
A	B	C		A	B	C		A	B	C
0	0	0		0	0	0		0	0	1
1	0	0		1	0	1		1	0	1
0	1	0		0	1	1		0	1	1
1	1	1		1	1	1		1	1	0

XOR				NOR		
A	B	C		A	B	C
0	0	0		0	0	1
1	0	1		1	0	0
0	1	1		0	1	0
1	1	0		1	1	0

Figure 2.18: Truth table representation for the logical functions AND, OR, NAND, NOR and XOR.

logical gate or system. For each of the logical gates we've studied so far, AND, OR, NAND, NOR and XOR, we've used a verbal expression to describe the function of the gate.

For example, with an AND gate we say, "The output of an AND gate is 1 if and only if both inputs are 1." This does get the logic across, but we need a better way to manipulate these gates and to design more complex systems. The method that we use is to create a truth table for the logic function. Figure 2.18 shows the truth tables for the five logic gates we've studied so far. The NOT gate is trivial so we won't include it here, and we've already seen the truth table for the tri-state gate, so it isn't included in the figure.

The truth table shows us the value of the output variable, C, for every possible value of the input variables, A and B. Since there are two independent input variables, the pair can take on any one of four possible combinations. Referring to Figure 2.18, we see that the output of the AND gate, C, is a logical 1 only when both A and B are 1. All other combinations result in C equal to 0. Likewise, the OR gate has its output equal to 1, if any one of the input variables, or both, is equal to 1. The truth table gives us a concise, graphical way to express the logical functions that we are working with.

Suppose that our AND gate has three inputs, or four inputs? What does the truth table look like for that? If we have three independent input variables: A, B, and C, then the truth table would have eight possible combinations, or rows, to represent all of the possible combinations of the input variables. If we had a 3-input AND gate, only one condition out of the eight, ($A = 1$, $B = 1$, $C = 1$) would product a 1 on the gate's output. For four input variables, we would need sixteen possible entries in our truth table. Thus, in general, our truth table must have 2^N entries for a system of N independent input variables. There's that pesky binary number system again. Figure 2.19 shows the truth tables and logic symbols for a 4-input AND gate and a 3-input OR gate. It is possible to find commercially available AND, NAND, OR and NOR circuits with 5 or more inputs. There are also versions of the NAND circuit that are designed to be able to expand the number of inputs to arbitrarily large values, although there are probably not many reasons to do so.

The XOR gate is an exception because it is only defined for two inputs. While we could certainly design a circuit with, for example, 5 inputs, A through E and 1 output, f, that has the property that $f = 0$ if and only if all the inputs are the same, it would not be an XOR gate. It would be an arbitrary digital circuit.

Consider the gray box of Figure 2.20. This is an arbitrary digital system that could be the control circuits of an elevator, a home heating and air conditioner, or fire alarm controller. Ultimately, we will be designing the circuit that is inside of the gray box as a circuit built from the logic gates that we've just discussed. Whatever it is, it is up to us to specify the logical relationships that each output variable has with the appropriate input variable or variables. Since the gray box has eight input variables, our truth table would have 256 entries to start with. In other words, we would design the system by first specifying the behavior of X, Y, and Z (output conditions) for each possible state of its inputs (a through h). Thus, if we were designing a heating system for a building and one of our inputs is derived from a temperature sensor that is telling us that the temperature in the room is lower than the temperature that we set on the thermostat, then it would be logical that this condition should trigger the appropriate output response (ignite burner in furnace).

A	B	C	D	f
0	0	0	0	0
1	0	0	0	0
0	1	0	0	0
1	1	0	0	0
0	0	1	0	0
1	0	1	0	0
0	1	1	0	0
1	1	1	0	0
0	0	0	1	0
1	0	0	1	0
0	1	0	1	0
1	1	0	1	0
0	0	1	1	0
1	0	1	1	0
0	1	1	1	0
1	1	1	1	1

A	B	C	f
0	0	0	0
1	0	0	1
0	1	0	1
1	1	0	1
0	0	1	1
1	0	1	1
0	1	1	1
1	1	1	1

Figure 2.19: Truth table and gate diagrams for a 4-input AND gate and a 3-input OR gate.

Referring to Figure 2-20, we might start out our design process by creating a spreadsheet with 256 rows in it and 11 columns. We would devote one column for each of the eight input variables and a separate column for each of the three output variables. Next, we would painstakingly fill in the table with all of the 256 possible combinations of the input variables, 00000000 through 11111111. Finally—and here's where the real engineering starts—for each row, we would have to decide how to assign a value to the output variables.

We'll go into the process of designing these systems in much greater detail in Chapter 3. For now, let's summarize this discussion by recognizing that we've used the truth table in two different, but related contexts:

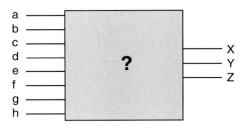

Figure 2.20: A digital system being designed. Each of the output variables X, Y, and Z is described in a separate truth table in terms of the combinations of the all possible states of the input variables.

1. The truth table could be used as a tabular format for describing the logical behavior of one of the standard gates (AND, OR, NAND, NOR, XOR or TS).
2. We can specify an arbitrarily complex digital system by using the truth table to describe how we want the digital system to behave when we finally create it.

Summary of Chapter 2

Here's what we've accomplished:

- Learned the basic logic gates, AND, OR and NOT and saw how more complex gates, such as NAND, NOR and XOR could be derived from them, and
- Learned that logic values that are dynamic, or change with time, may be represented on a graph of logic level or voltage versus time. This is called a *waveform*.
- Learned how CMOS logic gates are derived from electronic switching elements, MOSFET transistors.
- Learned how to describe the logical behavior of a gate or digital circuit as a *truth table*.

Chapter 2: *Endnotes*

[1] http://www.dnaftb.org/dnaftb/20/concept/index.html.

[2] J. Watson, *An Introduction to Field Effect Transistors*, published by Siliconix, Inc., Santa Clara, CA, 1970, p. 91.

[3] http://www.tomshardware.com.

[4] Fairchild Semiconductor Corp., Application Note 77, *CMOS, The Ideal Logic Family*, January, 1983.

Exercises for Chapter 2

1. Consider the simple AND circuit of Figure 1.14 and the OR circuit of Figure 2.5. Change the sense of the logic so that an open switch (no current flowing) is TRUE and a closed switch is FALSE. Also, change the sense of the light bulb so that the output is TRUE if the light is not shining and FALSE if it is turned on. Under these new conditions of negative logic, what logic function does each circuit represent?

2. Draw the circuit for a 2-input AND gate using CMOS transistors. Hint: use the diagram in Figure 2.16 as a starting point.

3. Construct the truth table for the following logical equations:

   a. $F = a * \bar{b} * \bar{c} + b * \bar{a}$

   b. $F = a * b + \bar{a} * c + b * c$

4. What is the effect on the signal between points A and B when the voltage at point X is raised to logic level 1?

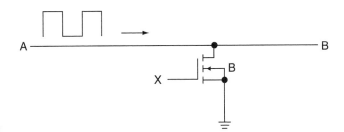

5. Construct a truth table for a *parity detection circuit*. The circuit takes as its input 4 variables, a through d, and has one output, X. Input d is a control input. If d = 1, then the circuit measures odd parity; if d = 0, the circuit measures even parity. Parity is measured on inputs a, b and c. Parity is odd if there are an odd number of 1's and parity is even if there is an even number of 1's. For example, if d = 1 and (a, b, c) = (1, 0, 0), then X = 1 because the parity is odd.

6. Draw the truth table that corresponds to the logic gate circuit shown, right:

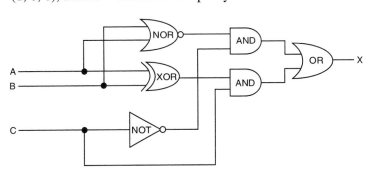

7. The gate circuit for the XOR function, equation $a \oplus b = X$, is shown in Figure 2.8. Given that you can also express the XOR function as:

$$a \oplus b = \sim [\sim (\overline{a} * b) * \sim (a * \overline{b})].$$

Redesign this circuit using only NAND gates.

8. Consider the circuit of Figure 2.3. Suppose that we replace the AND gate shown in the figure with (a) an OR gate and (b) and XOR gate. What would the output waveform look like for the case where input A = 0 and the case where input A = 1?

CHAPTER 3

Introduction to Asynchronous Logic

. . . .

Objectives

▶ *Use the principles of Boolean algebra to simplify and manipulate logical equations and turn them into logical gate designs;*

▶ *Create truth tables to describe the behavior of a digital system;*

▶ *Use the Karnaugh map to simplify your logical designs;*

▶ *Describe the physical attribute of logic signals, such as rise time, fall time and pulse width;*

▶ *Express system clock signals in terms of frequency and period;*

▶ *Manipulate time and frequency in terms of commonly used engineering units such as Kilo, Mega, Giga, milli, micro, and nano;*

▶ *Understand the logical operation of the Type 'D' flip-flop (D-FF);*

▶ *Describe how binary counters, frequency dividers, shift registers and storage registers can be built from D-FF's.*

. . . .

Introduction

Before we immerse ourselves in the rigors of logical analysis and design, it's fair to step back, take a breath and reflect on where all of this came from. We've may have been given the erroneous impression that logic sprang forth from Silicon Valley with the invention of the transistor switch.

We generally trace the birth of modern logical analysis to the Greek philosopher, Aristotle, who was born in 384 B.C., is generally acknowledged to be the father of modern logical thought. Thus, if we were to be completely accurate (and perhaps a bit generous), we might say that Aristotle is the father of the modern digital computer. However, if we were to look at the mathematics of the gates we've just been introduced to, we may find it difficult to make the leap to Aristolean Logic. However, one simple example might give us a hint.

Figure 3.1: Aristotle.

The basis of Aristotle's logic revolves around the notion of *deduction*. You may be more familiar with this concept from watching old Sherlock Holmes movies, but Sherlock didn't invent the idea of deductive reasoning, Aristotle did.

Aristotle proposes that a deductive argument has "things supposed," or a *premise* of the argument, and what "results of necessity" is the conclusion.[1]

One often cited example is:

1. Humans are mortal.
2. Greeks are human.
3. Therefore, Greeks are mortal.

We can convert this to a slightly different format:

1. Let A be the state of human mortality. It may be either TRUE or FALSE that human beings are mortal.
2. Let B be the state of Greeks. It may be TRUE or FALSE that natives of Greece are human beings.
3. Therefore, the only TRUE condition is that if A is TRUE AND B is TRUE then Greeks are mortal. Or, C (Greek mortality) = A * B.

We can trace our ability to manipulate logical expressions to the English mathematician and logician, George Boole (1816–1864).

"In 1854, Boole published an investigation into the Laws of Thought, on Which are founded the Mathematical Theories of Logic and Probabilities. Boole approached logic in a new way reducing it to a simple algebra, incorporating logic into mathematics. He pointed out the analogy between algebraic symbols and those that represent logical forms. It began the algebra of logic called *Boolean algebra*, which now finds application in computer construction, switching circuits etc."[2]

Figure 3.2:
George Boole.

Boolean algebra provides us with the toolset that we need to design complex logic systems with the certainty that the circuit will perform exactly as we intended it that it should. Also, as you'll soon see, the process of designing a digital circuit often leads to designs that are quite redundant and in need of simplification. Boolean algebra gives us the tools we need to simplify the circuit design and be confident that it will work as designed with the absolute minimum of hardware. As engineers, this is exactly the kind of analytical tool we need because we are most often tasked with producing the most cost-effective design possible.

In working with digital computing systems, there are two distinctly different binary systems with which we'll need to become familiar. Earlier in this course, we were introduced to the binary number system. This is convenient, because we need for our computer to be able to operate on numbers larger than one bit in width. The other component of this is the operation of the hardware as digital system. In order to understand how the numbers are manipulated, we need to learn the principles of binary logical algebra, or Boolean algebra.

Therefore, the first step in the process is to state some of the laws of Boolean algebra. In most cases, the laws should be obvious to you. In other cases, you might have to scratch your head a bit until you see that we're dealing with logical operations on variables that can only have two possible values, and not the addition and multiplication of variables that can take on a continuum of values. Also, be aware that there is no fixed convention for representing a NOT condition, and several variant representations may apply. This is partially historical because the basic ASCII set of printable characters did not give us a simple way to represent a NOT variable, that is a variable with a line over it, as shown in Chapter 2, Figure 2.4. Thus, you may see several different representations of the NOT condition. For example, /A , ~A , *A and A* are all commonly used ways of representing the condition, NOT A. I will avoid using the *A and A* notations because it is too easy to confuse this with the AND symbol. However, I will occasionally use ~A and /A interchangeably in the text for the NOT A condition if it makes the equation easier to understand. Most figures will continue to use the bar notation for the NOT condition. Sorry for the confusion.

Laws of Boolean Algebra

Laws of Complementation

• First law of complementation:	If $A = 0$ then $\overline{A} = 1$ and if $A = 1$ then $\overline{A} = 0$
• Second law of complementation:	$A * \overline{A} = 0$
• Third law of complementation:	$A + \overline{A} = 1$
• Law of double complementation:	$/(\overline{A}) = //A = A$

Laws of Commutation

• Commutation law for AND:	$A * B = B * A$
• Commutation law for OR:	$A + B = B + A$

Associative Laws

• Associative law for AND:	$A * (B * C) = C * (A * B)$
• Associative law for OR:	$A + (B + C) = C + (A + B)$

The associative laws enable us to combine three or more variables. The law tells us that we may combine the variables in any order without changing the result of the combinations. As we've seen in Chapter 2, this is the law that allows us, for example, to combine two, 2-input OR gates to obtain the logical equivalent of a single 3-input OR gate. It should be noted that when we say the "logical equivalent" we are neglecting any electrical differences. As we have seen, the timing behavior of a logic gate must be taken into account in any real system.

Distributive Laws

• First distributive law:	$A * (B + C) = (A * B) + (A * C)$
• Second distributive law:	$A + (B * C) = (A + B) * (A + C)$

The distributive laws look a lot like the algebraic operations of factoring and multiplying out.

Laws of Tautology

• First law of tautology: $A * A = A$	If $A = 1$ then $A * A = 1$. If $A = 0$, then $A * A = 0$. So the expression $A * A$ reduced to A.
• Second law of tautology: $A + A = A$	If $A = 1$, then $1 + 1 = 1$, If $A = 0$, then $0 + 0 = 0$. Again, the expression simply reduced to A.

Law of Tautology with Constants

•	$A + 1 = 1$
•	$A * 1 = A$
•	$A * 0 = 0$
•	$A + 0 = A$

Laws of Absorption

•	First law of absorption:	$A * (A + B) = A$
•	Second law of absorption:	$A + (A * B) = A$

This one is a bit trickier to see. Consider the expression: A * (A + B).

If A = 1, then this becomes 1 * (1 + B). By the law of tautology with constants, 1 + B = 1 so we are left with 1 + 1 = 1. If A = 0, the first expression now becomes 0 * (0 + B). Again, by the law of tautology with constants, this reduces to 0 * B, which has to be 0. Thus, in both cases, the expression reduced to the value of A, the value of B does not figure in the result.

DeMorgan's Theorems

•	Case 1: $(\overline{A * B}) = \overline{A} + \overline{B}$
•	Case 2: $(\overline{A + B}) = \overline{A} * \overline{B}$

DeMorgan's theorems are very important because they show the relationship between the AND and OR functions and the concepts of positive and negative logic. Also, DeMorgan's theorems show us that any logic function using AND gates and inverters (NOT gates) may be duplicated using OR gates and inverters. Also notice that the left side of both of the above equations are just the compound logic functions NAND and NOR, respectively.

Before we move on, we should discuss the relationship between DeMorgan's theorems and logic polarity. Up to now, we've adopted the convention that the more positive, or higher voltage signal was a 1, and the lower voltage, or more negative voltage was a 0. This is a good way to introduce the topic of logic because it's easier to think about TRUE/FALSE, 1/0 and HIGH/LOW in a consistent manner. However, while TRUE and FALSE have a logical significance, from an electrical circuit point of view it is rather arbitrary just how we define our logic.

This is not to say that the electrical circuit that is an AND gate would still be an AND gate if we swapped 1 and 0 in the electrical sense. It wouldn't. It would be an OR gate. Likewise, the OR gate would become an AND gate if we swapped 1 and 0 so that the lower voltage became a 1 and the higher voltage became a 0. You can verify this for yourself by reviewing the truth tables for the AND gate and OR gate in Lesson 1. You can see that if you take the truth table for the AND gate and swap all of the 1's with 0's and all of the 0's with 1's, you end up with the truth table for the OR gate (in negative logic). Try it again for the OR gate and you'll see that you now have an AND gate in negative logic.

An important point to keep in mind is that the same electronic circuit will give us the logical behavior of an AND gate if we adopt the convention that logical 1 is the more positive voltage. Thus, a logical 1 might be anything greater than about +3 volts and a logical 0 might be anything less than about 0.5 V. We call this *positive logic*. However, if we reverse our definition of what voltage represents a 1 and what voltage represents a 0, the same circuit element now gives us the logical OR function (*negative logic*).

Thus, in a digital system, we usually make TRUE and FALSE whatever we need it to be in order to implement the most efficient circuit design. Since we can't depend upon a consistent meaning for TRUE and FALSE, we generally use the term *assert*. When a signal is asserted, it becomes active. It may become active by changing from a logic level LOW to a logic level HIGH, or vice versa. As you'll see shortly when we discuss memory systems, most of the memory control signals are asserted LOW, even though the address of the memory cells and the data stored in them are asserted HIGH. You saw this in Chapter 2 when we discussed the logical behavior of the tri-state gate. Recall that the tri-state gate's output goes into the low-Z state when the output enable ($\overline{OE}$) input is asserted low. This does not mean that the $\overline{OE}$ signal is FALSE, or TRUE in the negative logic sense, it simply means that the signal become active in the low state.

In Figure 2.20, we considered the most general case of a logical system design. Each of 3 output variables is defined as a function of up to 8 input variables, i.e., $X = f(\,a, b, c, d, e, f, g, h)$, and so on. Note that output variables X, Y and Z may each be a separate function of some or all of the input variables, a through h. The problem that we must now solve is in four parts:

1. How do we use the truth table to describe the intended behavior of our digital system? This is just specifying your design.
2. Once we have the behavior we want (as defined by the truth table), how do we use the laws of Boolean algebra to transform the truth table into a Boolean algebraic description of the system?
3. When we can describe the design as a system of equations, can we use some of the rules of Boolean algebra to simplify the equations?
4. Finally, when we can describe the equations of our system in the simplest algebraic form, then how can we convert the equations to a real circuit design?

In a moment, we'll see how to slog through this, and using the rules of Boolean algebra, reduce it to a simpler form. However, if we knew beforehand that output Y only depended upon inputs c, d, and g, then we could immediately simplify the design task by limiting the truth table for Y to one with only 3 input variables instead of eight. As we'll soon see, the general case can be reduced to the simplified case, so either path gets you to the end point. It is often the case that you won't know beforehand that there is no dependence of an output variable on certain input variables; you only learn this after you go through all of the algebraic simplifications.

Figure 3.3, is an example of some arbitrary truth table design. It doesn't describe a real system, at least not one that I'm aware of. I just made it up to go through the simplification process.

Outputs E and F are the two dependent variables that are functions of input variables A through D. Since there are 4 input variables, there are 2^4, or 16, possible combinations in the truth table, representing all the possible combinations of the input variables. Also, note how each of the input variables are written in each column. Using a method like this insures that there are no missing or duplicate terms. Since this is a made-up example, there is no real relationship between the output variables and the input variables. This means that I arbitrarily placed 1's and 0's in the E and F columns to make the example look interesting. If this exercise was part of a real digital design problem, you, the designer, would consider each row of the truth table and then decide what should be the response of each of the dependent outputs do in response to that particular combination of inputs.

For example, suppose that we're designing a simple controller for a burglar alarm system. Let's say that output E controls a warning buzzer inside the house and output F controls a loud siren that can wake up the neighborhood if it goes off. Inputs A, B, and C are sensors that detect an intruder and input D is the button that controls whether or not the burglar alarm is active or not. If D = 0, the system is deactivated, if D = 1, the system is active and the sirens can go off.

A	B	C	D	E	F
0	0	0	0	0	1
1	0	0	0	0	0
0	1	0	0	0	0
1	1	0	0	0	0
0	0	1	0	1	0
1	0	1	0	1	0
0	1	1	0	0	1
1	1	1	0	0	0
0	0	0	1	0	0
1	0	0	1	0	0
0	1	0	1	0	0
1	1	0	1	0	0
0	0	1	1	0	0
1	0	1	1	0	0
0	1	1	1	0	0
1	1	1	1	1	0

$$E = \overline{A}*B*C*\overline{D} + A*\overline{B}*C*\overline{D} + A*B*C*D$$
$$F = \overline{A}*\overline{B}*\overline{C}*\overline{D} + \overline{A}*B*C*\overline{D}$$

Figure 3.3: Truth table for an example digital system design. Outputs E and F are the SUM of Products (minterm) representation of the truth table.

In this case, for all the conditions in the truth table where D = 0, you don't want to allow the outputs to become asserted, no matter what the state of inputs A, B, or C. If D = 1, then you need to consider the effect of the other inputs. Thus, each row of the truth table gives you a new set of conditions for which you need to independently evaluate the behavior of the outputs. Sometimes, you can make some obvious decisions when certain variables (D = 0) have global effects on the system.

However, suppose for a moment that the example of Figure 3.3 actually represented something real. Let's consider output variable F. The logical equation,

$$\underbrace{F = \overline{A} * \overline{B} * \overline{C} * \overline{D}}_{\text{term 1}} + \underbrace{\overline{A} * B * C * \overline{D}}_{\text{term 2}}$$

tells us that F will be TRUE under two different set of input conditions. Either the condition that all the inputs are FALSE (term 1) or A and D are FALSE, while B and C are TRUE (term 2) will cause the output variable F to be TRUE. How did we know this to be the case? We designed it that way! As the engineers responsible for this digital design, these are the two particular set of input conditions that can cause output F to be TRUE.

We call this form of the logical equation the *sum or products form, or minterm form.* There is an alternate form called the *maxterm form,* which could be described as a *product of sums.* The two forms can be converted from one to the other through DeMorgan's Theorems. For our purposes, we'll restrict ourselves to the minterm form.

Remember, this truth table is an example of some digital system design. In a way, it represents a shorthand notation for the design specification for the system. At this point we don't know why the columns for the dependent variables, *E* and *F*, have 1s or 0s in some rows and not others. That came out of the design of the system. We're engineers and that's the engineering design phase. What comes after the design phase is the implementation phase. So, let's tackle the implementation of the design.

Referring to Figure 3.3, we see that that output variable, E, is TRUE for three possible combinations of the input variables, A through D:

1. $\overline{A} * \overline{B} * C * \overline{D}$
2. $A * \overline{B} * C * \overline{D}$
3. $A * B * C * D$

We can express this relationship as a logical equation:

$$E = \overline{A} * \overline{B} * C * \overline{D} + A * \overline{B} * C * \overline{D} + A * B * C * D$$

The OR'ing of the three AND terms means that any one of the three AND terms will cause E to be TRUE. Thus, for the *combination*, we need to use AND. For the *aggregation*, we use OR. Likewise, we can express the value of *F* as

$$F = \overline{A} * \overline{B} * \overline{C} * D + \overline{A} * B * C * \overline{D}$$

At this point it would be relatively easy to translate these equations to logic gates and we would have our digital logic system done. Right? Actually, we're sort of right. We still don't know if the terms have any redundancies in them that we could possibly eliminate and make the circuit easier to build. The redundancies are natural consequences of the way we build the system from the truth table. Each row is considered independently of the other, so it is natural to assume that certain duplications will creep into our equations.

Using the laws of Boolean algebra, we could manipulate the equations and do the simplifications. However, the most common form of redundant term is $A * B + A * \overline{B}$.

It is easy to show that $A * B + A * \overline{B} = A$. How?

1. First Distributive Law: $A * B + A * \overline{B} = A * (B + \overline{B})$
2. Third Law of Complementation: $B + \overline{B} = 1$
3. Finally, $A * 1 = A$

Thus, if we can group the terms in a way that allows us to "factor out" a set of common AND terms and be left with an OR term appearing with its complement, then that term drops out.

The Karnaugh Map

In a classic paper, Karnaugh[3] (pronounced CAR NO) described a graphical method of simplifying the sum of products equation of the truth table without the need to resort to using Boolean algebra directly. The Karnaugh map is a graphical solution to the Boolean algebraic simplification:

$$A * B + A * \overline{B} = A$$

This simplification is one of the most commonly occurring ones because of the redundancies that are inherent in the construction of a truth table.

There are a few simple rules to follow in building the Karnaugh map (K-map). Figure 3.4 shows the construction of 3, 4 and 5 variable K-maps.

Refer to the K-map for 4 input variables. Note the vertical dark gray edges and the horizontal light gray edges. These edges are considered to be adjacent to each other. In other words, the map can be considered to be a cylinder wrapped either horizontally or vertically.

The method used to identify the columns may look strange to you but if you look carefully you'll see that as you move across the top of the map, the changes in the variables A and B are such that:

- Only one variable changes at a time,
- All possible combinations are represented,
- The variables in column 1 and column 4 are adjacent to each other.

The order for listing the variables shown in Figure 3.4 is not the only way to do it, it is just the way that I am most comfortable using and, through experience, is the way that I know I am drawing the map correctly. It is easy to have the correct set of variables listed, but to make a mistake in getting the order correctly listed, which will thus yield erroneous results. In other words, "Garbage in, garbage out."

Another point that should be noted is that the K-map yields the most simple form of the equation when the number of variables is 4 or less. For maps of 5 or more variables, the correct procedure is to use a 3-dimension map comprised of planes of multiple 4 variable maps. However, for the purposes of this discussion, we'll make it a point to note whenever the 5 variable maps require further simplification. In other words, it's easier to do a little Boolean algebra than it is to draw a 3D K-map.

We can summarize the steps necessary to simplify a truth table using the K-map process as follows:

1. The number of cells in the K-map equals the number of possible combinations of input variable states. For example,
   a. 4 input variables: A, B, C, D = 16 cells
   b. 3 input variables: A, B, C = 8 cells
   c. 5 input variables: A, B, C, D, E = 32 cells

Thus, the number of cells = $2^{(\text{NUMBER OF INPUT VARIABLES})}$

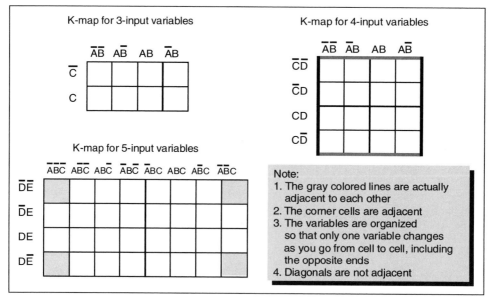

Figure 3.4: Format for building a Karnaugh map in 3, 4 and 5 variables.

2. Construct the K-map so that as you move across the columns or down the column, only one variable changes. Referring to Figure 3.5, note that the first and last columns are adjacent and the top and bottom rows are also adjacent. It is as if the map is actually wrapped on a cylinder. Note that the first and last cells along the diagonals are not considered to be adjacent.

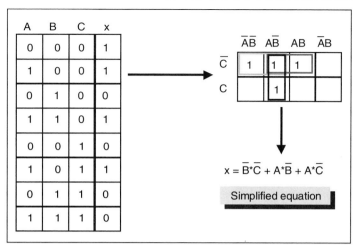

3. Construct a separate K-map for every output variable.

4. Place a "1" in every cell of the K-map that has a "1" in the corresponding row of the truth table.

Figure 3.5: Translating a truth table to a K-map. Each cell of the K-map represents one combination of the independent variables. A '1' is placed in each cell that contains a 1 in the corresponding row of the truth table.

5. Draw a loop around the largest possible number of *adjacent* cells that contain a 1. You can form loops of 2, 4, 8, 16, 32 and so on in adjacent cells.

6. You may form multiple loops and a cell in one loop may be in another loop, but each loop must contain at least one cell that is not contained in any other loop. Inspect the map for any loop whose terms are all enclosed in other loops and remove those loops.

7. Finally, simplify the loop by removing any variable that appears within that loop in both its complemented and un-complemented form. The simplified equation is now the ORing of the loops of the K-map where each loop represents the simplified minterm form.

Perhaps an example would help. Figure 3.5 demonstrates the process. We have a truth table with three independent input variables—A, B, and C—and one dependent output variable, x. Three input variables imply eight rows for the truth table. In Figure 2.5, there are four rows that have a 1 in the "x" column. We can thus write down the unsimplified logic equation for x:

$$x = \overline{A} * \overline{B} * \overline{C} + A * \overline{B} * \overline{C} + A * B * \overline{C} + A * \overline{B} * C$$

- Now, refer to Figure 3.5. The K-map corresponding to the truth table is shown on the right. Four cells contain a '1' term for the output variable 'x', in agreement with the truth table. We can draw three loops as follows:
- Light-gray loop around the term $\overline{A} * \overline{B} * \overline{C}$, and the term $A * \overline{B} * \overline{C}$
- Medium-gray loop around the term $A * \overline{B} * C$, and the term $A * \overline{B} * \overline{C}$
- Dark-gray loop around the term $A * \overline{B} * \overline{C}$, and the term $A * B * \overline{C}$

We can thus remove the variable A from the light-gray loop, B from the medium-gray loop and C from the dark-gray loop, respectively. The resulting equation: $x = \overline{B} * \overline{C} + A * \overline{B} + A * \overline{C}$ is the simplified version of our original equation.

Note that in this example, the cell corresponding to the state of the input variables $A * \overline{B} * \overline{C}$ is common to the three loops and it appears in each loop. However, each of the loops also contains one variable not contained in any other loop, so we can draw these three loops.

Before we go any further, it is reasonable to see if we could have achieved the same simplified logical equation by just doing the algebra the way George Boole intended us to. Below are the algebraic steps to simplifying the logical equation:

Step 1	$x = \overline{A} * \overline{B} * \overline{C} + A * \overline{B} * \overline{C} + A * B * \overline{C} + A * \overline{B} * \overline{C}$	From the truth table
Step 2	$x = \overline{A} * \overline{B} * \overline{C} + A * B * \overline{C} + A * \overline{B} * (C + \overline{C})$	First Law of Distribution
Step 3	$x = \overline{A} * \overline{B} * \overline{C} + A * B * \overline{C} + A * \overline{B}$	First Law of Complementation
Step 4	$x = \overline{A} * \overline{B} * \overline{C} + A * (B * \overline{C} + \overline{B})$	First Law of Distribution
Step 5	$x = \overline{A} * \overline{B} * \overline{C} + A * [(\overline{B} + \overline{C}) * (B + \overline{B})]$	Second Law of Distribution
Step 6	$x = \overline{A} * \overline{B} * \overline{C} + A * (\overline{B} + \overline{C})$	Law of Complementation
Step 7	$x = \overline{A} * \overline{B} * \overline{C} + A * \overline{B} + A * \overline{C}$	First Law of Distribution
Step 8	$x = \overline{B} * (\overline{A} * \overline{C} + A) + A * \overline{C}$	First Law of Distribution
Step 9	$x = \overline{B} * [(\overline{A} + A) * (\overline{C} + A)] + A * \overline{C}$	Second Law of Distribution
Step 10	$x = \overline{B} * (\overline{C} + A) + A * \overline{C}$	First Law of Distribution
Step 11	$x = \overline{B} * \overline{C} + \overline{B} * A + A * \overline{C}$	First Law of Distribution

Let's consider a slightly more involved example. Figure 3.6 shows a 4-input variable problem with two dependent variables for the outputs. Again, this is a made-up example, as far as I know it doesn't represent a real system. If we were actually trying to build a digital system, then the systems requirements would dictate the state of each output variable for a given set of input variables.

If you consider the K-map for variable 'X' you see that we are able to construct two simplifying loops. The medium-gray loop folds around the edges of the map because those edges are adjacent. In a similar way, the dark-gray loop folds around the top and bottom edges of the map. You might wonder why we couldn't also make a loop with the terms $A * B * \overline{C} * \overline{D}$ and $\overline{A} * B * \overline{C} * \overline{D}$.

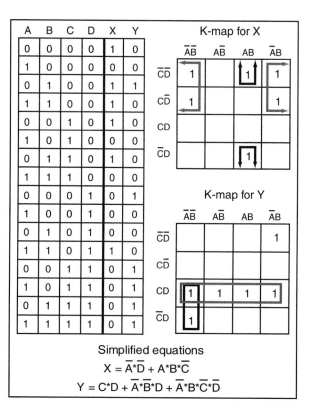

Figure 3.6: Simplifying a 4-variable truth table with two dependent variables.

We can't make another loop because the both terms are already in other loops. In order to make a third loop we would need to have one or more terms not contained in any other loop.

Using the K-map does not always result in the most simplification possible, but it comes pretty close. This is particularly true of K-maps larger than four variables. In fact, the five variable K-map should technically be represented as two four variable K-maps, one on top of the other. Remember that you may always try to use Boolean algebra and DeMorgan's Theorems to simplify your equations to their final form.

As a final step, let's convert our simplified logical equations to a real hardware gate implementation. We'll take the logical equation that we simplified in Figure 3.5 and convert it to its gate equivalent circuit. Figure 3.7 shows the implementation of the design using NOT, AND and OR gates. This is not the way that you "must" design it. It is just a convenient way to show the hardware design. The 3 input signals are shown along the top left of the figure. For each input variable, we also add a NOT gate because it will be convenient to assume that we'll need to use either the variable or its complement in the circuit design.

Also note that we use a black dot to indicate a point where two separate wires are physically connected together. We need to do this so we can differentiate wires that cross each other in our drawing, *but are not connected together*, from those wires that are connected together. The black dot serves that purpose for us.

The circuit in Figure 3.7 also shows the input variables, their complements, the combinatorial terms in the AND gate and the aggregation of the combinatorial terms using the OR gate. Notice that we needed three, 2-input AND gates and one, 3-input OR gate to implement the design. If we had a more complex problem we might choose to use AND and OR gates with more available inputs or use several levels of gates with fewer inputs. Thus, we could create an equivalent 7-input AND gate from three, 3-input AND gates.

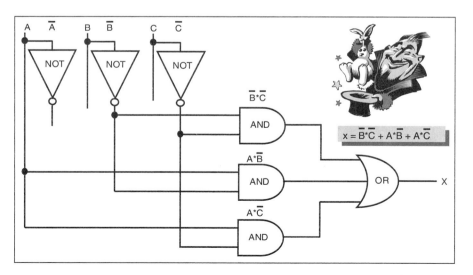

Figure 3.7: Circuit implementation of the logical equation $X = \overline{B} * \overline{C} + A * \overline{B} + A * \overline{C}$.

Does the circuit of Figure 3.7 actually agree with the original design of our truth table? It might be a good idea to do a quick check, just to build our confidence for the complexities to come. Let's do the first three terms of the truth table and we'll leave it as an exercise for you to do the remaining five terms.

1. *Term 1:* A = 0, B = 0, C = 0. Here the NOT gates for the variables B and C invert the inputs and feed the values B = 1 and C = 1 to the first AND gate. Since both inputs are '1', the output of this AND gate is 1. The output of this AND gate is the input to the OR gate. Since one of the inputs is '1', the output is also '1' and x = 1, just as required by the truth table.

2. *Term 2:* A = 1, B = 0, C = 0. According to the truth table, 'x' should also equal '1' for this situation. Since the first AND gate does not require variable A as an input, variable B and C are unchanged, so we also get x = 1 for this situation.

3. *Term 3:* A = 0, B = 1, C = 0. The first AND gate now gives us a '0' because the complement of B = 1 is $\overline{B}$ = 0. Thus, 0 AND 1 = 0. The second AND gate has A and $\overline{B}$ as its inputs. Since this condition has A = 0 and $\overline{B}$ = 0 as its inputs, it is also '0'. The third AND gate has A = 0 and $\overline{C}$ = 1 as its inputs. Again, it results in an output of '0'. Since all three inputs to the OR gate are '0', x = 0.

Before we leave the topic of logic gates and begin to consider systems that depend upon a synchronization or clock signal, let's examine how else we might build a digital system. The truth table is an interesting format because it looks very close in form to how a memory is formed. Figure 3.8 is the truth table example from Figure 3.3 but shown as if it was a memory device. We have four independent variables and two dependent variables.

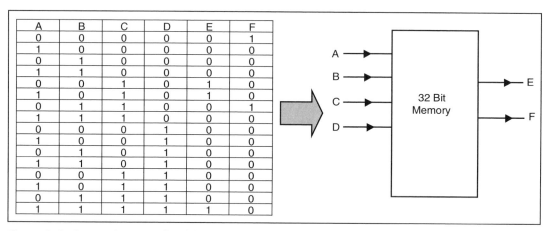

A	B	C	D	E	F
0	0	0	0	0	1
1	0	0	0	0	0
0	1	0	0	0	0
1	1	0	0	0	0
0	0	1	0	1	0
1	0	1	0	1	0
0	1	1	0	0	1
1	1	1	0	0	0
0	0	0	1	0	0
1	0	0	1	0	0
0	1	0	1	0	0
1	1	0	1	0	0
0	0	1	1	0	0
1	0	1	1	0	0
0	1	1	1	0	0
1	1	1	1	1	0

Figure 3.8: Converting a truth table to a memory image. The independent variables, A through D provides the address to the memory. The two dependent variables, E and F, are represented by the data in the memory cell corresponding to the address of the memory (row of the truth table).

Let's consider the implications of what we just did. Here we can imagine that we use a real memory device and fill it so that when we give it the appropriate address bit values (in this case, the appropriate combination of input variables A through D) the data out from the memory (the dependent variables *E* and *F*) is the circuit behavior that we have assigned to the truth table. Thus, we can implement logical systems either by creating an electrical circuit design using logical gates, or we can create a logically equivalent design by using a memory device to directly implement the contents of the truth table. In the case of memory as logic, we don't do any logical simplification as we would with a gate circuit design. Also, we might not get the speed that we need using one or the other method.

To make this point even stronger, let's redraw Figure 3.8 as Figure 3.9. The only difference is that we'll represent it as a real memory device. The independent variables (*A* through *D* in Figure 3.3) are represented as address bits, A0–A3, to the memory device. The dependent variables (*E* and *F* in Figure 3.3) are now the data bits stored in the various pairs of memory locations.

Thus, our memory needs to be able to hold a total of 32 bits, 16 bits for each of the two dependent variables. Each bit represents the state of that variable for the combination of states of the input variables forming the address of the memory cell.

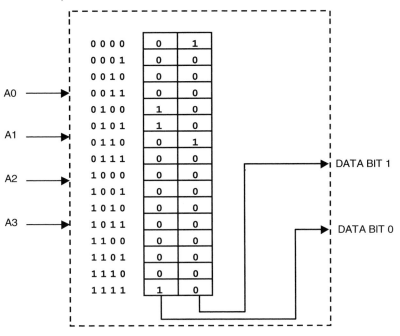

Figure 3.9: Converting a truth table to a memory image. The independent variables, A through D provides the address to the memory. The two dependent variables, E and F, are represented by the data in the memory cell corresponding to the address of the memory (row of the truth table).

Figures 3.8 and 3.9 represent an alternative way of creating a hardware implementation of the logical design. In the prior example, we used the laws of Boolean algebra and the K-maps to build a simplified set of logical equations that we could then implement as a combination of logic gates (also called *combinatorial logic*). Figure 3.8 shows that we could simply take the truth table, as is, and fill up a memory chip with all the possibilities and be done with it. It turns out that both methods are equally valid and are used where it's most appropriate. The use of memory as logic, called *microcode*, forms the basis for much of the control circuitry within a modern digital computer. We'll revisit this concept in a later chapter when we discuss *state machines*.

Clocks and Pulses

In Chapter 2, we first saw digital signals as waveforms. That is, we represented the logical signals as values that change over time. The waveform is a strip chart recorder view of the digital signal. The simplest digital signal that we might want to consider is the simple positive pulse of Figure 3.10, which schematically shows a single positive pulse with a pulse height of about 3 volts.

We might stop for a moment and ask, "What's a pulse?" You probably know the medical term for your "pulse." You feel, or should feel a pressure surge in your veins every time your heart beats. What you are feeling is the blood creating a pressure pulse as it flows because the pumping is discontinuous, and happens in discrete pulses of blood. You also can see this when you get an EKG done on your heart and you see the characteristic spikes of the electrical signals around your heart (Figure 3.11).

What is characteristic of any pulse is that the system goes from a relaxed state, such as low pressure, to an excited state (blood surge) and then back again. Electrically, we can describe a pulse as a signal going from low to high and then back to low, or vice versa. In other words, we could have a pulse such that the system goes from

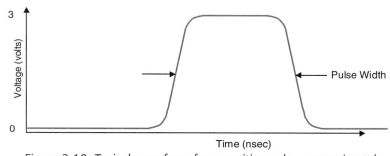

Figure 3.10: Typical waveform for a positive pulse, approximately 3 volts high.

asserted, to nonasserted, and then back to an asserted state. We call a pulse that goes from low to high and back to low a *positive pulse* a pulse that goes from high to low and back to high a *negative pulse*.

The pulse in Figure 3.10 is positive because it starts from the LOW state, goes HIGH, and then returns to the LOW state when it is completed. In this example, the width of the pulse, or *pulse width*, is a measure of the amount of time that the pulse exists, might be something like 50 nanoseconds (often abbreviated as 50 ns), or 50 billionths of a second. The pulse width is measure at a point mid way between the base level of the pulse and its nominal height. Thus, for a pulse that is 3 volts high, the pulse width would be measured at the 1.5 volts portion of the waveform.

Figure 3.12 shows a, more or less, "real pulse." The real pulse is what you might see if you had

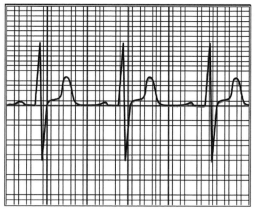

Figure 3.11: Part of an EKG showing the characteristics pulses of the electrical signals around the heart.

a really fast stopwatch, a really fast voltmeter and you could scribble like crazy (remember, this is a thought experiment). Real people, use analytical equipment called *oscilloscopes* to see this waveform. We'll look at some real oscilloscope waveforms later in the chapter. Notice how the gray line, which represents the change in voltage of the pulse over time, has some slope to it as it rises and falls. This is because the pulse can't change state infinitely rapidly. It takes some time for the voltage to rise to a 1, or fall back to 0. We call these times the *rise time* and *fall time*, respectively. Technically, for reasons that we don't need to consider, the rise and fall times are measure at the 10% and 90% points of the voltage.

Figure 3.12 shows how the pulse might be seen on the screen of an oscilloscope. The horizontal axis displays the elapsed time in some convenient units (nanoseconds, in this example) and the vertical axis displays how the voltage signal changes with time.

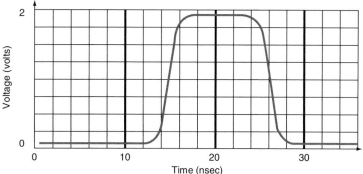

Figure 3.12: A positive pulse as it might appear on the display of an oscilloscope.

Figure 3.13 shows an actual rise time measurement from an R&D laboratory oscilloscope. The oscilloscope can perform a number of automatic measurements, such as rise time, fall time, pulse width, pulse height, frequency and period. The oscilloscope circuitry automatically analyzes the shape of the pulse waveform and locates the 10% and 90% points on the pulse and then calculates the time difference between those two points.

Before we move on to consider clocks, we should discuss one last point about gates. We've previously defined the *propagation delay* of a gate as time delay from a change at the input of the gate to the corresponding change at the output of the gate. Let's see what an actual propagation delay measurement might look like on our lab oscilloscope. Figure

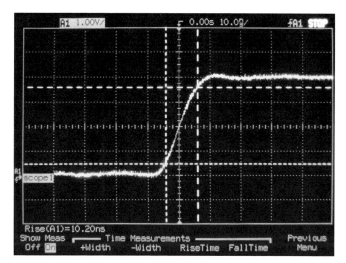

Figure 3.13: Oscilloscope rise time measurement. The vertical axis is 1 volt per division and the horizontal axis is 10 nanoseconds per division.

3.14 shows us what a propagation delay measurement might look like for a NOT gate. We connect our oscilloscope probes to the input and output of the NOT gate as shown in the figure. To make this measurement we start our oscilloscope trace a few nanoseconds before the input signal to the gate (the falling edge) occurs. This oscilloscope can simultaneously display the logic state of both the input and output waveforms so we see the input signal going low and then slightly later in time, the output signal goes high. In fact, if you look closely at

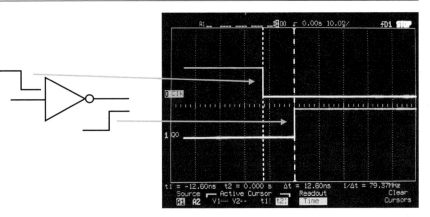

Figure 3.14: Oscilloscope trace of a propagation delay measurement. The upper trace represents a measurement probe on the input side of the NOT gate and the lower trace shows the output of the gate. The display shows that the output goes high 10.2 nanoseconds after the input goes low.

the oscilloscope display, you'll see that the time interval between the falling edge of the input and the riding edge of the output is 12.60 nanoseconds.

To this point, we've considered pulses as single events. However, just like the EKG shows, your heart pumps continuously, so the EKG displays a string of pulses.

In Figure 3.15, we once again go back to the idealized view of the waveform. Note the absence of any slope to the rising and falling edges of the waveform. We consider these pulses to have infinitely fast transitions from LOW to HIGH and HIGH to LOW. It doesn't hurt anything to make this assumption and it makes our diagrams easier to analyze and understand.

We refer to this continuous stream of pulses, such as the waveform shown in Figure 3.15, as a *clock*. The clock is not the same as the clock in your computer that gives you the time of day. The clock that we are concerned with in this context is a continuous stream of tightly regulated pulses, usually controlled by crystal oscillators. The crystals have the property that they can create a tuned circuit that resonates, or oscillates, at a very predictable and stable frequency. For example, even the cheapest digital wristwatch can keep time to better than 1 minute per month because it

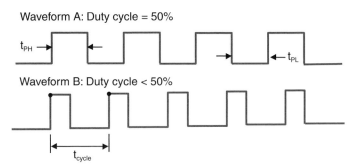

Figure 3.15: Examples of clock waveforms. When the pulse width of the high portion of the waveform equals the pulse width of the low portion of the waveform (waveform A), we have a 50% duty cycle. Waveform B has a duty cycle below 50%. The time difference between the two black dots on waveform B represent the period of the waveform.

uses a crystal that oscillates at approximately *32K Hz. We use the term Hertz, to represent cycles per second, or the oscillating frequency of a clock stream. The symbol is Hz. The unit, Hertz, was named in honor of the German physicist,* Heinrich Rudolf Hertz (1857–1894).

It's easy to get lost in the minutia of clock and waveforms, but let's stop for a second and consider the question, *"What really is a clock, anyhow?"*

A clock is a constant stream of pulses that is usually quite accurately spaced in time. Even an inexpensive digital watch only gains or loses a few seconds a month. Considering that its pulse is about 32 thousand beats per second, that's pretty impressive. We won't spend any time discussing how such accurate time signals are generated because we don't want to offend the Electrical Engineers, but we will try to understand what the clock is doing in our system.

Imagine that you are looking at the pendulum of an impressive old grandfather's clock. If you could start a stopwatch at the point where the pendulum just stops at one end of its travel and stop the watch when it travels to the side and returns, you would measure the period of the pendulum. The grandfather's clock uses the fact that the period of the pendulum's swing varies very little from cycle to cycle to advance the clock's time through a mechanism of gears. The point was to show you that in real systems then pendulum provides us with an accurate synchronization mechanism that we can use to drive the clock's internal timekeeping mechanism. Also, the action of the pendulum provides the source of synchronization for the entire clock. Each time it swings back and forth, it advances the gears that tell time.

Suppose it takes the pendulum 5 seconds to go from one extreme of its travel to the other and back again. This is 1 cycle every 5 seconds, so the period is 5 seconds. The number of cycles in one second is just the reciprocal of the period, or 0.2 Hz.

Within our digital systems, we use a clock signal to provide the same synchronization mechanism. In most computer systems, a single clock signal is distributed throughout the circuitry to provide one accurate timing source that all internal operations may be synchronized with. When you go into your local computer store to buy the latest PC, the salesperson will try to convince you to buy a PC with a 3.2 Gigahertz clock in order for you to get the maximum gaming experience. What are you really being sold? Just that this super PC has a faster clock. This means more things happen in one second, so it runs faster.

You're probably already familiar with the term "Hertz" and the abbreviation "Hz" if you've ever purchased a PC. The salesperson told you that this PC or that PC was a 3.0 "gigahertz" machine, and it was clearly better than your old 500 "megahertz" boat anchor. What you now understand is that the clock frequency of your old computer was 500 million cycles per second, or 500 MHz. The computer that is only $1,000 away from being in the trunk of your car has a clock frequency of 3 billion cycles per second, or 3 GHz. Since one billion is 1000 million, this is the same as saying that the new computer has a clock frequency of 3000 MHz, or roughly six times the clock speed of the old one.

What are we trying to show in Figure 3.16? Let's try to imagine that we're sitting on the clock signal input on a typical computer or microprocessor. You've got a really fast Radio Shack® voltmeter and you're measuring the voltage (logic levels) at the clock input. The clock voltage is low for a while (logic level 0), then it goes high for a while (logic level 1) and then low again, over and over. Each low-time interval is exactly the same as the previous one. Each high-time interval is also the same as every other one. The transitions from low to high and high to low are very fast and also equal. Thus, the time period for one complete cycle (low to high and back to low) is extremely repeatable.

Figure 3.16 is an actual oscilloscope display of a 2.5 MHz clock signal. Notice that the clock waveform of the figure is not as "clean" as our idealized waveforms, but it isn't too far off. The rising and falling edges are clearly visible in this view. We can also learn a valuable lesson from looking at the figure. Even though the clock waveform isn't as pretty as the idealized waveform, it functions correctly in our circuit. The reason is simple. In our digital world of 1's and 0's, nothing else counts. As long as the signal is less that than the zero threshold or greater than the threshold for a 1, the voltage will be properly interpreted.

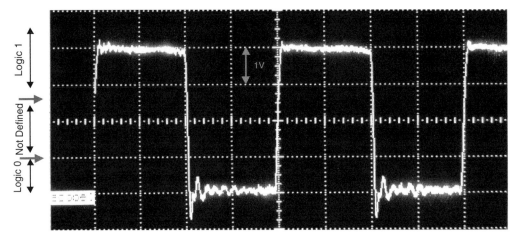

Figure 3.16: Oscilloscope image of a 2.5 MHz clock. Each vertical unit represents 1 volt and each horizontal unit represents 100 nanoseconds. Notice that as long as the signal is above the logic 1 threshold, or below the logic 0 threshold, it will be interpreted correctly.

Let's define some terms:
- *Frequency:* The number of clock pulses per unit time, usually seconds. The frequency is measured in Hertz, or Hz. One Hz is the same as one clock cycle per second.
- *Period:* The inverse of the frequency. It is the amount of time between two equal points on adjacent waveforms. The two black dots on waveform *B* in Figure 3.15 represent the period of the waveform, or the time for one cycle of the wave, t_{cycle}. The frequency of the clock equals the inverse of the period. A clock waveform with a period of one second has a frequency of 1 Hz.
- *Duty cycle:* Ratio of the amount of time the clock is HIGH to the period. A 50% duty cycle means that the clock is high for exactly ½ of the period, or the amount of time the clock is HIGH equals the amount of time the clock is LOW. We can also use the equation: Duty cycle = $(t_{PH} / (t_{PH} + t_{PL})) \times 100\%$.

Figure 3.17, shows the relationship between the common units of time (in computer lingo) and units of frequency.

Figure 3.17 also shows us that a clock frequency of 1 MHz has a clock period of 1 microsecond (μs) and a clock frequency of 1 GHz has a period of 1 nanosecond (1 ns). Thus, your 1-GHz Athlon computer has a clock that oscillates so quickly that light can only travel about 1 foot in the time it takes for one cycle of the clock to occur.

Let's review some useful relationships:

- A clock period of 1 ns has a frequency of 1 GHz.
- A clock frequency of 1 MHz has a period of 1 μsec (microsecond).
- A clock period of 1 msec (millisecond) has a frequency of 1 KHz.
- A clock period of 1 second has a frequency of 1 Hz.

- Common time (period) measurement
  - 1 millisecond (msec) = 10^{-3} sec
  - 1 microsecond (μsec) = 10^{-6} sec
  - 1 nanosecond (nsec) = 10^{-9} sec
  - 1 picosecond (psec) = 10^{-12} sec
  - 1 femtosecond (fsec) = 10^{-15} sec
- Figure of merit: The speed of light = 1 nsec/foot in free space
- Frequencies are the inverse of time
  - 1 kilohertz (KHz) = 10^{3} Hz (cycle per second)
  - 1 megahertz (MHz) = 10^{6} Hz
  - 1 gigahertz (GHz) = 10^{9} Hz
  - 1 terahertz (THz) = 10^{12} Hz

Figure 3.17: Common units of measurement of time and frequency. The speed of light in air is amazingly close to 1 nanosecond per foot. The speed of light through a conductor path on an integrated circuit is about half that, or about 6 inches per nanosecond.

Summary of Chapter 3

- Boolean algebra provides the rules for manipulating and simplifying logical equations.
- Truth tables provide a convenient method to describe the behavior of an arbitrary digital system in terms of the state of input variables and the resulting state of the output variables.
- The Karnaugh maps are a graphical method of simplifying the minterm form of truth tables.
- Digital systems are driven by clocks which are continuous streams of pulses and pulses may be describes in terms of their width, height, rise time and fall time.
- Frequency and period have an inverse relationship to on another.
- We use the engineering number system to describe the commonly occurring frequencies and times that we will be dealing with in digital systems.

Chapter 3: *Endnotes*

[1] Smith, Robin, "Aristotle's Logic," *The Stanford Encyclopedia of Philosophy (Fall 2003 Edition)*, Edward N. Zalta (ed.), http://plato.stanford.edu/archives/fall2003/entries/aristotle-logic/.

[2] J. J. O'Connor and E. F. Robertson, http://www.gap.dcs.st-and.ac.uk/.

[3] M. Karnaugh, *The Map Method for Synthesis of Combinational Logic Circuits,* taken from, *Computer Design Development Principle Papers,* Edited by Earl E. Swartzlander, Jr., ISBN 0-8104-5988-4, Hayden Book Company, Rochelle Park, NJ, 1976, p. 25.

[4] Gerald Williams, *Digital Technology, Second Edition*, Science Research Associates, Inc. ISBN 0-5742-1555-7, 1982.

Exercises for Chapter 3

1. Design a 1-bit full adder circuit. The full adder adds 2 input bits and a carry-in bit together and produces the sum and a carry-out bit. See the figure to the right. Create the truth table, Karnaugh maps, Simplified Boolean equations and a gate level diagram.

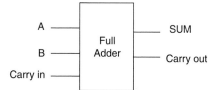

2. Prove the two cases of DeMorgan's Theorems using truth tables.

3. The circuit shown below is called a *Ring Oscillator.* It consists of 5 NOT gates connected as shown. Imagine that you are measuring the voltage level at point A. The *propagation delay* through each gate is exactly 10 ns. The propagation delay is defined as the time it takes the output to change when the input to the gate changes. Assume that at time t = 0 ns the voltage at point A goes from 0 to 1.

   a. Sketch the waveform that you would expect to see at point A.
   b. What is the period of oscillation for this circuit?
   c. What is the frequency of oscillation for this circuit?

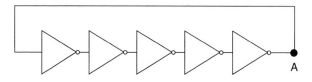

4. Design a truth table that has 4 address inputs, A, B, C and D and 1 output, x. Assume that A and B are the control inputs, C and D are arbitrary input variables. Fill in the truth table for x so that the design will implement different logic functions depending upon the state of the control inputs, A and B. The functions are as shown in the following table:

A	B	Output logic function of C and D (x)
0	0	NAND
0	1	XOR
1	0	NOR
1	1	AND

5. A *Priority Encoder* is a circuit whose output is the binary code of the most significant input that is turned on. Suppose that you have the circuit diagram as shown on the right.

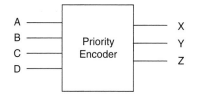

Here, A is the least significant input bit and D is the most significant input bit. X is the least significant output bit (2^0) and Z is the most significant output bit (2^2).

If all inputs are 0, all outputs are 0. The priority is determined by the most significant input bit that is 1. Create the truth table for this circuit and then use the Karnaugh map to simplify the truth table and draw the simplified circuit.

6. Assume a logic circuit that has as its inputs, two 4-bit binary numbers (A0 through A3 and B0 through B3) and as its output, a single binary output, Z. The output Z is TRUE (HIGH) if the two numbers are equal. Design the circuit that implements this equality tester.

7. You are the Chief Designer of the *Road to Nirvana* Hot Tub and Spa Company. Your assignment is to design a new digital spa controller to replace the old one that you adapted from a 1972 washing machine. You have the following input specification:

	Variable = 0	**Variable = 1**
Temperature indicator: A	*Water temperature is below the desired hot tub temperature*	*Water temperature is equal to or above the desired hot tub temperature*
Daily filter timer switch: B	*Circulation pump is off*	*Circulation pump is on*
Air blower switch: D	*Air blower is off*	*Air blower is on for soothing bubbles*
Key switch: E	*System is turned off*	*System is on*
Manual pump switch: F	*Circulation pump is off*	*Circulation pump is on*

Your logic controls the following outputs:

f = Pump Motor: 1 = On

g = Air Blower: 1 = On

h = Heater: 1 = On

See the figure, right:

How to proceed:

a. Create the truth table for the hot tub. Keep in mind that there may be several alternative ways to define the operation of the hot tub.

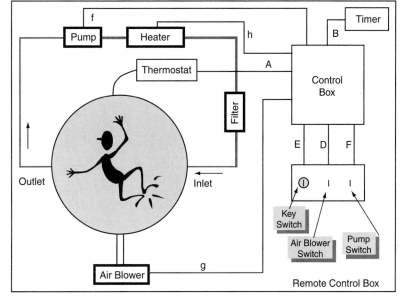

b. Use Karnaugh maps, DeMorgan's Theorems and the algebraic relationships to simplify your equations.

c. After you've simplified the equations as much as possible, draw the logic in terms of the gate structures.

8. Below is the truth table for a circuit called a *3-input by 8-output decoder*. Basically, it is used to assert LOW only one of its outputs at a time, based upon the binary value (0 to 7) of the input variables. The truth table is also shown. Using the Boolean algebraic methods and Karnaugh maps techniques you've learned, draw the simplest gate level diagram that you can to represent the circuit. Hint: Notice that the "active outputs" are low, not high. You can use this to your advantage to greatly simplify the logic of this problem. Note that there is a reason for calling the outputs $\overline{CS0}$ through $\overline{CS7}$. We'll see the reason when we study memory system organization.

A0	A1	A2	CS0	CS1	CS2	CS3	CS4	CS5	CS6	CS7
0	0	0	0	1	1	1	1	1	1	1
1	0	0	1	0	1	1	1	1	1	1
0	1	0	1	1	0	1	1	1	1	1
1	1	0	1	1	1	0	1	1	1	1
0	0	1	1	1	1	1	0	1	1	1
1	0	1	1	1	1	1	1	0	1	1
0	1	1	1	1	1	1	1	1	0	1
1	1	1	1	1	1	1	1	1	1	0

9. Assume that you have four variables, A, B, C and D defined as follows:

bool   A, B, C, D;

Consider the C control statement:

If (D)

    C = B;

else

    C = A;

Draw the gate equivalent circuit of this control circuit.

CHAPTER 4

Introduction to Synchronous Logic

Objectives

▶ Learn how logic gates are connect to create the flip-flop;
▶ Learn the different categories of flip-flops and their behavior;
▶ Learn about the different circuit configurations of the D-type flip-flop, including frequency dividers, counter, shift registers and storage registers;
▶ Learn how the D-type flip-flop is used to synchronize the transitions in state machines.

Up to now, we've limited our study of digital logic systems to *asynchronous* logic. Asynchronous means "not synchronized." In our system, this means that the change in the state of the output variable depends only on the state of the input variables and the combinatorial logic that links them. Change an input variable and the output variable would change to make the logic correct. There is no delay in process, other than the propagation delay through the combinatorial logic gates. However, within a computer, with millions of logic gates, we must be able to synchronize the change of logic state with some master signal (the clock) so that everything with the microprocessor can progress through a sequence of well-defined states. Therefore, let's now turn our attention to synchronous logic.

Look at the circuit configurations in Figure 4.1. Consider the upper circuit. Notice how the outputs of the gates are returned to the input of the opposite gate. We call this configuration *feedback*. Feedback is the screeching you hear when you place a microphone to close to the loudspeaker the amplified sound is coming from. Let's analyze the circuits in Figure 4.1. The circuit has two inputs, A and B and two outputs, Q and $\overline{Q}$. As you'll soon see, these outputs are always the complement of each other, but for now, we'll just call them Q and $\overline{Q}$.

According to Figure 4.1, inputs A and B and output Q are all at logic level '1' and output $\overline{Q}$ is

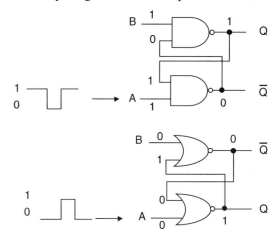

Figure 4.1: The Set/Reset (RS) Flip-Flop. The NAND gate design of the upper circuit is triggered by a negative going pulse. The NOR gate design of the lower circuit is triggered by a positive going pulse.

at logic '0'. This means that the upper NAND gate's two inputs are 1 and 0, the lower NAND gate's inputs are both 1. From the truth table for a NAND gate, we see that this circuit is stable because the inputs and outputs are all in their correct logic state. Now, let's perturb this stable system by applying a negative pulse to input A. Now things will change very rapidly. The output of the lower NAND gate goes to 1 because the inputs are now 1 and 0. The 1 is now applied to the input of the upper NAND gate, so the output goes to 0, since both inputs are 1. The output of the upper NAND gate is simultaneously applied to the input of the lower NAND gate, so both inputs are 0 and the output is 1. Finally, we remove the pulse that started it all. By removing the pulse, we are simply returning the signal at input A to its prior state. Notice the even though input A returns to 1, the outputs remain in their new state because the positive feedback from the upper gate forces the circuit to remain in the state, $Q = 0$ and $\overline{Q} = 1$.

It should be apparent to you that we could repeat the process with a negative going pulse on input *B* and the outputs would flip back to their original state. You should be able to repeat the analysis of the NAND gate in Figure 4.1 with the NOR gate. In this case, it is a positive going pulse that initiates the state transition, rather than a negative pulse.

Flip-Flops

If we apply a second negative pulse to input A of the NAND gate of Figure 4.1 the system remains unchanged because the output of the upper NAND gate is still 0. The only way to SET the circuit to the way it was is to provide a negative pulse at input B. Thus, we can see that 1 input is the SET input (setting $Q = 1$) and the other input is the RESET input (setting $Q = 0$). This type of circuit element is called a *flip-flop*, because the two outputs flip and flop back and forth like a playground teeter-totter. This particular type of flip-flop is an *RS flip-flop* because the two inputs alternately reset (R) or set (S) the outputs. We also refer to this type of behavior as *toggling* between two states, much like the toggle switch on the wall that controls the lights in a room. In that case, the switch toggles the lights on or off. Figure 4.2 is a schematic representation of the RS flip-flop as a unique circuit element. Here we've taken the two NAND or NOR gates from Figure 4.1 and drawn a different circuit symbol to represent them.

Figure 4.2 is a trivial example, but it introduces an important concept. In Chapters 1 through 3, we've been gradually increasing the complexity of the circuits being considered. For example, we looked at how an electronic switching element, a MOSFET transistor, could be configured as a simple inverter gate and how that basic configuration could be extended to more complex gates, such as NAND. We also saw how a compound gate, the XOR, is given a unique symbol in order to simplify the representation of circuits that contain it. Here, we are

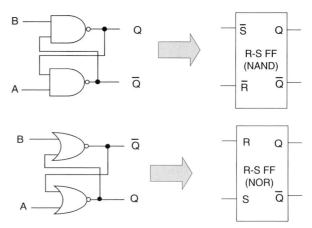

Figure 4.2: The set/reset (RS) flip-flop as a unique circuit symbol.

extending the concept as we begin to create new and more powerful circuit configurations out of the simpler building blocks. This is a process that we'll be continuing to use throughout this chapter and the rest of the book. Onward!

The RS flip-flop is important because it introduced the concept of *state dependency*. The state of the RS flip-flop's two outputs is not only dependent upon the state of the 2-input variables, A and B, it is also dependent upon the previous state of the outputs. This is very different behavior from what we observed with asynchronous, combinatorial logic. Now, the state of the outputs will be a function of the state of the inputs plus the previous state of the outputs. Even though the RS flip-flop is important from a "new concept" point of view, it has limited usefulness in real life. So let's spend some time and see if we can see how this concept is applied in more practical circuit configurations.

The RS flip-flop is an asynchronous device. Pulsing either the S (SET) or R (RESET) inputs will drive the Q and $\overline{Q}$ outputs accordingly. When the inputs are active, the outputs respond. There is no clock signal involved. Thus, we have to ask the question, "Where's the synchronization in all of this?" Good question. It isn't there yet. Let's make use of the gating properties of our NAND and NOR gates to introduce the concept of a clock. Figure 4.3 shows just such a flip-flop

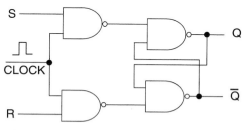

Figure 4.3: The clocked RS flip-flop.

design. We call this a *clocked RS flip-flop*. For simplicity, we'll only use the NAND-based designs, but you can easily convert it to a NOR design with the appropriate substitutions.

Introducing the additional two NAND gates now gives us a method of controlling the R and S inputs with a third input, the clock. Also, the R and S inputs are now active HIGH rather than active LOW with the RS flip-flop because of the inverting action of the two additional gates. Assume that Q = 0 and $\overline{Q}$ = 1. We wish to toggle the flip-flop to the opposite state. Assuming that the clock input is LOW, bringing the SET input HIGH will have no effect because the output of a NAND gate will go LOW only when both inputs are simultaneously HIGH. Thus, we can bring the SET input HIGH, but it isn't until the clock input also goes high will the other flip-flop pair toggle their states.

This is really getting close to what we want, but we still aren't there quite yet. We have introduced a clock synchronization mechanism, but it is a weak synchronization. The problem is that we are dependent upon the actual clock width for the degree of synchronization that we get. If the clock signal remains high for a length of time that is long compared to the time frames of the S and R inputs, then we've lost any synchronization that we hoped to gain. As long as the clock input is HIGH, we can continue to pulse the S and R inputs and cause the outputs to toggle continuously. Ideally, we would like to synchronize the flip-flop with either the rising or falling edge of the clock, rather than the level of the clock. Remember, the edge of a logic signal is the transition from HIGH to LOW or LOW to HIGH. For stability, we require that the transition occur very rapidly, independent of how long the signal remains in the HIGH or LOW state. Thus, even if we have a logic signal that changes state every 12 hours (the PM indicator on a digital clock) we still want that transition to occur in a few nanoseconds.

The circuit configuration of Figure 4.4 is perhaps the most famous flip-flop of all. It is called the *JK flip-flop*. According to Null and Lobur[1], the JK flip-flop was given its name in honor of Jack Kilby, an engineer with Texas Instruments Corporation who was one of the first inventors of the integrated circuit. Here, we've taken the basic clocked RS flip-flop and added two important new features:

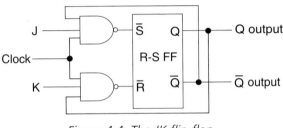

Figure 4.4: The JK flip-flop.

1. A second set of feedback loops were added from the Q and $\overline{Q}$ outputs to the inputs of the NAND gates,
2. A second set of inputs, J and K, were added to the NAND gates.

Conceptually, all that we've done is to add the additional feedback path to the circuit by way of converting the gating NAND gates from 2 input gates to 3 input gates. However, in the process we've come very close to building a JK flip-flop that actually works the way we want it to. Unfortunately, it doesn't work quite right. The problem is traceable back to the original problem with our clocked RS flip-flop. We were synchronizing with the level of the clock and not the edge of the clock. Here's how we want the JK flip-flop to work. Table 4.1 is the truth table for a JK flip-flop. The last condition, J = 1 and K = 1 is supposed to enable the flip-flop to toggle back and forth each time the clock is pulsed. As you'll soon see, this is a desirable circuit behavior for many reasons.

However, because we are still at the mercy of the duration of the clock, and when all three inputs are HIGH, we create an unstable state for the flip-flop. To see this, return to Figure 4.1 and analyze the circuit for the conditions when both inputs A and B go low at the same time. This type of behavior is also called a *race condition* and often leads to unstable and unpredictable circuit operation.

Table 4.1 Truth table for JK flip-flop.

J	K	Q after clock
0	0	no change
1	0	Q = 1
0	1	Q = 0
1	1	outputs toggle

Let's try to analyze the behavior of the circuit and see what might be causing the problem. Assume that the flip-flop is in the RESET state (Q = 0, $\overline{Q}$ = 1). J and K are both HIGH and the clock is LOW. Since we have connected the $\overline{Q}$ output back to one of the three inputs of the upper NAND gate, when the clock goes HIGH all three inputs will be HIGH and its output will go LOW. The LOW output will cause the RS portion of the flip-flop to toggle. That's good, so far, because that's the desired behavior.

As soon as the Q and $\overline{Q}$ outputs toggle, the output of the upper NAND gate will start to go HIGH again. However, the clock signal is still HIGH, so the lower NAND gate now goes LOW and toggles it back the other way. The circuit no exhibits one of two possible behaviors, neither of which can be predicted. Either both the Q and $\overline{Q}$ outputs go high and stay HIGH as long as the clock is HIGH, or the outputs toggle rapidly due to the instability introduced by the race condition. Only careful circuit analysis of the propagation delays and switching conditions will determine exactly how it will behave. In any event, it isn't what we want.

What we are lacking is a mechanism to shift control to the clock edge. The solution is to create a circuit with two gated RS flip-flops in sequence. Figure 4.5 is just the circuit that we need. While this might look quite a bit more complicated, we've only made a few modifications to the basic circuit configurations of Figures 4.3 and 4.4. Recall that the basic problem arose when we created the two feedback loops in Figure 4.4 in an attempt to create a circuit that would toggle when J = K = 1 and the clock was pulsed high.

In Figure 4.5 we've solved the race problem with the addition of the single NOT gate (gate #9) and by adding a second gated flip-flop. The new flip-flop design is called a *master-slave* JK flip-flop. Now, before you begin to get visions of chains and whips let me assure you that this is completely benign. The first grouping, the master, is used to bring in the data and stabilize it, and then the data is transferred to the slave, where it appears on the Q and $\overline{Q}$ outputs. The NOT gate and the slave portion of the circuit together

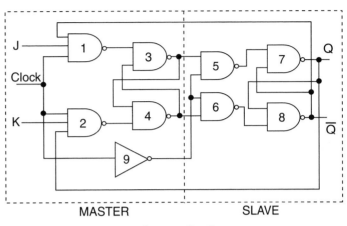

Figure 4.5: The master-slave JK flip-flop.

creates the circuit function that we've so far been lacking. When the clock signal goes HIGH on the master section of the flip-flop (gates 1 through 4) the output of the NOT gate goes low and locks the slave flip-flop, preventing any changes from occurring in its RS output section.

When the clock goes low again, the master flip-flop is disabled and its RS section cannot change state. However, the low going clock enables the slave flip-flop and it changes to agree with its gating inputs.

The two feedback loops that created all the previous race condition problems now operate properly because we've stopped the race condition. Each section of the flip-flop now operates on either the HIGH or LOW level of the clock, or on *alternate phases* of the clock. In other words, the race condition came about because the feedback signal from output to input was free to "race" around the circuit as long as the clock was HIGH. Now we've changed the circuit so that the feedback loop is effectively blocked because only ½ of the circuit is active at a time. The master-slave circuit configuration finally gives us the behavior that we need. The circuit is only active on the transitions of the clock and not on the level of the clock.

Figure 4.6 shows the JK master-slave flip-flop as a unique circuit block, and also shows its truth table. Note that now we use a down arrow to indicate that the outputs change state on the negative transition (HIGH to LOW) of the clock, or when the master portion transfers information to the slave portion. Also shown in Figure 4.6 is a slightly modified version of the circuit. By adding an inverter gate as shown, we limit the behavior of the JK flip-flop to only the conditions when J = 1 and K = 0, or vice versa. The inverter gate prevents the other two possible combinations from occurring, J = K = 0 and J = K = 1.

This new configuration, which we'll call the D *flip-flop*, is the most important version of the flip-flop families and we will focus our attention on this circuit almost exclusively through the rest of the book.

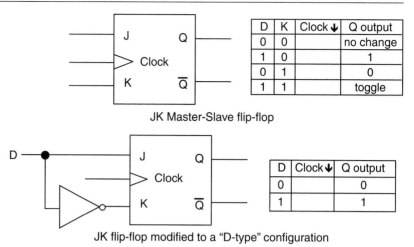

D	K	Clock ↓	Q output
0	0		no change
1	0		1
0	1		0
1	1		toggle

JK Master-Slave flip-flop

D	Clock ↓	Q output
0		0
1		1

JK flip-flop modified to a "D-type" configuration

If you look at the truth table for the D flip-flop in the lower portion of Figure 4.6 you might notice an interesting fact. When the clock goes low, Q output changes to agree with the D input. An alternative was of describing this is that, "On the falling

Figure 4.6: The master-slave JK flip-flop slightly modified to a 'D'-type configuration.

edge of the clock the data value appearing on the D input is stored in the cell and appears on the Q output." Now we know why it is called a "D" flip-flop. The 'D' means data. This is a memory cell.

Before we turn our exclusive attention to the D flip-flop (D-FF, for short), we need to make a few introductory remarks. As previously stated, even though we are able to express more complex logical functions in terms of more primitive gates, it doesn't necessarily follow that the actual design of the more complex function will be truly represented by the logical gate design. Always keep in mind that in most cases, there are circuit shortcuts that can be taken to reduce the overall complexity of the design. Therefore, we should always keep in mind that we are often looking at a logical, rather than actual, circuit implementation. Also, we may want to add additional functionality to a circuit block, or change slightly the behavior of the circuit we are dealing with. If you examine the data book of any of the digital logic integrated circuit manufactures, you'll see perhaps three or four different variations of the standard circuits. For example, you may find a JK flip-flop that changes state on the rising edge of the clock rather than the falling edge. The manufacturers do this to broaden the appeal of the part to the circuit design community who uses them.

For example, if the only JK FF that your company offers is a negative edge triggered FF and a large customer doesn't want to incur the added expense of inserting an additional NOT gate into their circuit just so they can convert a positive going clock edge to a negative going clock edge, then you have an unhappy customer. The customer can beat on your Marketing Department and convince them that you need to add a positive edge triggered JK FF to your product offerings, or they can take their business elsewhere, to a company that does offer the parts that they want.

The D-FF is a good example of this. Almost all D flip-flops in use today are positive edge triggered devices. Data is transferred from the D input to the Q output on the rising edge of the clock. In addition, the original functionality of the RS inputs has been retained to give an asynchronous method of setting or resetting the device. We may summarize the behavior of the D-FF as follows:

The logic level present on the D input just prior to the rising edge of the clock is transferred to the Q output on the low-to-high transition (rising edge) of the clock. The Q output will continue to retain the data until the next rising edge on the clock input.

Figure 4.7 illustrates the behavior of the D flip-flop circuit.

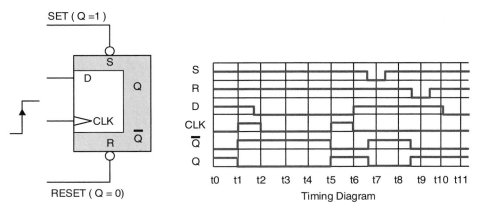

Figure 4.7: The D-type flip-flop. The shaded region represents the portion of the circuit that replicates the functionality of the RS flip-flop. The clear portion is the master-slave circuitry, which adds the clock and 'D' (data) inputs. The timing diagram for the 'D' flip-flop is on the right.

The timing diagram is just a way of looking at the relationship of several signals with respect to time. If you've ever seen a polygraph (lie detector) test administered on TV or in the movies (or for real!) then you've seen that the strip chart recorder is simultaneously plotting several body conditions at the same time. In other words, our timing diagram is just a polygraph test of a D-FF.

Notice the up-arrow, ↑, on the clock input. This indicates a rising edge. This means that only in the briefest instant in time just prior to the transition on the clock (CLK) input will the value on the D input be captured and transferred to the Q output. The symbol '▷' at the clock input of the 'D' flip-flop indicates that the device is *edge-triggered*, and is not sensitive to the absolute value of the logic level.

The R and S inputs use bubble notation on the inputs to indicate that they are *active,* or *asserted, LOW*. We can see this in the timing diagram in Figure 2.17. Just after time t6, $\overline{S}$ goes low. After a bit of a propagation delay, Q goes HIGH and $\overline{Q}$ goes LOW. Notice that no CLK input was present. $\overline{R}$ and $\overline{S}$ are asynchronous inputs. They can override the action of the clock. Similarly, $\overline{R}$ is asserted after t8 and the outputs switch back.

The D flip-flop in Figure 4.7 is comprised of two separate circuit functions. We know from the previous discussion that the device is actually a master-slave configuration. The $\overline{R}$ and $\overline{S}$ inputs, combined with the Q and $\overline{Q}$ outputs, form an RS flip-flop like the one we studied in Figure 4.1. These inputs can override the D and CLK inputs of the "D-type" portion of the circuit so that the outputs may be forced to any state at anytime with the assertion of the appropriate low-going $\overline{R}$ or $\overline{S}$ signal.

Figure 4.8 shows the logical implementation of a D-type flip-flop. Although the master-slave relationship is clearly visible, the circuit implementation is not as easy to discern as the JK example that we looked at earlier.

The circuit in Figure 4.8 uses six NAND gates to implement the design of the D flip-flop. We needed eight gates, plus two inverter gates, to convert the JK flip-flop to a D flip-flop. Also, notice that there is no feedback from the outputs to the inputs to implement the toggle function. In the case of the D-FF, it isn't one of its operational modes, so the circuit is not implemented in a way that adds the feedback.

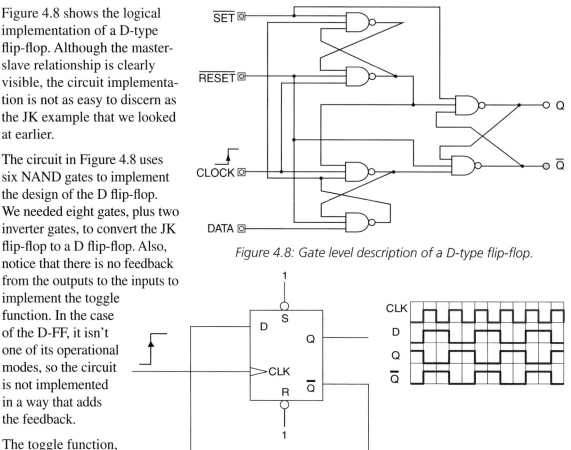

Figure 4.8: Gate level description of a D-type flip-flop.

The toggle function, that is having the outputs change state each time there is a pulse on the clock input, is a very useful feature to have in a digital system, so let's look at it in greater depth. Figure 4.9 shows a D-FF with the toggle function added back in. To add the toggle function, the $\overline{Q}$ output is returned to the D input. In the JK FF the toggle function was implemented be returning the Q output to the K input and the $\overline{Q}$ output to the J input, respectively.

Figure 4.9: 'D' flip-flop connected in a divide-by-two configuration.

Referring to the timing diagram in Figure 4.9, we see that when the flip-flop is configured in the toggle configuration, every two clock pulses results in one complete clock pulse on the Q or $\overline{Q}$ outputs. Each time we have a clock pulse it takes on flips the outputs, so we need two clock pulses to flip the outputs back and forth.

Another way of saying this is that the period of the waveform at the Q and $\overline{Q}$ outputs have a period of twice that of the clock itself. Since the frequency is the reciprocal of the period, the frequency of the waveform at the Q output is ½ that of the frequency of the waveform at the clock input. Thus, if the clock input has a frequency of 1 MHz, the Q output has a frequency of 1/2 MHz, or 500 KHz.

If we placed another one of these circuits to the right of the one in Figure 4.9 and feed the $\overline{Q}$ output into the clock input of the next one, the frequency would be halved again. We can see this in Figure 4.10. Referring to Figure 4.10, the waveforms for the input clock and the resulting changes in the Q1 and $\overline{Q1}$ are shown on the graph. This is identical to the circuit of Figure 4.9. Now, if we connect $\overline{Q1}$ so that it becomes the clock input for the second D-FF on the right,

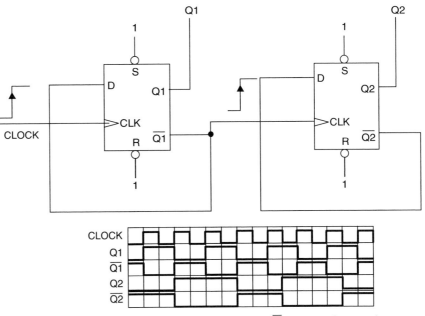

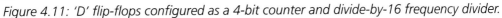

Figure 4.10: Two D-FF's connected in series. The $\overline{Q}$ output of one FF becomes the clock input for the next FF. The resulting transitions of the clock input and Q outputs are shown in the timing diagram.

then the circuit behavior is exactly the same, except that the period of all the resulting waveforms have been doubled (frequencies have been halved).

Clearly, we could continue to do this exercise with as many D-FF devices as we want to add. Each time we add a D-FF, the output period of that device is twice the period of the device that feeds it. In general, for a string of N D-FF's, the period of the Nth FF is simply 2^N times the period of the input clock to the first flip-flop in the chain. With enough flip-flops in a chain, we could easily reduce an input waveform of several megahertz to a leisurely 1 clock cycle per second.

Now let's go a step further in order to better see another property of this chain of D-FF's. In Figure 4.11, we see a chain of four D-FF's. The only other difference is that the $\overline{R}$ inputs are all tied

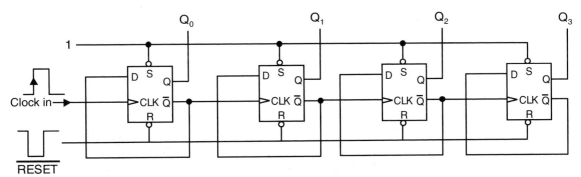

Figure 4.11: 'D' flip-flops configured as a 4-bit counter and divide-by-16 frequency divider.

79

together and presumably go to some location that can assert a RESET pulse at some time. Thus, whenever we apply a negative pulse to the four $\overline{R}$ inputs of this circuit, all of the Q outputs will be forced to go to 0, independent of the arrival of a clock signal. This is very convenient if we want to start the circuit running from a known state. In fact, that's exactly what the RESET button does on your PC.

In Figure 4.11, we show four of the circuits from Figure 4.9. Notice that we've tied all of the $\overline{S}$ inputs together and we're going to permanently hold them at a logic level of 1. Thus, we'll never have an occasion to "set" the Q outputs to 1. Tying an unused input to either logic level 1 or 0 is quite common in digital design and insures us that the circuit won't accidentally change state if that input were to see some noise. From our discussion of the RS flip-flop, we know that asserting the $\overline{R}$ input low will force the Q outputs to go to 0, thus giving us a known starting condition.

From our previous discussion, we can see that each successive D flip-flop, or stage, of the frequency divider divides the incoming clock by a factor of 2, so the output frequency, at the Q_3 output is one 16th of the frequency at the clock input, four stages to the left. Thus, a 16 MHz clock signal at the CLK input would be a 1 MHz clock signal at the Q_3 output. We can see the frequency division process in action by referring to the logic analyzer display of Figure 4.12. Unlike the oscilloscope display that shows us a high-fidelity view of the waveform, the logic analyzer display can show multiple waveforms at one time, but loses some information in the process. Thus, the logic analyzer view is more like Figure 3.15.

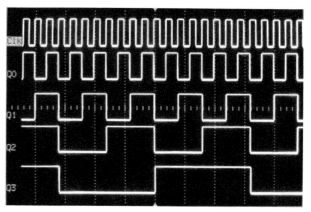

Figure 4.12: Logic analyzer display of a divide-by-16 circuit made from 4 'D' type flip-flops.

Let's look at Figure 4.12 in some detail. The clock is the uppermost waveform and the Q output of each successive D flip-flop is represented by the next lower waveform. Let's compare the CLK waveform with the Q0 waveform directly below it. Figure 4.13 is a magnified portion of Figure 4.12. The rising edges of the clock are labeled for ease of identification. Notice that on each rising edge of the CLK input, the Q0

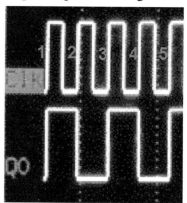

Figure 4.13: Magnified view of the clock input and Q0 output part of Figure 4.12.Assume that we've just asserted RESET and outputs Q_0 through Q_3 are all 0. There are no clock pluses coming in at the CLK input on the left. The circuit is quiet and life is good. Suddenly, there's a knock at the door. Oops, wrong book. Suddenly, a single clock pulse is received at the CLK input. Since the $\overline{Q}$ output = 1, the D input is 1, so after the clock pulse, the Q output of the leftmost D-flop goes to 1. The $\overline{Q}$ output goes from 1 to 0. That's a falling edge, so there is no change at the second D-flop.

output changes state. However, the D flip-flop requires rising edge to cause the output to change state, so the net effect is that it requires 2 input rising edges to cause the Q0 output to go through one complete cycle itself.

There's one more interesting feature of the circuit configuration of Figure 4.11 that we need to reiterate. We previously discussed that all of the $\overline{R}$ inputs are tied together so that asserting them will force the Q outputs to 0, independent of the state of the clock input. Thus, to start the system from a known state (all outputs = 0) we must assert the RESET input.

When the second clock pulse arrives at the CLK input, the first D-flop changes back and its output goes to 0. However, since its $\overline{Q}$ output now goes from 0 to 1, Q_1 now goes to 1. This is just what we would see on Figure 4.12 if we looked down a column on the waveform display after each CLK input arrives. As an exercise, make a truth table listing each clock pulse as a numeric entry down the leftmost column and the values of Q0 through Q3 as the three output columns. Start from clock pulse #0, or the RESET condition.

If you complete the suggested exercise, you'll quickly see that the circuit is also a binary counter that counts from 0000 to 1111 and then rolls around to 0000 on the 16th clock pulse. Not surprising, but we could continue to add counting stages to form a counter of any arbitrary length. This particular counting configuration is called a *ripple counter*, because the count ripples through the counter as each stage changes state. The fact that the count "ripples" through the counter means that we could be receiving the next pulse before the count change from the previous pulse completely ripples through the counter. Although the counter will still maintain an accurate count, it does limit our ability to accurately read the current count in the counter.

Up to now, we've been focusing on digital systems where each input or output variable occupies its own wire. If we want to transport 32 bits of data within a microprocessor, we need to create a bundle of 32 wires, or traces, in order to move the data around. This is called *parallel* data distribution because each individual data bit, or variable, travels in parallel with all the other data bits. However, there are some very good reasons to use *serial* data transmission methods. In serial data transmission, all of the data bits travel in sequence along a single wire, or at most, only a few wires. Some examples of parallel data transmission protocols that you might be familiar with are:

- Parallel port, LPT:
- PCI bus
- AGP bus
- IEEE 488 (HP-IB)

Examples of serial data transmission protocols are:

- RS-232 (COM ports, COM1..COM4)
- Ethernet
- USB
- Firewire

We'll now look at a circuit, built from the D flip-flops, that allows us to go from serial data formats to parallel data formats. Obviously, this is quite important because we must have a way to manipulate the data within the computer once the serial data is received. The circuit structure that we use is called a *shift register*. Consider Figure 4.14.

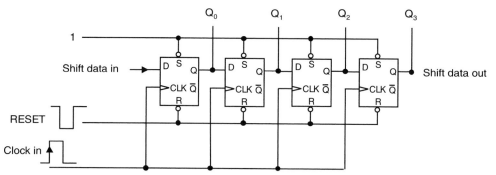

Figure 4.14: 4-bit shift register built from 'D' type flip-flops.

The RESET inputs (R) are tied together, just like the ripple counter of Figure 4.11. This allows us to simultaneously drive all of the outputs to 0. However, the big difference is that we have also tied the CLK (clock) inputs together and each Q output goes directly to the D input of the next stage. Thus, each D-flop will capture the value of the Q output to its left when the rising edge of the clock is applied to its CLK input, but because all of the CLK inputs are activated simultaneously, the data applied sequentially at the leftmost D-flop appears to move left-to-right through the shift register with each rising edge of the clock.

Let's see how this would look as a waveform. Suppose we want to send the number 6, or 0110 in binary, from one device to another along a serial cable. The method of translating the parallel data to serial is similar to the process of translating it back to parallel, but we won't consider that process here. We'll just assume that the data, 0110, now appears as the waveform traveling with a clock signal, as shown in Figure 4.15.

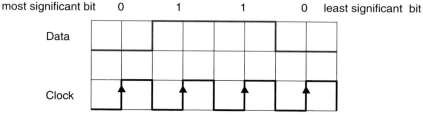

The data value is synchronized with the rising edge of the

Figure 4.15: Serial data transmission of the binary number 6.

clock, which may, or may not, be transmitted along with the data. Sometimes, a clock signal will be generated locally by the receiving device, rather than transmitted with the data. This is the way the COM ports on your computer operate. Other techniques involve combining the clock with the data and transmitted them together along a single wire. Special circuitry is used to recover the clock signals from the data at the receiving end. However, in all cases, a clock signal is required to synchronize the capturing of the data at the D input of the flip-flop. Thus, the 4 clock pulses each correspond to the receipt of one of the data bits.

Figure 4.16 shows the progression of the 4 data bits through the D-flops. Let's assume that in this example we are transmitting both the data and the clock along two separate wires, just as we've shown in Figure 2.24, remember that Figure 2.24 is actually a graph in time, so that if we had a very fast pencil and paper, each time that a rising edge of the clock occurs, we would record the state of the data on the other wire, we would see 0, 1, 1, 0, respectively. Although it may appear a little confusing at first, we can analyze the behavior as follows. Just prior to the clock edge at t1, the D input to the leftmost D-flop is 0 and all outputs have been RESET to 0. At t1, the 0's at the D inputs are transferred to the Q outputs, but since they were already 0, nothing appears to happen.

Now just prior to t2, the value at the leftmost D input becomes 1. Just after t2, Q0 goes to 1, to reflect the value at the D input. All other outputs are still 0 because we are still moving the first 0

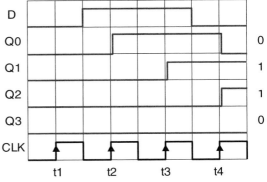

Figure 4.16: Propagation of the serial data through the 4-bit shift register.

through the chain. At t3, the third data bit, a 1, is clocked into the leftmost D-flop, and everything else moves to the right. Thus, just after t3, the Q0 through Q3 outputs are 1100, respectively. Finally, just after t4, the data bits shift one more time and the data is now completely stored in the shift register as a parallel value.

Storage Register

The final example of a D-flop circuit configuration that we're going to look at is perhaps the most important of all, the *storage register*. The storage register is a grouping of D-flops designed to receive and hold a digital data value. The value can be a bit, nibble, byte, word, long word, double word, and so on. The key point is that a single clock signal forces all the D-flops to store the data at their respective D inputs at the same time. In this sense, it is a parallel in/parallel out device.

In contrast to the storage register, the shift register of Figure 4.14 is a serial in/parallel out device. The reason that we need the storage register is that the data flowing through computers is extremely transitory, and we must find temporary storage locations where we can keep variables until we need them, or to store the results of instructions until we need to move them somewhere else.

Figure 4.17 shows eight D-flops in the circuit configuration of a storage register. We can see that any data present at the D inputs (the dark gray wires in the figure) will be stored by the flip-flops in the register after a rising clock edge. The data will then be available for the output data (light gray wires) to transmit. However, the key point is that the input data values (the signals on the dark gray wires) can now change, but the register is still holding the previous data stored in it.

In most computers, the number of D-flops that form the storage register matches the number of data bits in the computer's data path. Today, we use computers with data path widths from 4 bits wide to 128 bits wide. These registers are the key "data containers" in modern computers and microprocessors. You'll see in later chapters how, in many ways, the number and type of storage registers in a computer defines the architecture of that computer.

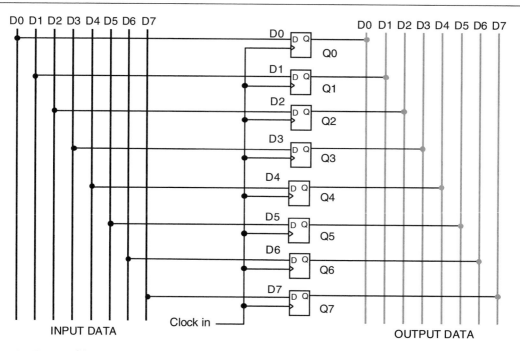

Figure 4.17: An 8-bit storage register. The input data (dark gray wires) is captured in the D-flops on the rising edge of the clock. The data then appears on the outputs (light gray wires). Registers such as this typically do not have SET and RESET inputs, so they have been omitted to simplify the diagram.

In order to reinforce the concept of the storage register as a unique circuit element instead of a collection of D-flops, we can redraw Figure 4.17 as shown in Figure 4.18.

Earlier in this chapter, we've repeatedly referred to the "state" of a flip-flop. By that we were really saying that in order to know what the new values of the Q and $\overline{Q}$ outputs will be, we had to know what the current value (or state) of the inputs and outputs are just prior to the arrival of the appropriate clock edge. What we're leading up to with all of this is the concept of a state machine.

The state machine represents all the possible combinations that the outputs and inputs of a digital system may exist in, as well as the paths that the system may take as it transitions through the different states of the state machine. Also, and perhaps most importantly, the state machine can change the path it takes because of changing conditions on its inputs. Unlike the asynchronous combinatorial logic that we studied earlier, state machines are synchronous devices. This means that a state machine can only change state on a clock edge. We can infer from this that somehow the D-flop and the state machine have some common lineage, and you're correct! We can best describe the behavior of a state machine with a table or a graphical picture (called a *state transition diagram*). Consider Figure 4.19.

The shaded circles in Figure 4.19 are the states of the system. We have two states in the system, 01 and 10, corresponding to the conditions $Q = 0, \overline{Q} = 1$ and $Q = 1, \overline{Q} = 0$, respectively. The arrows represent the possible state transitions and the numbers on the arrows represent the input condition at the time of the transition. Thus, when the device is in the state 01, it will remain in that state if D = 0, but transition to the 10 state if D = 1. Similarly, if it is in state 10, it will remain in that state if D = 1 when the clock signal comes, but will transition to the 01 state if D = 0.

The transition between states occurs on the rising edge of the clock, which is implied by the arrows. Even though the D-FF remains in a given state after the clock edge, we still use an arrow to show that it transitioned back to its current value.

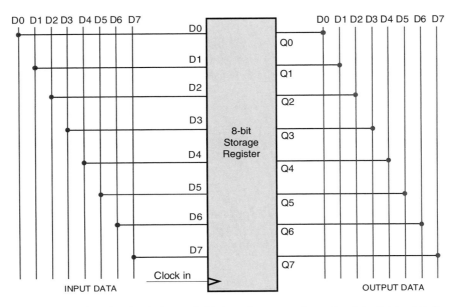

Figure 4.18: The circuit of Figure 2.26 is redrawn to show the 8 individual D-flops aggregated into a single device. The storage register is an integral part of most computer systems so it should be considered apart from the D-flop.

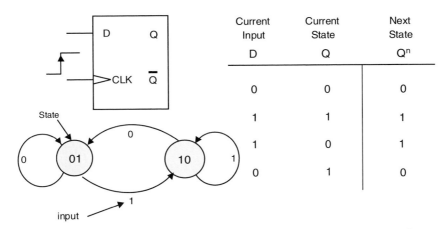

Current Input D	Current State Q	Next State Q^n
0	0	0
1	1	1
1	0	1
0	1	0

Figure 4.19: State transition diagram and truth table for a 'D' type flip-flop. The asynchronous SET and RESET inputs are omitted for clarity.

The truth table view (the right side of Figure 4.19) provides the same information, but is less intuitive then the state transition diagram. For simplicity, we've omitted the $\overline{Q}$ output of the D-FF in the truth table. The four rows of the table are the four possible state transitions that can occur. Note, however, that the output variable appears in two contexts, its value *before and after* the arrival of the rising edge of the clock pulse.

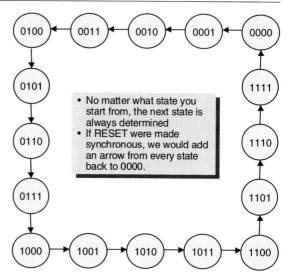

Figure 4.20 is the state transition diagram for the 4-bit counter example we considered in Figure 4.11. Since there are no other synchronous inputs to the counter, other than the clock, the diagram is very straightforward. No matter what state the counter is in, the next state is always predetermined. If the RESET input could be made synchronous (and it is easy to do, but we

Figure 4.20: State transition diagram for a 4-bit counter.

won't bother) then every bubble would have an arrow back to the 0000 state. These RESET arrows would have the label RESET = 0 on them to indicate that a 0 on the RESET input would force the counter to return to 0000 on the next clock pulse. All of the other arrows that indicate a transition to the next counter state would have the label RESET = 1, to indicate a normal state transition.

Earlier in Chapter 3, we considered that one possible way to implement asynchronous logic was to directly implement the truth table in a memory device, such as Figure 3.9. This would eliminate all of the gate design issues at once. We didn't discuss the reason for one system over the other, but we book-marked the idea for later elaboration. Let's elaborate! However, before we dive into the state machine it might be helpful to review the architecture of a computer memory. We'll be studying memories in much more detail in a later chapter, but for now, let's review what you already know about memories from your programming classes.

A memory device, such as a RAM or ROM memory chip contains a total number of *memory cells*. Each cell is capable of storing a single 1 or 0 value. We can imagine that the cells may be organized into various combinations to form memories of different *memory widths*. Once we know the total number of cells in the memory and the width of the memory we know how many *memory addresses* are required for this memory device. An example would probably help here. Suppose that we have a memory device that contains a total of 16,384 memory cells. Each memory cell stores a single *bit* of data. We want to design this memory device in such a way that each memory access will give us 8 bits of data. In other words, we want to store *chars* in this memory. Therefore, the number of unique addresses is 16,384 / 8 = 2,048. Thus, our memory device will contain:

- 16,384 total memory cells,
- 8 lines for reading and writing data to the memory,
- 2,048 uniquely addressable memory locations,
- 11 address lines for supplying 2,048 unique address combinations.

You might be wondering where the 11 address lines came from. Consider that we need to be able to address 2,048 unique address locations. Thus, the first address location should be given address 0000 and the last address location is given the address 2,047. It might seem strange that all zeroes is a valid address, but Computer Scientists are a hardy breed and we can deal with it!

In binary, the 2,048 unique addresses are represented as going from 000 0000 0000 to 111 1111 1111. Notice that we need 11 binary digits to represent all the possible addresses from 0000 to 2,047. It is a general rule that since 0000 is a valid address that the highest address number (all binary digits equal 1) will be one less than the number of unique address.

So, you can think of this particular memory device as being organized as 2,048 by 8. We can change our specification and say that we want to store 32-bit wide integers in this memory. Since there are more memory cells per address location, there must be fewer addresses to deal with. Since the memory is 4 times wider then before, we'll have 4 times fewer addresses. Our new memory device would have:

- 16,384 total memory cells
- 32 lines for reading and writing data to the memory
- 512 unique memory addresses
- 9 lines for supplying 512 unique address combinations.

So what does this have to do with state machines? Recall that Figure 3.9 was a truth table that looked suspiciously like a memory. The truth table in Figure 3.9 had four independent input variables, A through D and two dependent output variables, E and F. From our previous discussion, this is a memory with 4 address lines and a 2-bit wide output. The memory contains 32 memory cells.

The above discussion is not meant to imply that a memory is the only method that we could use to store the state transition information for our state machine. Since the memory is really an exact mapping of truth table we could use the methods that we learned earlier in this chapter. That is, build a Karnaugh map for each output variable and developed the simplified minterm equations for each. Then, we could construct the gate representation of the truth table and have our state machine. How exactly you choose to implement the design is determined by many factors. We'll put our trust in the hardware designer to do the right thing.

Now, suppose we wanted to really build a state machine. A microprocessor is a good example of a state machine. How would we do it? Well first, we would probably need to take a few more courses in Computer Science and Electrical Engineering, but we won't let that stop us. We'll look at the big picture. If we really wanted to build a microprocessor, we would have to design a state machine for its brain. The state machine would have thousands of possible states, with hundreds of input and output signals (variables). Such a state machine would be difficult to design using logic gates, but would be easier to design using a memory to implement the logical truth table. Consider Figure 4.22.

There are 4 input variables, A_{in} through D_{in}, and four output variables, A_{out} through D_{out}, respectively. The input variables form the address of a par-

Figure 4.21: Generic hardware designer[2].

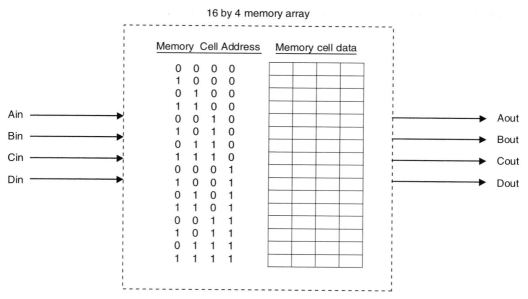

Figure 4.22: Organization of a 16 x 4 memory array as the truth table for a state machine.

ticular memory cell, 0000 through 1111 and output variables are just the data contained in the cell with the input address. This is shown in Figure 4.23.

In Figure 4.23, A0 through A3 are the addresses of the memory cells and D1 through D3 are the corresponding 4-bits of data stored in each cell. The outputs of the state machine, Q0 through Q3 drive out to the rest of the circuit, and are also routed back to provide the address for the next

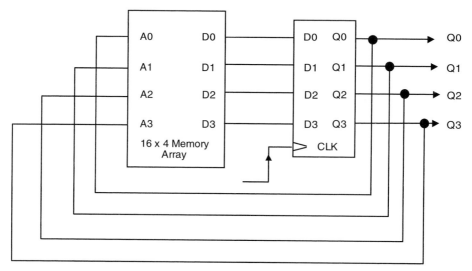

Figure 4.23: State machine implementation using a 16 x 4 memory array and a 4-bit storage register. The output variables are fed back to the input to provide the address value for the next state.

state. The 4-bit storage register provides the critical synchronization function. Since the data only transfers from the D inputs to the Q outputs on the positive clock edge, the register isolates the input and output transitions from each other and prevents the circuit from running in an uncontrollable fashion.

The 4-bit 'D' storage register provides exactly the same functionality in the state machine as the master-slave circuit did in the edge-triggered flip-flop. Without the D-FF to provide the appropriate synchronization, our circuit would race uncontrollably. To see why this is so, consider the circuit shown in Figure 4.24.

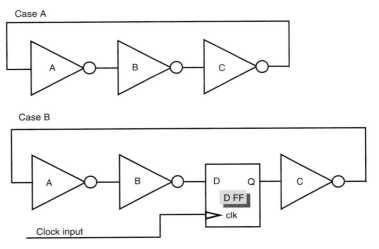

Figure 4.24: Case A: A circuit with a race condition. The signal that the output feeds back to the input is always the opposite polarity to the current input signal, the circuit can never establish a stable condition. Case B: The D-FF stabilizes the circuit and limits transitions to the rising edge of the clock.

In Case A, the circuit consists of three NOT gates (although any odd number of NOT gates greater than 1 would also work) connected in a loop. We can analyze the circuit as follows: Suppose that at a particular instant of time the input to gate A goes high. After an appropriate time lag, determined by the propagation delay of gate A (let's assume that this is 10 ns), the output of gate A goes low. Since this is the input to gate B the process repeats itself from gate B to gate C and back to gate A. So, 30 ns after the input to gate A goes high, the input changes state to low, and the process repeats itself again. After 60 ns, the input to gate 'A' goes high again and the cycle repeats itself. This is a classical a *race condition* and generally it's a bad thing to have happen.

Actually, it's not that bad if we want it to work that way. This is one of the ways we can generate clock pulses. We call this circuit an *oscillator*. With a few simple circuit manipulations we can add a crystal to the circuit and the frequency and periods of the oscillation become very precise. In the example circuit of Figure 4.24 the frequency would be the reciprocal of 60 ns, or approximately 16.7 MHz.

The circuit of Case B clearly stops the oscillations. Even though the value at the D input toggles with every rising edge of the clock, the circuit is stable. There is no uncontrolled signals changing state and all transitions in the system are limited to when the rising edge of the clock updates the D-FF.

What does this pleasant little diversion have to do with state machines? Well, everything! If we didn't have the D flip-flops in the circuit, then feeding the outputs of the memory back to the inputs could create a race condition and the state machine would be useless to us. With the D flip-flop in the circuit, we maintain control over the moment in time when we allow the outputs to feed back to the inputs and change the address of the memory cell. This is the state transition that we desire.

The design of this flip-flop isolates the current state (outputs Q0 through Qn) from the next state. Only during the briefest of times when the clock edge rises can the outputs of the flip-flops change state to agree with the D inputs. At that point, we have transitioned to the new state and the outputs of the current are fed back to the address inputs of the memory to provide the address for the next state. The data values stored in that address provide the values for the next state transition when the next rising edge of the clock occurs.

Thus, we can summarize this circuit architecture as follows:

- The memory cells hold all the possible states of the system just as the truth table contains all possible logical terms for the system.
- The output of the memory array (truth table) is the input to the D storage register and represents the next state of the state machine after the rising edge of the next clock pulse.
- The output of the D storage register is the current state of the system. The output is used to control external inputs and also provides the address to the memory array for the next state.
- The D storage register isolates the output of the state machine from the inputs to the state machine. State transitions can only occur on the rising edge of the clock signal.

Summary of Chapter 4

- The action of feedback creates a circuit element that has a state dependency. The state of the outputs depends not only on the state of the inputs but on a clock pulse and the present state of the outputs,
- The simplest form of this circuit, an RS flip-flop, is of limited usefulness.
- Once we add the additional functionality of the master-slave flip-flop, then we can overcome many of the instabilities that are inherent in an RS design.
- The JK flip-flop is the most versatile form of the device.
- A derivative design, the D-type flip-flop has the added property of a data storage element.
- The D-FF can be configured as a counter, a shift-register and a storage register.
- The D-FF is the key element in the implementation of a state machine because it limits the possible transitions of the system to only the instant in time during the rising edge of the clock.

Chapter 4: *Endnotes*

[1] Linda Null and Julia Lobur, *the essentials of Computer Organization and Architecture,* ISBN 0-7637-0444-X, Jones and Bartlett Publishers, Sudbury, MA, 2003, p. 116.

[2] The particular division of Hewlett-Packard®, where I was formerly employed for many years, had code names for the hardware and software designers. Hardware Designers were called "Toads," and Software Designers were called "Turkeys." I'm not sure where the naming convention came from; they just were Toads and Turkeys. Engineers who could design hardware and software were sometimes called "Capons," but we won't go there.

Over the years, I've made some additional observations about hardware and software types. Hardware Designers tend to wear Nike running shoes with untied laces, and Software Designers tend to wear sandals or Birkenstocks. I use to wear running shoes but recently I noticed that I'm wearing my Birkenstocks more and more. Could teaching in a CS Department be remolding my essence? Only time will tell.

Exercises for Chapter 4

(Note problems denoted by an asterisk are especially challenging.)

1. The circuit shown below consists of three D-type flip-flops. The black dots indicate those wires that are physically connected to each other. The $\overline{\text{RESET}}$ inputs are permanently tied to logic 1 and are never asserted. Before any clock pulses are received, all the $\overline{\text{SET}}$ inputs receive a negative pulse to establish the initial conditions for the circuit.

    Draw the state transition diagram for this circuit. Please denote the states in the order Q0, Q1 and Q2. Be sure to show the starting state, as indicated in the problem specification, and all subsequent states. Use arrows to indicate the transitions from one state to the next.

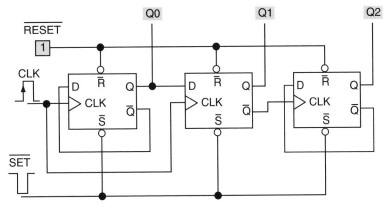

2. Shown below is a 3-bit counter made up of positive edge-triggered, D-type flip-flops:

    Assume a $\overline{\text{RESET}}$ pulse has just been asserted so that the outputs Q0, Q1 and Q2 are in the low state at time T0. Draw the timing diagram for outputs Q0, Q1 and Q2.

    For simplicity sake, you may assume that there is no delay through a flip-flop, so that the outputs change state coincidentally with the rising edge of the input clock.

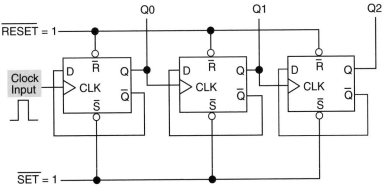

3. Consider the circuit shown below. Assume that a CLEAR pulse has been asserted. Create a truth table for this circuit, showing the state of the outputs after 4 clock pulses.

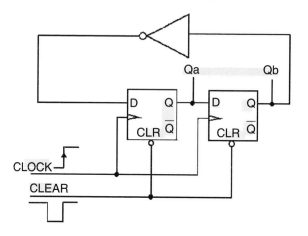

4. The figure below shows four D-type flip-flops and an exclusive OR (XOR) gate. A RESET pulse clears all of the Q outputs to zero. Wires that cross each other in the figure are only connected if there is a heavy black dot at the intersections. Create a truth table showing the state of outputs Q0 through Q3 after 4 clock pulses.

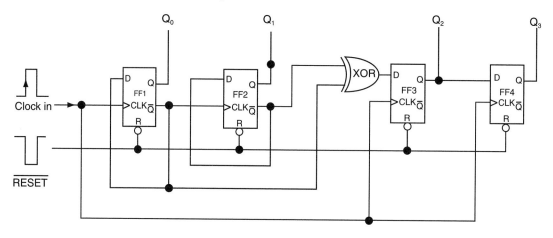

5. Consider the following circuit, comprised of 4, D-type flip-flops and an exclusive or (XOR) gate. Assume that a RESET pulse is issued to the circuit and the outputs A, B, C, D all go to zero. Create a truth table showing the state of the outputs after 8 clock pulses. Does the pattern repeat itself? If so, when?

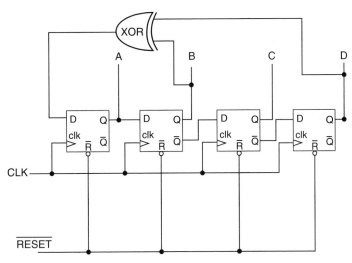

6. Consider a counter/frequency divider made up of eight stages of "D" flip-flops connected in series. Assume that the propagation delay through an individual flip-flop is 25 ns. In this case, the propagation delay is the time from the rising edge of the clock to the time that the output change has stabilized. What is the maximum possible input clock frequency that we can have before the last stage of the counter is still changing state when the next clock signal arrives. You may assume a square wave signal for the clock input.

7. One of the most common uses of the JK-FF is to build synchronous counting circuits. A synchronous counter differs from a ripple counter in that the clock goes to all of the FF stages in parallel and each stage either toggles or remains in its current state depending upon the state of the J and K inputs. Thus, the counter is finished counting much quicker than a ripple counter. Design a four-stage synchronous counter using JK flip-flops, such as the ones shown in Figure 4.6. *Hint: The only external gating that you'll need are three, 2-input AND gates.*

8*. The following circuit is called a synchronous counter with parallel load. It is particularly useful in computer applications. The LOAD/$\overline{\text{COUNT}}$ input determines if the circuit functions as a synchronous up-counter or as a parallel load register. For example, holding the LOAD/$\overline{\text{COUNT}}$ input HIGH and then pulsing the clock input loads the values D0 through D3 into the JK flip-flops. If the LOAD/$\overline{\text{COUNT}}$ input is then brought LOW, subsequent pulses on the clock input causes the counter to synchronously count up, commencing from the previously loaded value. Thus, if the value (D0, D1, D2, D3) = (0, 1, 1, 0) is loaded into the circuit with the LOAD/$\overline{\text{COUNT}}$ input HIGH, then the next clock pulse after LOAD/$\overline{\text{COUNT}}$ is brought low will cause the Q0 to Q3 outputs to count up to (1, 1, 1, 0).

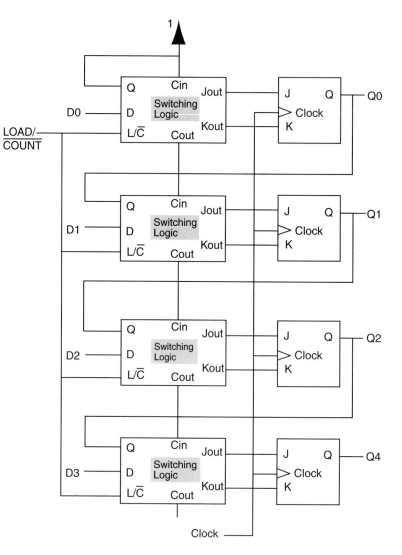

The box labeled "Switching Logic" contains the circuitry that implements the basic functionality of the synchronous counter with parallel load. Design the gate circuitry necessary to implement the switching logic for this device.

Hint: Solving the previous problem will help in understanding the requirements of this circuit design. Once that problem #7 is completed, it should be relatively straightforward to construct the truth table and K-map for the switching logic.

CHAPTER 5

Introduction to State Machines

Objectives

When you are finished with this chapter, you will be able to:
▶ *Describe the operation of state machines;*
▶ *Design a simple state machine;*
▶ *Explain how a state machine can be used to control the instruction sequence of a simple microprocessor.*

● ●

The topic of state machines in computer systems deserves much more discussion then we can give it in this book. However, having said that, let's press forward anyway and try to gain some insight into the topic. Although you have probably never thought about this, there is a symmetry that exists between hardware design implementations and software design implementations. As software developers, you are already familiar with what an algorithm is. You've probably written many different algorithms in your other programming classes, and perhaps, you may have even studied the properties of algorithms in a separate class. In this lesson, we'll see how you might create an algorithm using dedicated hardware, rather than a set of C or C++ instructions.

It is a fact that an algorithm may be solved in hardware as readily as in software. For example, you might be a "gamer" and enjoy playing video games on your PCs. You know that to really get into the game, you need a potent video card, one capable of high-speed image creation. If you don't have such a card, your game will still play, but it will be slower and less realistic then the game played with the hot card.

Another example is the 56K baud modems in your PC. The more expensive modems are fully self-contained and cost between $80 and $150. The less expensive "Win modems" must be used with the Windows® operating system and have a sufficiently fast PC to run properly. These modems might cost between $10 and $50. The difference is that the more expensive modem does the data translation in hardware while the Win modem does it in software.

Another good example can be found in the current crop of low-cost laser printers entering the market. Prior generations of laser printers contained high-performance hardware to convert the input data stream into a pattern of laser imprints on a photosensitive drum. The process of data conversion is called *rasterization*, and is similar to the method used to create an image on a CRT screen. Rasterization was handled in dedicated hardware by the laser printer, but as the computing power of PCs became more powerful, it made economic sense to move the rasterization back to the PC as part of the printer's driver code. The printer then simply transferred the raster data coming in into the laser bits on the screen. Of course, you could only use the printer with the appropriate software driver, and there was no way that the printer could be used as a stand-alone device.

In general, an algorithm solved in hardware is much faster, sometimes many orders of magnitude faster, then the same algorithm solved in software. Today, we have evolved a method of creating *application-specific integrated circuits*, or *ASICs*, as a way to create a hardware implementation of an algorithm. The video processor in your PC, the sound chip and the modem chip are all examples of ASICs in your PC, but they are even more prevalent throughout industry. In many of today's industrial and consumer applications, the ASICs handle the data processing and data manipulation, and the microprocessor handles the error processing and initialization.

In the previous chapter, we were introduced to the concept of the state of a system. State-based systems are a natural consequence of the introduction of the flip-flop. That is, a device whose outputs are not only dependent upon its inputs, but upon the current state of its outputs as well. Also, we introduced the concept of systems that can only change state when a synchronizing clock signal is present.

What we are leading to is that there is a class of digital system behaviors that a fundamentally sequential in their behavior. We'll soon look at how a computer steps its way through a finite sequence of steps as it fetches an instruction from memory, decodes the instruction, executes the instruction, writes back to memory any results and begins the sequence over again. This is a perfect example of such a sequential process.

How the computer is led through this sequence of steps is operation of its state machine engine. That engine can be controlled by data stored in memory, as we'll see in an example in this chapter, or it can be controlled by combinatorial logic, as we've previously learned to design. We'll also do an example of this type of design as well. Any sequential process that we can conceptually describe and implement as a synchronous digital system is called a *finite state machine*[1]. We can define a sequential (finite state) machine as a model that conforms to the following set of requirements:

- It can exist in a finite set S of states s.
- There is an initial state s_0 that is a member of set S.
- A finite set I of inputs i
- A next state function $d = d(s,i)$ that maps the present state values and the inputs into the next state in S.
- An output function, $f = f(s,i)$

Does the above formalism describe what we intuitively already know? Figure 4.20 shows the state transition diagram for a 4-bit binary counter. Does it conform to our definition of a finite state machine (FSM)?

- It can only exist in a set of 16 states, labeled 0000 through 1111. These 16 states define the set, S.
- There is an initial state, 0000, that is part of the set and it can be reached through initialization of the device, or as part of the sequence of steps.

- There are no inputs in this example, but that does not diminish the correctness of the analysis.
- The next state function is dependent upon the present state.
- The output of the counter is dependent upon the current state of the counter.

If *f* is a function only of its present state $f = f(s)$, as is the case with the 4-bit binary counter, we call that a *Moore machine*. If *f* is a function of both its current state and its inputs, $f = f(s,i)$, then it is called a *Mealy machine*. If the machine is in state *s* and receives the input *i,* the next state that it will go to is determined by $d = d(s,i)$

We can take the principles of FSM and apply them to the solution of process problems, such as the video card or modem examples. When we use these methods we are describing an *algorithmic state machine*. That is, we are applying the methods and hardware implementation techniques of FSM design to the solution of an algorithm.

While this all seems highly structured and mathematical (and it is) perhaps we can gain some insight into the process by implementing a FSM for a simple sequential process.

Figure 5.1 is a state diagram for an arbitrary hardware algorithm. After a reset pulse, the system is initialized to state 000. The dark gray arrows show the state transitions that will occur when the input variable, X = 1 and the light gray arrows show the state transitions that will occur when X = 0. Thus, if the system is in state 011 and X = 0, then the next state will be 100. However, if X = 1 when the clock arrives then the next state would be 010.

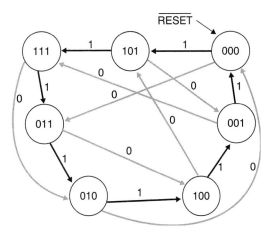

Figure 5.1: State diagram for an arbitrary sequential process.

Figure 5.2 is the implementation method for the algorithm. The output of the circuit are state variables Aout, Bout and Cout, as shown. Note that state 000 means Aout = 0, Bout = 0, Cout = 0, respectively. These variables also feed back to the truth table to become the input variables for the next state transition that will occur on the next clock. In this example, we will transfer

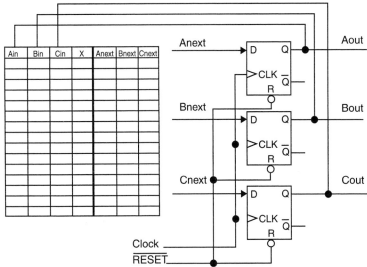

Ain	Bin	Cin	X	Anext	Bnext	Cnext

Figure 5.2: Implementation method for the sequential process of Figure 5.1. The truth table will be converted to combinatorial logic.

the requirements of the state diagram to the truth table and then create a hardware implementation in combinatorial logic for the truth table. When the combinatorial logic and the D-FF sequencer are combined, we should have a FSM that replicates our state diagram. Therefore, we will first complete the truth table, and then simplify it using K-Map techniques. Finally, we will implement the logical equations in hardware.

Figure 5.3 is the completed truth table. The first question to ask is, "Does this make sense?" Let's see. According to Figure 5.1 when the system is in its initial state, just after RESET, it can go to either state 101, if X = 1, or 011 if X = 0. Referring to the truth table in Figure 5.3, we see that when Ain, Bin, Cin = 000 and X = 0, Anext, Bnext, Cnext = 011 and when X = 1, Anext, Bnext, Cnext = 101, which does, indeed, agree with the state diagram.

Ain	Bin	Cin	X	Anext	Bnext	Cnext
0	0	0	0	0	1	1
1	0	0	0	1	0	1
0	1	0	0	0	0	0
1	1	0	0	X	X	X
0	0	1	0	1	1	1
1	0	1	0	0	0	1
0	1	1	0	1	0	0
1	1	1	0	0	1	0
0	0	0	1	1	0	1
1	0	0	1	0	0	1
0	1	0	1	1	0	0
1	1	0	1	X	X	X
0	0	1	1	0	0	0
1	0	1	1	1	1	1
0	1	1	1	0	1	0
1	1	1	1	0	1	1

Figure 5.3: Truth table for the sequential process example of Figure 5.1.

One important point to note is that there is one state 110, which is not part of the sequence process. One the truth table, we see this state represented as XXX. This is a shorthand way of saying that this state will never be reached (if the circuit is operating properly). What we can do is reserve the right to replace the X with either a 1 or 0 if it represents a possible simplification of the K-map. We can do this because, as you can see from the K-map, no other state leads us to the 110 state.

Before we begin working through the logic of the simplification process, we should stop and recall that the truth table is, in effect, a map of the contents of a memory device that could provide us with a circuit design for the sequential process. Figure 5.4 shows this alternative view of the logical design. Here, the inputs to the truth table, Ain, Bin, Cin, and X become the 4 address inputs to the memory. The data out bits, DATA0 through DATA2, correspond to the output variables of the truth table, or the next state for the process.

The K-maps for the variables and the simplified equations are shown in Figure 5.5. In this figure, the simplified equations for Anext includes an XOR term. This was added to further simplify a term $A * \overline{X} + \overline{A} * X$. It is fair to ask if any further simplifications could be made. Clearly, using the laws of Boolean algebra, there could be addition regroupings of some terms. However, from a gate complexity point of view, grouping terms may actually increase complexity because it could introduce an additional AND gate into the hardware implementation. Thus, it isn't always clear that the best algebraic solution leads us to the minimal hardware solu-

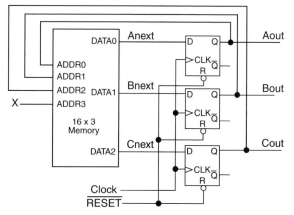

Figure 5.4 Memory-based solution to the sequential process.

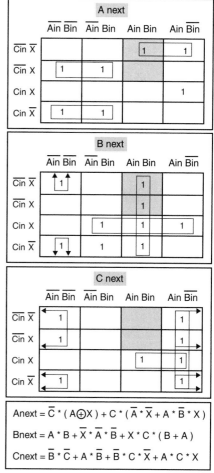

Figure 5.5: K-maps and simplified equations for variables Anext, Bnext and Cnext.

tion. There may be additional factors, such as available gates with the correct number of inputs, extra gates in the package needed elsewhere in the circuit, etc.

The hardware implementation is shown in Figure 5.6. Note how the $\overline{Q}$ outputs of the 'D' flip-flops were used as inverter gates for the truth table. Rather than simply using NOT gates to create the complements of Anext, Bnext, Cnext and X, the $\overline{Q}$ outputs were convenient sources of the complemented signals.

The circuit of Figure 5.6 meets the requirements of a finite state machine. That is:

- There is a set S of seven states that we sequence through in series, the transitions between the states occurs on the rising edge of the clock input to the D-FF's.
- We can establish an initial state, 000, with the RESET pulse, and the initial state is part of the set, S, of states.
- There is an input to the system, X, and three outputs, Aout, Bout and Cout.
- The next state function, $d = d(s,i)$ and the output function $f = f(s,i)$, are the same. In a later example, we'll see how these can be different functions entirely.

$$Anext = \overline{C} * (A \oplus X) + C * (\overline{A} * \overline{X} + A * \overline{B} * X)$$

$$Bnext = A * B + \overline{X} * \overline{A} * B + X * \overline{C} * (B + A)$$

$$Cnext = \overline{B} * \overline{C} + A * \overline{B} + \overline{B} * C * \overline{X} + A * C * X$$

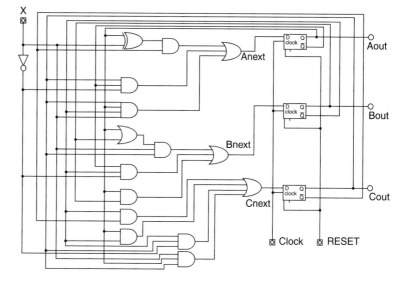

Figure 5.6: Hardware implementation of the sequence process from Figure 5.1.

Obviously, this example was completely made up and doesn't really represent a real-world digital design (although I'm sure you could buy some parts at a surplus electronics store and, with some simple instructions, build the circuit and watch it sequence through the states *ad nauseum*).

This is fun! Let's do another example. This time we'll do one that might be actually useful. Let's consider that we have a serial bit stream coming into our system. Suppose that we need to detect every time 3 or more successive 1 bits are received. Off hand, I don't know why you would want to do this, but given some time, I'm sure we could figure out a circuit need for such a circuit. Figure 5.7 is the state diagram for this circuit.

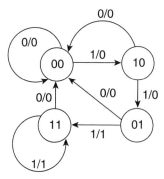

Here we have four states, 00, 10, 01 and 11, respectively. The arrows represent the state transitions on the clock edge. Associated with each arrow are two numbers separated by a forward slash. The first number is the state of the input bit at the time of the rising edge of the clock and the second number is the output signal, T, which goes TRUE if three or more successive 1's are detected in the bit stream.

Figure 5.7: State transition diagram for a circuit that detects 3 or more successive 1's in a serial bit stream.

In this case, the clock signal that we would use would likely be the clock that the system is using to synchronize with the data bit stream. Thus, a new data bit would appear on every rising edge of the clock. Every time that a 0 bit appears the circuit returns to the 00 state, waiting for the arrival of a 1 bit. With the arrival of the first 1 bit the circuit transitions to state 10. If a second 1 bit arrives it transitions to state 01. If the second bit to arrive is a 0 bit, the circuit returns to state 00. However, if a third 1 bit arrives, the circuit transitions to state 11 and the T output bit is asserted, signaling three successive 1 bits. Now, with as long as the incoming bits are 1's, the circuit remains in state 11 with the T bit set to 1. As soon as the next 0 bit arrives, the circuit returns to state 00. What is new in this example is the addition of an explicit output bit, T.

Figure 5.8 shows the truth table, K-map and simplified logical equations for the circuit.

The circuit diagram is shown in Figure 5.9. This problem is also an example of a FSM where the output bit, T, is a separate function of the inputs and the combinatorial logic of the problem. In fact, the state bits, A and B, do not have any role in the output condition, T. Clearly, the next state function, $d = d(s,i)$ and the output function $f = f(s,i)$, are different.

A	B	X	Anext	Bnext	T
0	0	0	0	0	0
0	0	1	1	0	0
1	0	0	0	0	0
1	0	1	0	1	0
0	1	0	0	0	0
0	1	1	1	1	1
1	1	0	0	0	0
1	1	1	1	1	1

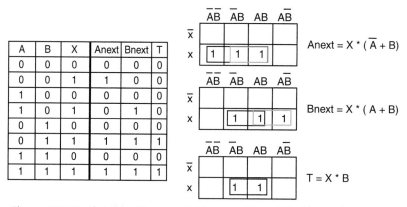

$Anext = X * (\overline{A} + B)$

$Bnext = X * (A + B)$

$T = X * B$

Figure 5.8: Truth table, Karnaugh Maps and simplified logical equations for the state diagram of Figure 5.7.

Another interesting observation that we could make is that this FSM is actually implementing an algorithm. It probably wouldn't take you very long to write an algorithm in C, C++ or Java that would mimic this hardware circuit in software. In fact, we have implemented an *algorithmic state machine, or ASM.*

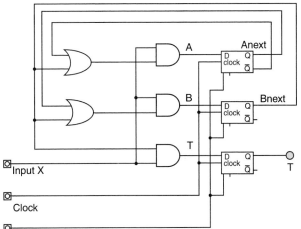

Algorithmic State Machine Example: A Traffic Intersection Controller

Let's walk through the gory details (at least once) of a real problem that might be typical of a range of problems that could be solved using Algorithmic State Machines[2, 3]. In this case, we'll consider the problem of designing the control circuitry for the traffic lights in a typical traffic intersection. Figure 5.10 shows the intersection. Here:

Figure 5.9: Hardware implementation of the state transition diagram of Figure 5.7.

- W = Signal light for westbound traffic
- E = Signal light for eastbound traffic
- N = Signal light for northbound traffic
- S = Signal light for southbound traffic

Since the N/S traffic is a busy four-lane road and the E/W road is only two lanes, we can assume that the N/S road has a considerably higher traffic flow then the E/W road. Therefore, we don't want to stop the N/S traffic at this intersection unless there is waiting E/W traffic. For example, how many times have you sat in your car at a red light at an intersection with perhaps 20 or 30 other cars, waiting for the light to change and there is no cross traffic in the intersection? But I digress…

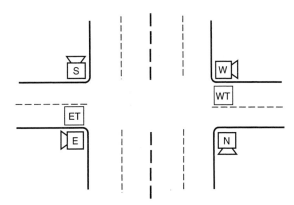

Figure 5.10: Traffic intersection with a four-lane N/S roadway and a two-lane E/W roadway. The E/W roadway has sensors to detect waiting traffic.

Here is a pseudo-code specification for our algorithm:

```
While (1)
{
  timer = 20 seconds;
  NS = green;
  EW = red;
  while ( timer != 0)
    wait;
  if (EW sensor != 0 )
  {
```

```
        timer = 5 seconds;
    NS = yellow;
        while ( timer != 0)
         wait;
        timer = 20 seconds;
        NS = red;
        EW = green;
        while ( timer != 0)
         wait;
           timer = 5 seconds;
        EW = yellow;
        while ( timer != 0)
         wait;
     }
  }
```

Notice that there is a new wrinkle in this problem. The times are different. We are nominally in the N/S state for 20 seconds, and then, if there is an EW car waiting, we have to transition for 5 seconds through a yellow light before turning on EW green for 20 seconds.

The statement, `while(1)`, is often used in embedded systems to identify the portion of the program code that will run forever. Once the system, such as a printer, is initialized, the printer runs forever (or until you turn off the power).

Let's look at the state transition diagram, shown in Figure 5.11. The state of the EW traffic sensor is an input to our algorithm and it can modify the behavior of the system. We can represent this condition by using the fact that there are two possible paths for the state machine to take once it is in the state NS GRN/EW RED. As long as neither of the E/W sensors detects a waiting car, the traffic control algorithm will remain in this state. However, if a waiting car is detected at the end of the time interval for NS GRN, then the state machine transitions to the state NS YEL/EW RED. Once in this state, the algoritm must sequence through NS RED/EW GRN and NS RED/EW YEL before it re-enters the state NS GRN/EW RED. There are no possible inputs in these other states that can change the sequence.

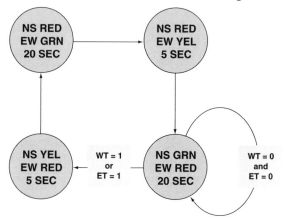

Figure 5.11: State transition diagram for the traffic intersection.

Again, assuming we could take care of the time requirements, you could easily change the pseudo code of this algorithm to a real program in C or C++. Of course, we'd need to supply an I/O port to drive the lights and a timer mechanism, but these are simple implementation details. Let's take a brief diversion from our algorithmic state machine design and introduce a new concept (just when you thought it was safe to go back into the water, bang!). Figure 5.12 shows us the state of the outputs represented as a timing diagram. This timing diagram shows us the state of the input signals that control the lights versus time. Notice in Figure 5.12 that our horizontal axis is calibrated in 5 second ticks. This just shows that the outputs can only change on the 5-second clock ticks.

The timing diagram of Figure 5.12 represents an alternative view of the system. In fact, this is what the controller outputs to the traffic light power circuits would look like on the screen of a logic analyzer. This kind of view is significant because it introduces us to another important concept, the idea of the states of the system represented as a set of *state vectors*. The state vectors are just the combination of all of the possible states of the inputs and outputs that our system might be in. Thus, if you create a table of all of the vectors, then you have an alternative way of representing all of the possible states of the system. Figure 5.13 is a representation of the timing diagram of Figure 5.12, but now represented as a state vector set.

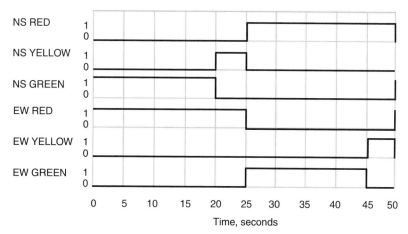

Figure 5.12: Traffic light sequence represented as a timing diagram.

At each clock transition, the state of the system can be described by the binary vector set or by the hexadecimal vector set. We have to be a bit careful here because the hexadecimal representation shouldn't be confused with a number. It's not. It is a collection of individual bits that are represented as an ensemble of bits, or in other words, a state vector.

The hexadecimal representation is convenient, as you'll see, for creating the actual memory image for the traffic controller, but the more accurate and informative representation would be to leave the description in terms of binary vectors. Unfortunately, it's just too convenient not to use hex.

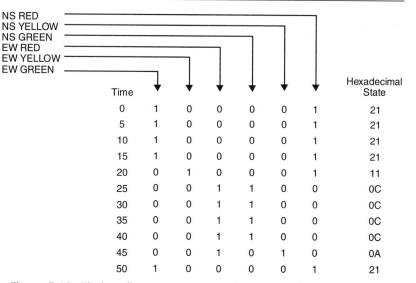

	Time							Hexadecimal State
	0	1	0	0	0	0	1	21
	5	1	0	0	0	0	1	21
	10	1	0	0	0	0	1	21
	15	1	0	0	0	0	1	21
	20	0	1	0	0	0	1	11
	25	0	0	1	1	0	0	0C
	30	0	0	1	1	0	0	0C
	35	0	0	1	1	0	0	0C
	40	0	0	1	1	0	0	0C
	45	0	0	1	0	1	0	0A
	50	1	0	0	0	0	1	21

Figure 5.13: Timing diagram represented as a set of state vectors. The Hexadecimal State representation on the right is a shorthand way of collecting the individual binary states of the six light signals. Thus, the hexadecimal numbers have no significance as unique number in their own right.

Let's now begin the actual process of designing the controller. As a first pass through, we'll ignore the E/W traffic sensors and just consider a simpler model, one with each direction getting 20 seconds of time with a 5 second yellow. Better to focus on the issues. Figure 5.14 shows the simplified algorithm converted to state variables. Also, we've redrawn it so that every state lasts 5 seconds in duration. By increasing the number of states, even though several of the states "seem to" represent the same condition, we can convert the problem to one of equal time intervals. Actually, the 4 states that each represents a 20 second long time interval are not the same. They just have the same output function.

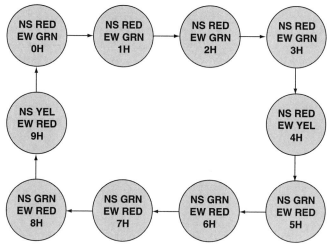

Now we've given each state a unique numerical identifier, numbering them from 0000 to 1001, or 0H to 9H. This may look suspiciously like a memory address to you, but my lips are sealed. I won't reveal the good part until the end. Let's specify the outputs and the binary weight of each bit. This is only for convenience sake, but we'll press on.

Figure 5.14 Algorithm for traffic intersection represented as a set of state variables. Each state is numbered in sequence and the period is 5 seconds long.

Consider the 6 outputs for the state 0000 (0 Hex).

- NS red is on: NSR or Data bit 0 (D0) = 1
- NS yellow is off: NSY or Data bit 1 (D1) = 0
- NS green is off: NSG or Data bit 2 (D2) = 0
- EW red is off: EWR or Data bit 3 (D3) = 0
- EW yellow is off: EWY or Data bit 4 (D4) = 0
- EW green is on: EWG or Data bit 5 (D5) = 1

Thus, the state variable for state 0000 is 100001, or state 0 hex = 21 hex. We also use the collective term data bits to represent the ensemble of state variables. Also, it will be convenient to give the individual bits that control the lights a collective identity so that we may deal with them as a group, rather than as individual entities. However, we should keep in mind that each output variable really is independent of the other. For example, in this instance, adding a number to the state value, 21 hex, has no logical significance. The next step is to convert our specification to a truth table.

Remember the 32-bit memory array that we looked at earlier. Let's see it again, but this time we'll expand the number of outputs to 6 in order to account for all 6 outputs for the traffic signals. Thus, we need a memory with a total of 96 memory cells, arranged as 16 by 6. Figure 5.15 shows the memory (truth table) layout for this project.

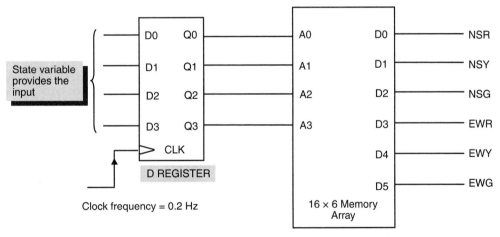

Figure 5.15: The state machine with the storage register added to provide the sequencing mechanism. This circuit still lacks the mechanism for sequencing through the various states in the system and for changing the algorithm if an E/W car is waiting.

Our clock generator is somewhere else in this system. Compared to the clock frequencies we've been discussing so far, 0.2 Hz (one cycle every 5 seconds is positively glacial, but that's the way were solving it). As an aside, let's consider the possibility of increasing the clock frequency to 1 Hz. This would have the apparent effect of increasing the number of states in the system, but there are other ways we might be able to redesign the system to handle that. However, one advantage of increasing the number of states is that we lower the response time of the system.

Suppose that we wanted to add an emergency vehicle detector to our intersection. That's a photosensitive device that is designed to detect the strobe lights on the tops of emergency vehicles. If a police car was rushing towards this intersection at 100 km per hour (~28 meters per second), then the police car could travel as much as ~139 meters (with a 5 second clock period) before all the lights in the intersection turned red. Clearly, there is room for improvements in our traffic intersection algorithm.

The output value from the D register is the address of the of the six memory cells that determine the data for the next state of the system after the rising edge of the clock pulse arrives. We can summarize the data by actually building the *state table*. This is just a convenient way of bringing all of the pieces together. Keep in mind that we still need to add a mechanism for sequencing through the various states in the system.

Figure 5.16 is the state table for our preliminary design. We call it preliminary because we still need to add the sequencing mechanism. That will come shortly. Referring to Figure 5.16, you can see that we've let the cat out of the bag. The numerical state identifiers, 0000 through 1001 are, indeed, the address in the READ ONLY MEMORY *ROM* that will become the truth table for our design. You're probably already familiar with the term "ROM" from your PC's BIOS ROM. The ROM is just a memory device that can be preprogrammed so that its data values are retained even if power is turned off. This is just what we need for our traffic controller design.

State					Outputs						
Q3	Q2	Q1	Q0	ROM Address	D5	D4	D3	D2	D1	D0	ROM Contents
0	0	0	0	0H	1	0	0	0	0	1	21H
0	0	0	1	1H	1	0	0	0	0	1	21H
0	0	1	0	2H	1	0	0	0	0	1	21H
0	0	1	1	3H	1	0	0	0	0	1	21H
0	1	0	0	4H	0	1	0	0	0	1	11H
0	1	0	1	5H	0	0	1	1	0	0	0CH
0	1	1	0	6H	0	0	1	1	0	0	0CH
0	1	1	1	7H	0	0	1	1	0	0	0CH
1	0	0	0	8H	0	0	1	1	0	0	0CH
1	0	0	1	9H	0	0	1	0	1	0	0AH
1	0	1	0	AH	X	X	X	X	X	X	Don't Care
1	0	1	1	BH	X	X	X	X	X	X	Don't Care
1	1	0	0	CH	X	X	X	X	X	X	Don't Care
1	1	0	1	DH	X	X	X	X	X	X	Don't Care
1	1	1	0	EH	X	X	X	X	X	X	Don't Care
1	1	1	1	FH	X	X	X	X	X	X	Don't Care

Figure 5.16 State table for the traffic controller showing the state variables (ROM addresses) and the data contained in each ROM address. The values identified as "X" are referred to as "Don't Cares" because the state machine will never go to these states.

Notice that our ROM actually has memory cells without any data in them. These are the "Don't Cares" at address 0A hex through 0F hex. This is because our design only has a total of 10 valid states (00 hex through 09 hex). Even though our memory has these extra states, we won't be bothered by them as long as our sequencer never takes us to them.

Adding the state sequencing mechanism is quite straightforward. All we need to do is increase the number of data outputs to our state table that we will feed back to the D storage register. Thus, we've generalized the concept of the state machine to show that it isn't necessary to feed all of the outputs back to the inputs. We only need to bring enough

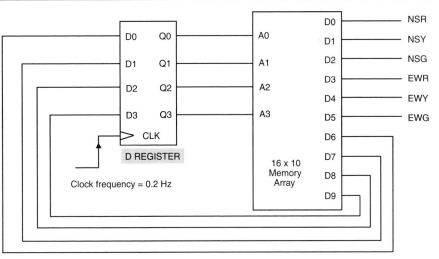

Figure 5.17: State machine with the addition of extra data bits in the ROM in order to provide a mechanism for sequencing through the states.

output states back to the inputs to provide a mechanism to sequence the state machine through its state. We call these additional outputs the *excitation outputs* because their function is to provide the sequencing (excitation) mechanism for the state machine. Figure 5.17 shows the new design.

Now, it will be necessary to expand the number of output bits in the ROM to provide the feedback to the D register for the sequencing system. Our ROM still has the same number of memory addresses but we've added 4 more output bits. Thus, the data in each ROM address is now 10 binary digits. Figure 5.18 shows the expanded state table.

Current State ROM Address	Next State				Outputs						ROM Contents
	D9	D8	D7	D6	D5	D4	D3	D2	D1	D0	
0H	0	0	0	1	1	0	0	0	0	1	061H
1H	0	0	1	0	1	0	0	0	0	1	0A1H
2H	0	0	1	1	1	0	0	0	0	1	0E1H
3H	0	1	0	0	1	0	0	0	0	1	121H
4H	0	1	0	1	0	1	0	0	0	1	151H
5H	0	1	1	0	0	0	1	1	0	0	18CH
6H	0	1	1	1	0	0	1	1	0	0	1CCH
7H	1	0	0	0	0	0	1	1	0	0	20CH
8H	1	0	0	1	0	0	1	1	0	0	24CH
9H	0	0	0	0	0	0	1	0	1	0	00AH
AH	X	X	X	X	X	X	X	X	X	X	Don't Care
BH	X	X	X	X	X	X	X	X	X	X	Don't Care
CH	X	X	X	X	X	X	X	X	X	X	Don't Care
DH	X	X	X	X	X	X	X	X	X	X	Don't Care
EH	X	X	X	X	X	X	X	X	X	X	Don't Care
FH	X	X	X	X	X	X	X	X	X	X	Don't Care

Figure 5.18: Modified state table now includes extra data bits for sequencing.

Finally, we are now in a position to go back and include the extra inputs for the EW traffic that we've neglected so far. Let's refresh our memory (it's been a long session) by reviewing the algorithm that we've previously developed. Figure 5.19 illustrates the state diagram that we developed and modified so that we may utilize the 5-second clock period for each state.

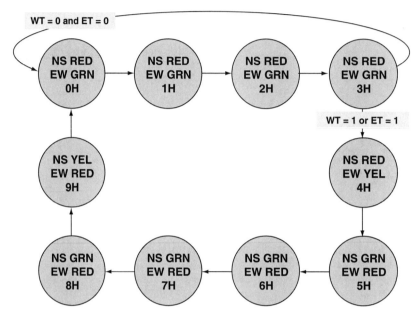

Figure 5.19: State transistion diagram for state machine including inputs for waiting EW traffic.

This new addition leads us to the state machine design in Figure 5.20.

The addition of these two independent inputs, ET and WT, which represent the in-the-road sensors for waiting eastbound and waiting westbound traffic, respectively, is extremely significant. The conditions of the independent inputs can affect the sequencing of the state machine. Thus, we've added a completely new dimension to our Algorithmic State Machine: the ability to change its state flow in response to a change in its input data. Do you recognize this new feature? You should. It's called a computer.

With the addition of independent inputs that are able to define new states for the state machine, our state machine is beginning to look like a decision-making device. Of course, it isn't really making a decision because we've preprogrammed into it exactly what it can do in response to the new data. The state machine in your computer is several thousand times more complex than this simple one, but you're beginning to get the picture. We call the state transition table (ROM) for the microprocessor the *microcode*. It is the microcode that gives the computer its personality. The computer's *instruction set architecture*, or ISA, is defined by the microcode.

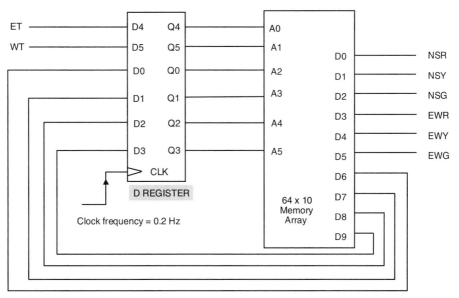

Figure 5.20: State machine design with independent inputs.

When a computer fetches an instruction from memory, the bit pattern that represents the instruction is the external combination of inputs to the microcode ROM. This pattern establishes the sequence of steps required to interpret and execute that instruction. Ever wonder why you get the infamous *Blue Screen of Death*? Sometimes (not always) it may be caused by an errant pointer that sends the program off to fetch an instruction from an illegal area of memory. The bit pattern there may be data values, or garbage, but not the bit pattern of a legal instruction. Since the microcode cannot interpret this pattern, the processor notifies the operation system and the program halts itself. Anyway, back to the problem at hand.

Let's re-examine our state ROM. We've added two more inputs, so the number of possible address combinations goes up to 64. Our state ROM is now 64 locations by 10 bits wide. Our simple traffic system controller has grown to 640 memory cells. It isn't hard to see why the microcode engine in a modern computer has millions of memory cells. Figure 5.21 is a table of the state ROM in all of its gory details.

Of course, the next logical step in the process would be to put back into our design the provisions to the emergency vehicle procedures. That would add one more input to the design, giving a total of 7 address bits to the ROM. Why stop there? We could add a left turn signal for the NS traffic. That would add at least one more output (NS turn light) and one more input (NS turn traffic waiting, NST). If we proceeded to develop the algorithm for this more complex situation, we'll probably see that the number of states in our system has increased beyond 16 so we'll need to add an additional output to feed back to the input for sequencing. That gives us a total of 11 outputs and 8 inputs to the ROM. If you have nothing to do some rainy weekend, why not create the state table for this real traffic intersection? I'll leave it as an optional, but highly instructive exercise for the motivated student.

	State			WT	ET		Next State					Outputs			
A5	A4	A3	A2	A1	A0	D9	D8	D7	D6	D5	D4	D3	D2	D1	D0
0	0	0	0	0	0	0	0	0	1	1	0	0	0	0	1
0	0	0	0	0	1	0	0	0	1	1	0	0	0	0	1
0	0	0	0	1	0	0	0	0	1	1	0	0	0	0	1
0	0	0	0	1	1	0	0	0	1	1	0	0	0	0	1
0	0	0	1	0	0	0	0	1	0	1	0	0	0	0	1
0	0	0	1	0	1	0	0	1	0	1	0	0	0	0	1
0	0	0	1	1	0	0	0	1	0	1	0	0	0	0	1
0	0	0	1	1	1	0	0	1	0	1	0	0	0	0	1
0	0	1	0	0	0	0	0	1	1	1	0	0	0	0	1
0	0	1	0	0	1	0	0	1	1	1	0	0	0	0	1
0	0	1	0	1	0	0	0	1	1	1	0	0	0	0	1
0	0	1	0	1	1	0	0	1	1	1	0	0	0	0	1
0	0	1	1	0	0	0	0	0	0	1	0	0	0	0	1
0	0	1	1	0	1	0	1	0	0	1	0	0	0	0	1
0	0	1	1	1	0	0	1	0	0	1	0	0	0	0	1
0	0	1	1	1	1	0	1	0	0	1	0	0	0	0	1
0	1	0	0	0	0	0	1	0	1	0	1	0	0	0	1
0	1	0	0	0	1	0	1	0	1	0	1	0	0	0	1
0	1	0	0	1	0	0	1	0	1	0	1	0	0	0	1
0	1	0	0	1	1	0	1	0	1	0	1	0	0	0	1
0	1	0	1	0	0	0	1	1	0	0	0	1	1	0	0
0	1	0	1	0	1	0	1	1	0	0	0	1	1	0	0
0	1	0	1	1	0	0	1	1	0	0	0	1	1	0	0
0	1	0	1	1	1	0	1	1	0	0	0	1	1	0	0
0	1	1	0	0	0	0	1	1	1	0	0	1	1	0	0
0	1	1	0	0	1	0	1	1	1	0	0	1	1	0	0
0	1	1	0	1	0	0	1	1	1	0	0	1	1	0	0
0	1	1	0	1	1	0	1	1	1	0	0	1	1	0	0
0	1	1	1	0	0	1	0	0	0	0	0	1	1	0	0
0	1	1	1	0	1	1	0	0	0	0	0	1	1	0	0
0	1	1	1	1	0	1	0	0	0	0	0	1	1	0	0
0	1	1	1	1	1	1	0	0	0	0	0	1	1	0	0
1	0	0	0	0	0	1	0	0	1	0	0	1	1	0	0
1	0	0	0	0	1	1	0	0	1	0	0	1	1	0	0

(continued)

0	1	1	1	1	0	1	0	0	0	0	0	1	1	0	0
0	1	1	1	1	1	1	0	0	0	0	0	1	1	0	0
1	0	0	0	0	0	1	0	0	1	0	0	1	1	0	0
1	0	0	0	0	1	1	0	0	1	0	0	1	1	0	0
1	0	0	0	1	0	1	0	0	1	0	0	1	1	0	0
1	0	0	0	1	1	1	0	0	1	0	0	1	1	0	0
1	0	0	1	0	0	0	0	0	0	0	0	1	0	1	0
1	0	0	1	0	1	0	0	0	0	0	0	1	0	1	0
1	0	0	1	1	0	0	0	0	0	0	0	1	0	1	0
1	0	0	1	1	1	0	0	0	0	0	0	1	0	1	0
1	0	1	0	0	0	0	0	0	0	0	0	1	0	0	1
1	0	1	0	0	1	0	0	0	0	0	0	1	0	0	1
1	0	1	0	1	0	0	0	0	0	0	0	1	0	0	1
1	0	1	0	1	1	0	0	0	0	0	0	1	0	0	1
1	0	1	1	0	0	0	0	0	0	0	0	1	0	0	1
1	0	1	1	0	1	0	0	0	0	0	0	1	0	0	1
1	0	1	1	1	0	0	0	0	0	0	0	1	0	0	1
1	0	1	1	1	1	0	0	0	0	0	0	1	0	0	1
1	1	0	0	0	0	0	0	0	0	0	0	1	0	0	1
1	1	0	0	0	1	0	0	0	0	0	0	1	0	0	1
1	1	0	0	1	0	0	0	0	0	0	0	1	0	0	1
1	1	0	0	1	1	0	0	0	0	0	0	1	0	0	1
1	1	0	1	0	0	0	0	0	0	0	0	1	0	0	1
1	1	0	1	0	1	0	0	0	0	0	0	1	0	0	1
1	1	0	1	1	0	0	0	0	0	0	0	1	0	0	1
1	1	0	1	1	1	0	0	0	0	0	0	1	0	0	1
1	1	1	0	0	0	0	0	0	0	0	0	1	0	0	1
1	1	1	0	0	1	0	0	0	0	0	0	1	0	0	1
1	1	1	0	1	0	0	0	0	0	0	0	1	0	0	1
1	1	1	0	1	1	0	0	0	0	0	0	1	0	0	1
1	1	1	1	0	0	0	0	0	0	0	0	1	0	0	1
1	1	1	1	0	1	0	0	0	0	0	0	1	0	0	1
1	1	1	1	1	0	0	0	0	0	0	0	1	0	0	1
1	1	1	1	1	1	0	0	0	0	0	0	1	0	0	1

Figure 5.21: State table for the traffic controller ROM. Note how the unused states, beginning with state 1010 have a data output equal to 09H, corresponding to the NS red and EW red lights being turned on. This is a safety feature in the unlikely event that the system enters an illegal state.

Figure 5.22 shows the important components and data paths within the circuit without getting bogged down in the minute details of actually making such a circuit work in the real world. The oscillator block produces the 0.2 Hz clock stream that we need to sequence the state machine. The Octal D flip-flop is an integrated circuit building block that contains eight D-FF's in a single package with a common clock input. Thus, all eight FF's are always clocked on the same rising clock edge.

The next two devices are the heart of our traffic controller. They contain the ROM code that we developed to implement the Algorithmic State Machine. Two separate ROMs were used because we need 10 outputs and most commercially available ROMs only have 8 outputs, so we need to partition the design among the two ROMs. For convenience, the 6 outputs of the upper ROM are used to control the lights and the 4 outputs of the lower ROM to sequence through the states. You can see that the 4 outputs from the lower ROM go back and become the inputs to the D-FF's. This clearly illustrates the separate functions of the sequencing function, $d(s,i)$, and the output function, $f(s,i)$.

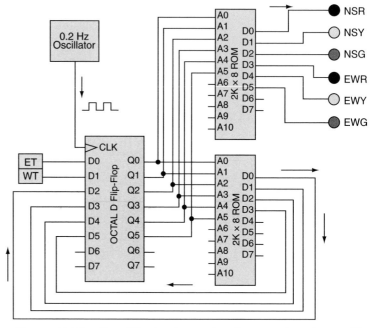

Figure 5.22: Simplified schematic diagram for the traffic intersection controller.

The ROMs in the use for this example each hold 16 K bits, organized as 2 K by 8. Since we're only using a total of 64 addresses in each ROM, you could argue that this is not the greatest engineering solution in the history of computer architecture. However, memory is relatively inexpensive, and, in this particular instance, a ROM with 16 K memory cells is less expensive then a much smaller one.

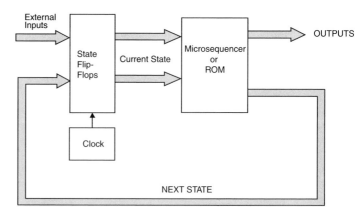

Figure 5.23: A generalized sequential digital machine.

Before we move on, let's take what we've learned about state machines and try to focus these principles on how a computer works. Figure 5.23 shows the basic sequencing mechanism of a digital computer. You can clearly see the elements of our Mealy machine. Conveniently, we've omitted pretty much everything else inside the computer, but they're just connected to the outputs of the microsequencer ROM. The external inputs to the microsequencer would include the machine language instructions as well as all of the other inputs, which can affect the program execution flow of the machine.

Suppose that we want to add two numbers together inside of our computer.

With respect to our newfound knowledge of state machine design, how might we do this? Let's assume that we want to add together two, 4-bit numbers. The circuit that adds two numbers together inside a digital computer is part of the *arithmetic and logic unit,* or *ALU.* You actually know enough about digital design to create the basic building block of the ALU. The basic adder circuit has 3 inputs (A, B and Carry In) and 2 outputs (SUM and Carry Out). We can summarize the behavior of a basic 2-bit adder circuit in the truth table shown as Table 5.1.

Table 5.1: Truth table for a 2-bit adder circuit.

A	B	Cin	Sum	Cout
0	0	0	0	0
1	0	0	1	0
0	1	0	1	0
1	1	0	0	1
0	0	1	1	0
1	0	1	0	1
0	1	1	0	1
1	1	1	1	1

Once you design this circuit and place the gates, you'll quickly see that you can greatly simplify it if you review the design of the exclusive OR gate. As you can see from the truth table, each stage of the adder simply adds the 2 input numbers plus any carry in from the previous stage and generates a sum and a carry out to the next stage. If we wanted to add two 32-bit numbers (*ints*) together, we would have to have 32 of these adders in a row. Anyway, let's see this in Figure 5.24.

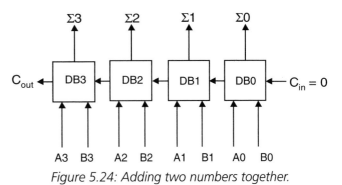

Figure 5.24: Adding two numbers together.

Now that we know how the numbers are actually added together, let's look at how a state machine might sequence through the operation to actually add the numbers together, get a result, and save it. Figure 5.25 shows this sequence schematically.

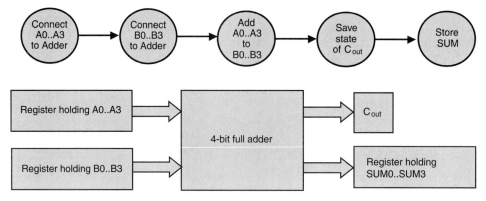

Figure 5.25: State machine sequence for adding two 4-bit numbers.

In this particular example, it would take us 5 clock cycles to complete the addition. Note that we've neglected some of the preliminary sequences, such as decoding the actual ADD instruction that got us to this point and how did the numbers actually get into the holding registers in the first place. Obviously, we're sweeping some additional material under the rug. We'll revisit this problem again in a later chapter in all of its gory details, so let's just focus on the concepts for now. We'll walk through the steps in the process.

1. The storage register holding the first operand, number A0..A3, is connected to the adder circuit so that the A inputs to the adder now see the first operand.
2. The storage register holding the second operand, number B0..B3, is next connected to the adder.
3. After the appropriate propagation delay (once the B input has stabilized), the result appears on the outputs of the adder.
4. The state of the Carry Out bit is save in the appropriate register. As you'll soon see, we'll also store some other results. For example, if the addition resulted in a sum of zero, we'd also store that information.
5. The sum is stored in an output register for further use.

Let's summarize what we've learned about state machines:

- The next state depends upon the current state, any inputs to the storage register and any outputs that are returned to the input of the storage register.
- A D-type register, comprised of D-type flip-flops, is used to synchronize the state transitions with the edge of a clock. The transition time for the clock must be much faster then any changes that may occur in the state machine. This synchronizes the state machine and prevents the circuit from "running away".
- A flow chart, or state chart for the ASM is the algorithm being implemented.
- The ROM-based state table is one way of implementing the truth table for the design. We could also follow the design steps and use the truth table to create the sum of products

(minterm) logical equations for each of the output variables. Once we have the truth table we can then generate the Karnaugh maps. The K-maps allow us to create a simplified gate design for the circuit.

The Algorithmic State Machine is the basis for almost all of today's computing engines. The Instruction Set Architecture of a modern computer is determined by its internal microcode ROM, which implements the state machine. The processor sequences through a series of states determined by:

- The instruction being executed
- The contents of internal registers
- The results of arithmetic or logical operations
- The type of memory accessing mode being used
- Asynchronous internal or external events (interrupts, RESET, error conditions)

Figure 5.26 shows this schematically.

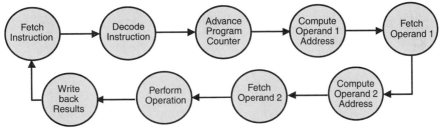

Figure 5.26: State sequencing in a simple microprocessor.

Modern Hardware Design Methodologies

Mead and Conway[4] proposed a new methodology for the design of *very large-scale integration (VLSI)* circuits. They describe a structured design system based upon a top-down approach to the development of a complex integrated circuit. They say,

> *The beginnings of a structured design methodology for VLSI systems can be produced by merging together in a hierarchy the concepts presented in this chapter. Designs are then done in a "top down" manner but with a full understanding by the architect of the successive lower levels of the hierarchy.*

> *To begin with, we plan our digital processing systems as combinations of register-to-register data transfer paths, controlled by finite state machines. Then the geometric shapes, relative sizes, and interconnection topologies of all subsystem modules are collectively planned so all modules will merge together snugly, with a minimum of space and time wasted by random interconnect wiring...*

> *A particularly uniform view of such a system of nested modules emerges if we view every module at every level as a finite state machine or data path controlled by a finite state machine. At the lowest level, elements such as the stack and register cells may be viewed as state machines with one feedback term (the output), two external inputs (the control signals) and a 1-bit state register. These rudimentary state machines are*

115

> *grouped in a structured manner to form portions of a state machine, or data path con-trolled by a state machine, at the next level of the hierarchy.*

Later[5] they go on to describe a method of designing the actual integrated circuits by creating a *computer-aided design* tool (CAD) that works like a macro assembler would work for software. If the basic circuit elements could be expressed as few standard building blocks, or cells, then a symbolic layout language of some kind could, using the macro assembler analogy, create an IC layout from these standard cells. They say,

> *The user defines symbols (macros) that describe the layout of the basic system cells. The locations and orientations of instances of these symbols are described in the language, as functions of the appropriate parameters. These symbolic descriptions may then be mechanically processed in a manner similar to the expansion of a macro assembly language program, to yield the intermediate form description of the system layout, which is analogous to machine code for generating output files.*

What Mead and Conway were describing were two concepts that should be very familiar to you. One is the idea of a structured approach to the design of the hardware. You might call this "software engineering". It starts with developing requirements documentation and then a set of formal specifications. From there, the various functional components (blocks) are defined and by a processes of top-down decomposition, the software progresses through the development process

The second concept will be familiar to you if you've studied how modern compilers convert a high-level language to an intermediate language (assembly language) and then to machine code. Here the low-level machine code is represented by the standard low-level cells that represent the transistor level circuits and interconnects between these cells.

What the authors were describing is what we today call *Silicon Compilers* and the process by which modern integrated circuits are design is called *silicon compilation*. Hardware circuit designers use a high-level development language, either Verilog or VHDL. VHDL is officially defined in IEEE Standard 1076-2001, *IEEE Standard VHDL Reference Manual*[6]. Verilog is considered to be "the other" hardware description language. Verilog started out as proprietary simulation language but was subsequently turned over to the IEEE and published as IEEE Standard 1364-1995, *IEEE Standard Description Language Based on the Verilog Hardware Description Language.*[7]

In 1981, a company, Silicon Compilers, Inc. was founded to *decouple the design description from the implementation*[8]. Their value proposition was to take care of the design implementation so that the designers could focus on the algorithm. To do this, they would create libraries and code modules which would be translatable to custom integrated circuit blocks, the same as if expert designers were handcrafting those circuits.

In 1985, the CMOS process became commercially available and this was the breakthrough needed to drive the ASIC industry. As we've seen, the CMOS gate is an almost ideal switch, and with it, the commercial viability of integrated circuits designed using silicon compilation was realized. Prior to the introduction of the CMOS process there was still too much hand-crafting needed to make an integrated circuit work properly. According to *Cheng,*[8] "power consumption went down, noise margin went up, and a 1 was a 1 and a 0 was a 0".

With the commercial viability of ASIC designs, the standardization of HDL tools emerged and logic synthesis, the design of digital hardware using silicon compilation, greatly accelerated. Silicon Compilers, Inc. merged with Silicon Design Labs in 1987 to form Silicon Compiler Systems Corporation. In 1990, SCS was acquired by Mentor Graphics, Inc., a leading supplier of Electronic Design Automation Tools.

You can see the impact of silicon compilation in Figure 5.27.

If you plot the data from Figure 5.27 on semi-logarithmic graph paper, you'll get an

Intel Microprocessors		
	Year of Introduction	Transistors
4004	1971	2,250
8008	1972	2,500
8080	1974	5,000
8086	1978	29,000
286	1982	120,000
386™ processor	1985	275,000
486™ DX processor	1989	1,180,000
Pentium® processor	1993	3,100,000
Pentium II processor	1997	7,500,000
Pentium III processor	1999	24,000,000
Pentium 4 processor	2000	42,000,000
Souce: Intel Corp.		

Figure 5.27: Growth in transistor count for the Intel family of microprocessors. From Cheng[9].

amazing close approximation to a straight line, which is a remarkable validation of Moore's Law. If we consider the number of designers required to design an integrated circuit, such as the 8086, with 29,000 transistors, versus the Pentium 4 processor, with 42,000,000 transistors, we must come to the conclusion that Intel could not possibly have used a Pentium 4 design team roughly 2000 times bigger than the 8086 design team. The conclusion we must draw is that efficiency of each designer to lay down transistors (or CMOS gates) onto silicon must account for the increase. So, just as C++ has freed the programmer from assembly language programming issues, so has silicon compilation freed the hardware designer from the low-level issues of the process of integrated circuit design.

As a person familiar with high-level languages, you should quickly become comfortable with the structure of the languages. In fact, there are relatively few signs that you are designing hardware, rather than writing software. Several commercial software companies have come to the same conclusion and today, there are commercial tools available that closely integrate the separate hardware and software design processes into a higher-level system view.

However, one remaining difference between the hardware and software development processes is the cost of fixing a defect. As software developers, you know that recompiling and rebuilding a software image might take a few days. The processes involved in distributing the new version of the code to the customers are also well established. You might post a new version to an FTP site in order to send out code updates.

The hardware developers do not have this luxury. Even though the design processes are converging, the hardware designer is still faced with the reality that hardware design is a complicated, expensive and time-consuming process with little or no margin for error. The cost of a hardware re-spin could easily be $500,000 with a time delay of three months or more. Thus, hardware design tools, even with silicon compilation, are heavily structured towards testing and design simulation. It is not unusual for the time required to develop a new ASIC to be equally divided between the actual time for design and the time required to thoroughly test the hardware in simulation. Of course, the hardware designers always have a Plan B in their back pockets. The universal solution to a hardware defect, is and will be for the foreseeable future, *fix it in software*.

Before we leave this topic, it is instructive to actually look at some VHDL code and compare it to the languages we are already familiar with. Following is an example of an adder circuit (*from Ashenden*[10]). We first need to declare an *entity*. This is analogous to declaring a variable.

```
entity adder is

    port ( a : in word ;

             b : in word ;

             sum : out word ) ;

    end entity adder ;
```

Next, we need to describe the internal operation of a module. In other words, we need to write the statements describing the behavior of the variables. This is done in the *architectural body* of the code.

```
architecture abstract of adder is

begin

    add_a_b : process ( a,b ) is

    begin

        sum <= a + b ;

    end process add_a_b ;

end architecture abstract;
```

The architecture body is named *abstract* and it contains a process *add_a_b*, which describes the operation of the entity. Like template functions in C++, this process assumes that the addition operator '+' has previously been defined for the addition of data type 'word'.

We could easily picture how this code snippet could compile to a circuit comprised of 1-bit adder primitives, as in Figure 5.24, except that it would be configured to add a 16-bit or larger variable called a 'word'.

Summary of Chapter 5

- We started Chapter 5 with a discussion of the generalized concept of an algorithm being either a solution based upon a set of software steps or a solution based in hardware.
- We looked at the definition of a finite state machine an saw how we define a state machine as either a Mealy Machine or a Moore Machine.
- We saw how the feedback of the current state, combined with a 'D' type flip-flop, allows us to synchronize and stabilize the state transitions.
- We examined how to use our knowledge of building truth tables to construct the hardware implementation of a state machine.
- We walked through the problem of building an algorithm into a state machine by constructing a memory-based implementation of a traffic intersection controller.
- Finally, we saw how hardware description languages are used to design hardware systems.

Chapter 5: *Endnotes*

[1] Thomas Richard McCalla, *Digital Logic and Computer Design,* ISBN 0-6752-1170-0, Merrill, New York, 1992, p. 265.

[2] Carver Mead and Lynn Conway, *Introduction to VLSI Systems,* ISBN 0-2010-4358-0, Addison-Wesley Publishing Company, Reading, MA, 1980, pp. 85–87.

[3] Claude A. Wiatrowski and Charles H. House, *Logic Circuits and Microcomputer Systems,* ISBN 0-0707-0090-7, McGraw-Hill Book Company, New York, 1980, pp. 1–11.

[4] Carver Mead and Lynn Conway, *Introduction to VLSI Systems,* ISBN 0-2010-4358-0, Addison-Wesley Publishing Company, Reading, MA, 1980, p. 89.

[5] Carver Mead and Lynn Conway, *Introduction to VLSI Systems,* ISBN 0-2010-4358-0, Addison-Wesley Publishing Company, Reading, MA, 1980, p. 98.

[6] Peter J. Ashenden, *The Designer's Guide to VHDL, Second Edition*, ISBN 1-55860-674-2, Morgan-Kaufmann Publishers, San Francisco, 2002, p. 671.

[7] Ashenden, *ibid.*, p. 677.

[8] Ed Cheng, http://bwrc.eecs.berkeley.edu/Seminars/Cheng%20-%209.27.02/Silicon%20Compilation,%2021%20years%20young.pdf.

[9] Cheng, *ibid.*

[10] Peter J. Ashenden, *The Designer's Guide to VHDL, Second Edition*, ISBN 1-55860-674-2, Morgan-Kaufmann Publishers, San Francisco, 2002, pp. 108–111.

Exercises for Chapter 5

1. The figure below is a state machine comprised of three D-type flip-flops and some logic gates.

You may assume that the RESET inputs to the three flip-flops have been asserted. Complete the truth table, shown below, for the state machine.

Draw the complete state transition diagram for this state machine. For simplicity, it is not necessary to include the effect of the RESET signal in your drawing. Hint: start with the partial state transition diagram shown, right.

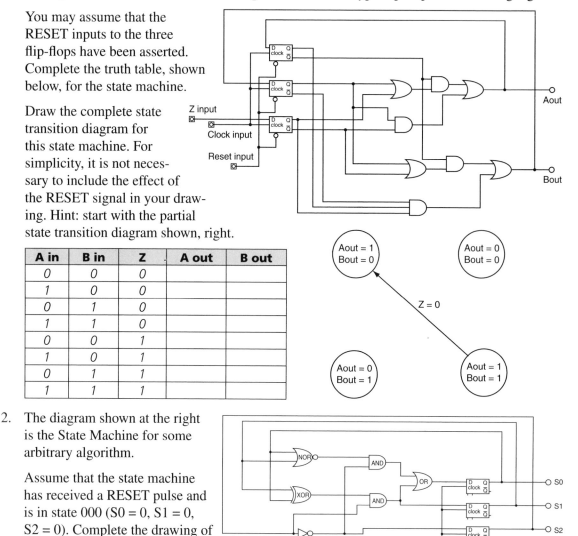

A in	B in	Z	A out	B out
0	0	0		
1	0	0		
0	1	0		
1	1	0		
0	0	1		
1	0	1		
0	1	1		
1	1	1		

2. The diagram shown at the right is the State Machine for some arbitrary algorithm.

Assume that the state machine has received a RESET pulse and is in state 000 (S0 = 0, S1 = 0, S2 = 0). Complete the drawing of the state transition diagram.

3. The circuit shown to the right consists of three D-type flip-flops. The black dots indicate those wires that are physically connected to each other. The $\overline{\text{RESET}}$ inputs are permanently tied to logic 1 and are never asserted. Before any clock pulses are received, the $\overline{\text{SET}}$ input receives a negative pulse to establish the initial conditions for the circuit. Draw the state transition diagram for this circuit.

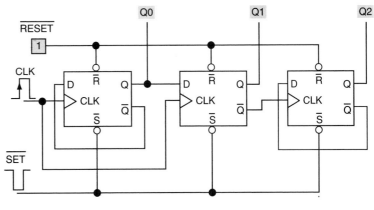

4. The figure, shown right, is a state diagram for a hardware algorithm.

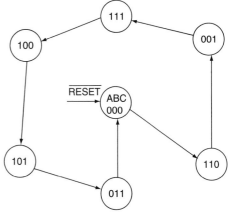

   After a reset pulse, the system is initialized to state 000. Shown below is the implementation method for the algorithm. The outputs of the circuit are state variables ABC, as shown. These variables also feed back to the truth table to become the input variables for the next state transition that will occur on the next clock. Draw the truth table that corresponds to this hardware algorithm. Simplify it, and then draw the entire circuit with the gate equivalent circuit replacing the truth table. Hint: Remember, there is one possible state that you can add to the truth table that does not appear in the algorithm. It may help you to include it in your truth table to get some added simplification.

5. You are the chief designer for the Happy Times Storm Door and Vending Machine Company. You've been given the task of redesigning the company's bestselling vending machine, a simple wall model that installs in convenience store restrooms. Since electrical power is available, you decide to replace the old mechanical model with a spiffy new electronic device.

   You decide to use a State Machine design format. The state machine is shown in the figure on the following page.

   The state machine can cycle through four possible states, S0–S3. In addition, it has two external inputs a and b, and one output, z. Input a represents a quarter being deposited in the coin slot. Input b represents a dime being deposited in the coin slot. The merchandise cost 30 cents and no change is given if the amount deposited is more than 30 cents. The output, $z = 1$, causes the merchandise to be dispensed.

   The four states are defined as follows:

   • S0 = Quiescent state, no money deposited.

- S1 = 10 cents deposited
- S2 = 20 cents deposited
- S3 = 25 cents deposited

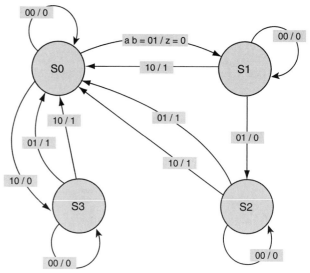

Assume that a customer deposits 10 cents ($ab = 01$). This takes it to state S1. This isn't enough money, so there is no merchandise dispensed ($z = 0$) in this transition. There are two possibilities. If another 10 cents is deposited, it goes to S2. Again, no merchandise is dispensed. However, if 25 cents is deposited, it dispenses the merchandise and returns to state S0.

If 25 cents is the first coin deposited, then it goes to S3, but no merchandise is dispensed. Any other coin being deposited will dispense the merchandise and return it to state S0.

It is not possible to deposit two coins at once, so the input, $ab = 11$, is not allowed. Also, the state transitions label 00/0 just means that nothing happens on that clock pulse. Complete the truth table, simplify the logical equations using the Karnaugh map and then draw the gate design for this circuit.

6. Design a state machine circuit that will detect the occurrence of the serial bit pattern 1001.

7. The circuit shown below is made up of four D-type flip-flops and two exclusive OR gates. Note that the SET input is not used. Only RESET may be asserted. Make a truth table showing:
   a. The state of outputs Q_0, Q_1, Q_2 and Q_3 after a $\overline{\text{RESET}}$ pulse.
   b. The state of the outputs after 16 clock pulses.
   c. How many clock pulses are required before the outputs recycle to the same pattern again?
   d. Redesign the circuit as a Finite State Machine. For simplicity, design it so that the pattern repeats itself after 8 clock pulses. Create a truth table for each state and for the next state, simplify it and draw the gate circuitry.

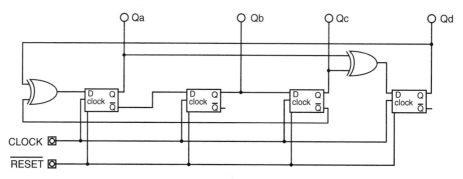

CHAPTER 6

Bus Organization and Memory Design

• •

Objectives

When you are finished with this chapter, you will be able to:

▶   *Understand the need for bus organization;*

▶   *Use the principles of tri-state logic to design bus oriented systems;*

▶   *Design the memory decoding circuitry for a modern microprocessor; and*

▶   *Design a memory system of any width or depth using the address, data and control I/O pins of modern memory circuits.*

• •

Bus Organization

In Chapter 2, we showed that the outputs of logic gates connect to the inputs of other logic gates. It is a general rule; outputs must go to inputs. You can simultaneously connect one output to a number of inputs so that when the output changes its state, all of the inputs connected to that output see the change at the same time (Of course, it still must be within the constraints imposed by the speed of light.). This is called a *one to N* circuit configuration.

You cannot, however, connect a number of inputs together unless they are tied to an output, because there needs to be a signal of some kind present to drive the inputs to either the 1 or 0 states. Without an output to drive them, inputs will tend to drift around, rattling from 1 or 0 and creating noisy signals in the computer. In general, all unused inputs are "tied" to either ground or the power supply voltage (Vcc). As you'll see shortly, a similar problem exists if we try to connect two or more outputs together. What do you think happens if an output that is in the logic 1 state is connected to an output that is in the logic 0 state? Do we end up with the average, 0.5? To see what might happen, you can take a 1.5 volt battery, such as a AA or AAA cell and, with a piece of wire, quickly short-circuit the positive terminal to the negative terminal of the battery. If you have good contact, you should see a spark and perhaps, even a puff of smoke. In general, computer designers do not like little puffs of smoke coming from inside of their computers. So hopefully, this experiment has convinced you that connecting outputs together is not a good idea.

As you might surmise from the above discussion, connecting two outputs together is similar to an electrical short circuit. In fact, we might actually damage the circuit elements because each one is trying to force the other to change, sort of like when I first got married. But that's another story.

Figure 6.1 introduces us to the basic dilemma of computer design. There are a lot of wires running around.

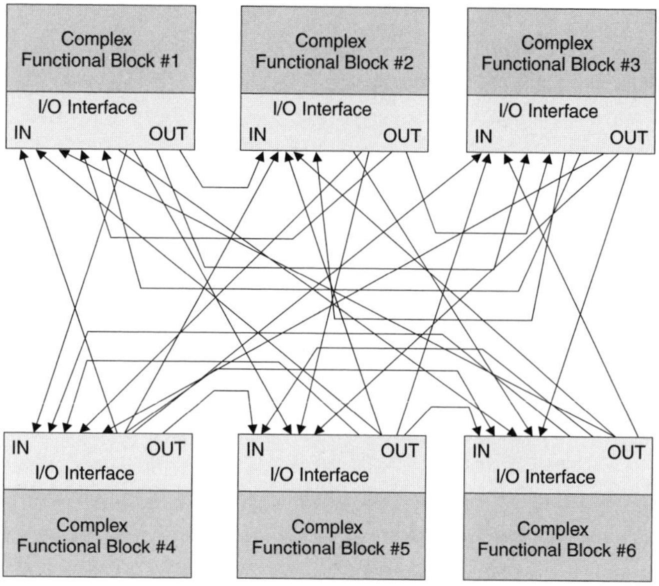

Figure 6.1 is supposed to show the "rat's nest" of wires that are required to interconnect six computer circuits to each other. In this example, we don't care what these functional blocks really are, we're just interested in how they're wired together. Also, the interconnection is designed to show just one data bit. We would have to multiply this maze of wires by 32 for a real computer system and modern computers have many more than six function blocks that need to be interconnected.

Each functional block connects its output signal to the five other blocks inputs. The arrowheads on the wires indicates which functional block is sending the signal. Within each functional block, there must be some kind of input/output (I/O) interface and control circuit so that the computer can synchronize which block is sending and which block

Figure 6.1: Computer organization based upon point-to-point wiring. The diagram shows the number of signal lines required to connect one data bit between six different internal components of a typical computer system. [NOTE: A color version of this figure is included on the DVD-ROM.]

is receiving, because it is customary that only one block sends and only one receives at any point in time. Thus, all the blocks might have 1s and 0s on their outputs, but at the right time, one of the blocks should be listening to only one of the possible outputs, which means that the I/O interface circuit must have some way to decide which functional block it's supposed to listen to, and which to ignore.

Figure 6.2 shows the I/O organization in somewhat more detail. The output from each block is simultaneously driven to all of the inputs of the other blocks in a 1" to 5" organization. On the input side, each block must have logic to decide which

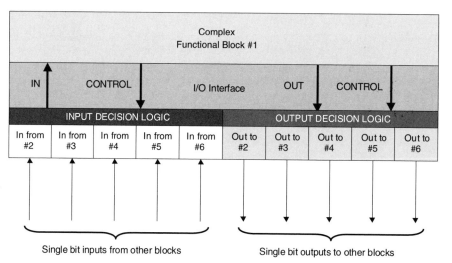

Figure 6.2: Detailed block view of circuitry necessary to implement a single bit communications protocol for six functional blocks using point-to-point wiring.

of the outputs it is going to accept at any point in time. Thus, it needs the decision logic, called a *multiplexer*, or MUX, that does a "5 to 1" reduction so that the correct data could be read by the input and then passed into the logic of the complex functional block. This is a lot of complexity. We'll be drowning in wires if we can't come up with a better solution. What would be ideal is if we could have the circuit connected as shown in Figure 6.3.

In Figure 6.3, we've managed to dramatically simplify our design. Only one wire comes out of each functional block and it carries the data out as well as the data in. Remember, this is still only 1 bit of data. This may be simple, but

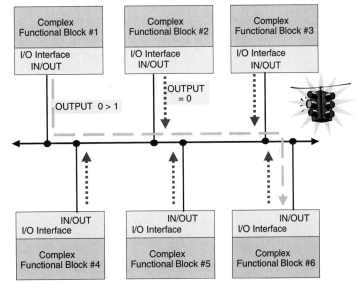

Figure 6.3: Using a bus protocol to interconnect the functional blocks. A problem has been created because inputs and outputs are tied together.

you're correct in observing that it won't work because we've got outputs and inputs all tied together. This is exactly the problem that we considered earlier. We're trying to send a 1 from block #1 to block #6 (gray dashed arrow). All of the other outputs are 0 (black dashed arrows) so the data cannot be sent. We need to be able to somehow manage the traffic flow (note the clever clip art) through the circuit. The ways in which we'll do it is by organizing our data paths into *busses,* and then make use of the fourth fundamental gate structure that we considered in Chapter 2, the tri-state buffer.

Busses were invented as a way to simplify the organization and flow of data within the computer system. We use busses to allow many devices to connect to the same data path at the same time. A bus is a grouping of similar signal. We'll look at busses in more detail in a moment, but for now, let's focus on figuring out how to connect those pesky signals together in the first place without blowing a fuse.

In most computers, there are three main busses called the *address bus*, the *data bus*, and the *status bus*. We'll be discussing these busses in more detail in the next chapter, but for now, let's consider the problem of the outputs once again. Figure 6.4 shows us the dilemma. The AND gate on the left has a 1 on both inputs, so its output is a 1, or 5 volts.

The logic gate on the right has a 1 and a 0 for its two inputs so its output is a 0, or 0 volts. The resulting signal on the bus is indeterminate. In order to fix the problem we need some way to separate the logic functions from the interface to the bus. We can do that by dividing the circuits of the logical devices that connect to the bus into two parts: the logic function and the bus interface unit. We saw the need for the interface circuit in Figures 6.1 and 6.2. However, in this case the bus interface unit will not be a multiplexer, which selected among the various input signals coming into the device; the interface logic will be much simpler than that. It will be a simple switch that connects or disconnects the outputs from the bus.

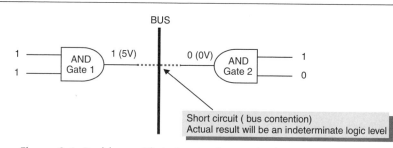

Figure 6.4: Problem with trying to tie two logic gate outputs to the same data bus. What is the resulting logic level that we would see on the bus?

Figure 6.5 shows this schematically. The interface to the bus is an electronic switch. As you already know, that electronic switch is the tri-state buffer. A bus control signal can rapidly activate the switch to either connect or disconnect the output from the bus. If all of the outputs of the various functional blocks are disconnected except

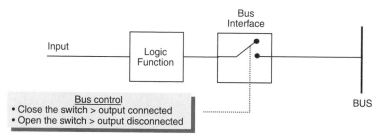

Figure 6.5: Mechanical switch representation of bus interface logic.

one, then the one output that is connected to the bus can "drive the bus" either high or low. Since it is the only talker and every other device is a "listener" (one to N) there will not be any conflicts from other output devices. All of the other outputs, which remain disconnected be their electronic switches, have no impact on the state of the bus signal.

As you know, with the tri-state buffers, we aren't really breaking the electrical continuity of the gates to the bus; we are simply making a rapid change in the electrical conductivity of a small portion of the circuit that resides between the logical output of the gate and the bus signal that it is connected to the circuit. In Figure 6.5, we simplify this bit of circuit magic by drawing the connect/disconnect portion of the circuit as if it was a physical switch, like the light switch on the wall of a room. However, keep in mind that we cannot open and close a real mechanical switch as nearly as quickly or as cleanly as we can go from high impedance (no signal flow) to low impedance (connect to the bus).

It is generally true that the bus control signal will be active low. Whatever logic state (1 or 0) we want to place on the bus, we must first bring the bus control for that gate, function block, memory cell and so forth, low to connect the gate to the bus so the output of the logical device can be connected to the bus. This signal has several names. Sometimes it is called *output enable* ($\overline{OE}$), *chip enable* ($\overline{CE}$) or *chip select* ($\overline{CS}$). These signals are not all the same, and, as we'll soon see, may perform different tasks within the device, but they all have the common property of disabling the output of the device and disconnecting it from the bus.

Referring back to Figure 2.6, the tri-state logic gate, we see that when the $\overline{OE}$, or bus control signal, is LOW, the output of the bus interface (BUS I/F) portion of the circuit follows the input.

When the $\overline{\text{CS}}$ signal is HIGH, the output of the BUS I/F goes to Hi-Z, effectively removing the gate from the circuit. One final point needs to be emphasized. The tri-state part of the circuit doesn't change the logical state of the gate, or memory device, or whatever else is connected to it. Its only function is to isolate the output of the gate from the bus so that another device may take control and send a signal on the bus.

Let's now look at how we can build our single bit bus into a real data bus. Refer to Figure 6.6. Here we see that the bus is actually a grouping of eight similar signals. In this case, it represents 8 data bits. So, if we were to take apart our VCR and look at the microprocessor inside of it, we might find an 8-bit micro-processor inside. The number of bits of the

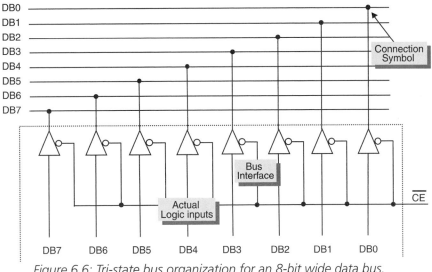

Figure 6.6: Tri-state bus organization for an 8-bit wide data bus.

data bus refers to the size of a number that we can fetch in one operation. Recall that 8-bits allow us to represent numbers from 0 to 255, or 2^8.

The bus is still made up of eight individual signal wires that are electrically isolated from each other. This is important to remember because, in order to keep our drawings simple, we'll usually draw a bus as one line, rather than 8, 16, or 32 lines. Each wire of the bus is labeled with the corresponding data bit, DB0 through DB7, with DB0 representing the number 2^0, or 1, and DB7 representing the number 2^7, or 128.

Inside the dotted line is our device that is connected to the bus. Each data bit inside the device connects with the corresponding data line on the bus through the tri-state buffer. Notice how the $\overline{\text{CE}}$ signal is connected to all of the tri-state buffers at once. By bringing $\overline{\text{CE}}$ LOW, all 8 individual data bits are simultaneously connected to the data bus. Also note that the eight tri-state buffers (drawn as triangles with the circle on the side) in no way change the value of the data. They simply connect the signals on the eight data bits, DB0 . . . DB7, inside of the functional block, to the corresponding data lines outside of the functional block. Don't confuse the tri-state buffer with the NOT gate. The NOT gate inverts the input signal while the tri-state buffer controls whether or not the signal is allowed to propagate through the buffer.

Figure 6.7 takes this concept one step further. Here we have four 32-bit storage registers connected to a 32-bit data bus. Notice how the data bits, D0 . . . D31 are drawn to show that the individual bits become the 32-bit bus. Figure 3.36 also shows a new logic element, the block labeled

"2:4 Address Decoder." Recall that we need some way to decide which device can put its data onto the bus because only one output may be on at a time. The decoder circuit does just that. Imagine that two signals, A0 and A1, come from some other part of

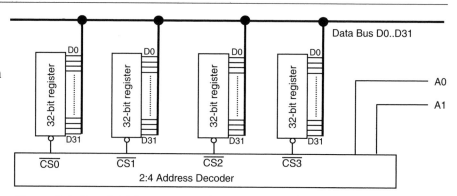

Figure 6.7: Four 32-bit storage registers connected to a bus. The 2:4 decoder takes the 2 input variables, A0 and A1, and selects only 1 of the 4 outputs, $\overline{CS0}..\overline{CS3}$ to enable.

the circuit and are used to determine which of the four, 32-bit registers shown in the figure should be connected to the bus at any given time. The two "address" input bits, labeled A0 and A1, give us four possible combinations. Thus, we can then create some relatively simple logic gate circuit to convert the combinations of our inputs, A0 and A1, to 4 possible outputs, the correct chip select bit, $\overline{CS0}$, $\overline{CS1}$, $\overline{CS2}$ or $\overline{CS3}$. This circuit design is a bit different from what you're already accustomed to because we want our "TRUE" output to go low, rather than go high. This is a consequence of the fact that tri-state logic is usually asserted with a low going signal.

Table 6.1 is the truth table for the 2:4 decoder circuit shown in Figure 6.7.

As we've previously discussed, it is common to see the circle drawn at the input to the tri-state gate. This means that the signal is *active low*. Just as the circle on the outputs of the inverter gate—NAND gate and NOR gate—indicate a signal inversion, the circle on the input to a gate indicates that the signal

Table 6.1: Truth table for a 2:4 decoder.

A0	A1	CS0	CS1	CS2	CS3
0	0	0	1	1	1
1	0	1	0	1	1
0	1	1	1	0	1
1	1	1	1	1	0

is asserted (TRUE) when it is LOW. This goes back to our earlier discussion about 1 and 0 being rather arbitrary in terms of what is TRUE and what is FALSE.

Before we move on to look at memory organization, let's revisit the algorithmic state machine from the previous chapter. With the introduction of the concept of busses in this chapter, we now have a better basis of understanding to see how the operation of a bus-based system might be controlled by the state machine. Figure 6.8 shows us a simplified schematic diagram of part of the control system of a computing machine. Each of the registers shown, Register A, Register B, Temporary Register and the Output Register have individual controls to read data into the register on a rising edge to the clock inputs, and an Output Enable ($\overline{OE}$) signal to allow the register to place data on the common data bus that connects all of the functional elements inside of the computer.

In order to place data into Register A we would first have to enable another source of the data, perhaps memory, and put the data onto the Data Bus. When the data is stable, the clock signal, clk_A

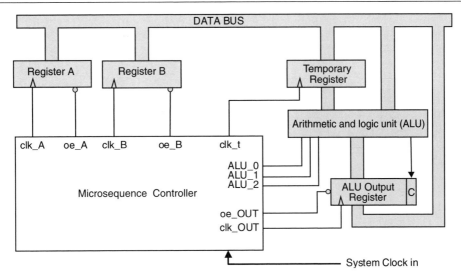

Figure 6.8 Schematic diagram of part of a computing engine. The registers are interconnected using tri-state logic and busses. The Microsequence Controller must sequence the Output Enable ($\overline{OE}$) of each register properly so that there are no bus contentions.

would go through a low to high transition and the data would be stored in the register. Remember that Register A is a just a collection of D-FF's with a common clock. Now suppose that we want to add the contents of register A and Register B, respectively. The arithmetic and logic unit (ALU) requires two input sources to add the numbers together. It has a temporary register and an output register associated with it because the ALU itself is an asynchronous gate design. That is, if the inputs were to change, the outputs would also immediately change because there is no D register there to synchronize its behavior.

Figure 6.9 shows a portion of the truth table that controls the behavior of the microsequence controller. In order to add two numbers together, such as the contents of A and B, we might issue the *assembly language instruction:*

ADD B,A

This instruction tells the computer to add together the contents of the A register and the B register and place the result back into register A. If there is any carry generated by the addition operation, it should be placed in the carry bit position, which is shown as the "C" bit attached to the output register of the ALU. Thus, we can begin to see that we would need several steps in order to add these numbers together. Let's describe the process in words:

1. Copy the data from memory in the A Register. This will be operand #2.
2. Copy operand #1 from memory into the B register.
3. Move the data from the A register into the temporary register.
4. Move the results of the addition in the ALU into the ALU Output Register and Carry bit.
5. Move the data from the Output register back into register A.

Referring to the portion of the truth table shown in Figure 6.9, we can follow the instruction's flow through the controller:

Clock pulse 1:	There is a low to high transition of the Register A input clock. This will store the data that is currently being held on the data bus in the register.
Clock pulse 2:	There is a low to high transition of the Register B input clock. This will store the data that is being held on the data bus into Register B.
Clock pulse 3:	The input to clock B returns to 0 and the Output Enable signal for Register A becomes active. This puts the data that was previously stored in Register A back onto the Data Bus. There is no effect of the clock signal to Register B going low because this is a falling edge.
Clock pulse 4:	There is a rising edge of the clock to the Temporary Register. This stores the data that was previously stored in Register A and is now on the Data Bus. The Output Enable signal of the A register is turned off and the Output Enable of Register B is turned on. At this point the data stored in Register B is on the Data Bus and the Data stored in register A is stored in the Temporary Register. The ALU input signals, ALU0, ALU1 and ALU2 are set for addition*, so the output of the ALU is the sum of A and B plus any carry generated.
Clock pulse 5:	The sum is stored in the Output Register on the rising edge of the clk_out signal.
Clock pulse 6:	The Output Enable of B register is turned off, the clock input to B is returned to 0 and the data in the Temporary Register is put on the Data Bus.
Clock pulse 7:	The Data is clocked into the A register and the Output Enable of the Temporary Register is turned off.
Clock pulse 8:	The system returns to the original state.

The ALU is a circuit that can perform up to eight different arithmetic or logical operations based upon the state of the three input variables, ALU_0..ALU_3. In this example, we assume that the ALU code, 000, sets the ALU up to perform an addition operation on the two input variables.

clock	clk_A	oe_A	clk_B	oe_B	clk_T	clk_OUT	oe_OUT	ALU_0	ALU_1	ALU_2	clk_A	oe_A	clk_B	oe_B	clk_T	clk_OUT	oe_OUT	ALU_0	ALU_1	ALU_2
1	0	1	0	1	0	0	1	0	0	0	1	1	0	1	0	0	1	0	0	0
2	1	1	0	1	0	0	1	0	0	0	0	1	1	1	0	0	1	0	0	0
3	0	1	1	1	0	0	1	0	0	0	0	0	0	1	0	0	1	0	0	0
4	0	0	0	1	0	0	1	0	0	0	0	1	0	0	1	0	1	0	0	0
5	0	1	0	0	1	0	1	0	0	0	0	1	0	0	0	1	1	0	0	0
6	0	1	0	0	0	1	1	0	0	0	0	1	0	1	0	0	0	0	0	0
7	0	1	0	1	0	0	0	0	0	0	1	1	0	1	0	0	1	0	0	0
8	1	1	0	1	0	0	1	0	0	0	0	1	0	1	0	0	1	0	0	0

Figure 6.9: Truth table for the Microsequence Controller of Figure 3.37 showing the before and after logic level changes with each clock pulse. The left-hand side of the table shows the state of the outputs before the clock pulse and the right hand side of the table shows the state of the outputs after the clock pulse. The areas shown in gray are highlighted to show the changes that occur on each clock pulse.

Now, before you go off and assume that you are a qualified CPU architect, let me warn you that it's a lot more involved then what we saw in this example. However, the principles are the same. For starters, we did not discuss how the instruction itself, ADD B, A actually got into the computer and how the computer figured out that the bit pattern which represented the instruction code was an ADD instruction and not something else. Also, we didn't discuss how we established the contents of the A and the B registers in the first place. However, for at least that portion of the state machine that actually does the addition process, it probably is pretty close to how it really works.

You've now seen two examples of how the state machine is used to sequence a series of logical operations and how these operations for the basis of the execution of an instruction in a computer. Without the concept of a bus-based architecture and the tri-state logical gate design, this would be very difficult or impossible to accomplish.

Memory System Design

We have already been introduced to the concept of the flip-flop. In particular we saw how the "D" type flip-flop could be used to store a single bit of information. Recall that the data present at the D input to the D-flop will be stored within the device and the stored value will be transferred to the Q output on the rising edge of the clock signal. Now, if the clock signal goes away, we would still have the data present on the Q output. In other words, we've just stored 1 bit of data. Our D-flop circuit is a memory cell.

Historically, there have been a large number of different devices used to store information as a computer's *random access memory*, or RAM. Magnetic devices, such as *core memories,* were very important until integrated circuits (IC) became widely available. Today, one IC memory device can hold 512 million bits of data. With this type of miniaturization, magnetic devices couldn't keep up with the semiconductor memory in terms of speed or capacity. However, memories based on magnetic storage have one ongoing advantage over IC memories—they don't forget their data when the power is turned off. Thus, we still have hard disk drives and tape storage as our secondary memory systems because they can hold onto their data even if the power is removed.

Many industry analysts are foretelling the demise of the hard drive. Modern FLASH memories, such as the ones that you may use in your PDA, digital camera or MP3 players are already at 1 gigabyte of storage capacity. FLASH memories retain their information even with power removed, and are faster and more rugged than disk drives. However, disk drives and tape drives still win on capacity and price per bit. At the time of this writing (Summer of 2004), a modern disk drive with a capacity of 160 gigabytes can be purchased for about $90 if you are willing to send in all of the rebate information and the company actually sends you back your rebate check. But that's another story for another time.

Hard disks and tape systems are *electromechanical systems*, with motors and other moving parts. The mechanical components cause them to be far less reliable and much slower than IC memories, typically 10,000 times slower than an IC memory device. It is for this reason that we don't use our hard disks for the main memory in our computer system because they're just too slow. However, as you'll see in a later lesson, the ability of the hard drive to provide almost limitless storage enables an operating system such as Linux or Windows to give the user the impression that every application that is opened on your desktop always has as much memory as it needs.

In this section, we will start from the D-flop as an individual device and see how we can interconnect many of them to form a memory array. In order to see how data can be written to the memory and read from the memory along the same signal path (although not at the same instant in time), consider Figure 6.10.

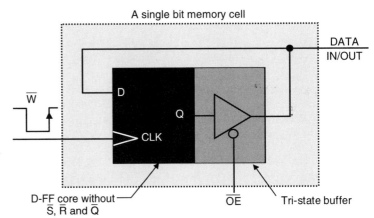

A single bit memory cell

D-FF core without S, R and Q

OE

Tri-state buffer

Figure 6.10 Schematic representation of a single bit of memory. The tri-state buffer on the output of the cell controls when the Q output may be connected to the bus.

The black box is just a slightly simplified version of the basic D flip-flop. We've eliminated the $\overline{S}$, $\overline{R}$ inputs and $\overline{Q}$ output. The dark gray box is the tri-state buffer, which is controlled by a separate $\overline{OE}$ (output enable) input. When $\overline{OE}$ is HIGH, the tri-state buffer is disabled, and the Q output of the memory cell is isolated (Hi-Z state) from the data lines (DATA I/O line). However, the Data line is still connected to the D input of the cell, so it is possible to write data to the cell, but the new data written to the cell is not immediately visible to someone trying to read from the cell until the tri-state buffer is enabled. When we combine the basic FF cell with the tri-state buffer, we have all that we need to make a 1-bit memory cell. This is indicated by the light gray box surrounding the two elements that we've just discussed.

The write signal is a bit misleading, so we should discuss it. We know that data is written into the D-FF on the rising edge of a pulse, which is indicated by the up-arrow on the write pulse ($\overline{W}$) in Figure 6.10. So why is the write signal, $\overline{W}$, written as if it was an active low signal? The reason is that we normally keep the write signal in a 1 state. In order to accomplish a write operation, the $\overline{W}$ must be brought low, and then returned high again. It is the low-to-high transition that accomplishes the actual data write operation, but since we must bring the write line to a low state in order to accomplish the actual writing of the data, we consider the write signal to be active low. Also, you should infer from this discussion that you would never activate the $\overline{W}$ line and the $\overline{OE}$ lines at the same time. Either you bring $\overline{W}$ low and keep $\overline{OE}$ high, or vice versa. They never are low at the same time. Now, let's return to our analysis of the memory array.

We'll take another step forward in complexity and build a memory out of tri-state devices and D-flops. Figure 6.11 shows a simple (well maybe not so simple) 16-bit memory organized as four, 4-bit nibbles. Each storage bit is a miniature D-flop that also has a tri-state buffer circuit inside of it so that we can build a bus system with it.

Each row of four D-FF's has two common control lines that provide the clock function (write) and the output enable function for placing data onto the I/O bus. Notice how the corresponding bit position from each row is physically tied to the same wire. This is why we need the tri-state control signal, $\overline{OE}$, on each bit cell (D-FF). For example, if we want to write data into row 2 of D-FF's the data must be place on the DB0 through DB3 from the outside device and the W2 signal must go high to store the data. Also, to write data into the cells, the $\overline{OE}$ signal must be kept in the HIGH state in order to prevent the data already stored in the cell from being placed on the data lines and corrupting the new data being written into a cell.

The control inputs to the 16-bit memory are shown on the left of Figure 6.11. The data input and output, or I/O, is shown on the top of the device. Notice that there is only one I/O line for each data bit.

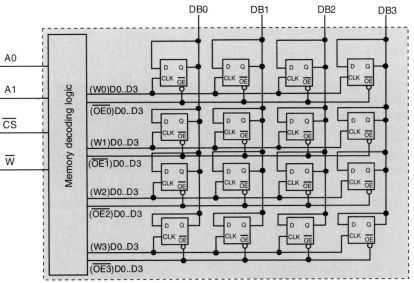

Figure 6.11: 16-bit memory built using discrete "D" flip-flops. We would access the top row of the four possible rows if we set the address bits, A0 and A1 to 0. In a similar vein, (A0, A1) = (1, 0), (0, 1) or (1, 1) would select rows 1, 2 and 3, respectively.

That's because data can flow in or out on the same wire. In other words, we've used bus organization to simplify the data flow into and out of the device. Let's define each of the control inputs:

A0 and A1	Address inputs used to select which row of the memory is being addressed for input or output operations. Since we have four rows in the device, we need two address lines.
$\overline{CS}$	Chip select. This active low signal is the master switch for the device. You cannot write into it or read from it if $\overline{CS}$ is HIGH.
$\overline{W}$	If the $\overline{W}$ line is HIGH, then the data in the chip may be read by the external device, such as the computer chip. If the $\overline{W}$ line is low, data is going to be written into the memory.

The signal $\overline{CS}$ (chip select) is, as you might suspect, the master control for the entire chip. Without this signal, none of the Q outputs from any of the sixteen D-FF's could be enabled, so the entire chip would remain in the Hi-Z state, as far as any external circuitry was concerned. Thus, in order to read the data in the first row, not only must (A0, A1) = (0, 0), we also need $\overline{CS} = 0$. But wait, there's more!

We're not quite done because we still have to decide if we want to read from the memory or write to it. If we want to read from it, we would want to enable the Q output of each of the four D-flops that make up one row of the memory cell. This means that in order to read from any row of the memory, we need the following conditions to be TRUE:

- READ FROM ROW 0 > (A0 = 0) AND (A1 = 0) AND ($\overline{CS}$ = 0) AND ($\overline{W}$ = 1)
- READ FROM ROW 1 > (A0 = 1) AND (A1 = 0) AND ($\overline{CS}$ = 0) AND ($\overline{W}$ = 1)

- READ FROM ROW 2 > (A0 = 0) AND (A1 = 1) AND ($\overline{CS}$ = 0) AND ($\overline{W}$ = 1)
- READ FROM ROW 3 > (A0 = 1) AND (A1 = 1) AND ($\overline{CS}$ = 0) AND ($\overline{W}$ = 1)

Suppose that we want to write four bits of data to ROW 1. In this case, we don't want the individual $\overline{OE}$ inputs to the D-flops to be enabled because that would turn on the tri-state output buffers and cause a conflict with the data we're trying to write into the memory. However, we'll still need the master $\overline{CS}$ signal because that enables the chip to be written to. Thus, to write four bits of data to ROW 1, we need the following equation:

WRITE TO ROW 1 > (A0 = 1) AND (A1 = 0) AND ($\overline{CS}$ = 0) AND ($\overline{W}$ = 0)

Figure 6.12 is a simplified schematic diagram of a commercially available memory circuit from NEC®, a global electronics and semiconductor manufacturer headquartered in Japan. The device is a μPD444008[1] 4M-Bit CMOS Fast Static RAM (SRAM) organized as 512 K × 8-bit wide words (bytes). The actual memory array is composed of an X-Y matrix 4,194,304 individual memory cells. This is just like the 16-bit memory that we discussed earlier, only quite a bit larger. The circuit has 19 address lines going into it, labeled A0 . . . A18. We need that many address lines because 2^{19} = 524,288, so 19 address lines will give us the right number of combinations that we'll need to access every memory word in the array.

The signal named $\overline{WE}$ is the same as the $\overline{W}$ signal of our earlier example. It's just labeled differently, but still required a LOW to HIGH transition to write the data. The $\overline{CS}$ signal is the same as our $\overline{CS}$ in the earlier example. One difference is that the commercial part also provides an explicit output enable signal (called $\overline{CE}$ in Figure 6.12) for controlling the tri-state output buffers during a read operation. In our example, the $\overline{OE}$ operation is implied by the state of the $\overline{W}$ input. In actual use, the ability to independently control $\overline{OE}$ makes for a more flexible part, so it is commonly added to memory chips such as

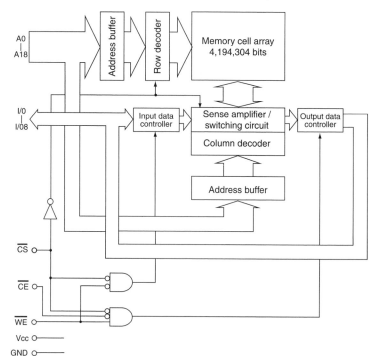

Truth Table

$\overline{CS}$	$\overline{CE}$	$\overline{WE}$	Mode	I/O	Supply current
H	x	x	Not selected	High impedance	Icc
L	L	H	Read	D$_{OUT}$	Icc
L	x	L	Write	D$_{IN}$	
L	H	H	Output Disable	High Impedance	

Remark x: Don't care

Figure 6.12: Logical diagram of an NEC μPD444008 4 M-Bit CMOS Fast Static RAM. Diagram courtesy of NEC Corporation.

this one. Thus, you can see that our 16-bit memory is operationally the same as the commercially available part.

Let's return to Figure 6.11 for a moment before we move on. Notice how each row of D-flops has two control signals going to each of the chips. One signal goes to the $\overline{OE}$ tri-state controls and the other goes to the CLK input. What would the circuit inside of the block on the left actually look like? Right now, you have all of the knowledge and information that you need to design it.

Let's see what the truth table would look like for this circuit. Figure 6.13 is the truth table.

You can see that the control logic for a real memory device, such as the µPD444008 in Figure 6.12 could become significantly more complex as the number of bits increases from 16 to 4 million, but the principles are the same. Also, if you refer to Figure 6.13 you should see that the decoding logic is highly regular and scalable. This would make the design of the hardware much more straightforward.

A0	A1	R/$\overline{W}$	$\overline{CS}$	W0	$\overline{OE0}$	W1	$\overline{OE1}$	W2	$\overline{OE2}$	W3	$\overline{OE3}$
0	0	0	0	0	1	1	1	1	1	1	1
1	0	0	0	1	1	0	1	1	1	1	1
0	1	0	0	1	1	1	1	0	1	1	1
1	1	0	0	1	1	1	1	1	1	0	1
0	0	1	0	1	0	1	1	1	1	1	1
1	0	1	0	1	1	1	0	1	1	1	1
0	1	1	0	1	1	1	1	1	0	1	1
1	1	1	0	1	1	1	1	1	1	1	0
0	0	0	1	1	1	1	1	1	1	1	1
1	0	0	1	1	1	1	1	1	1	1	1
0	1	0	1	1	1	1	1	1	1	1	1
1	1	0	1	1	1	1	1	1	1	1	1
0	0	1	1	1	1	1	1	1	1	1	1
1	0	1	1	1	1	1	1	1	1	1	1
0	1	1	1	1	1	1	1	1	1	1	1
1	1	1	1	1	1	1	1	1	1	1	1

Figure 6.13: Truth table for 16-bit memory decoder.

Data Bus Width and Addressable Memory

Before we move on to look at memory system designs of higher complexity, we need to stop and catch our breath for a moment, and consider some additional information that will help to make the upcoming sections more comprehensible. We need to put two pieces of information into their proper perspective:

1. Data bus width, and
2. Addressable memory.

The width of a computer's data bus determines the size of the number that it can deal with in one operation or instruction. If we consider embedded systems as well as desktop PC's, servers, workstations, and mainframe computers, we can see a spectrum of data bus widths going from 4 bits up to 128 bits wide, with data buses of 256 bits in width just over the horizon. It's fair to ask, "Why is there such a variety?" The answer is speed versus cost. A computer with an 8-bit data path to memory can be programmed to do everything a processor with a 16-bit data path can do, except it will take longer to do it. Consider this example. Suppose that we want to add two 16-bit numbers together to generate a 16-bit result. The numbers to be added are stored in memory and the result will be stored in memory as well. In the case of the 8-bit wide memory, we'll need to store each 16-bit word as two successive 8-bit bytes. Anyway, here's the algorithm for adding the numbers.

Case 1: 8-bit Wide Data Bus

1. Fetch lower byte of first number from memory and place in an internal storage register.
2. Fetch lower byte of second number from memory and place in another internal storage register.
3. Add the lower bytes together.
4. Write the low order byte to memory.
5. Fetch upper byte of first number from memory and place in an internal storage register.
6. Fetch upper byte of second number from memory and place in another internal storage register.
7. Add the two upper bytes together with the carry (if present) from the prior add operation.
8. Write the upper byte to the next memory location from the low order byte.
9. Write the carry (if present) to the next memory location.

Case 2: 16-bit Wide Data Bus

1. Fetch the first number from memory and place in an internal storage register.
2. Fetch the second number from memory and place in another internal storage register.
3. Add the two numbers together.
4. Write the result to memory.
5. Write the carry (if present) to memory.

As you can see, *Case 1* required almost twice the number of steps as *Case 2*. The efficiency gained by going to wider data busses is dependent upon the algorithm being executed. It can vary from as little as a few percent improvement to almost four times the speed, depending upon the algorithm being implemented.

Here's a summary of where the various bus widths are most common:

- 4, 8 bits: appliances, modems, simple applications
- 16 bits: industrial controllers, automotive applications
- 32 bits: telecommunications, laser printers, desktop PC's
- 64 bits: high end PCs, UNIX workstations, games (Nintendo 64)
- 128 bits: high performance video cards for gaming
- 128, 256 bits: next generation, very long instruction word (VLIW) machines

Sometimes we try to economize by using a processor with a wide internal data bus with a narrower memory. For example, the Motorola 68000 processor that we'll study in this class has a 16-bit external data bus and a 32-bit internal data bus. It takes two memory fetches to bring in a 32-bit quantity from memory, but once it is inside the processor it can be dealt with as a single 32-bit value.

Address Space

The next consideration in our computer design is how much addressable memory the computer is equipped to handle. The amount of externally accessible memory is defined as the *address space* of the computer. This address space can vary from 1024 bytes for a simple device to over 60 gigabytes for a high performance machine. Also, the amount of memory that a processor can address is independent of how much memory you actually have in your system. The Pentium processor in your PC can address over four billion bytes of memory, but most users rarely have more than 1 gigabyte of memory inside their computer. Here are some simple examples of addressable memory:

- A simple microcontroller, such as the one inside of your Mr. Coffee® machine, might have 10 address lines, A0 . . . A9, and is able to address 1024 bytes of memory ($2^{10} = 1024$).
- A generic 8-bit microprocessor, such as the one inside your burglar alarm, has 16 address lines, A0 . . . A15, and is able to address 65,536 bytes of memory ($2^{16} = 65,536$).
- The original Intel 8086 microprocessor that started the PC revolution has 20 address lines, A0 . . . A19, and is able to address 1,048,576 bytes of memory ($2^{20} = 1,048,576$).
- The Motorola 68000 microprocessor has 24 address lines, A0 . . . A23, and is able to address 16,777,216 bytes of memory ($2^{24} = 16,777,216$).
- The Pentium microprocessor has 32 address lines, A0 . . . A31, and is able to address 4,294,967,296 bytes of memory ($2^{32} = 4,294,967,296$).

As you'll soon see, we generally refer to addressable memory in terms of bytes (8-bit values) even though the memory width is greater than that. This creates all sorts of memory addressing ambiguities that we'll soon get into.

Paging

Suppose that you're reading a book. In particular, this book is a very strange book. It has exactly 100 words on every page and each word on each page is numbered from 0 to 99. The book has exactly 100 pages, also numbered from 0 to 99. A quick calculation tells you that the book has 10,000 words (100 words/page × 100 pages). Also, next to every word on every page is the absolute number of that word in the book, with the first number on page 0 given the address 0000 and the last number on the last page given the number 9,999. This is a very strange book indeed!

However, we notice something quite interesting. Every word on a page can be uniquely identified in the book in one of two ways:

1. Give the absolute number of the word from 0000 to 9,999.
2. Give the page number that the word is on, from 00 to 99 and then give the position of the word on the page, from 00 to 99.

Thus, the 45th word on page 36 could be numbered as 3644 in absolute addressing or as page = 36, offset = 44. As you can see, however we choose to form the address, we get to the correct word. As you might expect, this type of addressing is called *paging*. Paging requires that we supply two numbers in order to form the correct address of the memory location we're interested in.

1. *Page number* of the page in memory that contains the data,
2. *Page offset* of the memory location in that page.

Figure 6.14 shows such a scheme for a microprocessor (sometimes we'll use the Greek letter "mu" and the letter "P" together, µP, as a shorthand notation for microprocessor). The microprocessor has 20 address lines, A0 . . . A19, so it can address 1,048,576 bytes of memory. Unfortunately, we don't have a memory chip that is just the right size to match the memory address space of the processor. This is usually the case, so we'll need to add additional circuitry (and multiple memory devices) to provide enough memory so that every possible address coming out of the processor has a corresponding memory location to link to.

Since this memory system is built with 64 Kbyte memory devices, each of the 16 memory chips has 16 address lines, A0 through A15. Therefore, each of the address line of the address bus, A0 through A15, goes to each of the address pins of each memory chip.

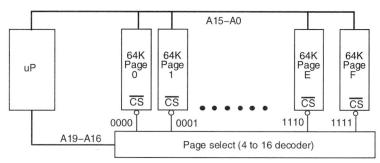

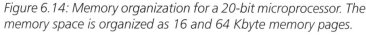

Figure 6.14: Memory organization for a 20-bit microprocessor. The memory space is organized as 16 and 64 Kbyte memory pages.

The remaining four address lines coming out of the processor, A16 through A19 are used to select which of the 16 memory chips we will be addressing. Remember that the four most significant address lines, A16 through A19 can have 16 possible combinations of values from 0000 to 1111, or 0 through F in hexadecimal.

Let's consider the microprocessor in Figure 6.14. Let's assume that it puts out the hexadecimal address 9A30D. The least significant address lines A0 through A15 from the processor go to each of the corresponding address inputs of the 16 memory devices. Thus, each memory device sees the hexadecimal address value A30D. Address bits A16 through A19 go to the page select circuit. So, we might wonder if this system will work at all. Won't the data stored in address A30D of each of the memory devices interfere with each other and give us garbage?

The answer is no, thanks to the $\overline{CS}$ inputs on each of the memory chips. Assuming that the processor really wants the byte at memory location 9A30D, the remaining four address lines coming out of the processor, A16 through A19 are used to select which of the 16 memory chips we will be addressing. Remember that the four most significant address lines, A16 through A19 can have 16 possible combinations of values from 0000 to 1111, or 0 through F in hexadecimal.

This looks suspiciously like the decoder design problem we discussed earlier. This memory design has a 4:16 decoder circuit to do the page selection with the most significant 4 address bits selecting the page and the remaining 16 address bits form the page offset of the data in the memory chips. Notice that the same address lines, A0 through A15, go to each of the 16 memory chips, so if the processor puts out the hexadecimal address E3AB0, all 16 memory chips will see the address 3AB0. Why isn't there a problem? As I'm sure you can all chant in unison by now it is the tri-state buffers which enable us to connect the 16 pages to a common data bus. Address bits A16 through A19 determine which one of the 16 $\overline{CS}$ signals to turn on. The other 15 remain in the HIGH state, so their corresponding chips are disabled and do not have an effect on the data transfer.

Paging is a fundamental concept in computer systems. It will appear over and over again as we delve further into the operation of computer systems. In Figure 6.14, we organized the 20-bit address space of the processor as 16, 64K byte pages. We probably did it that way because we were using 64K memory chips. This was somewhat arbitrary, as we could have organized the paging scheme in a totally different way; depending upon the type of memory devices we had available to us. Figure 6.15 shows other possible ways to organize the memory. Also, we could build up each page of memory from multiple chips, so the pages themselves might need to have additional hardware decoding on them.

Figure 6.15: Possible paging schemes for a 20-bit address space.

Page address	Page address bits	Page offset	Offset address bits	
NONE	NONE	0 to 1,048,575	A0 to A19	Linear address
0 to 1	A19	0 to 524,287	A0 to A18	
0 to 3	A19–A18	0 to 262,143	A0 to A17	
0 to 7	A19–A17	0 to 131,071	A0 to A16	
0 to 15	A19–A16	0 to 65,535	A0 to A15	Our example
0 to 31	A19–A15	0 to 32,767	A0 to A14	
0 to 63	A19–A14	0 to 16,383	A0 to A13	

It should be emphasized that the type of memory organization used in the design of the computer will, in general, be transparent to the software developer. The hardware design specification will certainly provide a memory map to the software developer, providing the address range for each type of memory, such as RAM, ROM, FLASH and so on. However, the software developer need not worry about how the memory decoding is organized.

From the software designer's point of view, the processor puts out a memory address and it is up to the hardware design to correctly interpret it and assign it to the proper memory device or devices.

Paging is important because it is needed to map the *linear address space* of the microprocessor into the physical capacity of the storage devices. Some microprocessors, such as the Intel 8086 and its successors, actually use paging as their primary addressing mode. The external address is formed from a page value in one register and an offset value in another. The next time your computer crashes and you see the infamous "Blue Screen of Death" look carefully at the funny hexadecimal address that might look like

BD48:0056

This is a 32-bit address in page-offset representation.

Disk drives use paging as their only addressing mode. Each disk is divided into 512 byte sectors (pages). A 4 gigabyte disk has 8,388,608 pages.

Designing a Memory System

You may not agree, but we're ready to put it all together and design a real memory system for a real computer. OK, maybe, we're not quite ready, but we're pretty close. Close enough to give it try. Figure 6.16 is a schematic diagram for a computer system with a 16-bit wide data bus.

First, just a quick reminder that in binary arithmetic, we use the shorthand symbol "K" to represent 1024, and not 1000, as we do in most engineering applications. Thus, by saying 256 K you really mean 262,144 and not 256,000. Usually, the context would eliminate the ambiguity; but not always, so beware.

The circuit in Figure 6.16 looks a lot more complicated than anything we've considered so far, but it really isn't very different than what we've already studied. First, let's look at the memory chips. Each chip has 15 address lines going into it, implying that it has 32K unique memory addresses because $2^{15} = 32,768$. Also, each chip has eight data input/output (I/O) lines going into

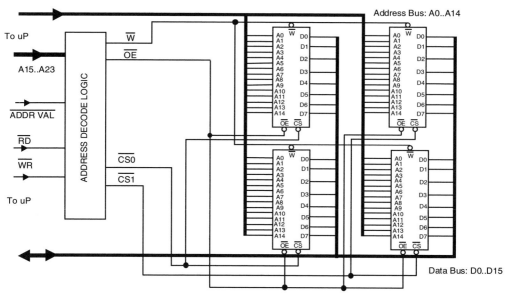

Figure 6.16: Schematic diagram for a 64 K × 16 memory system built from four 32 K × 8 memory chips.

it. However, you should keep in mind that the data bus in Figure 6.16 is actually 16 bits wide (D0…D15) so we would actually need two, 8-bit wide, memory chips in order to provide the correct memory width to match the width of the data bus. We'll discuss this point in greater detail when we discuss Figure 6.17.

The internal organization of the four memory chips in Figure 6.17 is identical to the organization of the circuits we've already studied except these devices contain 256 K memory cells and the memory we studied in Figure 6.11 had 16 memory cells. It's a bit more complicated, but the idea is the same. Also, it would have taken me more time to draw 256 K memory cells then to draw 16, so I took the easy way out.

This memory chip arrangement of 32 K memory locations with each location being 8-bits wide is conceptually the same idea as our 16-bit example in Figure 6.11 in terms of how we would add more devices to increase the size of our memory in both wide (size of the data bus) and depth (number of available memory locations). In Figure 6.11, we discussed a 16-bit memory organized as four memory locations with each location being 4-bits wide. In Figure 4.5, there are a total of 262,144 memory cells in each chip because we have 32,768 rows by 8 columns in each chip.

Each chip has the three control inputs, $\overline{OE}$, $\overline{CS}$ and $\overline{W}$. In order to read from a memory device we must do the following steps:

1. Place the correct address of the memory location we want to read on A0 through A14.
2. Bring $\overline{CS}$ LOW to turn on the chip.
3. Keep $\overline{W}$ HIGH to disable writing to the chip.
4. Bring $\overline{OE}$ LOW to turn on the tri-state output buffers.

The memory chips then puts the data from the corresponding memory location onto data lines D0 through D7 from one chip, and D8 through D15 from the other chip. In order to write to a memory device we must do the following steps:

1. Place the correct address of the memory location we want to read on A0 through A14.
2. Bring $\overline{CS}$ LOW to turn on the chip.
3. Bring $\overline{W}$ LOW to enable writing to the chip.
4. Keep $\overline{OE}$ HIGH to disable the tri-state output buffers.
5. Place the data on data lines D0 through D15. With D0 through D7 going to one chip and D8 through D15 going to the other.
6. Bring $\overline{W}$ from LOW to HIGH to write the data into the corresponding memory location.

Now that we understand how an individual memory chip works, let's move on to the circuit as a whole. In this example our microprocessor has 24 address lines, A0 through A23. A0 through A14 are routed directly to the memory chips because each chip has an address space of 32 K bytes. The nine most significant address bits, A15 through A23 are needed to provide the paging information for the decoding logic block. These nine bits tells us that this memory space may be divided up into 512 pages with 32 K address on each page. However, the astute reader will immediately note that we only have a total of four memory chips in our system. Something is definitely wrong! We don't have enough memory chips to fill 512 pages. Oh drat, I hate it when that happens!

Actually, it isn't a problem after all. It means that out of a possible 512 pages of addressable memory, our computer has 2 pages of real memory, and space for another 510 pages. Is this a problem? That's hard to say. If we can fit all of our code into the two pages we do have, then why incur the added costs of memory that isn't being used? I can tell you from personal experience that a lot of sweat has gone into cramming all of the code into fewer memory chips to save a dollar here and there.

The other question that you ask is this. "OK, so the addressable memory space of the µP is not completely full. So where's the memory that we do have positioned in the address space of the processor?" That's a very good question because we don't have enough information right now to answer that. However, before we attempt to program this computer and memory system, we must design the hardware so that the memory chips we do have are correctly decoded at the page locations they are designed to be at. We'll see how that works in a little while.

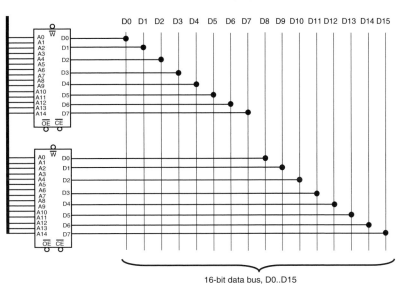

Figure 6.17: Expanding a memory system by width.

Let's return to Figure 6.16. It's important to understand that we really need two memory chips for each page of memory because our data bus is 16-bits wide, but each memory chip is only 8 data bits wide. Thus, in order to build a 16-bit wide memory, we need two chips. We can see this in Figure 6.17. Notice how each memory device connects to a separate group of eight wires in the data bus. Of course, the address bus pins, A0 through A14 must connect to the same wires of the address bus, because we are addressing the same address location both memory chips.

Now that you've seen how the two memory chips are "stacked" to create a page in memory that is 32 K × 16. It should not be a problem for you to design a 32 K × 32 memory using four chips.

You may have noticed that the microprocessor's clock was nowhere to be seen in this example memory design. Surely, one of the most important links in a computer system, the memory to processor, needs a clock signal in order to synchronize the processor to the memory. In fact, many memory systems do not need a clock signal to insure reliable performance. The only thing that needs to be considered is the timing relationship between the memory circuits and the processor's bus operation. In the next chapter, we'll look at a processor bus cycle in more detail, but here's a preview. The NEC μPD444008 comes in three versions. The actual part numbers are:

- μPD444008-8
- μPD444008-10
- μPD444008-12

The numerical suffixes, 8, 10 and 12, refer to the maximum *access time* for each of the chips. The access time is basically a specification which determines how quickly the chip is able to reliably return data once the control inputs have been properly established. Thus, assuming that the address to the chip has stabilized, $\overline{CS}$ and $\overline{OE}$ are asserted, then after a delay of 8, 10 or 12 nanoseconds (depending upon the version of the chip being used), the data would be available for reading into the processor. The chip manufacturer, NEC, guarantees that the access time will be met for the entire temperature range that the chip is designed to operate over. For most electronics, the commercial temperature range is 0 degrees Celsius to 70 degrees Celsius.

Let's do a simple example to see what this means. We'll actually look into this in more detail later on, but it can't hurt to prepare ourselves for things to come. Suppose that we have a processor with a 500 MHz clock. You know that this means that each clock period is 2 ns long. Our processor requires 5 clock cycles to do a memory read, with the data being read into the processor on the falling edge of the 5th clock cycle. The address and control information comes out of the processor on the rising edge of the first clock cycle. This means that the processor requires 4.5 × 2, or 9 ns to do a memory read operation. However, we're not quite done with our calculation. Our decoding logic circuit also introduces a time delay. Assume that it takes 1ns from the time the processor asserts the control and address signal to the time that the decoding logic to provide the correct signals to the memory system. This means that we actually have 8 ns, not 9 ns, to get the data ready. Thus, only the fastest version of the part (generally this means the most expensive version) would work reliably in this design.

Is there anything that we can do? We could slow down the clock. Suppose that we changed the clock frequency from 500 MHz to 400 MHz. This lengthens the period to 2.5 ns per clock cycle. Now 4.5 clock cycles take 11.25 ns instead of 9 ns. Subtracting 1 ns for the propagation delay through the decoding logic, we would need a memory that was 10.25 ns or faster to work reliably. That looks pretty encouraging. We could slow the clock down even more so we could use even cheaper memory devices. Won't the Project Manager be pleased! Unfortunately, we've just made a trade-off. The trade-off is that we've just slowed our processor down by 20%. Everything the processor does will now take 20% longer. Can we live with that? At this point, we probably don't know. We'll need to do some careful measurements of code execution times and performance requirements before we can answer the question completely; and even then we may have to make some pretty rough assumptions.

Anyway, the key to the above discussion is that there is no explicit clock in the design of the memory system. The clock dependency is implicit in the timing requirements of the memory-to-processor interface, but the clock itself is not required. In this particular design, our memory system is asynchronously connected to the processor.

Today, most PC memory designs are synchronous designs. The clock signal is an integral part of the control circuitry of the processor-to-memory interface. If you've ever added a memory "stick" to your PC then you've upped the capacity of your PC using *synchronous dynamic random access memory or SDRAM* chips. The printed circuit board (the stick) is a convenient way to mechanically connect the memory chips to the PC motherboard.

Figure 6.18 is a photograph of a 64 Megabyte (Mbyte) SDRAM memory module. This module holds 64 Mbytes of data organized as 1M × 64. There are a total of 16 memory chips on the module (front and back) each chip has a capacity of 32 Mbits, organized as 8M × 4. We'll look at the differences between asynchronous, or *static* memory systems and synchronous, *dynamic*, memory systems later on in this chapter.

Figure 6.18: 64 Mbyte SDRAM memory module.

Paging in Real Memory Systems

Our four memory chips of Figure 6.16 will give us two 32K × 16 memory pages. This leaves us 510 possible memory pages that are empty. How do we know where we'll have these two memory pages and where we will just have empty space? The answer is that it is up to you (or the hardware designer) to specify where the memory will be. As you'll soon see, in the 68000 system we want nonvolatile memory, such as ROM or FLASH to reside from the start of memory and go up from there. Let's state for the purpose of this exercise that we want to locate our two available pages of real memory at page 0 and at page 511.

Let's assume that the processor has 24 address bits. This corresponds to about 16M of addressable memory (2^{24} address locations). It is customary to locate RAM memory (read/write) at the top of memory, but this isn't required. In most cases, it will depend upon the processor architecture. In any case, in this example we need to figure out how to make one of the two real memory pages

Table 6.2: Page numbers and memory address ranges for a 24-bit addressing system.

Page Number (binary) A23................A15	Page number (hex)	Absolute address range (hex)
000000000	000	000000 to 007FFF
000000001	001	008000 to 00FFFF
000000010	002	010000 to 017FFF
000000011	003	018000 to 01FFFF
.	.	
.	.	
.	.	
.		
111111111	1FF	FF8000 to FFFFFF

respond to addresses from 0x000000 through 0x007FFF. This is the first 32 K of memory and corresponds to page 0. The other 32K words of memory should reside in the memory region from 0xFF8000 through 0xFFFFFF, or page 511. How do we know that? Simple, it's paging. Our total system memory of 16,777,216 words may be divided up into 512 pages with 32 K on each page. Since we have 9 bits for the paging we can divide the absolute address up as shown in Table 6.2.

We want the two highlighted memory ranges to respond by asserting the $\overline{CS0}$ or $\overline{CS1}$ signals when the memory addresses are within the correct range and the other memory ranges to remain unasserted. The decoder circuit for page 1FF is shown in Figure 6.19. The circuit for page 000 is left as an exercise for you.

Notice that there is new a signal called $\overline{ADDRVAL}$ (Address Valid). The Address Valid signal (or some other similar signal) is issued by the processor in order to notify the external memory that the current address on the bus is stable. Why is this necessary? Keep in mind that the addresses on the address bus are always changing. Just executing one instruction may involve five or more memory accesses with different address values. The longer an address stays around, the worse the performance of the processor will be. Therefore, the processor must signal to the memory that the current value of the address is correct and the memory may respond to it. Also, some processors may have two separate signals $\overline{RD}$ and $\overline{WR}$, to signify read and write operations, respectively. Other just have a single line R/W. There are advantages and disadvantages to each approach and we won't need to consider them here. For now, let's assume that our processor has two separate signals, one for a read operation and one for a write operation.

As you can see from Figure 6.16 and from the discussion of how the memory chips work in our system, it is apparent that we can express the logical conditions necessary to read and write to memory as:

MEMORY READ = $\overline{OE}$ * $\overline{CS}$ * WR
MEMORY WRITE = OE * $\overline{CS}$ * $\overline{WR}$

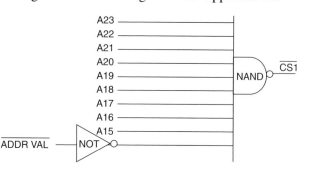

Figure 6.19: Schematic diagram for a circuit to decode the top page of memory of Figure 6.16.

In both cases, we need to assert the $\overline{CS}$ signal in order to read or write to memory. It is the control of the chip enable (or chip select) signal that allows us to control where in the memory space of the processor a particular memory chip will become active.

With the exception of our brief introduction to SDRAM memories, we've considered only *static RAM (SRAM)* for our memory devices. As you've seen, static RAM is derived from the D flip-flop. It is relatively simple interface to the processor because all we need to do is present an address and the appropriate control signals, wait the correct amount of time, and then we can read or write to memory. If we don't access memory for long stretches of time there's no problem because the feedback mechanism of the flip-flop gate design keeps the data stored properly as long as power is applied to the circuit. However, we have to pay a price for this simplicity. A modern SRAM memory cell requires five or six transistors to implement the actual gate design. When you're talking about memory chips that store 256 million bits of data, a six transistor memory cell takes up a lot of valuable room on the silicon chip (die).

Today, most high-density memory in computers, like your PC, uses a different memory technology called *dynamic RAM, or DRAM*. DRAM cells are much smaller than SRAM cells, typically taking only one transistor per cell. One transistor is not sufficient to create the feedback circuit that is needed to store the data in the cell, so DRAM's use a different mechanism entirely. This mechanism is called *stored charge*.

If you've ever walked across a carpet on a dry winter day and gotten a shock when you touched some metal, like the refrigerator, you're familiar with stored charge. Your body picked up excess charge as you walked across the carpet (now you represent a logical 1 state) and you returned to a logical 0 state when you got zapped as the charge left your body. DRAM cells work in exactly the same way. Each DRAM cell can store a small amount of charge that can be detected as a 1 by the DRAM circuitry. Store some charge and the cell has a 1, remove the charge and its 0. (However, just like the charge stored on your body, if you don't do anything to replenish the charge, it eventually leaks away.) It's a bit more complicated than this, and the stored charge might actually represent a 0 rather than a 1, but it will be sufficient for our understanding of the concept.

In the case of a DRAM cell, the way that we replenish the charge is to periodically read the cell. Thus, DRAM's get their name from the fact that we are constantly reading them, even if we don't actually need the data stored in them. This is the *dynamic* portion of the DRAM's name. The process of reading from the cell is called a *refresh cycle*, and must be carried out at intervals. In fact, every cell of a DRAM must be refreshed ever few milliseconds or the cell will be in danger of losing its data. Figure 6.20 shows a schematic representation of the organization of a 64 Mbit DRAM memory.

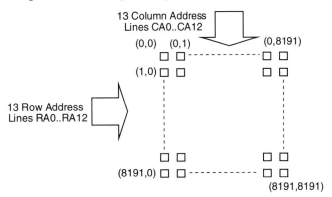

Figure 6.20: Organization of a 64 Megabit DRAM memory.

The memory is organized as a matrix with 8192 rows × 8192 columns (2^{13}). In order to uniquely address any one of the DRAM memory cells, a 26-bit address is required. Since we've already created it as a matrix, and 26 pins on the package would add a lot of extra complexity, the memory is addressed by providing a separate row address and a separate column address to the XY matrix. Fortunately for us, the process of creating these addresses is handled by the special chip sets on your PC's motherboard. Let's return to the refresh problem. Suppose that we must refresh each of the 64 million cells at least once every 10 milliseconds. Does that mean that we must do 64 million refresh cycles? Actually no; it is sufficient to just issue the row address to the memory and that guarantees that all of the 8192 cells in that row get refreshed at once. Now our problem is more tractable. If, for example the specification allows us 16.384 milliseconds to refresh 8192 rows in the memory, then we must, on average, refresh one row every 16.384×10^{-3} / 8.192×10^{3} seconds, or one row every two microseconds.

If this all seems very complicated, it certainly is. Designing a DRAM memory system is not for the beginning hardware designer. The DRAM introduces several new levels of complexity:

- We must break the full address down into a row address and a column address,
- We must stop accessing memory every microsecond or so and do a refresh cycle,
- If the processor needs to use the memory when a refresh also needs to access the memory, we then need some way to synchronize the two competing processes.

This makes the interfacing DRAM to modern processors quite a complex operation. Fortunately, the modern support chip sets have this complexity well in hand. Also, if the fact that we must do a refresh every two microseconds seems excessive to you, remember that your 2 GHz Athlon or Pentium processor issues 4,000 clock cycles every two microseconds. So we can do a lot of processing before we need to do a refresh cycle.

The problem of conflicts arising because of competing memory access operations (read, write and refresh) are mitigated to a very large degree because modern PC processors contain on-chip memories called *caches*. Cache memories will be discussed in much more detail in a later chapter, but for now, we can see the effect of the cache on our off-chip DRAM memories by greatly reducing the processor's demands on the external memory system.

As we'll see, the probability that the instruction or data that a processor requires will be in the cache is usually greater than 90%, although the exact probability is influenced by the algorithms being run at the time. Thus, only 10% of the time will the processor need to go to external memory in order to access data or instructions not in the cache. In modern processors, data is transmitted between the external memory systems and the processor in *bursts*, rather than one byte or word at a time. Burst accesses can be very efficient ways to transfer data. In fact, you are probably already very familiar with the concept because so many other systems in your PC rely on burst data transfers. For example, you hard drive transfers data to memory in bursts of a sector of data at a time. If your computer is connected to a 10Base-T or 100Base-T network then it is processing packets of 256 bytes at time. It would be just too inefficient and wasteful of the system resources to transmit data a byte at a time.

SDRAM memory is also design to efficiently interface to a processor with on-chip caches and is specifically designed for burst accesses between the memory and the on-chip caches of the processor. Figure 6.21 is an excerpt from the data sheet for an SDRAM memory device from Micron Technology, Inc.®, a semiconductor memory manufacturer located in Boise, ID. The timing diagram is for the MT48LC128MXA2[2] family of SDRAM memories. The devices are 512 Mbit parts organized as 4, 8 or 16-bit wide data paths. The 'X' is a

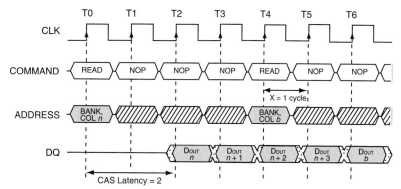

Figure 6.21: Timing diagram of a burst memory access for a Micron Technology Inc. part number MT48LC128MXA2 SDRAM memory chip. Diagram courtesy of Micron Technology.

placeholder for the organization (4, 8 or 16 bit wide). Thus, the MT48LC128M4A2 is organized as 32 M × 4, while the MT48LC128M16A2 is organized as 8 M × 16.

These devices are far more complicated in their operation then the simple SRAM memories we've looked at so far. However, we can see the fundamental burst behavior in Figure 6.21.

The fields marked *COMMAND, ADDRESS* and *DQ* are represented as bands of data, rather than individual bits. This is a simplification that allows us to show a group of signals, such as 14 address bits, without having to show the state of each individual signal. The band is used to show where the signal must be stable and where it is allowed to change. Notice how the signals are all synchronized to the rising edge of the clock. Once the READ command is issued and the address is provided for where the burst is to originate, there is a two clock cycle latency and sequentially stored data in the chip will then be available *on every successive clock cycle.* Clearly, this is far more efficient then reading one byte at a time.

When we consider cache memories in greater detail, we'll see that the on-chip caches are also designed to be filled from external memory in bursts of data. Thus, we incur a penalty in having to set-up the initial conditions for the data transfer from external memory to the on-chip caches, but once the data transfer parameters are loaded, the memory to memory data transfer can take place quite rapidly. For this family of devices the data transfer takes place at a maximum clock rate of 133 MHz.

Newer SDRAM devices, called *double data rate, or DDR* chips, can transfer data on both the rising and falling edges of the clock. Thus, a DDR chip with a 133 MHz clock input can transfer data at a speedy 266 MHz. These parts are designated, for reasons unknown, as PC2700 devices. Any SDRAM chip capable of conforming to a 266 MHz clock rate are PC2700.

Modern DRAM design takes many different forms. We've been discussing SDRAM because this is the most common form of DRAM in a modern PC. Your graphics card contains video DRAM. Older PC's contained *extended data out*, or *EDO DRAM*. Today, the most common type of SDRAM is DDR SDRAM. The amazing thing about all of this is the incredibly low cost of this type of memory. At this writing (summer of 2004), you can purchase 512 Mbytes of SDRAM for about 10 cents per megabyte. A memory with the same capacity, built in static RAM would cost well over $2,000.

Memory-to-Processor Interface

The last topic that we'll tackle in this chapter involves the details of how the memory system and the processor communicate with each other. Admittedly, we can only scratch the surface because there are so many variations on a theme when there are over 300 commercially available microprocessor families in the world today, but let's try to take a general overview without getting too deeply enmeshed in individual differences.

In general, most microprocessor-based systems contain three major bus groupings:

- Address bus: A unidirectional bus from the processor out to memory.
- Data bus: A bi-directional bus carrying data from the memory to the processor during read operations and from the processor to memory during write operations.
- Status bus: A heterogeneous bus comprised of the various control and housekeeping signals need to coordinate the operation of the processor, its memory and other peripheral devices. Typical status bus signals include:
  a. RESET,
  b. interrupt management,
  c. bus management,
  d. clock signals,
  e. read and write signals.

This is shown schematically in Figure 6.22 for the Motorola®* MC68000 processor. The 68000 has a 24-bit address bus and a 16-bit external data bus. However, internally, both address and data can be up to 32 bits in length. We'll discuss the interrupt system and bus management system later on in this section.

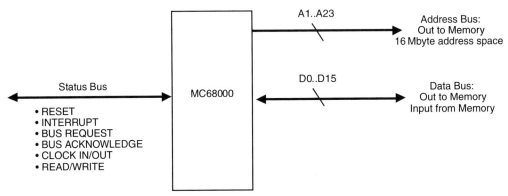

Figure 6.22: Three major busses of the Motorola 68000 processor.

---

* The Motorola Corporation has recently spun off its Semiconductor Products Sector (SPS) to form a new company, *Freescale®, Inc.* However, old habits die hard, so we'll continue to refer to processors derived from the 68000 architecture as the Motorola MC68000.

The Address Bus is the aggregate of all the individual address lines. We say that it is a *homogeneous bus* because all of the individual signals that make up the bus are address lines. The address bus is also unidirectional. The address is generated by the processor and goes out to memory. The memory does not generate any addresses and send them to the processor over this bus.

The Data Bus is also homogeneous, but it is bidirectional. Data goes out from memory to the processor on a read operation and from the processor to memory on a write operation. Thus, data can flow in either direction, depending upon the instruction being executed.

The Status Bus is *heterogeneous*. It is made up of different kinds of signals, so we can't group them in the same way that we do for address and data. Also, some of the signals are unidirectional, some are bidirectional. The Status Bus is the "housekeeping" bus. All of the signals that are also needed to control system operation are grouped into the Status Bus.

Let's now look at how the signals on these busses work together with memory so that we may read and write. Figure 6.23 shows us the processor side of the memory interface.

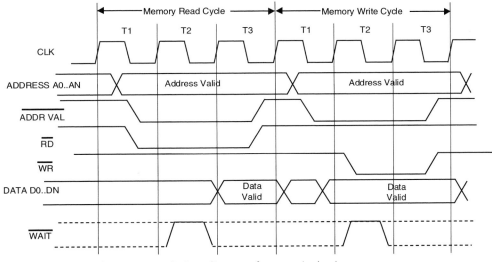

Figure 6.23: Timing diagram for a typical microprocessor.

Now we can see how the processor and the clock work together to sequence the accessing of the memory data. While it may seem quite bewildering at first, it is actually very straightforward. Figure 6.23 is a "simplified" timing diagram for a processor. We've omitted many additional signals that may present or absent in various processor designs and tried to restrict our discussion to the bare essentials.

The Y-axis shows the various signals coming from the processor. In order to simplify things, we've grouped all the signals for the address bus and the data bus into a "band" of signals. That way, at any given time, we can assume that some are 1 and some are 0, but the key is that we must specify when they are valid. The crossings, or X's in the address and data busses is a symbolic way to represent points in time when the addresses or data on the busses may be changing, such as an address changing to a new value, or data coming from the processor.

Since the microprocessor is a state machine, everything is synchronized with the edges of the clock. Some events occur on the positive going edges and some may be synchronized with the negative going edges. Also, for convenience, we'll divide the bus cycles into identifiable time signatures called "T states." Not all processors work this way, but this is a reasonable approximation of how many processors actually work. Keep in mind that the processor is always running these bus cycles. These operations form the fundamental method of data exchange between the processor and memory. Therefore, we can answer a question that was posed at the beginning of this chapter. Recall that the state machine truth table for the operation, ADD B, A, left out any explanation of how the data got into the registers in the first place, and how the instruction itself got into the computer.

Thus, before we look at the timing diagram for the processor/memory interface, we need to remind ourselves that the control of this interface is handled by another part of our state machine. In algorithmic terms, we do a "function call" to the portion of the state machine that handles the memory interface, and the data is read or written by that algorithm.

Let's start with a READ cycle. During the falling edge of the clock in T1 the address becomes stable and the $\overline{\text{ADDR VAL}}$ signal is asserted LOW. Also, the $\overline{\text{RD}}$ signal goes LOW to indicate that this is a read operation. During the falling edge of T3 the READ and ADDRESS VALID signals are de-asserted indicating to memory that that the cycle is ending and the data from memory is being read by the processor. Thus, the memory must be able to provide the data to the processor within two full clock cycles (all of T2 plus half of T1 and half of T3).

Suppose the memory isn't fast enough to guarantee that the data will be ready in time. We discussed this situation for the case of the NEC static RAM chip and decided that a possible solution would be to slow the processor clock until the access time requirements for the memory could be guaranteed to be within specs. Now we will consider another alternative. In this scenario, the memory system may assert the $\overline{\text{WAIT}}$ signal back to the processor. The processor checks the state of the $\overline{\text{WAIT}}$ signal on the on the falling edge of the clock during T2 cycle. If the $\overline{\text{WAIT}}$ signal is asserted, the processor generates another T2 cycle and checks again. As long as the $\overline{\text{WAIT}}$ signal is LOW, the processor keeps marking time in T2. Only when $\overline{\text{WAIT}}$ goes high will the processor complete the bus cycle. This is called a *wait state*, and is used to synchronize slower memory to faster processors.

The write cycle is similar to the read cycle. During the falling edge of the clock in T1 the address becomes valid. During the rising edge of the clock in T2 the data to be written is put on the data bus and the write signal goes low, indicating a memory write operation. $\overline{\text{WAIT}}$ signal has the same function in T2 on the write cycle. During the falling edge of the clock in T3 the $\overline{\text{WR}}$ signal is de-asserted, giving the memory a rising edge to store the data. $\overline{\text{ADDR VAL}}$ also is de-asserted and the write cycle ends.

There are several interesting concepts buried in the previous discussion that require some explanation before we move on. The first is the idea of a state machine that operates on both edges of the clock, so let's consider that first. When we input a single clock signal to the processor in order to synchronize its internal operations, we don't really see what happens to the internal clock.

Many processors will internally convert the clock to a *2-phase clock*. A timing diagram for a 2-phase clock is shown in Figure 6.24.

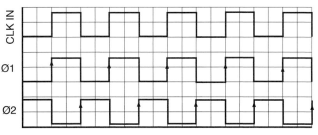

Figure 6.24: Figure 4.9: A two-phase clock.

The input clock, which is generated by an external oscillator, is converted to a 2-phase clock, labeled $\phi1$ and $\phi2$. The two clock phases now 180 degrees out of phase from each other, so that every rising or falling edge of the CLK IN signal generates an internal rising clock edge. How could we generate a 2-phase clock? You actually already know how to do it, but there's a piece of information that we first need to place in context. Figure 6.25 is a circuit that can be used to generate a 2-phase clock.

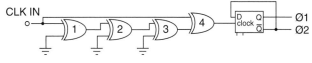

Figure 6.25: A two-phase clock generation circuit.

The 4 XOR gates are convenient to use because there is a common integrated circuit part which contains 4 XOR gates in one package. This circuit makes use of the propagation delays that are inherent in a logic gate. Suppose that each XOR gate has a propagation delay of 10 ns. Assume that the clock input is LOW. One input of XOR gates 1 through 3 is permanently ties to ground (logic LOW). Since both inputs of gate 1 are LOW, its output is also LOW. This situation carries through to gates 2, 3 and 4. Now, the CLK IN input goes to logic state HIGH. The output of gate #4 goes high 10 ns later and toggles the D-FF to change state. Since the Q and $\overline{Q}$ outputs are opposite each other, we conveniently have a source of two alternating clock phases by nature of the divide-by-two wiring of the D-FF.

After a propagation delay of 30 ns the output of gate #3 also goes HIGH, which causes the output of XOR gate #4 to go LOW again because the output of an XOR gate is LOW if both inputs are the same and HIGH if the inputs are different. At some time later, the clock input goes low again and we generate another 30 ns wide positive going pulse at the output of gate #4 because for 30 ns both outputs are different. This cause the D-FF to toggle at both edges of the clock and the Q and $\overline{Q}$ outputs give us the alternating phases that we need. Figure 6.26 shows the relevant waveforms.

This circuit works for any clock frequency that has a period greater than 4 XOR gate delays. Also, by using both outputs of the D-FF, we are guaranteed a two-phase clock output that is exactly 180 degrees out of phase from each other.

Now we can revisit Figure 6.23 and see the other subtle point that was

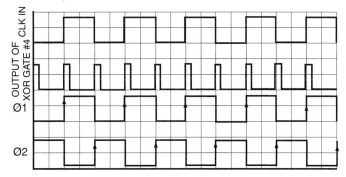

Figure 6.26: Waveforms for the 2-phase clock generation circuit.

buried in the diagram. Since we are apparently changing states on the rising and falling edges of the clock, we now know that the internal state machine of the processor is actually using a 2-phase clock and each of the 'T' states is, in reality, two states. Thus, we can redraw the timing diagram for a READ cycle as a state diagram. This will clearly demonstrate the way in which the WAIT state comes into play. Figure 6.27 shows the READ phase of the bus cycle, represented as a state diagram.

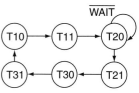

Figure 6.27: State diagram for a processor READ cycle.

Referring to Figure 6.27 we can clearly see that in state T20 the processor tests the state of the $\overline{\text{WAIT}}$ input. If the input is asserted LOW, the processor remains in state T20, effectively lengthening the total time for the bus cycle. The advantage of the wait state over decreasing the clock frequency is that we can design our system such that a wait penalty is incurred only when the processor accesses certain memory regions, rather than slowing it for all operations. We can now summarize the entire bus READ cycle as follows:

- T10: READ cycle begins. Processor outputs new memory address for READ operation.
- T11: Address is now stable and $\overline{\text{AD VAL}}$ goes LOW. $\overline{\text{RD}}$ goes low indicating that a READ cycle is beginning.
- T20: READ cycle continues.
- T21: Process samples $\overline{\text{WAIT}}$ input. If asserted T21 cycle continues.
- T30: READ cycle continues.
- T31: READ cycles terminates. $\overline{\text{AD VAL}}$ and $\overline{\text{RD}}$ are de-asserted and processor inputs the data from memory.

Direct Memory Access (DMA)

We'll conclude Chapter 6 with a brief discussion of another form of memory access called DMA, or direct memory access. The need for a DMA system is a result of the fact that memory system and the processor are connected to each other by busses. Since the bus is the only path in and out of the system, conflicts will arise when peripheral devices, such as disk drives or network cards have data for the processor, but the processor is busy executing program code.

In many systems, the peripheral devices and memory share the same busses with the processor. When a device, such as a hard disk drive needs to transfer data to the processor, we could imagine two scenarios.

Scenario #1

1	Disk drive:	"Sorry for the interrupt boss, I've got 512 bytes for you."
2	Processor:	"That's a big 10-4 little disk buddy. Gimme the first byte."
3	Disk drive:	"Sure boss. Here it is."
4	Processor:	"Got it. Gimme the next one."
5	Disk drive:	"Here it is."

Repeat steps 4 and 5 for 510 more times.

Scenario #2

1	Disk drive:	"Yo, boss. I got 512 bytes and they're burning a hole in my platter. I gotta go, I gotta go." (BUS REQUEST)
2	Processor:	"OK, ok, pipe down lemme finish this instruction and I'll get off the bus. OK, I'm done, the bus is yours, and don't dawdle, I'm busy." (BUS GRANT)
3	Disk drive:	"Thanks boss. You're a pal. I owe you one. I've got it." (BUS ACKNOWLEDGE)
4	Disk drive:	"I'll put the data in the usual spot." (Said to itself)
5	Disk drive:	"Hey boss! Wake up. I'm off the bus."
6	Processor:	Thanks disk. I'll retrieve the data from the usual spot."
7	Disk drive:	"10-4. The usual spot. I'm off."

As you might gather from these two scenarios, the second was more efficient because the peripheral device, the hard disk, was able to take over memory control from the processor and write all of its data in a single burst of activity. The processor had placed its memory interface in a tri-state condition and was waiting for the signal from the disk drive that it could return to the bus. Thus, the DMA allows other devices to take over control of the busses and implement a data transfer to or from memory while the processor idles, or processes from a separately cached memory. Also, given that many modern processors have large on-chip caches, the processor looses almost nothing by turning the external bus over to the peripheral device. Let's take the humorous discussion of the two scenarios and get serious for a moment. Figure 6.28 shows the simplified DMA process. You may also gather that I shouldn't quit my day job to become a sitcom writer, but that's a discussion for another time.

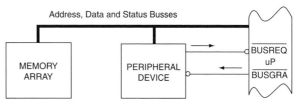

Figure 6.28: Schematic representation of a DMA transfer.

In the simplest form, there is a *handshake* process that takes place between the processor and the peripheral device. A handshake is simply an action that expects a response to indicate the action was accepted. The process can be described as follows:

- The peripheral device requests control of the bus from the processor by asserting the BUS REQUEST ($\overline{\text{BUSREQ}}$) signal input on the processor.
- When processor completes present instruction cycle, and no higher level interrupts are pending, it sends out a BUS GRANT ($\overline{\text{BUSGRA}}$), giving the requesting device permission to begin its own memory cycles.
- Processor then idles, or continues to process data internally in cache, until BUSREQ signal goes away

Summary of Chapter 6

- We looked at the need for bus organization within a computer system and how busses are organized into address, data and status busses.
- Carrying on our discussion of the previous chapter, we saw how the microcode state machine would work with the bus organization to control the flow of data on internal busses.

- We saw how the tri-state buffer circuit enables individual memory cells to be organized into larger memory arrays.
- We introduced the concept of paging as a way to form memory addresses and as a method to build memory systems.
- We looked at the different types of modern memory technology to understand the use of static RAM technology and dynamic RAM technology.
- Finally, we concluded the overview of memory with a discussion of direct memory access as an efficient way to move blocks of data between memory and peripheral devices.

Chapter 6: *Endnotes*

[1] http://www.necel.com/memory/pdfs/M14428EJ5V0DS00.pdf.

[2] http://download.micron.com/pdf/datasheets/dram/sdram/512MbSDRAM.pdf.

[3] Ralph Tenny, *Simple Gating Circuit Marks Both Pulse Edges*, Designer's Casebook, Prepared by the editors of *Electronics,* McGraw Hill, p. 27.

Exercises for Chapter 6

1. Design a 2 input, 4-output memory decoder, given the truth table shown below:

Inputs		Outputs			
A	B	O1	O2	O3	O4
0	0	0	1	1	1
1	0	1	0	1	1
0	1	1	1	0	1
1	1	1	1	1	0

2. Refer to Figure 6.11. The external input and output (I/O) signals defined as follows:

A0, A1: Address inputs for selecting which row of memory cells (D-flip flops) to read from, or write to.

$\overline{CE}$: Chip enable signal. When low, the memory is active and you read from it or write to it.

R/$\overline{W}$: Read/Write line. When high, the appropriate row within the array may be read by an external device. When low, an external device may write data into the appropriate row. The appropriate row is defined by the state of address bits A0 and A1.

DB0, DB1, DB2, DB3: Bidirectional data bits. Data being written into the appropriate row, or read from the appropriate row, as defined by A0 and A1, are defined by these 4 bits.

The array works as follows:

A. To read from a specific address (row) in the array:
   a. Place the address on A0 and A1
   b. Bring $\overline{CE}$ low
   c. Bring R/$\overline{W}$ high.
   d. The data will be available to read on D0..D3

B. To write to a specific address in the array:
   a. Place the address on A0 and A1
   b. Bring $\overline{CE}$ low
   c. Bring R/$\overline{W}$ low.
   d. Place the data on D0..D3.
   e. Bring R/$\overline{W}$ high.

C. Each individual memory cell is a standard D flip-flop with one exception. There is a tri-state output buffer on each individual cell. The output buffer is controlled by the $\overline{CS}$ signal on each FF. When this signal is low, the output is connected to the data line and the data stored in the FF is available for reading. When this output is high, the output of the FF is isolated from the data line so that data may be written into the device.

Consider the box labeled "Memory Decoding Logic" in the diagram of the memory array. Design the truth table for that circuit, simplify it using K-maps and draw the gate logic to implement the design.

3. Assume that you have a processor with a 26-bit wide address bus and a 32-bit wide data bus.
   a. Suppose that you are using memory chips organized as 512 K deep × 8 bits wide (4 Mbit). How many memory chips are required to build a memory system for this processor that completely fills the entire address space, leaving no empty regions?
   b. Assuming that we use a page size of 512 K, complete the following table for the first three pages of memory:

4. Consider the memory timing diagram from Figure 6.23. Assume that the clock frequency is 50 MHz and that you do not want to add any wait states to slow the processor down. What is the slowest memory access time that will work for this processor?

5. Define the following terms in a few sentences:
   a. Direct Memory Access
   b. Tri-state logic
   c. Address bus, data bus, status bus

6. The figure shown below is a schematic diagram of a memory device that will be used in a memory system for a computer with the following specifications:
   - 20-bit address bus
   - 32-bit data bus
   - Memory at pages 0, 1 and 7

   a. How many addressable memory locations are in each memory device?

   b. How many bits of memory are in each memory device?

   c. What is the address range, in hex, covered by each memory device in the computer's address space? You may assume that each page of memory is the same size as the address range of one memory device.

   d. What is the total number of memory devices required in this memory design?

   e. Why would a memory system design based upon this type of a memory device not be capable of addressing memory locations at the byte-level? Discuss the reason for your answer in a sentence or two.

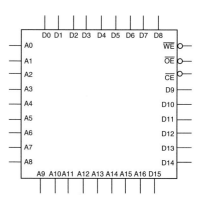

7. Assume that you are the chief hardware designer for the *Soul of the City Bagel and Flight Control Systems Company*. Your job is to design a computer to memory sub-system for a new, automatic, galley and bagel maker for the next generation of commercial airliners now being designed. The microprocessor is the AB2000, a hot new chip that you've just been waiting to have an opportunity to use in a design.

The processor has a 16-bit address bus and a 16-bit data bus as shown in the figure, below (For simplicity, the diagram only shows the relevant processor signals for this exercise problem). Also shown in the figure are the schematic diagrams and pin designations for the ROM and SRAM chips you will need in your design.

The memory subsystem has three status signals that you must use to design your memory array:
a. $\overline{WR}$: Active low, it indicates a write to memory cycle
b. $\overline{RD}$: Active low, it indicates a read from memory cycle
c. $\overline{ADVAL}$: Active low, it indicates that the address on the bus is stable and may be considered to be a valid address

The memory chips occupy the following regions of memory:
a. ROM from 0x0000 through 0x3FFF
b. SRAM from 0xC000 through 0xFFFF
c. Other stuff (not your problem) from 0x4000 through 0xBFFF

a. Design the memory decoder circuit that you will need for this circuit. It must be able to properly decode the memory chips for their respective address ranges and also decode the processor's $\overline{WR}$, $\overline{RD}$ and $\overline{ADVAL}$ signals to create $\overline{OE}$, $\overline{CE}$, and $\overline{WR}$. Hint, refer to the text sections on the memory system to get the equations for $\overline{OE}$, $\overline{CE}$ and $\overline{WR}$.

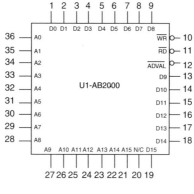

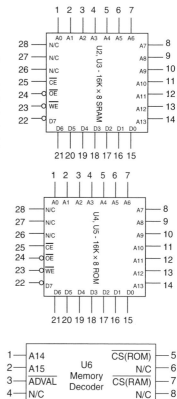

Assume that the circuit that you design will be fabricated into a memory interface chip designed, U6. Since the printed circuit designer needs to know the pin designations for all the parts, you supply the specification of the pin functions for U6 shown, below. You will then design the memory decoder to these pin designations. Finally, for simplicity, you decide not to implement byte addressability. All memory accesses will be word-width accesses.

b. Create a net list for your design. A sample net list is shown in the figure, below. The net list is just a table of circuit interconnections or "nets". A net is just the common connection of all I/O pins on the circuit packages that are connected together. Thus, in this design, you might have a net labeled "ADDR13", which is common to: Pin #23 on chip number U1 (shorthand U1-23), U2-14, U3-14, U4-14, U5-14. The net list is used by printed circuit board manufacturers to actually build your PC board. A start for your net list is shown below:

Net name					
addr0	U1-36	U2-1	U3-1	U4-1	U5-1
addr1	U1-35	U2-2	U3-2	U4-2	U5-2
addr2	U1-34	U2-3	U3-3	U4-3	U5-3
addr3	U1-33	U2-4	U3-4	U4-4	U5-4
•					
•					
data0	U1-14	U2-15	U4-15		

8. Complete the analysis for each of the following situations:

   a. A microprocessor with a 20-bit address range using eight memory chips with a capacity of 128K each: Build a table showing the address ranges, in hexadecimal, covered by each memory chip.

   b. A microprocessor with a 24-bit address range and using four memory chips with a capacity of 64K each: Two of the memory chips occupy the first 128K of the address range and two chips occupy the top 128K of the address range. Build a table showing the address ranges, in hexadecimal, covered by each of the chips. *Notice that there is a big address range in the middle with no active memory in it.* This is fairly common.

   c. A microprocessor with a 32-bit address range and using eight memory chips with a capacity of 1M each (1M = 1,048,576): Two of the memory chips occupy the first 2M of the address range and 6 chips occupy the top 6M of the address range. Build a table showing the address ranges, in hexadecimal, covered by each of the chips.

   d. A microprocessor with a 20-bit addressing range and using eight memory chips of different sizes. Four of the memory chips have a capacity of 128K each and occupy the first 512K consecutive addresses from 00000 on. The other four memory chips have a capacity of 32K each and occupy the topmost 128 K of the addressing range.

CHAPTER 7

Memory Organization and Assembly Language Programming

· ·

Objectives

When you are finished with this lesson, you will be able to:
▶ *Describe how a typical memory system is organized in terms of memory address modes;*
▶ *Describe the relationship between a computer's instruction set architecture and its assembly language instruction set; and*
▶ *Use simple addressing modes to write a simple assembly language program.*

· ·

Introduction

This lesson will begin our transition from hardware designers back to software engineers. We'll take what we've learned so far about the behavior of the hardware and see how it relates to the instruction set architecture (ISA) of a typical microprocessor. Our study of the architecture will be from the perspective of a software developer who needs to understand the architecture in order to use it to its best advantage. In that sense, our study of assembly language will be a metaphor for the study of the architecture of the computer. This perspective is quite different from that of someone who wants to be able to design computer hardware. Our focus throughout this book has been on the understanding of the hardware and architectural issues in order to take make our software developments complement the design of the underlying hardware.

We're first going to investigate the architecture of the Motorola 68000 processor. The 68K ISA is a very mature architecture, having first come out in the early 1980s. It is a fair question to ask, "Why am I learning such an old computer architecture? Isn't it obsolete?" The answer is a resounding, "No!" The 68K architecture is one of the most popular computer architectures of all time and derivatives of the 68K are still being designed into new products all the time. Palm PDA's still use a 68K derivative processor. Also, a completely new processor from Motorola, the ColdFire® family, uses the 68K architecture, and today, it is one of the most popular processors in use. For example, the ColdFire is used in many inkjet printers.

In subsequent chapters, we will also look at two other popular architectures, the Intel X86 processor family and the ARM family. With these three microprocessor architectures under our belts, we'll then be familiar with the three most popular computer architectures in the world today. Much of what we'll be discussing as we begin our investigation will be common to all three processor families, so studying one architecture is as good as another when it comes to gaining an understanding of basic principles.

Another reason for starting with the 68000 is that the architecture lends itself to the learning process. The memory addressing modes are straightforward, and from a software developer's point of view, the linkages to high level languages are easy to understand.

We'll start by looking at the memory model again and use that as a jumping off point for our study of how the computer actually reads and writes to memory. From a hardware perspective, we already know that because we just got finished studying it, but we'll now look at it from the ISA perspective. Along the way, we'll see why this sometimes strange way of looking at memory is actually very important from the point of view of higher level languages, such as C and C++. Ladies and gentlemen, start your engines...

Memory Storage Conventions

When you start looking at a computer from the architectural level you soon realize that relationship between the processor and memory is one of the key factors in defining the behavior and operational characteristics of the machine. So much of how a computer actually accomplishes its programming tasks revolves around its interface to memory. As you'll soon see when we begin to program in assembly language, most of the work that the computer does involves moving data in and out of memory. In fact, if you created a bar graph of the assembly language instructions used in a program, whether the original program was written in C++ or assembly language, you'll find that the most used instruction is the MOVE instruction. Therefore, one of our first items of business is to build on our understanding of how memory systems are organized and look at the memory system from the prospective of the processor.

Figure 7.1 is a fairly typical memory map for a 68K processor. The processor can uniquely address 16,777,216 (16M) bytes of memory. It has 23 external word address lines and 2-byte selector control signals for choosing one or the other of the two bytes stored in a word location. The external data bus is 16-bits wide, but the internal data paths are all 32 bits wide. Subsequent members of the 68K family, beginning with the 68020, all had 32-bit external data buses. In Figure 7.1 we see that the first 128 K words (16-bit) of memory occupy the address range from 0x000000 to 0x01FFFE. This will typically be the program code and the vector tables. The vector table occupies the first 256 long words of memory. This is the reason that nonvolatile, read-only memory occupies the lower address range. The upper address range usually contains the RAM memory. This is volatile, read/write memory where variables are stored. The rest of the memory space contains the address of I/O devices. All the rest of the memory in Figure 7.1 is empty space. Also, we'll see in a moment why the addresses for the last word of ROM and the last word of RAM are 0x01FFFFE and 0xFFFFFE, respectively.

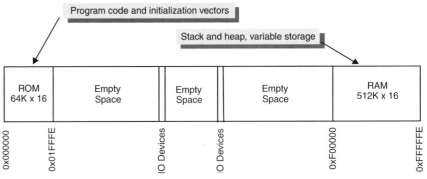

Figure 7.1: Memory map for a 68K-based computer system.

Since 8-bits, or a byte, is the smallest data quantity that we usually deal with, storing byte-sized characters (chars) in memory is straightforward when we're dealing with a memory that is also 8-bits wide. However, our PCs all have memories that are 32 bits wide and lots of computers are in use today that have 16 bit data path widths. As an extreme example, we don't want to waste 24 bits of storage space simply because we want to store an 8-bit quantity in a 32-bit memory, so computers are designed to allow byte addressing. Think of byte addressing in a 32-bit word as an example of paging with the page address being the 32-bit word and the offset being the four possible byte positions in the page. However, there is one major difference between byte addressing and real paging. With byte addressing, we do not have a separate word address that corresponds to a page address. Figure 7.2 shows this important distinction.

We call this type of memory storage *byte packing* because we are literally packing the 32-bit memory full of bytes. This type of addressing introduces several ambiguities. One, is quite serious, and the others are merely new to us. We discuss the serious one in a moment. Referring to Figure 7.2 we see that the byte at memory address FFFFF0 (char) and the 32-bit long word (int) at FFFFF0 have the same address. Isn't this a disaster waiting to happen? The answer is a definite "maybe." In C and C++, for example, you must declare a variable and its type before you can use it. Now you see the reason for it. Unless the compiler knows the type of the variable stored at address FFFFF0, it doesn't know the type of code it must generate to manipulate it. Also, if you want to store a char at FFFFF0, then the compiler has to know how much storage space to allocate to it.

Also, notice that we cannot access 32-bit words at addresses other than those that are divisible by 4. Some processors will allow us to store a 32-bit value at an odd boundary, such as byte addresses 000003-000006, but many others will not. The reason is that the processor would have to make several additional memory operations to read in the entire value and then do some extra work to reconstruct the bytes into the correct order. We call this a *nonaligned access* and it is generally quite costly in terms of processor performance to allow it to happen. In fact, if you look at the memory map of how a compiler stores objects and structures in memory, you'll often see spaces

Long Word Address

000000	Byte 0 – Address 000000	Byte 1 – Address 000001	Byte 2 – Address 000002	Byte 3 – Address 000003
000004	Byte 0 – Address 000004	Byte 1 – Address 000005	Byte 2 – Address 000006	Byte 3 – Address 000007
000008	Byte 0 – Address 000008	Byte 1 – Address 000009	Byte 2 – Address 00000A	Byte 3 – Address 00000B
00000C	Byte 0 – Address 00000C	Byte 1 – Address 00000D	Byte 2 – Address 00000E	Byte 3 – Address 00000F
000010	Byte 0 – Address 000010	Byte 1 – Address 000011	Byte 2 – Address 000012	Byte 3 – Address 000013

•
•
•

FFFFF0	Byte 0 – Address FFFFF0	Byte 1 – Address FFFFF1	Byte 2 – Address FFFFF2	Byte 3 – Address FFFFF3
FFFFF4	Byte 0 – Address FFFFF4	Byte 1 – Address FFFFF5	Byte 2 – Address FFFFF6	Byte 3 – Address FFFFF7
FFFFF8	Byte 0 – Address FFFFF8	Byte 1 – Address FFFFF9	Byte 2 – Address FFFFFA	Byte 3 – Address FFFFFB
FFFFFC	Byte 0 – Address FFFFFC	Byte 1 – Address FFFFFD	Byte 2 – Address FFFFFE	Byte 3 – Address FFFFFF

Figure 7.2: Relationship between word addressing and byte addressing in a 32-bit wide word.

in the data, corresponding to intentional gaps so as not to create regions of data containing non-aligned accesses.

Also notice that when we are doing 32-bit word accesses, address bits A0 and A1 aren't being used. This might prompt you to ask, "If we don't use them, what good are they?" However, we do need them when we need to access a particular byte within the 32-bit words. A0 and A1 are often called the *byte selector* address lines because that is their main function. Another point is that we really only need byte selectors when we are writing to memory. Reading from memory is fairly harmless, but writing changes everything. Therefore, you want to be sure that you modify only the byte you are interested in and not the others. From a hardware designer's perspective having byte selectors allows you to qualify the write operation to only the byte that you are interested in.

Many processors will not explicitly have the byte selector address lines at all. Rather, they provide signals on the status bus which are used to qualify the WRITE operations to memory. What about storing 16-bit quantities (a *short* data type) in 32-bit memory locations? The same rules apply in this case. The only valid addresses would be those addresses divisible by 2, such as 000000, 000002, 000004, and so on. In the case of 16-bit word addressing, the lowest order address bit, A0, isn't needed. For our 68K processor, which has a 16-bit wide data bus to memory, we can store two bytes in each word of memory, so A0 isn't used for word addressing and becomes the byte selector for the processor.

Figure 7.3 shows a typical 32-bit processor and memory system interface. The READ signal from the processor and the CHIP SELECT signals have been omitted for clarity. The processor has a 32-bit data bus and a 32-bit address bus. The memory chips represent one page of RAM somewhere in the address space of the processor. The exact page of memory would be determined by the design of the Address Decoder logic block. The RAM chips each have a capacity of 1 Mbit and are organized as 128K by 8.

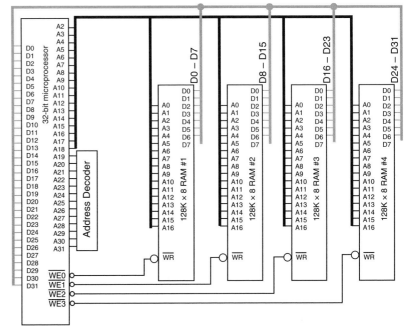

Figure 7.3: Memory organization for a 32-bit microprocessor. Chip select and READ signals have been omitted for clarity.

Since we have a 32-bit wide data bus and each RAM chip has eight data I/O lines, we need four memory chips per 128K wide page. Chip #1 is connected to data lines D0 through D7, chip #2 is connected to data lines D8 through D15, chip #3 is connected to data lines D6 through D23 and chip #4 is connected to data lines D24 to D31, respectively.

The address bus from the processor contains 30 address lines, which means it is capable of address 2^{30} long words (32-bit wide). The additional addressing bits needed to address the full address space of 2^{32} bytes are implicitly controlled by the processor internally and explicitly controlled through the 4 WRITE ENABLE signals labeled $\overline{WE0}$ through $\overline{WE3}$.

Address lines A2 through A18 from the processor are connected to address inputs A0 through A16 of the RAM chips, with A2 from the processor being connected to A0 on each of the 4 chips, and so on. This may seem odd at first, but it should make sense to you after you think about it. In fact, there is no special reason that each address line from the processor must be connected to the same address input pin on each of the memory devices. For example, A2 from the processor could be connected to A14 of chip #1, A3 of chip #2, A8 of chip #3 and A16 of chip #4. The same address from the processor would clearly be addressing different byte addresses in each of the 4 memory chips, but as long as all of the 17 address lines are from the processor are connected to all 17 address lines of the memory devices, the memory should work properly.

The upper address bits from the processor, A19 through A31 are used for the page selection process. These signals are routed to the address decoding logic where the appropriate $\overline{CHIP\ SELECT}$ signals are generated. These signals have been omitted from Figure 7.3. There are 13 higher order address bits in this example. This gives us 2^{13} or 8,192 pages of memory. Each page of memory holds 128K 32-bit wide words, which works out to be 2^{30} long words. In terms of addresses, each page actually contains 512 Kbytes, so the byte address range on each page goes from byte address (in HEX) 00000 through 7FFFF. Recall that in this memory scheme, there are 8,192 pages with each page holding 512 Kbytes. Thus, page addresses go from 0000 through 1FFF. It may not seem obvious to you to see how the page addresses and the offset addresses are related to reach other but if you expand the hexadecimal addresses to binary and lay them next to each other you should see the full 32-bit address.

Now, let's see how the processor deals with data sizes smaller than long words. Assume that the processor wants to read a long word from address ABCDEF64 and that this address is decoded to be on the page of Figure 7.3. Since this address is on a 32 bit boundary, A0 and A1 = 0, and are not used as part of the external address that goes to memory. However, if the processor wanted to do a word access of either one of the words located at address ABCDEF64 or ABCDEF66, it would still generate the same external address. When the data was read into the processor, the ½ of the long word that was not needed would be discarded. Since this is a READ operation, the contents of memory are not affected.

If the processor wanted to read any one of the 4 bytes located at byte address ABCDEF64, ABCDEF65, ABCDEF66 or ABCDEF67, it would still perform the same read operation as before. Again, only the byte of interest would be retained and the others would be discarded.

Now, let's consider a write operation. In this case, we are concerned about possibly corrupting memory, so we want to be sure that when we write a quantity smaller than a long word to memory, we do not accidentally write more than we intend to. So, suppose that we want to write the byte at memory location ABCDEF65. In this case, only the $\overline{WE1}$ signal would be asserted, so only that byte position could be modified. Thus, to write a byte to memory, we only activate one of the 4 WRITE ENABLE signals. To write a word to memory we would active either $\overline{WE0}$ and $\overline{WE1}$ together or $\overline{WE2}$ and $\overline{WE3}$. Finally, to write a long word, all four of the WRITE ENABLE lines would be asserted.

What about the case of a 32-bit word stored in a 16-bit memory? In this case, the 32-bit word can be stored on any even word boundary because the processor must always do two consecutive memory accesses to retrieve the entire 32-bit quantity. However, most compilers will still try to store the 32-bit words on natural boundaries (addresses divisible by 4). This is why assembly language programmers can often save a little space or speed up an algorithm by overriding what the compiler does to generate code and tweaking it for greater efficiency.

Let's get back on track. For a 32-bit processor, address bits A2...A31 are used to address the 1,073,741,824 possible long words, and A0...A1 address the four possible bytes within the long word. This gives us a total of 4,294,967,296 addressable byte locations in a 32-bit processor. In other words, we have a byte addressing space of 4 GB. A processor with a 16-bit wide data bus, such as the 68K, uses address lines A1–A23 for word addressing and A0 for byte selection.

Combining all of this, you should see the problem. You could have an 8-bit byte, a 16-bit word or a 32-bit word with the same address. Isn't this ambiguous? Yes it is. When we're programming in a high-level language, we depend upon the compiler to keep track of these messy details. This is one reason why *casting* one variable type to another can be so dangerous. When we are programming in a low-level language, we depend upon the skill of the programmer to keep track of this.

Seems easy enough, but it's not. This little fact of computer life is one of the major causes of software bugs. How can a simple concept be so complex? It's not complex, it's just ambiguous. Figure 7.4 illustrates the problem. The leftmost column of Figure 7.4 shows a string (aptly named "string") stored in an 8-bit memory space. Each ASCII character occupies successive memory locations.

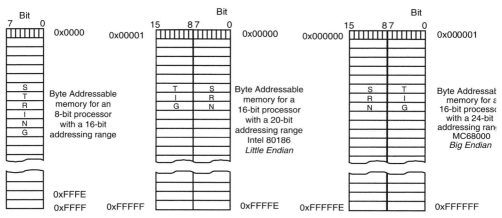

Figure 7.4: Two methods of packing bytes into 16-bit memory words. Placing the low order byte at the low order end of the word is called Little Endian. Placing the low order byte at the high order side of the word is called Big Endian.

The middle column shows a 16-bit memory that is organized so that successive bytes are stored right to left. The byte corresponding to A0 = 0 is aligned with the low order portion, DB0 . . . DB7, of the 16-bit word and the byte corresponding to A0 = 1 is aligned with the high order portion, DB8..DB15, of the 16-bit word. This is called Little Endian organization. The rightmost column stores the characters as successive bytes in a left to right fashion. The byte position corresponding to A0 = 0 is aligned with the high order portion of the 16-bit word. This is called Big Endian organization. As an exercise, consider how the bytes are stored in Figure 7.2. Are they big or little Endian?

Motorola and Intel, chose to use different endian conventions and Pandora's Box was opened for the programming world. Thus, C or C++ code written for one convention would have subtle bugs when ported to the other convention. It gets worse than that. Engineers working together on projects misinterpret specifications if the intent is one convention and they assume the other. The ARM architecture allows the programmer to establish which type of "endianess" will be used by the processor at power-up. Thus, while the ARM processor can deal with either big or little endian, it cannot dynamically switch modes once the endianess is established. Figure 7.5 shows the difference between the two conventions for a 32-bit word packed with four bytes.

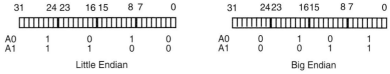

Figure 7.5: Byte packing a 32-bit word in Little Endian and Big Endian modes.

If you take away anything from this text, remember this problem because you will see it at least once in your career as a software developer.

Before you accuse me of beating this subject to death, let's look at it one more time from the hardware perspective. The whole area of memory addressing can be very confusing for novice programmers as well as seasoned veterans. Also, there can be ambiguities introduced by architectures and manufacturer's terminology. So, let's look at how Motorola handled it for the 68K and perhaps this will help us to better understand what's really going on, at least in the case of the Motorola processor, even though we have already looked at the problem once before in Figure 7.3. Figure 7.6 summarizes the memory addressing scheme for the 68K processor.

The 68K processor is capable of directly addressing 16 Mbytes of memory, requiring 24 "effective" addressing lines. Why? Because $2^{24} = 16,777,216$. In Figure 7.6 we see 23 address lines. The missing address line, A0, is synthesized by two additional control signals, $\overline{LDS}$ and $\overline{UDS}$.

For a 16-bit wide external data bus, we would normally address bit A0 to be the byte selector. When A0 is 0, we choose the even byte, and when A0 = 1, we choose the odd byte. The *endianness* of the 68K is Big Endian, so that the even byte is aligned with D8 through D15 of the data bus. Referring to figure 4.3.1 we see that there are two status bus signals coming out of the processor, designate $\overline{UDS}$, or *Upper Data Strobe*, and $\overline{LDS}$, or *Lower Data Strobe*.

When the processor is doing a byte access to memory, then either $\overline{LDS}$ or $\overline{UDS}$ is asserted to indicate to the memory which part of the word is being accessed. If the byte at the even address

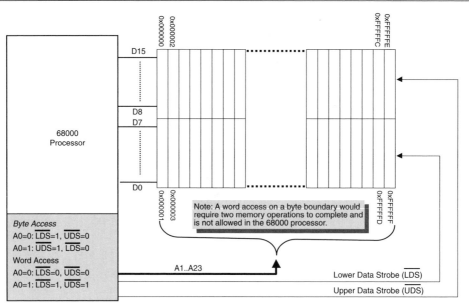

Figure 7.6: Memory addressing modes for the Motorola 68K processor

is being accessed (A0 = 0), then $\overline{\text{UDS}}$ is asserted and $\overline{\text{LDS}}$ stays HIGH. If the odd byte is being accessed (A0 = 1), then $\overline{\text{LDS}}$ is asserted and $\overline{\text{UDS}}$ remains in the HIGH, or OFF, state. For a word access, both $\overline{\text{UDS}}$ and $\overline{\text{LDS}}$ are asserted. This behavior is summarized in the table of Figure 7.6.

You would normally use $\overline{\text{LDS}}$ and $\overline{\text{UDS}}$ as gating signals to the memory control system. For example, you could the circuit shown in Figure 7.7 to control which of the bytes are being written to.

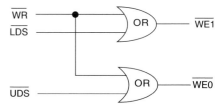

Figure 7.7: Simple circuit to control the byte writing of a 68K processor

You may be scratching your head about this circuit. Why are we using OR gates? We can answer the question in two ways. First, since all of the signals are asserted LOW, we are really dealing with the negative logic equivalent of an AND function. The gate that happens to be the negative logic equivalent of the AND gate is the OR gate, since the output is 0 if and only if both inputs are 0.

The second way of looking at it is through the equivalence equations of DeMorgan's Theorems. Recall that:

$$(\overline{A * B}) = \overline{A} + \overline{B} \quad (1)$$
$$(\overline{A + B}) = \overline{A} * \overline{B} \quad (2)$$

In this case, equation 1 shows that the OR of $\overline{A}$ and $\overline{B}$ would be equivalent to using positive logic A and B and then obtaining the NAND of the two signals.

Now, suppose that you attempted to do a word access at an odd address. For example, suppose that you wrote the following assembly language instruction:

```
move.w D0,$1001     * This is a non-aligned access!
```

This instruction tells the processor to make a copy of the word stored in internal register D0 and store the copy of that word in external memory beginning at memory address $1001. The processor would need to execute two memory cycles to complete the access because the byte order requires that it bridge the two memory locations to correctly place the bytes. Some processors are capable of this type of access, but the 68K is one of the processors that can't. If a nonaligned access occurs, the processor will generate an exception and try to branch to some user-defined code that will correct the error, or at least, die gracefully.

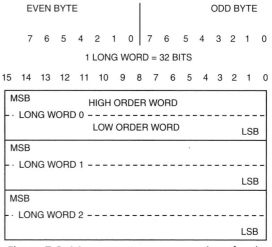

Figure 7.8: Memory storage conventions for the Motorola 68000 processor

Since the 68K processor is a 32-bit wide processor internally, but with only a 16-bit wide data bus to external memory, we need to know how it stores 32-bit quantities in external memory as well. Figure 7.8 shows us the convention for storing long words in memory.

Although Figure 7.8 may seem confusing, it is just a restatement of Figure 7.2 in a somewhat more abbreviated format. The figure tells us that 32-bit data quantities, called *longs* or *long words* are stored in memory with the most significant 16 bits (D16 – D31) stored in the first word address location and the least significant 16 bits (D0 – D15) stored in the next highest word location. Also, even byte addresses are aligned with the high order portion of the 16-bit word address (Big Endian).

Introduction to Assembly Language

The PC world is dominated by an instruction set architecture (ISA) first defined by Intel over twenty-five years ago. This architecture, called X86 because of its family members, has the following lineage:

$$8080 \rightarrow 8086 \rightarrow 80186 \rightarrow 80286 \rightarrow 80386 \rightarrow 80486 \rightarrow \text{Pentium}$$

The world of embedded microprocessors—that is microprocessors used for a single purpose inside a device, such as a cell phone—is dominated by the Motorola 680X0 ISA:

$$68000 \rightarrow 68010 \rightarrow 68020 \rightarrow 68030 \rightarrow 68040 \rightarrow 68060 \rightarrow \text{ColdFire}$$

ColdFire unites a modern processor architecture, called RISC, with backward compatibility with the original 68K ISA. (We'll study these architectures in a later lesson.) Backward compatibility is very important because there is so much 68K code still around and being used. The Motorola 68K instruction set is one of the most studied ISAs around and your can find an incredible number of hits if you do a Web search on "68K" or "68000."

Every computer system has a fundamental set of operations that it can perform. These operations are defined by the *instruction set* of the processor. The reason for a particular set of instructions is due to the way the computer is organized internally and designed to operate. This is what we would call the architecture of the computer. The architecture is mirrored by the assembly language instructions that it can execute, because these instructions are the mechanism by which we access the computer's resources. The instruction set is the atomic element of the processor. All of the complex operations are achieved by building sequences of these fundamental operations.

Computers don't read assembly language. They take their instructions in machine code. As you know, the machine code defines the entry point into the state machine microcode table that starts the instruction execution process. Assembly language is the *human-readable* form of these machine language instructions. There is nothing mystical about the machine language instructions, and pretty soon you'll be able to understand them and see the patterns that guide the internal state machines of the modern microprocessor. For now, we'll focus on the task of learning assembly language. Consider Figure 7.9.

Instead of writing a program in machine language as:	We write the program in assembly language as:		
00000412 307B7048	MOVEA.W	(TEST_S,PC,D7),A0	*We'll use address indirect
00000416 327B704A	MOVEA.W	(TEST_E,PC,D7),A1	*Get the end address
0000041A 1080	MOVE.B	D0,(A0)	*Write the byte
0000041C B010	CMP.B	(A0),D0	*Test it
000041E 67000008	BEQ	NEXT_LOCATION	*OK, keep going
00000422 1600	MOVE.B	D0,D3	*copy bad data
00000424 61000066	BSR	ERROR	*Bad byte
00000428 5248	ADDQ.W	#01,A0	*increment the address
0000042A B0C9	CMPA.W	A1,A0	*are we done?

Figure 7.9: The box on the right is a snippet of 68K code in assembly language. The box on the left is the machine language equivalent.

Note: We represent hexadecimal numbers in C or C++ with the prefix '0x'. This is a standardization of the language. There is no corresponding standardization in assembly language, and different assembler developers represent hex numbers in different ways. In this text we'll adopt the Motorola convention of using the '$' prefix for a hexadecimal number.

The machine language code is actually the output of the assembler program that converts the assembly language source file, which you write with a text editor, into the corresponding machine language code that can be executed by a 680X0 processor. The left hand box actually has two columns, although it may be difficult to see that. The left column starts with the hexadecimal memory location where the machine language instruction is stored. In this case the memory location $00000412 holds the machine language instruction code 0x307B7048. The next instruction begins at memory location 0x00000416 and contains the instruction code 0x327B704A. These two machine language instructions are given by these assembly language instructions.

```
MOVEA.W (TEST_S,PC,D7),A0
MOVEA.W (TEST_E,PC,D7),A1
```

Soon you'll see what these instructions actually mean. For now, we can summarize the above discussion this way:

- Starting at memory location $00000412, and running through location $00000415, is the machine instruction code $307B7048. The assembly language instruction that corresponds to this machine language data is **MOVEA.W (TEST_S,PC,D7),A0**

- Starting at memory location $00000416, and running through location $00000419, is the machine instruction code $327B704A. The assembly language instruction that corresponds to this machine language data is **MOVEA.W (TEST_E,PC,D7),A1**

Also, for the 68K instruction set, the smallest machine language instruction is 16-bits long (4 hex digits). No instruction will be smaller than 16-bits long, although some instructions may be as long as 5, 16-bit words long.

There is a 1:1 correspondence between assembly language instructions and machine language instructions. The assembly language instructions are called *mnemonics*. They are designed to be a shorthand clue as to what the instruction actually does. For example:

MOVE.B	*move a byte of data*
MOVEA.W	*move a word of data to an address register*
CMP.B	*compare the magnitude of two bytes of data*
BEQ	*branch to a different instruction if the result equals zero*
ADDQ.W	*add (quickly) two values*
BRA	*always branch to a new location*

You'll notice that I've chosen a different font for the assembly language instructions. This is because fonts with fixed spacing, like "courier", keep the characters in column alignment, which makes it easier to read assembly language instructions. There's no law that you must use this font, the assembler probably doesn't care, but it might make it easier for you to read and understand your programs if you do.

The part of the instruction that tells the computer what to do, is called the *opcode* (short for "operation code"). This is only one half of the instruction. The other half tells the computer how and where this operation should be performed. The actual opcode, for example MOVE.B, is actually an opcode and a modifier. The opcode is MOVE. It says that some data should be moved from one place to another. The modifier is the ".B" suffix. This tells it to move a byte of data, rather than a word or long word. In order to complete the instruction we must tell the processor:

1. where to find the data (this is called operand 1), and
2. where to put the result (operand 2).

A complete assembly language instruction must have an opcode and may have 0,1 or 2 operands. Here's your first assembly language instruction. It's called NOP (pronounced *No op*). It means do nothing. You might be questioning the sanity of this instruction but it is actually quite useful. Compilers make very good use of them. The NOP instruction is an example of an instruction that takes 0 operands.

The instruction CLR.L D4 is an example of an instruction that takes one operand. It means to clear, or set to zero, all 32 bits (the ".L" modifier) of the internal register, D4.

The instruction MOVE.W D0,D3 is an example of an instruction that takes two operands. Note the comma separating the two operands, D0 and D3. The instruction tells the processor to move 16 bits of data (the ".W" modifier) from data register D0 to data register D3. The contents of D0 are not changed by the operation. All assembly language programs conform to the following structure:

Column 1	Column 2	Column 3	Column 4 . . . .
LABEL	OPCODE	OPERAND1,OPERAND2	*COMMENT

Each instruction occupies one line of text, starting in column 1 and going up to 132 columns.

1. The LABEL field is optional, but it must always start in the first column of a line. We'll soon see how to use labels.
2. The OPCODE is next. It must be separated from the label by white space, such as a TAB character or several spaces, and it must start in column 2 or later.
3. Next, the operands are separated from the opcode by white space, usually a tab character. The two operands should be separated from each other by a comma. There is no white space between the operands and the comma.
4. The comment is the last field of the line. It usually starts with an asterisk or a semi-colon, depending upon which assembler is being used. You can also have comment lines, but then the asterisk must be in column 1.

Label

Although the label is optional, it is a very important part of assembly language programming. You already know how to use labels when you give a symbolic name to a variable or a constant. You also use labels to name functions. In assembly language we commonly use labels to refer to the memory address corresponding to an instruction or data in the program. The label must be defined in column 1 of your program. Labels make the program much more readable. It is possible to write a program without using labels, but almost no one ever does it that way. The label allows the assembler program to automatically (and correctly!) calculate the addresses of operands and destinations. For example, consider the following snippet of code in Figure 7.10.

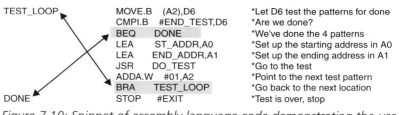

Figure 7.10: Snippet of assembly language code demonstrating the use of labels.

The code example of Figure 7.10 has two labels, TEST_LOOP and DONE. These labels correspond to the memory locations of the instructions, "MOVE.B (A2),D6" and "BRA TEST_LOOP" respectively. As the assembler program converts the assembly language instructions into machine language instructions it keeps track of where in memory each instruction will be located. When it encounters a label as an operand, it replaces the label text with the numeric value. Thus, the instruction "BEQ DONE" tells the assembler to calculate the numeric value necessary to cause the program to jump to the instruction at the memory location corresponding to the label "DONE" if the test condition, equality, is met. We'll soon see how to test this equality. If the test fails, the branch instruction is ignored and the next instruction is executed.

Comments

Before we get too far offshore, we need to make a few comments about the proper form for commenting your assembly language program. As you can see from Figure 7.10 each assembly language instruction has a comment associated with it. Different assemblers handle comments in different ways. Some assemblers require that comments that are on a line by themselves have an asterisk '*' or a semicolon ';' as the first character on the line. Comments that are associated with instructions or assembler directives might need the semicolon or asterisk to begin the comment, or they might not need any special character because the preceding white space defines the location of the comment block. The important point is that assembly language code is not self-document-ing, and it is easy for you to forget, after a day or so goes, exactly what you were trying to do with that algorithm.

Assembly code should be profusely commented. Not only for your sanity, but for the people who will have to maintain your code after you move on. There is no reason that assembly code cannot be as easy to read as a well-document C++ program. Use equates and labels to eliminate magic numbers and to help explain what the code is doing. Use comment blocks to explain what sections of an algorithm are doing and what assumptions are being made. Finally comment each instruc-tion, or small group of instructions, in order to make it absolutely clear what is going on.

In his book, *Hackers: Heroes of the Computer Revolution,* Steven Levy[1] describes the coding style of Peter Samson, an MIT student, and early programmer,

> …*Samson, though, was particularly obscure in refusing to add comments to his source code, explaining what he was doing at a given time. One well-distributed program Samson wrote went on for several hundreds of assembly language instructions, with only one com-ment beside an instruction which contained the number 1750. The comment was* RIPJSB, *and people racked their brains about its meaning until someone figured out that 1750 was the year that Bach died, and that Samson had written an abbreviation for Rest In Peace Johann Sebastian Bach.*

Programmer's Model Architecture

In order to program in assembly language, we must be familiar with the basic architecture of the processor. Our view of the architecture is called the *Programmer's Model* of the processor. We must understand two aspects of the architecture:

1. the instruction set, and
2. the *addressing modes.*

The addressing modes of a computer describe the different ways in which it accesses the operands, or retrieves the data to be operated on. Then, the addressing modes describe what to do with the data after the operation is completed. The address modes also tell the processor how to calculate the destination of a nonsequential instruction fetch, such as a branch or jump to a new location. Addressing modes are so important to the understanding of the computer we'll need to study them a bit before we can create an assembly language program. Thus, unlike C, C++ or JAVA, we'll need to develop a certain level of understanding for the machine that we're programming before we can actually write a program.

Unlike C or C++, assembly language is not portable between computers. An assembly language program written for an Intel 80486 will not run on a Motorola 68000. A C program written to run on an Intel 80486 *might be able to run* on a Motorola 68000 once the original source code is recompiled for the 68000 instruction set, but differences in the architectures, such as big and little endian, may cause errors to be introduced.

It might be worthwhile and stop for a moment to reflect on why, as a programmer, it is important to learn assembly language. Computer science and programming depends upon a working knowledge of the processor, its limitations and its strengths. To understand assembly language is to understand the computing engine that your code is running on. Even though high-level languages like C++ and JAVA do a good job of abstracting the low level details, it is important to keep in mind that the engine is not infinitely powerful and that its resources are not limitless.

Assembly language is tightly coupled to the design of the processor and represents the first level of simplification of the binary instruction set architecture into a human readable form. Generally there is a 1:1 match between the assembly language instruction and the binary or hexadecimal instruction that results. This is very different from C or C++, where one C statement, may generate hundreds of lines of assembly code.

It is still true that you can generally write the tightest code in assembly language. While C compilers have gotten pretty good, they're not perfect. A program written in assembly language will often use less space and run faster than a program with the same functionality written in C. Many games are still coded in assembly language for exactly that reason.

The need for efficient code is especially true of interrupt handlers and algorithms written for specialized processors, such as *digital signal processors (DSPs)*. Many experience programmers will argue that any code that must be absolutely deterministic cannot be written in C, because you cannot predict ahead of time the execution time for the code generated by the compiler. With assembly language you can control your program at the level of a single clock cycle. Also, certain parts of the run time environment must be written in assembly because you need to be able to establish the C runtime environment. Thus, boot-up code tends to be written in assembly. Finally, understanding assembly language is critically important for debugging real time systems. If you've ever been program in a programming environment such as Microsoft's Visual C++® and while you are debugging your code you inadvertently step into a library function, you will then find yourself knee-deep in x86 assembly language.

Motorola 68000 Microprocessor Architecture

Figure 7.11 is a simplified schematic diagram of the 68K architecture. Since Motorola has a large number of family members this particular architecture is also referred to as CPU16. CPU16 is a subset of the CPU32 architecture, the ISA for the 68020 and later processors.

Let's briefly identify some of the important functional blocks that we'll later be using.

- *Program counter:* Used to hold the address of the next instruction to be fetched from memory. As soon as the current instruction is decoded by the processor the program counter (PC) is updated to point to the address of the next sequential instruction in memory.
- *General registers:* The 68K processor has 15 general-purpose registers and two special-purpose registers. The general purpose registers are further divided into eight data registers, D0 . . . D7 and seven address registers, A0 . . . A6. The data registers are used to hold and manipulate data variables and the address registers are used to hold and manipulate memory addresses. The two special purpose registers are used to implement two separate stack pointers, A7 and A7'. We'll discuss the stack a bit later in the text.

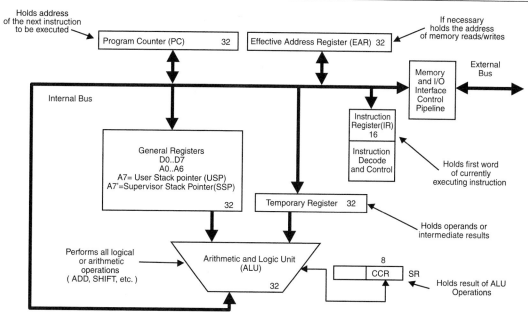

Figure 7.11: Architecture of the Motorola 68K processor.

- *Status register:* The status register is a 16-bit register. The bits are used to describe the current state of the computer on an instruction-by-instruction basis. The condition code register, CCR, is part of the status register and holds the bits that directly relate to the result of the last instruction executed.
- *Arithmetic and logic unit (ALU):* The ALU is the functional block where all of the mathematical and logical data manipulations are performed.

Figure 7.12 is the *Programmer's Model* of the 68K architecture. It differs from Figure 7.11 because it focuses only on the details of the architecture that is relevant to writing a program.

In this text, we're not going to deal with the status register (SR), or the supervisor stack pointer (A7'). Our world from now on

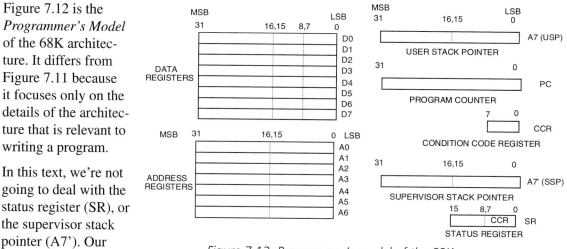

Figure 7.12: Programmer's model of the 68K processor

will focus on D0 . . . D7, A0 . . . A6, A7, CCR and the PC. From these registers, we'll learn everything that we need to know about this computer architecture and assembly language programming.

Condition Code Register

The condition code register (CCR) deserves some additional explanation. The register is shown in more detail in Figure 7.13.

	DB7	DB6	DB5	DB4	DB3	DB2	DB1	DB0
CCR				X	N	Z	V	C

Figure 7.13: Condition Code Register. The shaded bits are not used.

The CCR contains a set of five condition bits whose value can change with the result of each instruction being executed. The exact definition of each of the bits is summarized below.

- *X BIT (extend bit):* used with multi-precision arithmetic
- *N BIT (negative bit):* indicates that the result is a negative number
- *Z BIT (zero bit):* indicates that the result is equal to zero
- *V BIT (overflow):* indicates that the result may have exceeded the range of the operand
- *C BIT (carry bit):* indicates that a carry was generated in a mathematical operation

The importance of these bits resides with the family of test and branch instructions such as BEQ, BNE, BPL, BMI, and so on. These instructions test the condition of an individual *flag*, or CCR bit, and either take the branch if the condition is true, or skip the branch and go to the next instruction if the condition is false. For example, BEQ means *branch equal*. Well, equal to what? The BEQ instruction is actually testing the state of the *zero flag*, Z. If Z = 1, it means that the result in the register is zero, so we should take the branch because the result is equal to zero. So *branch equal* is an instruction which tells the processor to take the branch if there was a recent operation that resulted in a zero result.

But BEQ means something else. Suppose we want to know if two variables are equal to each other. How could we test for equality? Simple, just subtract one from the other. If we get a result of zero (Z = 1), they're equal. If they're not equal to each other we'll get a nonzero result and (Z = 0). Thus, BEQ is true if Z = 1 and BNE (branch not equal) is true if Z = 0. BPL (branch plus) is true if N = 0 and BMI (branch minus) is true if (N = 1).

There are a total of 14 conditional branch instructions in the 68K instruction set. Some test only the condition of a single flag, others test logical combinations of flags. For example, the BLT instruction (branch less than) is define by: $BLT = N * \overline{V} + \overline{N} * V$ ($N \oplus V$) (not by bacon, lettuce, and tomato).

Effective Addresses

Note: Even though the 68K processor has only 24 address lines external to the processor, internally it is still a 32-bit processor. Thus, we can represent addresses as 32-bit values in our examples.

Let's now go back and look at the format of the instructions. Perhaps the most commonly used instruction is the MOVE instruction. You'll find that most of what you do in assembly language is moving data around. The MOVE instruction takes two operands. It looks like this:

```
MOVE.W    source(EA),destination(EA)
```

For example:

```
MOVE.W    $4000AA00,$10003000
```

tells the processor to *copy* the 16-bit value stored in memory location 0x4000AA00 to memory location 0x10003000.

Also, the MOVE mnemonic is a bit misleading. What the instruction does is to overwrite the contents of the destination operand with the source operand. The source operand isn't changed by the operation. Thus, after the instruction both memory locations contain the same data as $4000AA00 did before the instruction was executed.

In the previous example, the source operand and the destination operand are addresses in memory that are exactly specified. These are *absolute addresses*. They are absolute because the instruction specifies exactly where to retrieve the data from and where to place it. Absolute addressing is just one of the possible addressing modes of the 68K. We call these addressing modes *effective addresses*. Thus, when we write the general form of the instruction:

MOVE.W   source(EA),destination(EA)

we are saying that the source and destination of the data to move will be independently determined by the effective addressing mode for each.

For example:

MOVE.W   D0,$10003000

moves the contents of one of the processor's eight internal data registers, in this case data register D0, to memory location $10003000. In this example, the source effective address mode is called *data register direct* and the destination effective address mode is absolute.

We could also write the MOVE instruction as

MOVE.W   A0,D5

which would move the 16-bit word, currently stored in address register A0 to data register D5. The source effective address is *address register direct* and the destination effective address is data register direct.

Suppose that the contents of address register A0 is $4000AA00 (we can write this in a shorthand notation as <A0> = $4000AA00). The instruction

MOVE.W   D5,(A0)

moves the contents of data register D5 to memory location $4000AA00. This is an example of *address register indirect* addressing. You are probably already familiar with this addressing mode because this is a pointer in C++. The contents of the address register become the address of the memory operation. We call this an *indirect addressing* mode because the contents of the address register are not the data we want, but rather the memory address of the data we want. Thus, we aren't storing the data directly to the register A0, but indirectly by using the contents of A0 as a pointer to its ultimate destination, memory location $4000AA00. We indicate that the address register is being used as a pointer to memory by the parentheses around the address register. Suppose that <A1> = $10003000 and <A6> = $4000AA00. The instruction

MOVE.W   (A1),(A6)

would copy the data located in memory location $10003000 to memory location $4000AA00. Both source and destination effective address modes are the address register indirect mode.

Let's look at one more example.

MOVE.W   #$234A,D2

would place the hexadecimal number, $234A directly into register D2. This is an example of the *immediate address mode*. Immediate addressing is the mechanism by which memory variables are initialized. The pound sign (#) tells the assembler that this is a number, not a memory location. Of course, only the source effective address could be an immediate address. The destination can't be a number—it must be a memory location or a register.

The effective address (EA) specifies how the operands of the instruction will be accessed. Depending upon the instruction being executed, not all of the effective addresses may be available to use. We'll discuss this more as we go. The type of effective address that will be used by the instruction to access one or more of the operands is actually specified as part of the instruction. Recall that the minimum size for an opcode word is 16-bits in length. The opcode word provides all of the information that the processor needs to execute the instructions. However, that doesn't mean that the 16-bit opcode word, by itself, contains all the information in the instruction. It *may* contain enough information to be the entire instruction, like *NOP*, but usually it only contains enough information to know what else it must retrieve, or *fetch*, from memory in order to complete the instruction. Therefore, many instructions may be longer than the op-code word portion of the instruction.

If we think about the microcode-based state machine that drives the processor, this all begins to make sense. We need the op-code word to give us our entry point into the microcode. This is what the processor does when it is decoding an instruction. In the process of executing the entire instruction, it may need to go out to memory again to fetch additional operands in order to complete the instruction. It knows that it has to do these additional memory fetches because the path through the state machine, as defined by the op-code word, predetermines it.

Consider the form of the opcode word for the MOVE instruction shown in Figure 7.14

opcode	dst EA	src EA
15 12 11	6 5	0

Figure 7.14: Machine language instruction format for the MOVE instruction

The op-code field may contain three types of information about the MOVE instruction within the field defined by the four data bits, DB15 – DB12. These possibilities are defined as follows:

- 0001 = MOVE.B
- 0011 = MOVE.W
- 0010 = MOVE.L

The source EA and destination EA are both 6-bit fields that are each further subdivided into two, 3-bit fields each. One 3-bit field, called the *register field*, can take on a value from 000 to 111, corresponding to one of the data registers, D0 through D7, and address register A0 through A7. The other 3-bit field is the *mode field*. This field describes which effective addressing mode is being used for the operation. We'll return to this point later on. Let's look at a real example. Consider the instruction

MOVE.W   #$0A55,D0

What would this instruction look like after it is assembled? It may help you to refer to an assembly language programming manual, such as the *Programmer's Reference Manual*[3] to better understand what's going on. Here's what we know. It's a MOVE instruction with the size of the data transfer being 16-bits, or one word. Thus, the four most significant bits of the opcode word (D15 – D12) are 0011. The source effective address (D5 – D0) is immediate, which decodes as 111 100. The destination effective address (D11 – D6) is data register D0, which decodes as 000 000. Putting this all together, the machine language opcode word is:

0011 000 000 111 100 , or $303C

The translation from 0011 000 000 111 100 (binary) to $303C (hex) may have seemed a bit confusing to you because the bit fields of the instruction usually don't align themselves on nice 4-bit boundaries. However, if you proceed from right to left, grouping by four, you'll get it right every time.

Is that all the elements of the instruction? No, because all we know at this point is that the source is an immediate operand, but we still don't know what the immediate number is. Thus, the opcode word tells the processor that it has to go out to memory again and fetch another part of the source operand. Since the effective addressing mode of the destination register D0, is Data Register Direct, there is no need to retrieve any additional information from memory because all the information that is needed to determine the destination is contained in the op-code word itself. Thus, the complete instruction in memory would be $303C 0A55. This is illustrated in Figure 7.15.

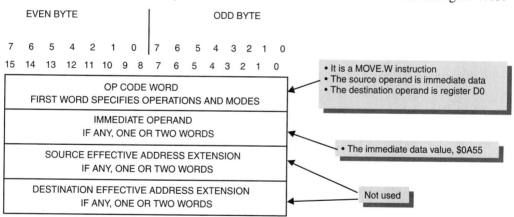

Figure 7.15: Memory storage of the instruction MOVE.W #$0A55,D0

Now, suppose that the instruction uses two absolute addresses for the source effective address and the destination effective address

MOVE.W $0A550000,$1000BB00

This machine language op-word would decode as 0011 001 111 111 001, or $33F9. Are we done? Again, the answer is no. This is because there are still two absolute addresses that need to be fetched from memory. The complete instruction looks like

$33F9 0A55 0000 1000 BB00.

The complete instruction takes up a total of five 16-bit words of memory and requires the processor to make five separate memory-read operations to completely digest the instruction. This is illustrated in Figure 7.16.

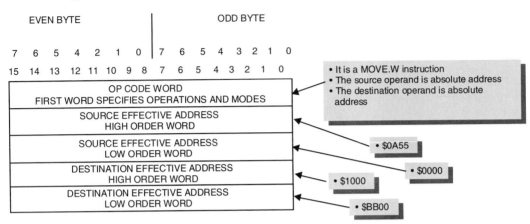

Figure 7.16: Memory storage of the instruction `MOVE.W $0A550000,$1000BB00`

Word Alignment

The last topic that we need to cover in order to prepare ourselves for the programming tasks ahead of us is that of word alignment in memory. We touched on this topic earlier in the lesson but it is worthwhile to review the concept. Recall that we can address individual bytes in a word by using the least significant address bit, A0, to indicate which byte we're interested in. However, what would happen if we tried to access a word or a long word value on an odd byte boundary? Figure 7.17 illustrates the problem.

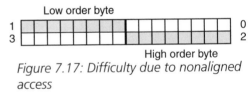

Figure 7.17: Difficulty due to nonaligned access

In order to fetch a word located on an odd-byte boundary the processor would have to make two 16-bit fetches and then discard parts of both words to correctly read the data. Some processors can do this operation. It is an example of the *nonaligned access* mode that we discussed earlier. In general, this type of access is very costly in terms of processor efficiency. The 68K cannot perform this operation and it will generate an internal exception if it encounters it. The assembler will not catch it. This is a run time error. The cure is to never program a word or long word operation on a byte (odd) memory boundary.

Reading the Programmer's Reference Manual

The most daunting task that you'll face when you set out to write an assembly language program is trying to comprehend the manufacturer's data book. The Motorola 68K *Programmer's Reference Manual* is written in what we could generously call "a rather terse style." It is a reference for people who already know how to write an assembly language program rather than a learning tool for programming students who are trying to learn the methods. From this point on you will need the *Programmer's Reference Manual,* or a good book on 68K assembly language programming, such as *Clements*[2] textbook. However, the textbook isn't particularly efficient when it comes to writing a program and needing a handy reference, such as the Programmer's Reference Manual. Fortunately, it is easy to get a free copy of the reference manual, either from Freescale, or their corporate website (see the chapter references for the URL).

So, assuming that you have the book in front of you, let's take a quick course in understanding what Motorola's technical writing staff is trying to tell you. The following text is similar to the layout of the reference pages for an instruction.

1. *Heading*:   **ADDI**   Add Immediate **ADDI**

2. *Operation:*   Immediate Data + Destination → Destination

3. *Syntax:*   ADDI #<data>,<ea>

4. *Attributes:*   Size = (Byte, Word, Long)

5. *Description:* Add the immediate data to the destination operand and store the result in the destination location. The size of the operation may be specified to be byte, word or long. The size of the immediate data matches the operation size.

6. Condition Codes:

X	N	Z	V	C
●	●	●	●	●

N   Set if the result is negative. Cleared otherwise.

Z   Set if the result is zero. Cleared otherwise.

V   Set if an overflow is generated. Cleared otherwise.

C   Set if a carry is generated. Cleared otherwise.

X   Set the same as the carry bit.

7. *Instruction format:*

	15	14	13	12	11	10	9	8	7	6	5	4	3	2	1	0
OP-CODE	0	0	0	0	0	1	1	0	Size		Mode	Effective Address				Register
Immediate Field	Word data								Byte data							
	Long data word (includes previous word)															

8. *Instruction fields:*

Size filed – Specifies the size of the operation:

DB 7	DB 6	Operation Size
0	0	Byte operation
0	1	Word operation
1	0	Long word operation

9. *Effective Address field:*

Specifies the destination operand. Only data alterable addressing modes are allowed as shown:

Addr. Mode	Mode	Register		Addr. Mode	Mode	Register
Dn	000	reg. num:Dn		(XXX).W	111	000
An	---	not allowed		(XXX).L	111	001
(An)	010	reg. num:An		#<data>	---	not allowed
(An)+	011	reg. num:An				
-(An)	100	reg. num:An				
(d_{16},An)	101	reg. num:An		(d_{16},PC)	---	not allowed
(d_8,An,Xn)	110	reg. num:An		(d_8,PC,Xn)	---	not allowed

10. *Immediate field:* (Data immediately following the op-code word)

If size = 00, then the data is the low order byte of the immediate word.

If size = 01, then the data is the entire immediate word.

If size = 10, then the data is the next two immediate words.

Let's go through this step-by-step

1.	Instruction and mnemonic: ADDI—Add Immediate
2.	Operation: Immediate Data + Destination→Destination Add the immediate data to the contents of the destination effective address and place the result back in the destination effective address.
3.	Assembler syntax: ADDI #(data),<ea> This explains how the instruction is written in assembly language. Notice that there is a special opcode for an immediate add instruction, ADDI rather than ADD, even though you still must insert the # sign in the source operand field.
4.	Attributes: Byte, word or long word (long) You may add a byte, word or long word. If you omit the .B, .W or .L modifier the default will be .W.
5.	Description: Tells what the instruction does. This is your best hope of trying to understand what the instruction is all about since it is the only real English grammar on the page.
6.	Condition codes: Lists the condition codes (flags) that are affected by the instruction.
7.	Instruction format: How the machine language instruction is created. Note that there is only one effective address for this instruction.
8.	Effective Address Field: The effective address modes that are permitted for this instruction. It also shows the mode and register values for the effective address. Note that an address register is not allowed as the destination effective address, nor is an immediate data value. Also notice that the two addressing modes involving the Program Counter, PC, are also not allowed to be used with this instruction.

Flow Charting

Assembly language program, due to their highly structured nature, are good candidates for using flow charts to help plan the structure of a program. In general, flow charts are not used to plan programs written in a high-level language such as C or C++, however, flow charts are still very useful for assembly language program planning.

In creating a flow chart, we use a rectangle to represent an "operation." The operation could be one instruction, several instructions, or a subroutine (function call). The rectangle is associated with doing something. We use a diamond to represent a decision point (program flow control), and use arrows to represent program flow. Figure 7.18 is a simple flow chart example of a computer starting up a program.

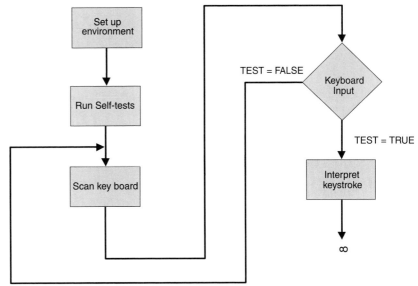

Figure 7.18: Flow chart for a simple program to initialize a system.

The flow chart starts with the operation "set-up environment." This could be a few instructions or tens of instructions. Once the environment is established the next operation is to "run the self-tests." Again, this could mean a few instructions or a lot of instructions, we don't know. It depends upon the application. The decision point is waiting for a keyboard input. The program scans the keyboard and then tests to see if a key has been struck. If not, it goes back (loops) and tries again. If a key has been struck, the program then interprets the keystroke and moves on.

Flow-charting is a very powerful tool for helping you to plan your program. Unfortunately, most students (and many professional programmers) take the "code hacking" approach and just jump in and immediately start to write code, much like writing a novel. This probably explains why most programmers wear sandals rather than basketball sneakers with the laces untied, but that's another issue entirely.

Writing an Assembly Language Program

Remember Leonardo DiCaprio's famous line in the motion picture *Titanic,* "I'm king of the world!" In assembly language programming you are the absolute monarch of the computer world. There is nothing that can stop you from making incredible coding blunders. There are no type checks or compiler warnings. Just you and the machine, *mano a mano.* In order to write a program in assembly language, you must be continuously aware of the state of the system. You must keep in mind how much memory you have and where it is located; the locations of your peripheral devices and how to access them. In short, you control everything and you are responsible for everything.

For the purposes of being able to write and execute programs in 68K assembly language, we will not actually use a 68K processor. To do that you would need a computer that was based on a 68K family processor, such as the original Apple MacIntosh® We'll approach it in a different way. We'll use a program called an *instruction set simulator (ISS)* to take the place of the 68K processor.

Commercially available ISSs can cost as much as $10,000 but ours is free. We're fortunate that two very good simulators have been written for the 68000 processor. The first simulator, developed by Clements[4] and his colleagues at the University of Teesside, in the United Kingdom, may be downloaded from his website.

Kelley[5] developed a second simulator, called *Easy68K*. This is a much newer simulator and has extensive debugging support that is lacking in the Easy68K version. Both simulators were designed to run under the Windows® operating system Also, Easy68K was compiled for 32-bit Windows, so it has a much better chance of running under Windows 2000, XP and so on, although, the Teesside simulator seems to be reasonably well-behaved under the more modern versions of Windows.

The simulators are closer to integrated design environments (IDE). They include a text editor, assembler, simulator and debugger in a package. The simulators are like a computer and debugger combined. You can do many of the debugger operations that you are use to, such as

- peek and poke memory
- examine and modify registers
- set breakpoints
- run from or to breakpoints
- single-step and watch the registers

In general, the steps to create and run an assembly language program are simple and straight forward.

1. Using your favorite ASCII-only text editor, or the one included with the ISS package, create your source program and save it in the old DOS 8.3 file format. Be sure to use the extension .x68. For example, save your program as *myprog.x68*. The editor that comes with the program will automatically append the x68 suffix to your file.

2. Assemble the program using the included assembler. The assembler will create an absolute machine language file that you may run with the simulator. If your program assembles without errors, you'll then be able to run it in the simulator. The assembler actually creates two files. The absolute binary file and a *listfile*. The listfile shows you your original source program and the hexadecimal machine language file it creates. If there are any errors in your source file, the error will be shown on the listfile output.

In the Teesside assembler, the assembly language program runs on a simulated computer with 1M byte of memory occupying the virtual address range of $00000 . . . $FFFFF. The *Easy68K* simulator runs in a full 16M address space. The Teesside simulator package is not without some minor quirks, but in general, it is well-behaved when running under Windows.

To program in assembly language you should already be a competent programmer. You should have already had several programming classes and understand programming constructs and data structures. Assembly language programming may seem very strange at first, but it is still programming. While C and C++ are free form languages, assembly language is very structured. Also, it is up to you to keep track of your resources. There is no compiler available to do resource allocation for you.

Don't be overwhelmed by the number of opcodes and operands that are part of the 68K, or any processor's ISA. The real point here is that you can write fairly reasonable programs using just a subset of the possible instructions and addressing modes. Don't be overwhelmed by the number of instructions that you might be able to use. Get comfortable writing programs using a few instructions and addressing modes that you understand and then begin to integrate other instructions and addressing modes when you need something more efficient, or just want to expand your repertoire.

As the programming problems that you will be working on become more involved you'll naturally look for more efficient instructions and effective addressing modes that will make your job easier and result in a better program. As you'll soon discover when we look at different computer architectures, very few programmers or compilers make use of all of the instructions in the instruction set. Most of the time, a small subset of the instructions will get the job done quite nicely. However, every once in a while, a problem arises that was just made for some obscure instruction to solve. Happy coding.

Pseudo Opcodes

There are actually two types of opcodes that you can use in your programs. The first is the set of opcodes that are the actual instructions used by the 68000 processor. They form the 68K ISA. The second set is called *pseudo-ops*, or pseudo opcodes. These are placed in your program just like real opcodes, but they are really instructions to the assembler telling it how to assemble the program. Think of pseudo-ops just as you would think of compiler directives and #define statements.

Pseudo-ops are also called *assembler directives*. They are used to help make the program more readable or to provide additional information to the assembler about how you want the program handled. It is important to realize that commercially available, industrial-strength assemblers are every bit as complex as any modern compiler. Let's look at some of these pseudo-opcodes.

- **ORG** *(set origin):* The ORG pseudo-op tells the assembler where in memory to start assembling the program. This is not necessarily where the program might be loaded into memory, it is only telling the assembler where you intend for it to run. If you omit the **ORG** statement, the program will be assembled to run starting at memory location $00000. Since the memory range from $00000..$003FF is reserved for the system vectors, we will generally "**ORG**" our program to begin at memory address $00400. Therefore, the first line of you program should be

```
<label>    ORG    $400    <*comment>
```

The next pseudo-op directive is placed at the end of your source file. It has two functions. First, it tells the assembler to stop assembling at this point and, second, it tells the simulator where to load the program into memory.

- **END** *(end of source file):* Everything after END is ignored. Format:

```
<no label>    END    <address>
```

Note that the **ORG** and **END** directives are complementary. The **ORG** directive tells the assembler how to resolve address references; in effect, where you intend for the program to run. The **END** directive instructs the loader program (part of the simulator) where to place the program code in memory. Most of the time the addresses of **ORG** and **END** will be the same, but they don't have to be. It is quite possible for a program to be loaded in one place and then relocated to another when it is time to run. For our purposes, you would generally start your program with:

ORG $400   and end it with: END $400

- **EQU** *(equate directive):* The equate pseudo-op is identical to the #define in C. It allows you to provide a symbolic name for a constant value. The format is

```
<label>    EQU    <expression>    <*comment>
```

The expression may be a mathematical expression, but most likely it will just be a number. You may also use the equate directive to create a new symbolic value from other symbolic values. However, the values of the other symbols must be known at the time the new symbol is being evaluated. This means that you cannot have forward references.

The equate directive, like the "#define" directives in C and C++, are instruction to the C assembler to substitute the numeric value for the symbolic name in your source file. For example,

```
Bit0_test    EQU    $01    * Isolate data bit 0

ANDI.B    #Bit0_test,D0    * Is bit 0 in D0 = 1?
```

will be more meaningful to you than:

```
ANDI.B    #$01,D0    * Magic numbers
```

especially after you haven't looked at the code for several days.

- **SET** *(set symbol):* *SET* is like *EQU* except that set may be used to redefine a symbol to another value later on. The format is

```
<label>    SET    <expression>
```

Data Storage Directives

The next group of pseudo-ops is called *data storage directives*. Their purpose is to instruct the assembler to allocate blocks of memory and perhaps initialize the memory with values.

- **DC** *(define constant):* Creates a block of memory containing the data values listed in the source file. The format is:

```
<label>    DC.<SIZE>    <item>,<item>,....
```

Example

```
error_msg    DC.B    'Error 99',$0D,$0A,$00    *error message
```

A text string that is placed inside of single quotation marks is interpreted by the assembler as a string of ASCII characters. Thus, this directive is equivalent to writing

```
error_msg    DC.B    $45,$72,$72,$6F,$72,$20,$39,$39,$0D,$0A,$00
```

As you can see, using the single quotes makes your intention much more understandable. In this case, the memory location associated with the label, *error_msg*, will contain the first byte of the string, $45.

- **DCB** *(define constant block):* Initialize a block of memory to the same value. The length is the number of bytes, word or long words. The format is

<label>   DCB.<size>   <length>,<value>

- **DS** *(define storage):* Generates an un-initialized block of memory. Use this if you need to define a storage area that you will later use to store data. The format is:

<label>   DS.<size>   <length>

- **OPT** *(set options):* Tells the assembler how you want the assembler to create your program code in places where you have not explicitly directed it and on how you want the listfile formatted.

The only option that is worth noting here is the **CRE** option. This option tells the assembler to create a list of cross-references on the listfile. This is an invaluable aid when it comes time to debug your program. For example, examine the block of code in case 1.

CASE 1: Listfile without the CRE option

```
Source file: EXAMPLE.X68
Defaults: ORG $0/FORMAT/OPT A,BRL,CEX,CL,FRL,MC,MD,NOMEX,NOPCO
 1
 2 ***********************
 3 *
 4        * This is an example of
 5        * not using cross-references
 6 *
 7 ***********************
 8 00000400 ORG $400
 9
10 00000400 103C0000    START:      MOVE.B #00,D0
11 00000404 66FA        TEST:       BNE     START
12 00000406 B640        COMPARE:    CMP.W   D0,D3
13 00000408 6BFC        WAIT:       BMI     COMPARE
14 00000400                         END     $400
Lines: 14, Errors: 0, Warnings: 0.
```

CASE 2: Listfile with the CRE option set

```
Source file: EXAMPLE.X68
Defaults: ORG $0/FORMAT/OPT A,BRL,CEX,CL,FRL,MC,MD,NOMEX,NOPCO
1
2 *******************
3 *
4 * This is an example of using cross-
```

```
 5 * references
 6 ********************
 7
 8                            OPT  CRE
 9 00000400                   ORG  $400
11 00000400 103C0000 START:   MOVE.B    #00,D0
12 00000404 66FA     TEST:    BNE       START
13 00000406 B640     COMPARE: CMP.W     D0,D3
14 00000408 6BFC     WAIT:    BMI       COMPARE
15 00000400                   END       $400
Lines: 15, Errors: 0, Warnings: 0.

SYMBOL TABLE INFORMATION
Symbol-name Type Value Decl Cross reference line numbers
COMPARE LABEL 00000406 13 14.
START LABEL 00000400 11 12.
TEST LABEL 00000404 12 * * NOT USED * *
WAIT LABEL 00000408 14 * * NOT USED * *
```

Notice how the **OPT CRE** directive creates a symbol table of the labels that you've defined in your program, their value, the line numbers where they are first defined and all the line numbers that refer to them. As you'll soon see, this will become an invaluable aid for your debugging process.

Analysis of an Assembly Language Program

Suppose that we want to implement the simple C assignment statement, $Z = Y + 24$, as an assembly language program. What would the program look like? Let's examine the following simple program. Assume that $Y = 27$. Here's the program.

```
ORG       $400                *Start of code
          MOVE.B    Y,D0      *Get the first operand
          ADDI.B    #24,D0    *Do the addition
          MOVE.B    D0,Z      *Do the assignment
          STOP      #$2700    *Tell the simulator to stop
          ORG       $600      *Start of data area
Y         DC.B      27        *Store the constant 27 in memory
Z         DS.B      1         *Reserve a byte for Z
          END       $400
```

Notice that when we use the number 27 without the "$" preceding it, the assembler interprets it as a decimal number. Also notice the multiple **ORG** statements. This defines a code space at address $400 and a data space at address $600. Here's the assembler listfile that we've created. The comments are omitted.

```
1 00000400                    ORG       $400
2 00000400    103900000600    MOVE.B    Y,D0
3 00000406    06000018        ADDI.B    #24,D0
```

```
4 0000040A      13C000000601     MOVE.B      D0,Z

5 00000410      4E722700         STOP        #$2700

6 * Comment line

7 00000600                       ORG         $600

8 00000600      1B          Y:   DC.B        27

9 00000601      00000001    Z:   DS.B        1

10 00000400                      END         $400
```

Let's analyze the program line by line:

Line 1: The **ORG** statement defines the starting point for the program.

Line 2: **MOVE** the byte of data located at address $600 (Y) into the data register D0. This moves the value 27 ($1B) into register D0.

Line 3: Adds the byte number 24 ($18) to the contents of D0 and store the result in D0.

Line 4: Move the byte of data from D0 to memory location $601.

Line 5: This instruction is used by the simulator to stop program execution. It is an artifact of the ISS and would not be in most real programs.

Line 6: Comment

Line 7: The **ORG** statement resets the assembler instruction counter to begin counting again at address $600. This effectively defines the data space for the program. If no **ORG** statement was issued, the data region would begin immediately after the **STOP** instruction.

Line 8: Defines memory location $600 with the label "Y" and initializes it to $1B. Note that even though we use the directive **DC**, there is nothing to stop us from writing to this memory location and changing its value.

Line 9: Defines memory location $601 with the label "Z" and reserves 1 byte of storage. There is no need to initialize it because it will take on the result of the addition.

Line 10: **END** directive. The program will load beginning at address $400.

Notice that our actual program was only three instructions long, but two of the instructions were **MOVE** instructions. This is fairly typical. You'll find that most of your program code will be used to move variables around, rather than do any operations on them.

Summary of Chapter 7

- How the memory of a typical computer system may be organized and addressed
- How byte addressing is implemented and the ambiguity of the Big Endian versus Little Endian methods of byte addressing
- The basic organization of assembly language instructions and how the opcode word is interpreted as part of a machine language instruction
- The fundamentals of creating an assembly language program
- The use of pseudo opcodes to control the operation of the assembler program
- Analyzing a simple assembly language program.

Chapter 7: *Endnotes*

[1] Steven Levy, *Hackers: Heroes of the Computer Revolution,* ISBN 0-385-19195—2, Anchor Press/Doubleday, Garden City, 1984, p. 30.

[2] Alan Clements, *68000 Family Assembly Language,* ISBN 0-534-93275-4, PWS Publishing Company, Boston, 1994.

[3] Motorola Corporation, *Programmer's Reference Manual,* M68000PM/AD REV 1. This is also available on-line at: http://e-www.motorola.com/files/archives/doc/ref_manual/M68000PRM.pdf.

[4] Alan Clements, http://a.clements@uk.ac.tees.

[5] Charles Kelley, http://www.monroeccc.edu/ckelly/tools68000.htm.

Exercises for Chapter 7

1. Does the external and internal widths of a processor's data bus have to be the same? Discuss why they might differ.

2. Explain why the Address bus and Data bus of a microprocessor are called homogeneous and the Status bus is considered to be a heterogeneous bus.

3. All of the following instructions are either illegal or will cause a program to crash. For each instruction, briefly state why the instruction is in error.
   a. MOVE.W $1000,A3
   b. ADD.B   D0,#$A369
   c. ORI.W   #$55AA007C,D4
   d. MOVEA.L   D6,A8
   e. MOVE.L $1200F7,D3

4. Briefly explain in a few sentences each of the following terms or concepts.
   a. Big Endian/Little Endian:
   b. Nonaligned access:
   c. Address bus, data bus, status bus:

5. Consider the following 68000 assembly language program. What is the *byte value* stored in memory location $0000A002 after the program ends?

```
* System equates

foo       EQU          $AAAA
bar       EQU          $5555
mask      EQU          $FFFF
start     EQU          $400
memory    EQU          $0000A000
plus      EQU          $00000001
magic     EQU          $2700


* Program starts here

ORG       start
          MOVE.W       #foo,D0
```

```
        MOVE.W      #bar,D7
        MOVEA.L     #memory,A0
        MOVEA.L     A0,A1
        MOVE.B      D0,(A0)
        ADDA.L      #plus,A0
        MOVE.B      D7,(A0)
        ADDA.L      #plus,A0
        MOVE.W      #mask,D3
        MOVE.W      D3,(A0)
        MOVE.L      (A1),D4
        SWAP        D4
        MOVE.L      D4,D6
        MOVE.L      D6,(A1)
        STOP        #magic
        END         start
```

6. Examine the code fragment listed below. What is the longword value stored in memory location $4000?

```
Start   LEA         $4000,A0            *Initialize A0
        MOVE.L      #$AAAAFFFF,D7       *Initialize D7 and D6
        MOVE.L      #$55550000,D6
        MOVE.W      D7,D0               *Load D0
        SWAP        D6                  *Shell game
        SWAP        D0
        MOVE.W      D6,D0               *Load D0
        MOVE.L      D0,(A0)             *Save it
```

7. Examine each of the following Motorola 68,000 assembly language instructions. Indicate which instructions are correct and which are incorrect. For those that are incorrect, write brief explanation of why they are incorrect.

```
    a.  MOVE.L      D0,D7
    b.  MOVE.B      D2,#$4A
    c.  MOVEA.B     D3,A4
    d.  MOVE.W      A6,D8
    e.  AND.L       $4000,$55AA
```

8. Four bytes of data are located in successive memory locations beginning at $4000. Without using any additional memory locations for temporary storage, reverse the order of the bytes in memory.

9. Examine the following snippet of 68000 assembly language code. What are the *word* contents of D0 after the ADDQ.B instruction? Hint, all logical operations are bit by bit.

```
MOVE.W      #$FFFF,D1
MOVE.W      #$AAAA,D0
EOR.W       D1,D0
ADDQ.B      #01,D0          *What are the word contents of D0?
```

10. Consider the following snippet of assembly code. What is the longword value stored in register D1 at the conclusion of the fourth instruction?

```
START       MOVE.L        #$FA865580,D0
            MOVE.L        D0,$4000
            LSL.W         $4002
            MOVE.L        $4000,D1            * <D1> = ?
```

11. This problem is designed to expose you to the assembler and simulator. At the end of this introduction is a sample program. You should create an assembly source file and then assemble it without any errors. Once it assembles properly, you should then run it in the simulator of your choice. It is best to single-step the simulator from the starting point of the program. The ultimate objective of the exercise is to answer this question, "What is the WORD VALUE of the data in memory location $4000 when the program is just about to loop back to the beginning and start over again?"

```
*******************************************************************
*
*  My first 68000 Assembly language program
*
*******************************************************************
*  Comment lines begin with an asterisk
*  Labels, such as "addr1" and "start", if present, must begin in column
*  1 of a line.
*  OP Codes, such as MOVE.W or Pseudo OP Codes, such as EQU, must begin
*  in column two or later.
*  Watch out for comments, if the text spills over to the next line and
*  you forget to use an asterisk, you'll get an assembler error.


*******************************************************************
*
*  Beginning of EQUates section, just like #define in C
*
*******************************************************************
addr1       EQU        $4000
addr2       EQU        $4001
data2       EQU        $A7FF
data3       EQU        $5555
data4       EQU        $0000
data5       EQU        4678
data6       EQU        %01001111
data7       EQU        %00010111


*******************************************************************
*
*  Beginning of code segment. This is the actual assembly language
```

```
* instructions.
*
* * * * * * * * * * * * * * * * * * * * * * * * * * * * * * * * * * * * * * * * * * * * * * * * * * * * * * * * * * * * * *

              ORG      $400         * Program starts at $400
   start      MOVE.W   #data2,D0    * Load D0
              MOVE.B   #data6,D1    * Load D1
              MOVE.B   #data7,D2    * load D2
              MOVE.W   #data3,D3    * load D3
              MOVEA.W  #addr1,A0    * load address register
              MOVE.B   D1,(A0)+     * transfer byte to memory
              MOVE.B   D2,(A0)+     * transfer second byte
              MOVEA.W  #addr1,A1    * load address
              AND.W    D3,(A1)      * Logical AND

   *Stop here. The next instruction shows how a label is used

              JMP    start         * Program loops forever
              END    $400          * Stop assembly here
```

Comments on the program:

1. The "EQU", "END $400" and "ORG $400" instructions are the pseudo-op instructions. They are instructions for the assembler, they are not 68000 machine instructions and do not generate any 68000 code.

2. Also note that the default number system is decimal. To write a binary number, such as 00110101 you would write it with the percent sign, %00110101. Hexadecimal numbers, such as 72CF, are written with the dollar sign, $72CF.

3. Notice that every instruction is commented.

4. Look at the code. Notice that most of the instructions are just moving data around. This is very common in assembly language programming.

5. The instruction, **AND.W D3,(A1)** is an example of another address mode called *indirect* addressing. In C++ we would call this a pointer. The parentheses around A1 mean that the value contained in the internal register A1 should be considered to be a memory address, and not a data value. This instruction tells the computer to take the 16-bit quantity stored in register D3 and do a logical AND with the 16-bit value stored in memory at the address pointed to by the A1 register and then put the results back into the memory location pointed to by A1.

6. The instruction, **MOVE.B D1,(A0)+**, is an example of *indirect addressing with post increment.* The instruction moves a byte of data (8-bits) from internal data register D1 to the memory location pointed to by the contents of address register A0. In this way it is similar to the previous instruction. However, once the instruction is executed, the contents of A0 are incremented by one byte, so A0 would be pointing to the next byte of memory.

CHAPTER 8

Programming in Assembly Language

Objectives

When you are finished with this lesson, you will be able to:

▶ *Deal with negative and real numbers as they are represented in a computer;*

▶ *Write programs with loop constructs;*

▶ *Use the most common of the addressing modes of the 68000 ISA to write assembly language programs;*

▶ *Express the standard program control constructs of the C++ language in terms of their assembly language equivalents;*

▶ *Use the program stack pointer register to write programs containing subroutine calls.*

Introduction

In Chapter 7, we were introduced to the basic structure of assembly language programming. In this lesson we'll continue our examination of the addressing modes of the 68K and then examine some more advanced assembly language programming constructs, such as loops and subroutines with the objective of trying to use as much of these new methods as possible in order to improve our ability to program in assembly language.

Since you already understand how to write programs in high-level languages, such as C, C++ and Java, we'll try to understand the assembly language analogs of these programming constructs in terms of the corresponding structures that you're already familiar with, such as DO, WHILE, and FOR loops. Remember, any code that you write in C or C++ will ultimately be compiled down to machine language, so there must be an assembly language analog for each high-level language construct.

One of the best ways to understand assembly language programming is to closely examine an assembly language program. It actually doesn't help very much to work through the various instructions themselves, because once you're familiar with the Programmer's Model of the 68K and the addressing modes, you should be able find the instruction that you need in the reference manual and develop your program from there. Therefore, we'll focus our efforts on trying to understand how programs are designed and developed, and then look at several example programs to try to understand how they work.

Numeric Representations

Before we move deeper into the mysteries of assembly language we need to pick-up a few loose ends. In particular, we need to examine how different types of numbers are stored within a computer. Up to now, we've only looked at positive number from the perspective of the range of the number in terms of the number of binary bits we have available to represent it.

The largest positive number that you can represent is given by:

$$2^N - 1$$

where N is the number of binary bits.

Thus, we can represent a range of positive integers as follows:

Number of binary bits	Maximum value	Name	C equivalent
4	15	nibble	----
8	255	byte	char
16	65,535	word	short
32	4,294,967,295	long	int
64	18,446,744,073,709,551,616	long long	----

Obviously, there are two missing numeric classes: negative numbers and real numbers. Let's look at negative numbers first; then we'll move onto real numbers.

It may seem strange to you but there are no circuits inside of the 68K that are used to subtract one number from another. Subtraction, however, is very important, not only because it is one of the four commonly used arithmetic operations (add, subtract, multiply, and divide), but it is the basic method to use to determine if one value is equal to another value, greater than the other value, or less than the other value. These are all comparison operations. The purpose of a comparison instruction, such as CMP or CMPA, is to set the appropriate flags in the condition control register (CCR), but not change the value of either of the quantities used in the comparison. Thus, if we executed the instruction:

```
test    CMP.W  D0,D1          * Does <D0> = <D1> ?
```

the zero bit would be set to 1 if <D0> = <D1> because this is equivalent to subtracting the two values. If the result of the subtraction is equal to zero, then the two quantities must be equal. If this doesn't click with you right now, that's OK. We'll discuss the CCR in much more detail later on in this lesson.

In order to subtract two numbers in the 68K architecture it is necessary to convert the number being subtracted to a negative number and then the two numbers may be added together. Thus:

$$A - B = A + (-B)$$

The 68K processor always assumes that the numbers being added together may be positive or negative. The algorithm that you write must utilize the carry bit, C, the negative bit, N, the overflow bit, V, and the extended bit, X, to decide upon the context of the operation. The processor also uses negative numbers to branch backwards. It does this by adding a negative number to the contents of the program counter, PC, which effectively causes a backwards branch in the program flow. Why this is so deserves a bit of a diversion. Recall that once an op-code word is decoded the computer knows how many words are contained in the instruction currently being executed. One of the first things it does is to advance the value in the PC by that amount so that the value contained in the PC is now the address of the next instruction to be executed.

Branch instructions cause the processor to execute an instruction out-of-sequence by simply adding the operand of the branch instruction to the value in the PC. If the operand, called a *displacement*, is positive, the branch is in the forward direction. If the displacement is a negative number, the branch is backwards.

In order to convert a number from positive to negative we convert it to it's *two's complement* representation. This is a two-step process.

Step 1: Complement every bit of the byte, word or long word. To complement the bits, you change every 1 to a 0 and every 0 to a 1. Thus, the complement of $55 is $AA because Complement(01010101) = 10101010. The complement of 00 if $FF, and so on.

Step 2: Add 1 to the complement. Thus, –$55 = $AB.

Two's complement is a version a method of subtraction called *radix complement*. This method of subtraction will still work properly if you are working base, 2, 8, 10, 16 or any other base system. It has the advantage of retaining a single method of arithmetic (addition) and dealing with subtraction by converting the subtrahend to a negative number.

We define the radix complement of a number, N, represented in a particular base (radix) containing d digits as follows:

radix complement = $r^d - N$ for nonzero values of N and = 0 if N = 0. In other words, –0 = 0

Let's see how this works for the decimal number 4,934. Thus, $r^d - N$ = $10^4 - 4,934 = 5,066$

So 5,066 equals –4,934. Let's subtract 4,934 from 8013 and see if it actually works. First, we'll just subtract the numbers in the old fashioned way:

$$8,013 - 4,934 = 3,079$$

Now, we'll add together 8,013 and the radix complement of 4,934:

$$8,013 + 5,066 = 13,079 = >1\ 3079$$

Clearly, the least significant 4 digits are the same, but we're left with a 1 in the most significant digit position. This is an artifact of the radix complement method of subtraction and we would discard it. Why we are left with this number can be seen if we just multiply out what we've done.

If x = a – b, then $x = a - (r^d - b) = r^d + (a - b)$. Thus, we are left with a numeric result in two parts: the radix to the power of the number of digits and the actual result of the subtraction. By discarding the radix to the power portion of the number, we are left with answer we want. Thus, two's complement arithmetic is just radix complement arithmetic; for the base two number system.

In order to convert a negative number back to its positive representation, perform the two's complement transformation in exactly the same way. The most significant bit of the number is often called the *sign bit* because a negative number will have a 1 in the most significant bit position, but it isn't strictly a sign bit, because the negative of +5 isn't –5.

In any case, this is a minor point because the N flag is always set if result of the operation places a 1 in the most significant bit position for that byte, word or long word. Examine the following snippet of code.

Code Example

```
        ORG        $400
start   MOVE.B     #00,D0
        MOVE.B     #$FF,D0
        MOVE.W     #$00FF,D0
        MOVE.W     #$FFFF,D0
        MOVE.L     #$0000FFFF,D0
        MOVE.L     #$FFFFFFFF,D0
        END        $400
```

Tracing the code in a simulator shows that the N bit is set each time the most significant bit is a 1 for the operation being performed. Thus, for a byte operation, DB7 = 1. For a word operation, DB15 = 1. For a long word operation, DB31 = 1.

Below is the simulator trace for this program. The N bit and the contents of register D0 are highlighted in gray. Also notice that the simulator represents numbers as positive or negative as well. Thus, $FF is shown as –1 in instruction 2, but is shown in the register as $FF.

By creating two's complement negative numbers, all arithmetic operations are converted to addition. However, when we use negative numbers our range is cut in half. Thus,

1. Range of 8-bit number is –128 to +127 (zero is positive)
2. Range of 16-bit numbers is –32,768 to +32,767
3. Range of 32-bit numbers is –2,147,483,648 to +2,147,483,647

```
PC=000400  SR=2000  SS=00A00000  US=00000000  X=0
A0=00000000 A1=00000000 A2=00000000 A3=00000000  N=0
A4=00000000 A5=00000000 A6=00000000 A7=00A00000  Z=0    Program start
D0=00000000 D1=00000000 D2=00000000 D3=00000000  V=0
D4=00000000 D5=00000000 D6=00000000 D7=00000000  C=0
---------->MOVE.B #0,D0


PC=000404  SR=2004  SS=00A00000  US=00000000  X=0
A0=00000000 A1=00000000 A2=00000000 A3=00000000  N=0
A4=00000000 A5=00000000 A6=00000000 A7=00A00000  Z=1    After instruction
D0=00000000 D1=00000000 D2=00000000 D3=00000000  V=0    MOVE.B #0,D0
D4=00000000 D5=00000000 D6=00000000 D7=00000000  C=0
---------->MOVE.B #-1,D0


PC=000408  SR=2008  SS=00A00000  US=00000000  X=0
A0=00000000 A1=00000000 A2=00000000 A3=00000000  N=1
A4=00000000 A5=00000000 A6=00000000 A7=00A00000  Z=0    After instruction
D0=000000FF D1=00000000 D2=00000000 D3=00000000  V=0    MOVE.B #-1,D0
D4=00000000 D5=00000000 D6=00000000 D7=00000000  C=0
---------->MOVE.W #255,D0
PC=00040C  SR=2000  SS=00A00000  US=00000000  X=0
A0=00000000 A1=00000000 A2=00000000 A3=00000000  N=0
A4=00000000 A5=00000000 A6=00000000 A7=00A00000  Z=0    After instruction
```

```
D0=000000FF D1=00000000 D2=00000000 D3=00000000 V=0   MOVE.W #255,D0
D4=00000000 D5=00000000 D6=00000000 D7=00000000 C=0
---------->MOVE.W #-1,D0


PC=000410 SR=2008 SS=00A00000 US=00000000 X=0
A0=00000000 A1=00000000 A2=00000000 A3=00000000 N=1
A4=00000000 A5=00000000 A6=00000000 A7=00A00000 Z=0   After instruction
D0=0000FFFF D1=00000000 D2=00000000 D3=00000000 V=0   MOVE.W #-1,D0
D4=00000000 D5=00000000 D6=00000000 D7=00000000 C=0
---------->MOVE.L #65535,D0


PC=000416 SR=2000 SS=00A00000 US=00000000 X=0
A0=00000000 A1=00000000 A2=00000000 A3=00000000 N=0
A4=00000000 A5=00000000 A6=00000000 A7=00A00000 Z=0   After instruction
D0=0000FFFF D1=00000000 D2=00000000 D3=00000000 V=0   MOVE.L #65535,D0
D4=00000000 D5=00000000 D6=00000000 D7=00000000 C=0
---------->MOVE.L #-1,D0


PC=00041C SR=2008 SS=00A00000 US=00000000 X=0
A0=00000000 A1=00000000 A2=00000000 A3=00000000 N=1
A4=00000000 A5=00000000 A6=00000000 A7=00A00000 Z=0   After instruction
D0=FFFFFFFF D1=00000000 D2=00000000 D3=00000000 V=0   MOVE.L #-1,D0
D4=00000000 D5=00000000 D6=00000000 D7=00000000 C=0
```

What happens when your arithmetical operation exceeds the range of the numbers that you are working with? As software developers I'm sure you've seen this bug before. Here's a simple example which illustrates the problem. In the following C++ program *bigNumber* is a 32-bit integer, initialized to just below the maximum value positive value for an integer, +2,147,483,647. As we loop and increment the number 10 times, we will eventually exceed the maximum allowable value for the number. We can see the result of the overflow in the output box.

Unfortunately, unless we write an error handler to detect the overflow, the error will go undetected. We will have the same problem with two's complement arithmetic.

Two's complement notation makes the hardware easier but developing the correct mathematical algorithms can be a minefield for the unwary programmer! Algorithms for arithmetic operations must carefully set and use the condition flags in order to insure that the operation is correct.

For example, what happens when the sum of an addition of two negative numbers (bytes) is less than –128? You need to consider the state of the overflow bit overflow bit. In this case, V = 1 if a two's complement addition overflows, but V = 0 if no overflow occurs. For

Source Code	Program Output
#include <iostream.h>	The big number equals 214748361
int main(void)	The big number equals 214748362
{	The big number equals 214748363
int bigNumber = 2147483640 ;	The big number equals 214748364
for (int i = 1; i <= 10 ; i++)	The big number equals 214748365
cout << "The big number equals"	The big number equals 214748366
<< bigNumber + i << endl ;	The big number equals 214748367
return 0 ;	The big number equals -2147483648
}	The big number equals -2147483647
	The big number equals -2147483646

an n bit signed number, V=1 indicated that the true result is *greater than* $2^{n-1} - 1$, or *less than* $- 2^{n-1}$. Without the overflow flag, positive results will be interpreted as negative numbers, and *vice versa*.

The expression **N XOR V** always gives the correct sign of a two's complement result. This might not be so obvious. Suppose that you added $6A and $7C as bytes. These are two positive numbers. If we add these two numbers together we get $E6. However, $E6 is a negative number. The computer interprets this as -$1A, which is clearly incorrect. The overflow flag would have been set to 1 and the negative flag would have been set to 1, so we at least know that the sign of the result is a positive number and that an overflow has occurred. How would you remedy the situation? The same way that you do in C or C++: use a larger variable type. Now, if we add $006A and $007C we get $00E6, but the number is properly interpreted as a positive number. If you were only adding positive numbers together then you would also need to watch the Carry Bit to tell you if you overflowed on the addition.

We can see how using two's complement to represent signed numbers maps the range of the unsigned number to the range of the signed numbers. If we map the corresponding signed and unsigned numbers, we can see how the negative numbers are represented.

Unsigned 32-bit number	0.......... 2,147,483,647 | 2,147,483,648.......... 4,294,967,295
Signed 32-bit number	0.......... 2,147,483,647 | − 2,147,483,648............................−1

Thus, the maximum value that the unsigned number can reach is equal to −1 as a two's complement and the number that is one digit greater than the maximum value of the signed number is equal to the maximum negative number in the range of the signed numbers.

Real numbers, or numbers containing a whole part and a fractional part, can only be approximated within most computers. You are all familiar with the *float* and *double* data types. These can only approximate a real number. In our computer, the question becomes one of, "How accurately do you want to approximate the real number? I'm sure that you, as software professionals, would ever write this kind of an *if* statement because you know that the test could erroneously pass or fail do to progressive round-off and conversion errors.

```
float a;
float b;
if ( b < a)
     [ do this ];
else
     [do that];
```

Let's see how real numbers are represented in a modern computer. Just as we represent the fractional part of a decimal number as the base, 10, to progressively larger negative powers of 10 as we move to the right, away from the decimal point, we would use the same method for fractional numbers in any base. Thus, the binary number 10100101 ($A5) is equivalent to:

	2^7	2^6	2^5	2^4	2^3	2^2	2^1	2^0	•	2^{-1}	2^{-2}	2^{-3}	2^{-4}		
case 1	1	0	1	0	0	1	0	1	•					X	2^0
case 2		1	0	1	0	0	1	0	•	1				X	2^1
case 3			1	0	1	0	0	1	•	0	1			X	2^2
case 4				1	0	1	0	0	•	1	0	1		X	2^3
case 5					1	0	1	0	•	0	1	0	1	X	2^4

Thus, the binary integer 10100101 can be represented as the real number:

$$1010.0101 \times 2^4$$

While the representation might look a bit strange at first, it is just the mantissa and exponent notation that you already are familiar with for decimal numbers. You can see from the above example that we limited ourselves to 4 decimal places. If continued the process for another step without adding another decimal digit, our number would have become:

$$101.010 \times 2^5$$

Thus, when we convert back to an integer value, our original number, $A5 becomes $A4. If you're an Accountant that's probably OK, but not so for Computer Scientists and Engineers!

Today, we represent floating point numbers in an industry standard form, called IEEE-754. The 1985 standard is maintained and published by the Institute of Electrical and Electronic Engineers (IEEE). The standard describes the format for representing single and double precision floating point numbers in computer systems. The standard is almost universally accepted because many of the higher performance microprocessors contain on-chip floating-point units, or *FPUs*. Without a standard for representing floating point numbers, the representation of floating point numbers would be left to the microprocessor companies and much of the software libraries would not be portable across platforms.

The IEEE-754 standard defines single, double and quad formats for floating point numbers with each standard requiring 32, 64 and 128 bits, respectively. Figure 8.1 illustrates the partitioning of the bit fields for each of the formats.

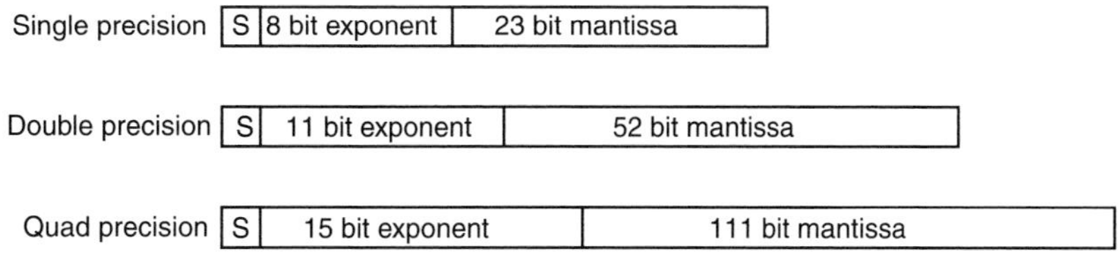

The actual representation of the numbers is not obvious and there are several areas of the format that need further explanation. First, it is interesting to note that the floating point representation of a

Single precision	S	8 bit exponent	23 bit mantissa	
Double precision	S	11 bit exponent	52 bit mantissa	
Quad precision	S	15 bit exponent	111 bit mantissa	

Figure 8.1: IEEE-754-1985 floating point number representations.

number uses the sign and magnitude representation for a number, rather than the integer representation of two's complement. The sign bit, S, is 0 for a positive number and 1 for a negative number.

The mantissa is always adjusted so that it is a number between 1 and 2. Thus, the mantissa form of the mantissa is similar to the way that we try to represent a decimal number in scientific notation. In general, a number in scientific notation is has a mantissa that is greater than 1 and less than 10, so that we have one digit to the left of the decimal point and then a fractional part.

In the IEEE notation, a base 2 number is adjusted so that the mantissa always has a 1 to the left of the fractional part, but since the 1 is always there, it is omitted from the numeric fields in order to gain an extra bit of precision in the mantissa field. Thus, the single precision number actually requires 33 bits to represent it, but only 32 bits are stored because there is always a 1 to the left of the decimal point.

The exponent also requires a bit of investigation. There is no sign bit for the exponent, so it might first appear that we are using two's complement for the exponent. This is a pretty good guess and the method used is similar in principle to using a two's complement representation. Rather than two's complement, the range of the bit field is offset by a bias value. In the case of an 8-bit exponent, the bias value is 127 (01111111). So, if the 8-bit value of the exponent is 10000001, the value of the exponent would be calculated as:

$$129 - 127 = 2.$$

For a double precision number the bias value is 1023 and for a quad precision number the bias value is 32,767. We can thus combine these parts and define a floating point number, N, as follows:[1]

$$N = -1^S \times 1.F \times 2^{E-B}$$

Here, S is the sign bit, F is the fractional part of the mantissa, E is the exponent and B is the bias. Thus, for a single precision number, we can represent the exponent in the range of 2^{-127} to 2^{128}.

Branching and Program Control

Most programs only execute five to seven instructions before taking a branch (nonsequential program fetch). Most branches are usually paired with the test instructions (CMP instructions) within the program so that a condition is tested and then the branch immediately follows the test. This pairing of the test and branch is a general rule because you don't want to lose the state of the flag, which is set by the test instruction, by executing another instruction. However, while pairing the test instruction with the branch instruction is the most common pairing, it isn't always necessary. Consider the follow two code snippets.

```
        Snippet #1                        Snippet #2
         MOVE.W   #05,D0                    MOVE.W   #05,D0
loop     SUBI.W   #01,D0           loop     SUBI.W   #01,D0
         CMPI.W   #00,D0                    BNE      loop
         BNE      loop
```

Both code snippets work equally well, but snippet #2 saves 1 instruction because the zero flag would be automatically set by the SUBI.W instruction. It is redundant to test the state of the flag with the CMPI.W instruction since it is already set by the subtraction instruction that precedes it. Ok, you might think that this is a bit extreme, but we worry about these things in assembly language. One of the primary reasons to program in assembly language is to gain absolute cycle by cycle control over the system. Shaving one redundant instruction could translate to speeding-up a time-critical routine by a few microseconds. Is that important? It's hard to say right now, but I can predict with some certainty that it might very well be important at some time in your future as a software developer.

The state of flags inside the condition code register, or CCR, (Z,N,V,X,C) determines if a particular branch will be taken. Sometimes, branches are taken as a result of exception processing, such as interrupts, bus errors, trap instructions or other errors. *TRAP* instructions are important because they are commonly used to interface the operating system, O/S, to the user's application code. We'll be using the *TRAP* instruction later on when we interface assembly language program to the keyboard and display for user I/O. The TRAP instructions provide a well-controlled method of accessing the operating system services of the simulator.

Status bits (*flags*) in the CCR may or may not change state with each the execution of each instruction. It is customary to place the test instruction for the branch immediately in front of the branch instruction so that no other instruction will possibly alter the flag value. The following table summarizes the meaning of the CCR register flags.

Bit	Definition	Meaning
Z	ZERO	Set to 1 if result = 0, set to 0 if result is not zero
N	NEGATIVE	Bit equals the most significant bit (MSB) of the result
C	CARRY	Set to 1 if carry is generated out of the MSB of the operands for an addition. Also set to 1 if a borrow is generated. Set to 0 otherwise.
V	OVERFLOW	Set to 1 if there was an arithmetic overflow. This implies that the result cannot be represented by the operands. Otherwise set to 0.
X	EXTENDED	Transparent to data movement instructions. When affected by arithmetic instructions, it is set the same as the carry.

The branch instruction tests the state of the appropriate CCR flags, either individually or through a logical combination of the flags. If the logical condition evaluates to TRUE the branch is taken. Recall that the program counter (PC) is always pointing to the address of the next instruction to be executed. Adding or subtracting an *offset value* to the contents of the PC before the next instruction is fetched from memory, causes the processor to fetch the next instruction from a different location. The operand of a branch instruction is an 8-bit or 16-bit offset, called a *displacement*. Therefore, if the branch test condition evaluates to true, the displacement is added to the current value in the PC and this becomes the address of the next instruction.

In other words: <PC> + displacement →PC

The form of instruction is `Bcc <displacement>`. Here "*cc*" is shorthand notation for the condition code of the actual branch. You must replace *cc* with the proper test condition. Figure 8.2 is a summary of all of the conditional branch instructions and the way that they are evaluated.

Figure 8.3 is a segment of code containing three examples of conditional branches. The first branch test is a result of the instruction:

```
CMP.L   (A0),D0    * Read it back
```

Here, register A0 is a pointer to a location in memory. The value in that memory location is compared with the contents of D0 and if they're equal, the branch is taken (BEQ) to the instruction indicated by the label **addr_ok**.

The second branch test is a result of the comparison instruction

```
CMPI.B    #max_cnt,(A3) * Have we hit 4 bad locations yet?
```

Here, the immediate value, **max_cnt**, is compared with the contents of the memory location pointed to by A3. If they are equal to each other the branch is taken to the instruction defined by the label done.

The third branch test is a result of the comparison instruction

```
CMPA    A0,A1          * Have we hit the last address yet?
```

Unlike the CMP instruction, the CMPA instruction is used when we need to compare the value of an address register. Here, we compare the contents of two address registers, A0 and A1, and that the

Figure 8.2: Summary of the conditional branch instructions and their meanings instructions and the way that they are evaluated.

Bcc	MEANING	LOGICAL TEST
BCC	Branch if CARRY clear	$C = 0$
BCS	Branch if CARRY set	$C = 1$
BEQ	Branch if result equals zero	$Z = 1$
BNE	Branch if result does not equal zero	$Z = 0$
BGE	Branch if result is greater or equal	$N * V + \overline{N} * \overline{V} = 1$
BGT	Branch if result is greater than	$N * V * \overline{Z} + \overline{N} * \overline{V} * \overline{Z} = 1$
BHI	Branch if result is HI	$\overline{C} * \overline{Z} = 1$
BLE	Branch if result is less than or equal	$Z + N * \overline{V} + \overline{N} * V = 1$
BLS	Branch if result is low or the same	$C + Z = 1$
BLT	Branch if result is less than	$N * \overline{V} + \overline{N} * V = 1$
BMI	Branch if result is negative	$N = 1$
BPL	Branch if result is positive	$N = 0$
BVS	Branch if the resulted caused an overflow	$V = 1$
BVC	Branch if no overflow resulted	$V = 0$

```
tst_loop  MOVE.L   D0,(A0)            * Load value into memory
          CMP.L    (A0),D0            * Read it back
          BEQ      addr_ok            * Test passed, they're the same
not_ok    MOVE.W   A0,(A4)            * Save the bad address location
          ADDQ.B   #1,(A3)            * Increment the counter
          CMPI.B   #max_cnt,(A3)      * Have we hit 4 bad locations yet?
          BEQ      done               * Yes, quit test
addr_ok   ADDQ.L   #inc_addr,A0       * Increment Address pointer
          CMPA     A0,A1              * Have we hit the last address yet?
          BGE      tst_loop           * No, keep testing
          BRA      next_test          * Yes, go to the next test
done      STOP     #exit_pgm          * Quit back to simulator
```

If the contents of the memory location currently being pointed to by A3 equals the number, max_cnt, then the Z flag will be set and the branch will be taken to the instruction labeled, "done"

If the contents of the memory location currently being pointed to by A0 equals the contents of register D0, then the Z flag will be set and the branch will be taken to the instruction labeled, "addr_ok"

Figure 8.3: Code segment showing conditional and unconditional branches.

branch if A0 is greater than or equal to A1. You'll find that there are a number of instructions that are used only for address registers, or only for data registers. Finally, the **branch always** instruction (**BRA**) is an unconditional branch back to another instruction (not shown in this segment).

Addressing Modes

Up to now, we've focused on becoming comfortable with some of the fundamentals of assembly language programming. In order to do that, we've neglected a systematic study of the addressing modes of the 68K instruction set architecture. Before we go any further into other aspects of the architecture, such as stacks and subroutines, we'll need to at least have a single once-over of the primary addressing modes.

You're probably familiar with most of these right now from some of our examples. The following list of addressing modes is based upon the effective address field of the opcode word. Thus, we'll have a "mode/register" combination where that is appropriate; or a "mode/subclass" combination if we need to further subdivide the mode. This will become clearer as we discuss the modes.

Mode 0, Data Register Direct

Source or destination is a data register (D0 . . . D7)

Mode 1, Address Register Direct

Source or destination is an address register (A0 . . . A6)

Register direct addressing is the simplest of the addressing mode. The source or destination of an operand is a data register or an address register and the contents of the specified source register provide the source operand. Similarly, if a register is a destination operand, it is loaded with the value specified by the instruction. The following examples all use register direct addressing for source and destination operands.

* **MOVE.B D0,D3**: Copies the source operand in register D0 to register D3
* **SUB.L   A0,D3**: Subtract the source operand in register A0 from register D3
* **CMP.W   D2,D0**: Compare the source operand in register D2 with register D0
* **ADD.W   D3,D4**: Add the source operand in register D3 to register D4

Registers are the most precious resource you have in a computer. They are part of the architecture of the computer and they have the fastest access time in operations. Most arithmetic and logical operations must use one of the data registers or one of the address registers as one of the operands in the calculation. For example, the **ADD** instruction must use a data register as either the source or destination operand. The other operand may be one of several effective address modes.

Register direct addressing is also efficient because it uses short instructions. It takes only three bits to specify one of eight data registers. Register direct addressing is fast because the external memory does not have to be accessed. In general, programmers use register direct addressing to hold variables that are frequently accessed (i.e., scratchpad storage).

Good compilers make use of register direct addressing to increase performance. In fact, the declaration in C or C++ uses the keyword "register":

```
register int foo = 0;
```

tells the compiler that, if possible, you would like to keep "foo" available as a register variable rather than as a memory location. The compiler might not be able to grant your wish, but you've requested it.

It's pretty hard for assembly language programmers to keep track of registers and their contents, and to make the most effective use of them. This will become more apparent when we look at the architecture of the RISC processor. RISC processors have large register sets (the AMD 29000 has 256 registers) to give compilers plenty of fast, local storage.

Mode 2, Address Register Indirect

The address register, A0 through A6, contains the memory address of the source or destination of the effective address. This is a pointer in C. We also call this indirect addressing because the contents of the address register is not the data we're interested in, it is a pointer to the data, so it allows us to indirectly access memory.

Therefore, with indirect addressing the contents of the address register is the address of the data we are trying to access.

Until now, we have always directly specified the source or destination memory location. Indirect addressing allows us to manipulate addresses by using the contents of an address register to specify the effective address of the operand. If you examine the assembly language code produced by a C compiler, you'll see that pointers directly translate to address register indirect addressing modes. Some processor architectures, such as those in the Intel 80X86 family, use indirect addressing almost exclusively.

In *address register indirect addressing*, the instruction specifies that one of the 68000's address registers, A0 through A6, holds the memory address of the data we need. In writing the assembly language instruction, parentheses mean *the contents of.* For example, the instruction **MOVE.B (A0),D0** instructs the processor to load data register D0 with *the contents of* the memory location pointed to by address register A0. The source address register contains the address of the operand. The processor then accesses the operand *pointed to* by the address register and finally, the contents of the address register pointed to by A0 are copied to the data register. We can see this in Figure 8.4.

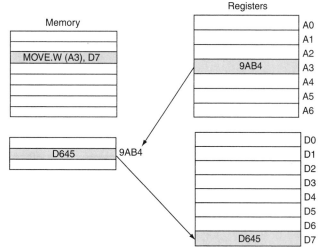

Figure 8.4: Example of the Address Register Indirect addressing mode.

We can also interpret address register indirect addressing in terms of the behavior of the state machine. Consider these two snippets of code:

Snippet #1		Snippet #2	
MOVE.W	$0600,D0	MOVEA	#$600,A0
MOVE.W	D0,$4000	MOVEA	#$4000,A1
		MOVE.W	(A0),(A1)

In Snippet #1 the opcode word is decoded to show that the source EA is an absolute word. This means that the processor must do a few things:

Go out to memory and read the next word of the instruction. It reads in $0600.

1. Place $0600 on the address bus and read the contents of the memory location at address $0600 into D0.

In the next instruction, it sees that the destination EA is an absolute address:

1. Go out to memory and read the next word of the instruction, It reads in $4000.
2. Place $4000 on the address bus and write the data stored in D0 out to address $4000.

The key is that the state machine must use the effective address mode in order to determine the value of the address location of the source or destination. A large part of how it sequences through the state machine is basically the determination of these real memory addresses from the address mode calculations. Once it has determined the real address value, the sequence to read or write to memory is always the same.

In snippet #2, the task seems slightly more complex, but it actually is much more efficient. The two MOVEA instructions are used to load the address registers, A0 and A1 with their initial values, then the process is much more efficient.

1. Place the contents of address register A0 onto the address bus and read the contents of the memory location into a temporary holding register inside of the processor.
2. Place the contents of register A1 onto the address bus and write the contents of the temporary holding register out to that memory location.

Notice that in the above example once we loaded starting values into the address register we no longer concerned ourselves with what those values were. We could do all sorts of incrementing and decrementing of address values and our memory loads and stores would be handled appropriately.

I hope that you can see that address register indirect is a very powerful addressing mode. In fact, you'll probably use it more often than any other mode. The reason it is so important is that indirect addressing enables us to compute a memory location during program execution instead of being fixed when the program is assembled. Thus, if we are able to compute an address, then we can easily move up and down through tables and structures.

Mode 7, Subclass 4: Immediate Addressing

The source value, preceded by the # sign, is the data. It is not a memory address. An immediate operand is also called a literal operand. We use the immediate addressing mode most often to initialize variables, or to provide commonly used numbers in the program. Keep these two important points in mind when trying to use immediate addressing.

1. The immediate addressing mode can only be used to specify a *source operand* because you cannot store a number (data) in a number.
2. You must place the pound sign, #, in front of the numeric value as an indicator to the assembler that this is an immediate value. Otherwise, it may be interpreted as a memory location. This may be a hard bug to find because you won't get an assembler error for specifying a memory location instead of a number.

For example, the instruction

```
MOVE.B #4,D0
```

uses a literal source operand and a register direct destination operand. The literal source operand, 4, is part of the instruction. The destination register, D0, is addressed using register direct addressing. Each operand of the instruction can use a different addressing mode. The effect of the instruction is to copy the literal value 4 to the data register, D0.

Mode 7, Subclass 000: Absolute Addressing (Word)

The memory location is explicitly specified as a 16-bit word. However, since the full address of the 68K architecture is 32 bits long, this addressing mode uses a 16-bit *sign extended address*. The most significant bit of the address operand is used to fill all the upper address bits from

A15 . . . A31. If MSB = 0, then the address will be the lower 32k (A0 . . . A14). If MSB = 1, then the address will be the upper 32K.

For example, if the 16-bit address is $B53C, or binary 1011 0101 0011 1100, then the "sign extended" 32-bit address is $FFFFB53C. If the 16-bit address is $7ABC, or binary 0111 1010 1011 1100, then the sign extended address is $00007ABC.

You might be wondering, "Why bother?" Well, for one thing, the absolute short, or word, form of the address saves memory space and is faster because it means that there is one less extension word fetch to make from memory. Also, most ROM code is placed in low memory and RAM code would be located in high memory. For example, the 68K Exception Vector Table is located in first 256 long words of memory.

Mode 7, Subclass 001: Absolute Addressing (Long)

The memory location is explicitly specified as a 32-bit word. The actual address of the operand is contained in the two words following the instruction word. After reading the two address words from memory, the data to be operated on is then read from the external 32-bit memory address formed by concatenating the high-order word and low-order word to form the full memory address.

Since the 68K has only 24 external address bits, there exists a potential problem called *aliasing*. Aliasing is a result of inadvertently duplicating an address. Remember that a full 32-bit address may contain 256, 24-bit pages. The best solution is to try to keep all of the upper address bits, A24 through A31, equal to zero.

In direct or absolute addressing, the instruction provides the address of the operand in memory. Direct addressing requires two memory accesses to fetch the complete instruction. The first is to access the instruction and the second is to access the actual operand. For example, the instruction CLR.B 1234 clears (sets to zero) the contents of memory location 1234.

Even though it seems pretty straightforward, absolute addressing is not used very often. In all but the most simple computer systems we need the flexibility to move code around in memory without regard to the absolute location that the code resides at. Absolute addressing prevents the code from being relocated and it slows the processor because multiple memory accesses are required in order to determine the memory address of the operand.

Mode 3, Address Register Indirect with Postincrement

The address register, A0 . . . A6, contains the address of the source or destination of the effective address. After the instruction is executed the contents of the address register is incremented by one. This is the 68K method of implementing the stack POP operation.

Mode 4, Address Register Indirect with Predecrement

The address register, A0 . . . A6, contains the address of the source or destination of the effective address. Before the instruction is executed the contents of the address register is decremented by one. This is the 68K method of implementing the stack PUSH operation. We'll discuss the stack PUSH and POP operation in a later section.

Mode 3 and mode 4 are examples of a general addressing method called *auto-incrementing*. If the addressing mode is specified as (A0)+, the *contents of the address register are incremented after* they have been used. For example: Suppose that the address register, A3, contains the value $9AB4. When used as an indirect pointer it points to memory location $00009AB4. Assume that we're going to execute the instruction:

```
ADD.L      (A3)+,D4
```

In words, this instruction will add the source effective address to the destination effective address and store the results in the destination effective address. The contents of register A3, <A3> = $9AB4. Thus, the long word content of memory location $9AB4 is added to the content of register D4 and the result is stored back into D4. The instruction is completed by incrementing the contents of address register A3. How much will it be incremented by? If you say 1, keep it to yourself. Since the operation is on a long word quantity, <A3> →<A3 + 4> , or $9AB8. Thus, the size of the incrementing operation matches the size of the operand being manipulated.

If the instruction after the **ADD.L (A3)+,D4** *instruction resulted in* a branch back of the program, then the result would keep adding the contents of successive memory locations to D0. Auto-incrementing instructions are valuable because so much of memory is data in ordered lists, and as you'll see in a moment, they are also used to implement stack-based addressing operations.

Let's examine the use of the address register indirect with post-incrementing (whew!) instruction mode with a piece of example code. The following fragment of code uses address register indirect addressing with post-incrementing to add together five numbers stored in consecutive memory locations. Note the use of the instruction, LEA (load effective address). This instruction is designed expressly for the purpose of placing an address into an address register. The program adds together 5 byte number values stored in memory.

Example

```
        ORG        $400
        MOVE.B     #5,D0        *Five numbers to add
        LEA        Table,A0     *A0 points to the numbers
        CLR.B      D1           *Clear the sum
  Loop  ADD.B      (A0)+,D1     *REPEAT
        SUB.B      #1,D0
                   1            *Add number to the total

        BNE        Loop         *UNTIL all numbers added
        STOP       #$2700


                                *Some dummy data
 Table  DC.B       1,4,2,6,5

        END        $400
```

The mode 3 and mode 4 address modes automatically increment the value in the address register after operand is fetched from memory. The value in the register increments by one, two, or four bytes if the corresponding operation is on bytes, words or long words, respectively.

Let's summarize the primary addressing modes.

- *Register direct addressing* is used for variables that can be held in registers.
- *Literal (immediate) addressing* is used for constants that do not change. Its primary use is to initializes variables.

- *Direct (absolute) addressing* is used to directly specify the address of variables that reside in memory. Although conceptually simple, it prevents programs from being relocatable.
- *Address register indirect addressing* is used when an address needs to be computed, or sequentially addressed.
- *Address register indirect with postincrement or predecrement* is used for sequential data manipulation or PUSH and POP operations to the stack.

The only difference between register direct addressing and direct addressing is that the former uses registers to store operands and the latter uses memory.

Let's look at a simple assembly language program which summarizes most of the concepts that we've just discussed. Assume that we have a sequence of bytes residing in successive memory locations beginning at a memory location labeled "data". Figure 8.5 is a flow chart of the algorithm.

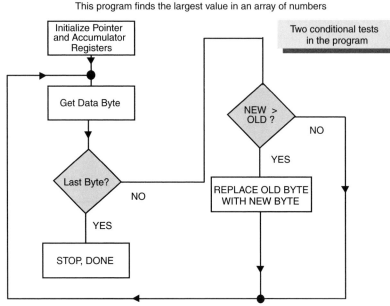

Figure 8.5: Flow chart for an algorithm to find the largest number in a sequence of numbers but the NULL byte must be present.

In the program we assume that the string of bytes is terminated by a byte containing 00. We'll call this the NULL byte. It's the same termination method used for character strings in C. The program assumes that there can be a zero length string of numbers in memory, but the NULL byte must be present.

Here's the program:

```
*****************************************************************************
* Program to find the largest value in an array of bytes
*
* The array starts at memory location $1000 and is terminated by a null byte, $00
*
*****************************************************************************
1   start     ORG     $400            * Program origin
2             LEA     data,A0         * Use A0 as a pointer
3             CLR.B   D0              * D0 will record the largest byte, zero it
4   next      MOVE.B  (A0)+,D1        * Repeat loop, read a byte
5             BEQ     exit            * Is this the null byte?
6             CMP.B   D0,D1           * D1 - D0, If New > Old
7             BLE     end_test        *
```

```
8                 MOVE.B   D1,D0          * Then OLD = New
9    end_test     BRA      next           * Go back and get the next byte
10   exit         STOP     #$2700         * Return to simulator
11
12                ORG      $1000          * Data region
13   data         DC.B     12,13,5,6,4,8,4,10,0 * Sample data
14                END      $400           * Program terminator and entry point
```

Let's analyze the program. Note that the numbers in the leftmost column shouldn't be placed in an actual program. They were placed there to aid in identifying each line of instructions. Also notice the header block enclosed in asterisks before the actual program code. This is used to describe the algorithm, much the way a comment block is placed at the beginning of a function in C++.

1. The ORG pseudo-opcode identifies the beginning of the program.
2. LEA (Load Effective Address) initializes address register A0, with the memory address of the data. In effect, we are assigning a pointer to the array variable, *data*. We'll keep the pointer in the A0 register. In an earlier code snippet we used MOVEA instead of LEA to load a value into an address register. If you are setting up an address register for later use, LEA is the preferred instruction.
3. CLR.B is used to zero the contents of register D0 so that we can properly test for a larger value. Also, in case the array is zero length, we won't report a wrong value.
4. MOVE.B copies the byte pointed to by A0 into register D1. Since we are using the address register indirect with post incrementing mode, the contents of A0 automatically increments to point to the next byte in the string *after the data is retrieved.*
5. BEQ tests to see if the byte we retrieved is the NULL byte. If it is, the branch is taken and we go to the exit point of the program. If this test fails we fall through to the next instruction.
6. CMP.B checks to see if the new byte is larger than the currently largest byte.
7. BLE executes the result of the previous CMP.B instruction. If the new byte is not larger than the current byte in D0, it jumps over the next instruction and returns to the beginning of the loop. If it is larger, the next instruction is executed. Notice how lines 6 and 7 pair an instruction that compares two value with an instruction that branches based upon the result of the comparison.
8. MOVE.B replaces the value in D0 with the new largest value.
9. BRA always returns to the start of the loop.
10. STOP brings us back to the simulator.
11. ORG relocated the instruction counter to begin assembling code to be loaded in memory at address $1000 and above.
12. DC.B is the pseudo-opcode to load the sample data into memory at the memory location defined with the label "data".
13. END tells the assembler to stop assembling and to load the program to run at $400.

Assembly Language and C++

Up to now, we've been considering assembly language in isolation, as if we're learning to program a new and cryptic programming language. That's sort of true. However, let's not lose sight of the fact that many of the compilers of high-level languages output assembly language. The assembly language output of these compilers is then assembled to object code. This means that the coding constructs that we've learned in our other programming classes must also translate to a comparable assembly language construct.

C and C++ already have built-in constructs for changing the program flow. **IF/ELSE**, **WHILE**, **DO/WHILE**, **SWITCH**, **FOR**, and function calls all can change the flow of the program. In assembly language we must build our own constructs using the assembly language instructions that are available to us, **Branch**, **Jump**, **Jump to Subroutine** (function call).

Let's look at how some very familiar C++ constructs survive when we look under the hood and see what a compiler might do with them. First, recall that the assembly language comparison instructions (**CMP**, **CMPI** and **CMPA**) subtract one operand from the other in order to set the appropriate condition code flags, but do not save the result. Thus, if <D0> = 10 and <D1> = 10, the instruction:

CMP.B D0,D1

will subtract the contents of data register D1 from the contents of data register D0. Since both registers contain the number 10, the subtraction will equal zero, with a result that Z=1. However, unlike the actual subtraction instruction,

SUB.B D0,D1

the contents of D0 and D1 are unchanged by the operation. Let's compare two code segments. The first, shown below, is a simple C++ *IF* construct:

```
int a = 3, b = 5;
if( a == b)
    { Execute this code };
else
    { Execute this code }
```

A compiler might convert this to this assembly language code segment:

```
          MOVE.L    #3,D0
          MOVE.L    #5,D1
          CMP.L     D0,D1
          BEQ       equal
not_equal {This code executes}
equal     {This code executes}
```

Loop construct in assembly language are similar to their C++ brethren. It is instructive to look at the loop construct in some detail because loops are one of the more difficult assembly language coding structures to wrap your mind around. Let's look at a simple *FOR* loop construct:

```
for ( int counter = 1; counter < 10 ; counter++ )
      {Execute these statements}
```

```
          MOVE.L #1,D0                    *D0 is the counter
          MOVE.L #10,D1                   *D1 holds terminal value
for_loop  CMP.B  D0,D1                    *Do the test
          BEQ    next_code                *Are we done yet?
          { Execute some other loop instructions}*Increment the counter
          ADDQ.B        #1,D0
          BRA           for_loop          *Go back
next_code{ Execute the instructions after the loop }
```

The *DO/WHILE* construct is also commonly used in C++. The form of the instruction is:

```
DO
    { Execute these statements }
WHILE ( The test condition is true }
```

The assembly language analog is shown below:

```
          MOVEA.W   #start_addr,A2    *Initalize the loop conditions
test_loop JSR       subroutine        *This is a function call in C++
          CMPI.B    #test_value,(A2)+ * Compare the new value
          BNE       test_loop         *The test condition is still true
          {Next set of instructions  * { This code after the loop}
```

Notice that the function call, *JSR* subroutine, happens at least once because this is a *DO/WHILE* loop. The instruction,

`CMPI.B #test_value,(A2)+`

is the heart of the equality test. Here the value, *test_value*, is tested against the value in memory pointed to by address register A2. After the instruction is executed, the value contained in register A2 is automatically incremented to point to the next memory location. Thus, automatically incrementing our memory pointer, A2, gives us the next value to test when we re-enter the loop.

Let's do one more exercise in the analysis of an assembly language algorithm. In this case, we'll take a very simple assembly language coding example and successively refine it to make it as efficient as possible. In this case, efficiency will be measured by code compactness and execution speed. Here's the problem statement:

> *Write the data value, $FF, into memory locations $1000 through $1005, inclusive*

Example #1 Brute Force Method

```
*****************************************************************
* This program puts FF in memory locations $1000 through $1005
*****************************************************************
* System equates

load_val    EQU     $FF           * Byte to load
pgm_start   EQU     $400          * Program runs here
stack       EQU     $2000         *Put stack here


            ORG     pgm_start
            LEA     stack,SP          *Initialize the stack pointer
            MOVE.B  #load_val,$1000 *Load first value
            MOVE.B  #load_val,$1001 *Load second value
            MOVE.B  #load_val,$1002 *Load third value
            MOVE.B  #load_val,$1003 *Load fourth value
```

Before we analyze the program, there is an instruction that might look strange to you. The instruction:

`LEA    stack,SP     *Initialize the stack pointer`

places the address of a variable that we've defined as *stack* (memory location $2000) into something called *SP*. SP is the way that our assembler designates register A7 or A7', the stack pointer.

We'll discuss the stack pointer in more detail later. For now, we'll accept on faith the need to locate the stack pointer somewhere in high memory, and we should do it as one of the first things that we do in our program. Also note that we use the *load effective address,* LEA, instruction to do this.

Now, let's analyze the program. The program in example #1 runs, but it is far from efficient. All of our destination operands for the MOVE.B instructions are absolute addresses. We're wasting memory space for the instructions because we are exactly specifying the memory location for each data transfer. Suppose that we were moving 600,000 bytes instead of 6 bytes. Clearly that scenario would not work for this algorithm.

We can improve on Example #1 by using a data register to hold the data value, $FF, that we are moving to memory and we can also use an address register to point to the memory when we want to store the data. This saves us time and space because the data, $FF, is now available to us in a register, rather than having to retrieve it from memory each time. The disadvantage is that we must initially load some registers, which requires additional instructions. However, the overhead of loading the registers will be offset by the speed gained by using shorter instructions that execute in less time.

Example #2 Using Registers

```
*****************************************************************
* This program puts FF in memory locations $1000 through $1005
*****************************************************************

* System equates
load_val        EQU         $FF         *Byte to load
pgm_start       EQU         $400        *Program runs here
stack           EQU         $2000       *Put stack here
start_addr      EQU         $1000       *First address to load

*Program starts here
                ORG         pgm_start
                LEA         stack,SP        *Initialize the stack pointer
                LEA         start_addr,A0   *Set up A0 as the pointer
                MOVE.B      #load_val,D0    *Put it in a data register
                MOVE.B      D0,(A0)         *Load first value
                ADDA.W      #01,A0          *Point to the next address
                MOVE.B      D0,(A0)         *Load second value
```

This is better because each instruction that moves data only depends upon the contents of registers. Can we do better? Yes, we can eliminate the instruction,

```
                ADDA.W      #01,A0
```

by using the auto-incrementing address mode, which is officially known as *address register indirect with post-incrementing.*.

Example #3 Auto Incrementing

```
*****************************************************************
* This program puts FF in memory locations $1000 through $1005
*****************************************************************

* System equates
```

```
load_val         EQU          $FF              *Byte to load
pgm_start        EQU          $400             *Program runs here
stack            EQU          $2000            *Put stack here
start_addr       EQU          $1000            *First address to load


*Program starts here

                 ORG          pgm_start
                 LEA          stack,SP             *Initialize the stack
pointer
                 LEA          start_addr,A0        *Set up A0 as the pointer
                 MOVE.B       #load_val,D0         *Put it in a data register
                 MOVE.B       D0,(A0)+             *Load first value
                 MOVE.B       D0,(A0)+             *Load second value and in-
crement
                 MOVE.B       D0,(A0)+             *Load third value and in-
crement
                 MOVE.B       D0,(A0)+             *Load fourth value and in-
crement
                 MOVE.B       D0,(A0)+             *Load fifth value
                 MOVE.B       D0,(A0)              *Load last value
                 STOP         #$2700               *Return to the simulator
                 END          pgm_start            *Stop assembling here
```

Is example #3 as good as we can do? This is an interesting question that requires some in-depth analysis. Why? Because for a simple program with 6 data write statements, trying to make the code more compact by using a loop construct might actually decrease performance rather than improve it. However, if we were moving a lot of data, this in-line algorithm would clearly not work.

The question about an in-line algorithm versus a loop construct is a very interesting one from the point of view of computer performance. Computers run most efficiently when they can execute large blocks of in-line code without needing to branch or loop. Compilers will often convert *for* loops to in-line code (if the number of iterations is a manageable value) in order to improve performance. We'll see the reason for this when we discuss pipelining in a later lesson.

Anyway, let's put in a loop construct and see if it improves things.

Example #4 The DO/WHILE Loop Construct

```
*****************************************************************
* This program puts FF in memory locations $1000 through $1005
*****************************************************************
* System equates

load_val         EQU          $FF              *Byte to load
pgm_start        EQU          $400             *Program runs here
stack            EQU          $2000            *Put stack here
start_addr       EQU          $1000            *First address to load
end_addr         EQU          $1005            *Last address to load
```

```
                ORG             pgm_start

                LEA             stack,SP        *Initialize the stack pointer
                LEA             start_addr,A0   *Set up A0 as the pointer
                LEA             end_addr,A1     *A1 will keep track of the
end
                MOVE.B          #load_val,D0    *Put it in a data register
loop            MOVE.B          D0,(A0)+        *Load value and increment
                CMPA.W          A1,A0           *Are we done yet?
                BLE             loop            *No, go back
                STOP            #$2700          *Return to the simulator
                END             pgm_start       *Stop assembling here
```

We've managed to decrease our algorithm from 10 to 8 actual instructions (not counting pseudo-ops). That's better, but for this simple algorithm the value of the loop is hidden by the overhead or creating it. However, if the loop was going to execute 1,000,000 times or so, then writing the algorithm as in-line code would be prohibitive. Let's look at this algorithm with a *for loop* construct.

Example #5 The FOR Loop Construct

```
*****************************************************************
* This program puts FF in memory locations $1000 through $1005
*****************************************************************

* System equates

load_val        EQU             $FF             * Byte to load
pgm_start       EQU             $400            * Program runs here
stack           EQU             $2000           *Put stack here
start_addr      EQU             $1000           *First address to load
loop_ctr        EQU             6               *Number of time through the loop

                ORG             pgm_start

                LEA             stack,SP        *Initialize the stack pointer
                LEA             start_addr,A0   *Set up A0 as the pointer
                MOVE.B          #loop_ctr,D1    *D1 will keep track
                MOVE.B          #load_val,D0    *Put it in a data register
loop            MOVE.B          D0,(A0)+        *Load value and increment
                SUBQ.B          #01,D1          *Decrement counter
                BNE             loop            *Are we done yet?
                STOP            #$2700          *Return to the simulator
                END             pgm_start       *Stop assembling here
```

Counting instructions tells us that the *for* loop and the *do/while* loop constructs give us identical results. However, to really know if one would give us an improvement over the other we would really need to look at each instruction and count the actual number of machine cycles used. Then, and only then, would we know if the *for loop* was more or less efficient than the *do/while* loop.

One thing we can say at this point is that we're pretty good where we are. Can we get better? Yes, but it isn't obvious how we can do this. The answer lies with a more complex instruction, the DBcc instruction. Refer to the Motorola Programmer's Reference Manual for a complete discussion of this instruction. It's a difficult instruction to master, but it combines in one instruction the two instructions:

```
SUBQ.B          #01,D1          *Decrement counter
BNE             loop            *Are we done yet?
```

By using the DBcc instruction we can make our program even more compact.

Example #6 Using the DBcc Instruction

```
************************************************************
* This program puts FF in memory locations $1000 through $1005
************************************************************

* System equates

load_val        EQU         $FF         * Byte to load
pgm_start       EQU         $400        * Program runs here
stack           EQU         $2000       *Put stack here
start_addr      EQU         $1000       *First address to load
loop_ctr        EQU         6           *Number of time through the loop

                ORG         pgm_start

                LEA         stack,SP        *Initialize the stack pointer
                LEA         start_addr,A0   *Set up A0 as the pointer
                MOVE.B      #loop_ctr,D1    *D1 will keep track
                MOVE.B      #load_val,D0    *Put it in a data register
loop            MOVE.B      D0,(A0)+        *Load value and increment
                DBF         D1,loop         * Decrement loop counter and
branch if D1 = -1
                STOP        #$2700          *Return to the simulator
                END         pgm_start       *Stop assembling here
```

Well this is probably as good as it gets. The DBF instruction is very complex and difficult to master, but once you've mastered it, you can add it to your programmer's toolbox and use it when you need to shave a cycle or two. The instruction works this way. First, it either tests a condition code flag (DBcc) or is forced to always be false (DBF). Each time through the loop, the data register (in this case D1) is decremented, if the condition is false. The loop is exited under two conditions:

1. If the condition is true, or
2. The value in the data register $= -1$.

Thus, the **DBF D1,loop** instruction keeps branching back to "loop" and decrementing D1 until $<D1> = -1$, then it exits the loop. Since we're using the version of the instruction that always tests to false, we're always going to branch back.

Let's review what we've just seen. In this sequence we've gone from a very straight-forward solution (take this data and put it here) to a much more compact and elegant solution. In the process we went from simple instructions and addressing modes to more complex instructions, addressing modes and algorithmic structures. In order to make the program as compact as possible, we used a rather complex instruction, DBF, in order to combine a register decrement operation with a test and branch operation. The following table shows the number of clock cycles and execution times (assuming a 16 MHz clock frequency) for the various instructions that we've used.

Instruction	Clock Cycles	Instruction Time (microseconds)
MOVE.B #$FF,$1000	28	1.75
MOVE,B D0,$1000	20	1.25
ADDA.W #01,A0	12	0.75
MOVE.B D0,(A0)	8	0.5
MOVE.B D0,(A0)+	8	0.5

There is a factor of 3.5 in the number of clock cycles required to move the byte value $FF into each memory location depending upon whether the data is loaded from memory each time and written to the specific memory location versus using the data and address registers to hold and manipulate the values.

Stacks and Subroutines

The concept of a stack is fundamental to how the computer manages its data and addresses. The stack is an example of a LAST-IN/FIRST-OUT or LIFO data structure. There are two special registers in the 68K architecture, A7 and A7' that are dedicated to stack operations. A7 is the *user stack pointer,* and is referenced as SP, not A7, in assembly language instructions. A7' is the *supervisor stack pointer*.

In real life, the supervisor stack pointer would be reserved for operating system use, along with a set of special instructions that are usable in *supervisor mode.* Normally, our programs would run at a lower priority than the operating system. We would run our programs in *user mode* and would automatically access the A7 register for our stack pointer rather than the supervisor stack pointer. However, since the programs that we're running on the simulator are rather simple and don't require the use of an operating system (other than your PCs operating system), we're always in supervisor mode and we'll be using the supervisor stack pointer for our operations. Both the supervisor mode and user mode refer to the stack pointer as SP, which stack pointer you get is determined by what mode you're in. Figure 8.6 summarizes the operation of the stack.

The reasons for the pre-decrement and post-increment addressing modes are now apparent. The stack pointer is always pointing to the memory address that is the top of the stack. When additional data needs to be placed on the stack, the pointer must first decrement in order to point to the next available space, so that when the data storage occurs, the data is moved to free memory. The post-increment mode accesses the current memory location on the stack *and then* increments, so the SP is pointing to the previous item on the stored on the stack.

Now that we have showed how the stack implements the LIFO structure, we can move on to examine subroutines. Plain and simply, subroutines in assembly language are the equivalent of function calls in C. When the program encounters the *jump to subroutine (JSR)* instruction, the processor automatically *pushes* the long word address of the *next instruction* onto the stack and then jumps to the location specified in the operand. Since the program counter is always pointing to the next instruction to be fetched, the action of the JSR instruction is to first move the contents of the program counter to the current location of the stack pointer and then load the program counter with the new operand address. This effectively causes the program to "jump" to a new location in memory.

The *return from* subroutine (RTS) instruction is placed at the end of the subroutine. It will cause the stack to pop the return address back into the PC and the next instruction will be executed from the point in the program where the program jumped to the subroutine.

When you write a program in C or C++ the compiler manages all of the housekeeping

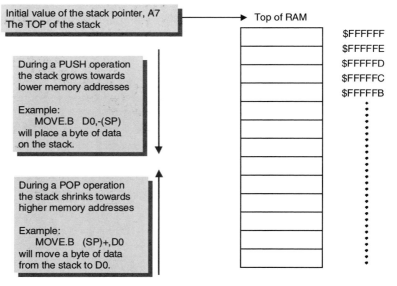

Figure 8.6: Operation of the 68000 stack pointer register, A7.

that needs to go on when you do a procedure or a function call. In assembly language it is up to the programmer to:

- make sure that all subsequent stack pushes and pops line up,
- that all resources used by the subroutine (registers and memory) are properly saved before the subroutine uses them and restored when the subroutine returns, and restored when the subroutine returns,
- decide on a mechanism for parameter passing between the subroutines and the main program.

In general, a subroutine will likely need to use register resources that are being used by the main program, or by other subroutines. The 68K processor has a mechanism to manage this. The **MOVEM** instruction is designed to quickly specify a list of registers that should be saved onto the stack upon the entry into the subroutine and restored upon exit from the subroutine. By saving the registers that you intend to use in the subroutine upon entry, and then restore them upon exit, the subroutine is able to freely use the registers without corrupting the data that is being used by other routines. Thus, it is generally not necessary to save all of the registers on entry into the subroutine, but it can't hurt. You can always streamline the code once you decide which registers you aren't using in your subroutine. Also, you will likely need to move parameters in and out of the subroutines, just as you do in C. However, it is up to you to decide how to do this, since there is no compiler there to force a convention upon you.

Thus, on entry to the subroutine, use the:

```
MOVEM    <register list>,-(SP)
```

to PUSH registers onto the stack, and on exit from the subroutine, use the

```
MOVEM (SP)+,<register list>
```

to POP registers from the stack.

One caveat (Darn those caveats!). It is very convenient to use registers to pass parameters into a subroutine and to use registers to return results from a subroutine. The C/C++ keyword *return* generally causes the compiler to put the result of the function call into a designated register and then return from the subroutine. Here's the "gotcha". If you intend to return a result in a register then you better not have that particular register as member of the group of registers that you save and restore upon subroutine entry and exit. Why? Because when you restore the register on exit you will write over the result that you wanted to return. Many students have burned the midnight oil trying to find that particular bug.

Creating a register list is made easy by the *REG* pseudo-opcode assembler directive. REG means *register range*. It allows a list of registers to be defined. The format is:

`<label>    REG    <register list>`

Registers may be specified as a single register—A0 or D0—separated by slashes (i.e., A1/A5/A7/ D1/D3), or register ranges may also be specified, such as A0–A3. For example, this will define a register list called "save_reg"

`save_reg    REG    A0-A3/A5/D0-D7`

Registers are automatically saved or restored in a fixed order, the ranges specified in the REG directive does not specify the order. Refer to the **MOVEM** instruction in your programmer's manuals for a complete, although incomprehensible, explanation of the **MOVEM** instruction. Figure 8.7 summarizes the storing and restoring of resources when we enter and exit a subroutine.

Subroutines may be nested, just like function calls in C. It is important to keep in mind that subroutines require careful stack management because the return path depends upon having the correct sequences of return addresses stored on the stack. Subroutines should always return to the point where they were invoked, the next instruction after the JSR instruction. However, the *jump table* is one of the few exceptions to this rule. We'll study the jump table in a later lesson. Figure 8.8 illustrates the nesting of subroutines.

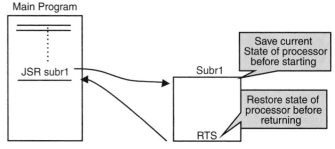

Figure 8.7: Resource management with subroutines.

Subroutines result in efficient coding because the same block of code can be reused many times. The alternative is to have all the code written in-line. Here are some guidelines for writing and using subroutines.

1. You must have the user stack established before attempting to execute a JSR instruction.
2. Locate the subroutines after the main part of your program, but before your data storage area.
3. Each subroutine should have a comment block header listing:
    a. Subroutine name
    b. What it does

c. Registers used and saved
d. Parameters input and returned

4. The first instruction line of the subroutine must have label with the name of the subroutine. This is how the assembler is able to insert the effective address of the destination operand.

5. Save the registers that you'll use on entry to the subroutine.

6. Restore registers on exit from the subroutine.

7. Always return to point in the program where subroutine was called. In other words, don't push a new return address onto the stack and don't jump or branch from the subroutine to somewhere else, leaving the return address on the stack.

8. Nesting of subroutines is permitted, just like function calls in C.

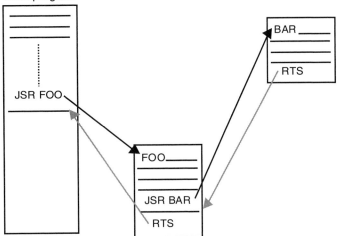

Figure 8.8: Nesting subroutines. The RTS instruction should always be used to return to the instruction after the last JSR instruction.

A Sample Program: The Gory Details

We've spent a good bit of time analyzing the primary addressing modes and examining some programming concepts. It's time to get our hands dirty and look at two programs in some greater detail. The program is a complete program that should put together all of the concepts that we've learned so far.

For the next example program, let's construct a program that really does something useful. It checks the integrity of the memory in the system. This is a pretty common program and is widely used in many computer systems, especially at boot-up time when the system is going through its self-check. You PC does this every time you turn it on or press RESET. Here's the plan. Let's walk through a sample memory test program to see all of the elements that we've discussed so far used in a real example. Our goals will be to:

- examine the structure of the assembly language program;
- see how comments are used;
- see how equates (EQU) are used to define constants;
- see how labels are used to define memory locations;
- look at how the pseudo op directives (ORG, DS, DC, END) are used in practice;
- see how the load effective address (LEA) command is used;
- observe the various addressing modes used in the program;
- see how the compare instruction is combined with the branch instruction to test and modify program flow;

- observe the header block on the subroutine; and,
- see how the subroutine is used and how parameters are passed.

As a way of introduction, let's discuss what the program is doing. It has been hard-coded to always test the memory region from $2000 to $6000, although the starting and ending addresses are somewhat arbitrary and the "equates" section allows us to easily change the region to be tested.

The program writes a byte value out to memory and then reads it back immediately. The value read back should be the same as the value written. If not, there could be a bad memory location, or perhaps a broken or shorted memory line. As a way to determine the cause of the problem, we use four different byte values: $00, $FF, $55 and $AA. These represent all data lines low, all data lines high, even data lines low and odd data lines high, and even data lines high, odd data lines low, respectively. We'll also keep track of the number of failures we detect by storing up to 10 address locations that failed the memory test.

The first section of the program includes the system equates. Consider this to be your header file. This is where you keep all of your #define statements. Notice that almost every constant or initial value is defined here and given a symbolic name. Also notice that the maximum number of bad locations, *maxcnt*, was defined as a decimal number because its intent will be clearer that way. The assembler will convert it to hexadecimal for us.

```
*************************************************************
*
* Memory test program
*
* This is a program to test memory from byte address $2000
* to byte address $6000. It uses four test patterns, 00,
* $FF, $AA, $55
* and it can store up to 10 bad address locations.
*
*************************************************************
*
* System Equates
*
end_test    EQU    $11      * Test pattern terminator
test1       EQU    00       * First Test Pattern
test2       EQU    $FF      * Second test pattern
test3       EQU    $55      * Third Test Pattern
test4       EQU    $AA      * Fourth test pattern
st_addr     EQU    $2000    * Starting Address of Test
end_addr    EQU    $6000    * Ending address of test
stack       EQU    $7000    * Stack location
maxcnt      EQU    10       * Maximum number of bad addresses
```

The highlighted areas illustrate the use of the *load effective address*, **LEA,** command to establish an address in the stack pointer or address register. The subroutine call instruction, JSR, is italicized. The two symbolic variables, **tests** and **bad_cnt** are data storage locations at the end of the program.

```
                ORG      $400             * Start of program
start           LEA      stack,SP         * Initialize the stack pointer
                CLR.B    D0               * Initialize D0
                CLR.B    bad_cnt          * Initialize the bad address counter
                LEA      tests,A2         * A2 points to the test patterns
                LEA      bad_addr,A3      * Pointer to bad count storage
test_loop       MOVE.B   (A2)+,D6         * Let D6 test the patterns for done
                CMPI.B   #end_test,D6     * Are we done?
                BEQ      done             * Yes, quit
                LEA      st_addr,A0       * Set up the starting address in A0
                LEA      end_addr,A1      * Set up the ending address in A1
                JSR      do_test          * Go to the test
                MOVE.B   bad_cnt,D7       * Get the current count
                CMPI.B   #maxcnt,D7       * Have we max'ed out yet?
                BGE      done             * Quit program
done            STOP     #$2700           * Return to the simulator
```

Notice that we saved the registers that we used in the subroutine. We did not need to save A0, A1 and A2 because they were used to pass in the address parameters that we were using.

```
****************************************************************
*
* Subroutine do_test
*
* This subroutine does the actual testing.
* A0 holds the starting address.
* A1 holds the ending address. A2 points to the test pattern to use in
* this test.
* This routine will test the memory locations from A0 to A1 and put the
* address of any failed memory locations in bad_addr and will also
* increment the count in bad_cnt. If the count exceeds 10
* the test will stop
*
****************************************************************
do_test         MOVEM.W  A3/D1/D7,-(SP)   * Save the registers
                MOVE.B   (A2),(A0)        * Write the byte
                MOVE.B   (A0),D1          * Use D1 to hold the value written
                CMP.B    (A2),D1          * Do the comparison
                BNE      error_byte       * Update counter
                BRA      next_test        * OK, test again
error_byte      MOVE.W   A0,(A3)+         * Store address and increment ptr
                ADDI.B   #01,bad_cnt      * Increment the bad count location
                MOVE.B   bad_cnt,D7       * Have we max'd out?
                CMPI.B   #maxcnt,D7       * Check it
                BGE      exit             * Return, we're done.
next_test       ADDA.W   #01,A0           * Increment A0
                CMPA.W   A0,A1            * Test if we're done
                BGE      do_test          * go back and test the next addr
```

```
                MOVEM.W   (SP)+,A3/D1/D7 * Restore the registers
    exit        RTS                      * return to test program
```

The data storage region contains out test patterns and the reserved memory for holding the count and the bad addresses. Notice that the **END** directive comes at the end of all of the source code, not just the program code. The value defined by **end_tests** is similar to the *NULL* character that we use to terminate a string in C. Each time through the test loop we check for this character to see if the program is done.

```
* Data storage region

    tests       DC.B  test1,test2,test3,test4,end_tests   * tests
    bad_cnt     DS.W  1                   * counter for bad    locations
    bad_addr    DS.W  10                  * save space for 10  locations
                END   $400                * end of program and load address
```

Suggested Exercise

Carefully read the code and then build a flow chart to describe how it works. Next, create a source file and run the program in the simulator. In order to test the program, change the ending address for the test to something reasonably close to the beginning, perhaps 10 or 20 bytes away from the start. Next, assemble the program and, using the list file, set a breakpoint at the instruction in the subroutine where the data value is written to memory. Using the trace instruction write the data value to memory, but then change the value stored in memory before starting to trace the program again. In other words, force the test to fail. Watch the program flow and confirm that it is behaving the way you would expect it to. If you are unsure about why a particular instruction or addressing mode is used, review it in your notes or refer to your *Programmer's Reference Manual*. Finally, using this program as a skeleton, see if you can improve on it using other addressing modes or instructions. Please give this exercise a considerable amount of time. It is very fundamental to all of the programming concepts that we've covered so far.

Summary of Chapter 8

Chapter 8 covered:

- How negative and real numbers are represented and manipulated within a computer.
- Branches and the general process of conditional code execution based upon the state of the flags in the CCR.
- The primary addressing modes of the 68K architecture
- High level language loop constructs and their analog in assembly language.
- Using subroutines in assembly language programming.
- A detailed walk-through of an assembly language program to test memory.

Chapter 8: *Endnotes*

[1] Alan Clements, *68000 Family Assembly Language,* ISBN 0-5349-3275-4, PWS Publishing Company, Boston, 1994, p. 29

Exercises for Chapter 8

1. Shown below on the right is a schematic diagram of a 7-segment display. The table on the left represents the binary code that displays the corresponding digits on the display. Thus, to illuminate the number '4' on the display, you would set DB1, DB2, DB5 and DB6 to logic level 1, and all the other data bits to logic level 0.

   - The display is memory-mapped to byte address $1000.
   - There is a hardware timer located at address $1002.
   - The timer is started by writing a 1 to DB4. DB4 is a write-only bit and reading from it always gives DB4=0.
   - When the timer is started DB0 ($\overline{\text{BUSY}}$) goes low and stays low for 500 milliseconds. After 500 milliseconds, the timer times-out and DB0 goes high again.
   - DB0 is read-only and writing to it has no effect on the timer. All other bit positions may be ignored. The timer control register is shown schematically, right:

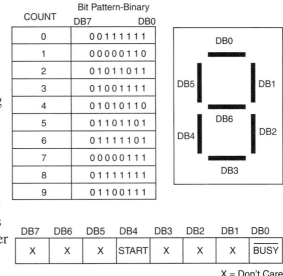

COUNT	Bit Pattern-Binary DB7 DB0
0	0 0 1 1 1 1 1 1
1	0 0 0 0 0 1 1 0
2	0 1 0 1 1 0 1 1
3	0 1 0 0 1 1 1 1
4	0 1 0 1 0 1 1 0
5	0 1 1 0 1 1 0 1
6	0 1 1 1 1 1 0 1
7	0 0 0 0 0 1 1 1
8	0 1 1 1 1 1 1 1
9	0 1 1 0 0 1 1 1

DB7	DB6	DB5	DB4	DB3	DB2	DB1	DB0
X	X	X	START	X	X	X	BUSY

X = Don't Care

Write a short 68K assembly language *subroutine* that will *count down to zero* from the number passed into it in register D0.B. The current state of the count down is shown on the seven-segment display. The count down rate is one digit every two seconds.

> *Notes:*
> - *This is a subroutine. There is no need to ORG your program, set-up a stack pointer or use and END pseudo-op.*
> - *You may assume that number passed-in is in the range of 1 to 9. You do not have to do any error checking. The subroutine is exited when the counter reaches 0.*

2. Assume that some external device has transmitted a sequence of byte values to your computer. Along with the sequence of bytes the external device transmits a *checksum value* that you will use to determine if the byte sequence that you received is exactly the same as the byte sequence that was transmitted. To do this you will calculate the same checksum for the byte stream that you received and then compare it with the checksum that was transmitted to you. If they are equal, then it is extremely likely that there was no error in transmission.

Part A: Write an assembly language subroutine, *not a program*, which will calculate a checksum for a sequence of bytes located in successive memory locations. The checksum is simply a summation of the total value of the bytes, much like summing a column of numbers. The checksum value is a 16-bit value. Any overflow or carry beyond 16-bits is ignored.

1. The information that the subroutine needs is passed into the subroutine as follows:
   a. Register A0 = Pointer to the byte string in memory, represented as a *long word*.
   b. Register D0 = Checksum passed to the subroutine for comparison, represented as a *word* value.
   c. Register D1 = Length of byte sequence, represented as a *word* value.
2. Any overflow or carry generated by the checksum calculation past 16 bits is ignored. Only the *word* value obtained by the summation is relevant to the checksum.
3. If the calculated checksum agrees with the transmitted value, then address register A0 returns a pointer to the start of the string.
4. If the checksum comparison fails, the return value in A0 is set to 0.
5. With the exception of address register A0, all registers should return from the subroutine with their original values intact.

Part B: What is the probability that if there was an error in the byte sequence, it wouldn't be detected by this method?

3. What is the value in register D0 after the highlighted instruction has completed?

```
00000400 4FF84000      START     LEA       $4000,SP
00000404 3F3C1CAA                MOVE.W    #$1CAA,-(SP)
00000408 3F3C8000                MOVE.W    #$8000,-(SP)
0000040C 223C00000010            MOVE.L    #16,D1
00000412 203C216E0000            MOVE.L    #$216E0000,D0
00000418 E2A0                    ASR.L     D1,D0
0000041A 383C1000                MOVE.W    #$1000,D4
0000041E 2C1F                    MOVE.L    (SP)+,D6
00000420 C086                    AND.L     D6,D0
00000422 60FE          STOP_HERE BRA       STOP_HERE
```

4. Write a *subroutine* that conforms to the following specification. The subroutine takes as its input parameter list the following variables:

   - A longword memory address in register A0, where A0 points to the first element of a sequence of 32-bit integers already present in memory.
   - A 32-bit longword value in register D1, where the value in D1 is a search key.
   - A positive number between 1 and 65,535 in register D0, where the value in D0 determines how many of the integer elements in the sequence pointed to by A0 will be searched.
   - The subroutine returns in register D2 the value zero, if there is no match between the search key and the numbers in the sequence being searched; or the numeric value of the memory location where the first match is found

Note that once a match occurs there is no need to keep searching and you should assume that you have no knowledge of the state of the rest of the program that is calling your subroutine.

5. The memory map of a certain computer system consists of ROM at address 0000 through 0x7FFF and RAM at address 0x8000 through 0xFFFF. There is a bug in the following snippet of code. What is it?

Code Snippet:
```
MOVE.W      $1000,D0
MOVE.W      D0,$9000
LEA.W       $2000,A1
MOVEA.W     A1,A2
MOVE.W      D0,(A2)
```

6. Write a short 68K assembly language program that will add together two separate 64-bit values together and stores the result. The specifications are as follows:
   a. Operand 1: High order 32-bits stored in memory location $1000
   b. Operand 1: Low order 32-bits stored in memory location $1004
   c. Operand 2: High order 32-bits stored in memory location $1008
   d. Operand 2: Low order 32-bits stored in memory location $100C

   Store the result as follows:

   a. High order 32-bits in memory location $1020
   b. Low order 32-bits in memory location $1024

   Any carry out generated by the high order addition in memory location $101F.

7. The diagram shown below represents a circular array of eight lights that are connected to an 8-bit, memory mapped, I/O port of a 68K-based computer system. Each light is controlled by a corresponding bit of the I/O port. Writing a 1 to a bit position will turn on the light, writing a 0 will turn it off.

   Write a short 68K assembly language program that will turn on each lamp in succession, keep it on for two seconds, turn it off and then turn on the next lamp. The specifications are as follows:

   - The 8-bit wide parallel I/O port is mapped at memory address $4000.
   - There is a 16-bit wide time-delay port located at memory address $8000. DB0 through DB11 represent a count-down timer that can generate a time delay. Writing a value to DB0-DB11 will cause DB15 to go from low to high. The timer then counts down from the number stored in DB0-DB11 to zero. When the timer reaches zero, DB15 goes low again and the timer stops counting. Each timer tick represents 1 millisecond. Thus, writing $00A to the timer will cause the timer to count down for 10 milliseconds.
   - DB15 is a read-only bit, writing a value to it will not change it or cause any problems.

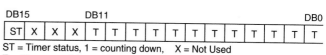

ST = Timer status, 1 = counting down, X = Not Used
T = Timer countdown value

- There is no interrupt from the timer, you will need to keep examining the memory location to see when the timer stops counting.

8. Write a short 68K assembly language subroutine that will send a string of ASCII characters to a serial port according to the following specification:

   - The serial port is memory mapped as two successive byte locations at address $4000 and $4001. The actual port for sending and receiving characters is at address $4000 and the status port is located at address $4001.
   - Data bit 0 (DB0) of the Status Port is assigned to indicate the state of the Transmit Buffer. When the serial device is ready to transmit the next character TBE = 1, or the signal, Transmit Buffer Empty is true. When the Transmit Buffer is empty the next character may be sent. Writing a character to address $4000 starts the serial data transmission and sets TBE = 0. The next character may be sent when TBE=1.
   - On entry, the subroutine should save the values of any registers that it may use and restore these registers on exit.
   - The location of the string to print is passed into the subroutine in register A0.
   - The string is terminated with the null character, $FF.
   - The stack pointer has already been initialized, there is no need to establish a stack pointer
   - Assume that this is a polling loop, there is no interrupt occurring. You must continually test the state of TBE in order to know when to send the next character.

9. Write a program that fills all of memory between two specified addresses with the word pattern $5555 This is similar to what the block fill (BF) command does in the instruction set simulator, (ISS), but you will do it with 68K assembly language code. The memory region that you will fill is $2000 to $20FF, inclusive.

10. Write a memory test program that will be capable of testing memory *as words* (16-bits at a time) in the region from $00001000 to $0003FFFF. It should test up to, and including $0003FFFF, but not past it. A memory test program works as follows:

    - Fill the every memory location in the region being tested with the word data value.
    - Read back each memory location and compare it to the data value that you wrote.
    - Compare the value that you read back with the value that you wrote. If they don't agree, then you have a bad memory location.
    - Start testing the memory using two different test patterns: $FFFF and $AAAA.
    - After you complete a memory test with one of the patterns, complement the bits (change 1's to 0's and 0's to 1's) and repeat the test with the new pattern. Thus, you'll be cycling through the test a total of 4 times using these patterns.
    - Repeat the test one more time using the starting test pattern $0001. Use the ROL.L instruction to move the 1 bit one position to the left each time through the memory test until you've run the memory test 16 times, shifting the 1 to the left each time you repeat the test.
    - Complement the test pattern that you just used and repeat the above bit shifting test.
    - Here are the particulars for the assignment:
      - The program should be ORG'ed to run at memory address $00000400.
      - The program tests the memory from $00001000 to $0003FFFF, inclusive
      - The stack pointer should be located at $000A0000.

- The starting memory test patterns are: $FFFF, $AAAA and $0001
- The test will fill all the memory region of interest with one of the test patterns. Next, it reads the pattern back and compares the value read to the value written. If you write the program in a way that writes a word to memory and then reads it back immediately then you are not "adhering to the specifications".
- The test is repeated for each of the two starting test patterns, their complement and the shifted bit pattern and finally, its complemented bit pattern.
- If an error is detected, the address of the memory location where the error occurred, the data written and the data read back is stored in memory variables.
- If more than one error occurs the program should store the total error count (number of bad locations found) and keep only the address and data information for the last error detected.
- You should allow for a count of up to 65,535 bad memory locations.

Discussion:

This program encompasses many of the fundamental aspects of assembly language programming. If you study the program, you'll see that it lends itself to using a subroutine to actually do the memory test. Imagine if you wrote it in C, what would it look like? How would you pass the testing parameters to the function?

- Don't forget to initialize the stack pointer!
- The program has several loops in it. What are they? How will you define the regions of memory to test? How do you know when you've done all of the tests? How do you know if you wrote all of the memory locations that you need to?
- Be sure that you understand how to use the pseudo-ops EQU, ORG, CRE, DC.L, DC.B, DC.W and DS.L, DS.W, DS.B, END
- Understand the instructions JRS, RTS, LEA and the addressing modes, (An) and (An)+.
- This program can be done in less than 50 instructions, but you have to know what you are trying to do. The Easy68K simulator counts cycles. The program that runs in the least number of clock cycles, even if it has more instructions, is generally the more efficient one.
- Try coding it in stages. Once you've completed the flow chart for the program, write the assembly code for each block and test it. How do you test it? Well, you can assemble it. If it assembles properly, you've made some progress. Next, run it in the simulator and verify that it is doing what you want it to.
- When you test your program, one good programming trick is to use the EQUates to change the region of memory that your testing to only a few words, not the entire space. That way you can quickly walk through the code. So instead of testing from $00001000 to $0003FFFF, you test from $00001000 to $0000100A.
- Check to see what happens if a memory location has bad data. Change the data value in a memory location using the simulator after your program has filled it with the test pattern. Did your program catch it? Did it deal with it? The Easy68K simulator has a nice memory interface that you can access through the view window.
- This program is almost entirely comprised of three instructions, MOVE, Bcc, and CMP (CMPA). The addressing mode will probably be address register indirect because you'll constantly be writing to successive memory locations and reading from successive

memory locations. The address register is an ideal place to hold the starting address, the ending address and the address of where you currently are in memory. To point to the next memory location, increment the address register's contents. You can do this explicitly by adding to it, or implicitly with the (An)+ addressing mode.

- Think about how you might terminate the test when you're using the ROL instruction. You could set up a counter, and then count 32 times. That would work. However, look closely at the ROL instruction. Where does the "1" bit go to after it shifts out of the MSB position? What instruction would test for that condition?

The general structure of your program should be as follows:

a. Comment header block: Tells what your program does.

b. System equates: Define your variables.

c. ORG statement: Your program starts here:

d. Main program code: Everything except subroutines are here.

e. STOP $#2700 instruction: This ends the program gracefully and kicks you back into the simulator.

f. Subroutine header block: All subroutines should have their own header block

g. Subroutine label: Each subroutine should have a label on the first instruction. Otherwise you can't get there from here.

h. Subroutine code: This does the work. Don't forget to keep track of what registers are doing the work and how parameters are passed.

i. RTS: Every subroutine has to return eventually.

j. Data area: All of your variables defined with the DC and DS pseudo ops are stored here.

k. END: The last line of the program should be END $400. This tells the assembler that the program ends here and that it should load at $400.

If all else fails, review the memory test programming example in the chapter. Finally, note that if you try to run your program under the various forms of Windows you may notice strange behavior. The program might seem to run very quickly if you run it in a small region of memory, but then seem to die if you run it in a larger region of memory. This is not your program's fault. It is a problem with Windows when it runs a console application.

Windows monitors I/O activity in a console window. When it doesn't see any input or output in the window, it severely limits the amount of CPU cycles given to the application tied to the window. Thus, as soon as your program begins to take some time to run, Windows throttles it even more.

There are several ways around it, depending upon your version of Windows. You could try hitting the ENTER key while your program is running. It won't do anything in the window, but it will fool the operating system into keeping your program alive.

Another trick is to open the PROPERTIES menu for the window and play with the sensitivity settings so that Windows doesn't shut you down. This worked in Win98SE at Win2000 at school.

CHAPTER 9

Advanced Assembly Language Programming Concepts

. . . .

Objectives

When you are finished with this lesson, you will be able to:

▶ *Program in assembly language using all addressing modes and instructions of the 68K processor architecture;*
▶ *Describe how assembly language instructions and addressing modes support high level languages;*
▶ *Disassemble memory images back to the instruction set architecture;*
▶ *Describe the elements and functions of a single-board computer system.*

. . . .

Introduction

Now that we're sufficiently grounded in most of the 68K programming fundamentals, let's move deeper into subject by examining more closely some additional instructions and addressing modes. The addressing modes that we'll examine now are more obscure from an assembly language programmer's point of view, but critically important if you're programming for the 68K family using C or C++ as your development language of choice.

Since the overwhelming majority of programmers write in C or C++, having addressing modes that support the high-level language constructs are an important consideration for computer designers. The addressing modes we'll study now allow us to implement data structures and to write code that is position independent, or *relocatable*.

Relocatable code can run anywhere in the address space of the processor because there are no *absolute references* to memory locations containing instruction or data. Being able to be loaded anywhere in the address space is important since operating systems have to be able to manage tasks and memory in such a way that a task (program) might have to run from address $A30000 one time and $100000 another time. Also, programs that are written to be position-independent tend to be more efficient in terms of memory size and speed because the destinations of jumps and fetches are determined by the contents of the registers, rather than having to be retrieved from memory as part of the instruction.

Now before we attack the advanced addressing modes you might be wondering about the idea of relocatable code and absolute address references. Look at the following code segment:

```
start    LEA      $4000,A0
         MOVE.W   D0,(A0)
```

Are we making an absolute address reference? Absolutely (sorry)! However, once we have an address in a register, even if it was initially derived from an absolute reference, we have the ability to modify that address according to where that program is loaded in memory.

Advanced Addressing Modes

Mode 5, Address Register Indirect with Displacement

A signed (positive or negative value), 16-bit displacement is added to the contents of the address register to form the effective address. Thus, the effective address, (EA) = (An) +/– 16-bit displacement. For example if <A6> = $1000, the instruction:

`MOVE.L   $400(A6),D0`

would fetch the long word located at memory address $1400 and copy it into data register D0. This addressing mode is very important because it is used to locate local variables in C functions.

This is a "gotcha" in the above instruction. The form of the above instruction is represented in the *Programmer's Reference Manual* as d_{16}`(An)`, which would lead you to believe that the value in the last example, $400, is a displacement value. However, most assembler programs do not expect you to calculate a displacement yourself. The assembler expects that you will insert a label or an absolute memory reference and it will calculate the displacement value for you. Thus, the number $400 might not be interpreted as a displacement of $400, but rather, as the memory location that you wish to calculate the displacement to. Therefore, if your assembler program is giving you errors, such as, *displacement too large* or *out-of-range error*, then it is likely expecting an absolute address or label, rather than an offset.

When you make a function call in C or C++, the compiler establishes a *stack frame* using one of the address registers as the stack pointer. The locations of the different variables are identified by their displacements from the pointer. Thus, if A6 was pointing to the beginning of a stack frame for a function, **foo()**, the local variable being fetched is located $400 bytes from the pointer.

Mode 6, Address Register Indirect with Index

The contents of the address register is added to the contents of an index register (A0 . . . A6 or D0 . . . D6) plus an 8-bit displacement.

$$EA = (An) + (Xn) + d_8$$

If <A5> = $00001000 and <D3> = $AAAA00C4, the instruction

`MOVE.W $40(A5,D3.W),D4`

would fetch the word contents of memory location $00001104 and copy the data into register D4. To see this, try running this code fragment in the simulator.

Example

```
org     $400
lea     $00001000,A5
move.l  #$AAAA00C4,D3
move.w  #$AAAA,D1
move.w  D1,$40(A5,D3.W)
stop    #$2700
end     $400
```

At first, you might think that the effective address would be $AAAA1104. However, the index register for this instruction, D3, is using only the word value, not the long word value, to calculate the effective address.

This addressing mode may seem very strange to you, but assume for a moment that you've created an array of compound data types (structure) in C. Each data type would have some fixed offset from the beginning of the structure. In order to access a particular data element of a particular structure, you must index into the array using D3 and then find the particular element with the fixed offset of $40.

Mode 7, Subclass 2: Program Counter with Displacement

A signed, 16-bit displacement is added to the current contents of the program counter.

$$EA = (PC) + d_{16}$$

This is an example of the general class of addressing modes known as *PC relative*. Please don't mistake it to mean something related to a PC, such as a Palm Pilot®. PC relative addressing is the most important addressing mode for generating relocatable code. In PC relative addressing the effective address of the operand is computed by adding the *sign extended* value in the extension word to the current value of the PC. The resulting address is then placed on the address lines and the data is fetched from external memory.

For example, assume that the <PC> = $D7584420

$7AFE = $\underline{0}$111 1010 1111 1110. Sign extending the most significant bit (underlined) to 32 bits gives us 0000 0000 0000 0000 0111 1010 1111 1110.

EA = $D7584420 + $00007AFE = $D758BF1E

If <PC> = $00000400 at the point in the program when the instruction **MOVE.W $100(PC),D4** is executed, the word contents of memory location $500 will be fetched from memory and copied to D4. However, if this code segment was relocated to another place in memory and run again it would still be fetched correctly, as long as the data to be fetched was located $100 bytes away from the instruction.

The reason is that the current value of the program counter will be used to calculate the effective address. The PC always points to the next instruction to be fetched from memory, regardless of where that program code is residing in memory. As long as the effective address can be calculated as being located a fixed distance from the current value of the PC, everything works properly.

The same holds true for JUMP and JUMP TO SUBROUTINE instructions. Up to now, we've always considered the destinations of the JMP and JSR instructions to be absolute. However, consider the following two code examples:

Example: Case 1

```
JSR    foo    *assembler generates absolute address for foo
```

Example: Case 2

```
        JSR    foo(PC)      *assembler generates relative address

  foo   {More instructions here}
        RTS            *Return from subroutine
```

In Case 1, the assembler calculates the absolute address for the label, **foo**. When the instruction is executed, the address of the next instruction is placed on the stack and the absolute address of **foo** is placed in the PC. The next instruction fetched from memory is the instruction located at address, **foo**.

In Case 2, the assembler calculates the displacement (distance) from the current value of the PC to **foo**. When the instruction is executed, the address of the next instruction (current value of the PC) is placed on the stack and the displacement is added to the current value in the PC and the sum is returned to the PC. The next instruction is fetched from the location at address, **foo**.

However, only Case 2 would allow the program (main code plus subroutines) to be moved to another location in memory without having to readjust all of the addresses.

Mode 7, Subclass 3: Program Counter with Index

The contents of the program counter is added to the contents of an index register (A0 . . . A6 or D0 . . . D6) plus an 8-bit displacement. Thus, EA = (PC) + (Xn) + d_8.

Example

MOVE.W $40(PC,D3.L),D6

This addressing mode is exactly the same as address register indirect with displacement, except that the PC is used as the base register for the address calculation, rather than an address register.

In summary, indexing (using another register to provide a variable displacement) is a powerful method of implementing algorithms based upon tables, strings, arrays, lists, etc. The displacement value provides a constant offset from the current value of the program counter to the start of the table (base address). The index register provides the variable pointer into the table. By making it PC relative, the entire table automatically moves with the program. This is very useful in compiler operations that need to address a variable with a constant offset (displacement) in a structure.

68000 Instructions

Within each category of instruction there are several variants. For example, the ADD instruction family consists of:

- ADD: Adds an effective address to a data register or data register to an effective address.
- ADDA: Adds an effective address to an address register
- ADDI: Adds an immediate value to an effective address
- ADDQ: Adds an immediate data value from 1 to 8 to an effective address
- ADDX: Adds a data register value to a data register value and includes the value of the X bit.

As you know, the effective address may be some, or all of the addressing modes we've studied so far. Obviously, there are a number of ways to formulate an instruction to do what you want. The conclusion is that learning to program in assembly language is like learning to speak by using a dictionary. Learn a few simple words first, and then broaden your vocabulary to add richness and efficiency. Most algorithms can be written in many ways, but the most efficient algorithms may be hard to code without lots of programming experience. First get your algorithm to run, then try to tune it up!

MOVE Instructions

* **MOVE** (Move data): Copies data from source to destination. You should know this one by now.

MOVEA: (Move address): Copies data into an address register. The destination is always an address register and the data size must be word or long word. Consider the formats of the MOVE and MOVEA instructions below:

MOVE instruction:

15	14	13 12	11 9	8 6	5 3	2 0
0	0	SIZE	Destination Register	Destination Mode	Source Mode	Source Register

MOVEA instruction:

15	14	13 12	11 9	8		6	5 3	2 0
0	0	SIZE	Destination Register	0	0	1	Source Mode	Source Register

They look different until you realize that bits 8,7,6 are just defining a mode 1 addressing mode, which is the address register direct addressing mode. So we still don't really have a reason for a unique mnemonic. The reason is that we need to have a different representation is the size of the operation. Since the MOVEA instruction can only move word or long word values into an address register, we prevent the situation of inadvertently creating an illegal op-code.

MOVEM (Move multiple registers): The MOVEM instruction can only be used to move registers to memory and vice versa. It is most often used with the postincrementing and predecrementing addressing modes. Predecrementing is used to transfer registers to a stack structure in memory and Postincrementing is used to transfer data from the memory stack back to the registers. The stack pointer could be the SP register for a system stack, or one of the other address registers for a user stack frame. The order of the register list is unimportant because the 68000 always writes to memory in the order A7 to A0, then D7 to D0. It always reads from memory in the order D0 to D7, then A0 to A7.

Logical Instructions

AND

The **AND** instruction performs the bit-wise, logical AND. For example, if <D0> = $3795AC5F and <D1> = $B6D34B9D before the instruction **AND.W D0,D1** is executed, then <D0> = $3795AC5F and <D1>= $B6D3081D after the instruction. Note that only the lower 16-bits of each register are used in the operation. To see why this is the result, let's do the bit-wise AND in binary:

Example

```
$3795AC5F = 0011 0111 1001 0101 1010 1100 0101 1111
                        AND
$B6D34B9D = 1011 0110 1101 0011 0100 1011 1001 1101
                       EQUALS
$3691081D = 0011 0110 1001 0001 0000 1000 0001 1101
```

A simple trick to remember is this:

- Any hex digit **AND'ed** with F returns the digit
- Any hex digit **AND'ed** with 0 returns 0

ANDI

ANDI: AND immediate data. Example, **ANDI.B   #$5A,D7**

Other Logical Instructions

- **OR**: Perform the bit-wise logical *OR*
- **ORI**: OR immediate data
- **EOR**: Perform the bit-wise logical *Exclusive OR*
- **EORI**: *Exclusive OR* the immediate data
- **NOT**: Perform the bit-wise logical complement of the operand

The **NOT** instruction takes a single operand. If <D3> = $FF4567FF, then the instruction **NOT. B   D3** would change the data in D3 such that, <D3> = $FF456700. Note that you cannot have a *NOT Immediate* instruction because **NOTI.B   #$67** has no meaning.

Shift and Rotate Instruction

This group of instructions is rather involved in that they have different rules depending upon how many positions the bits are shifted and whether the effective address is a register or memory. However, shift and rotate instructions are very important as part of the overall need to manipulate bits and maneuver parts of data words to make them more accessible for arithmetic and logical operations.

For example, suppose that you have read four ASCII characters in from a modem. The characters are $31, $41, $30, $30. Furthermore, the characters are supposed to be a 4-digit hexadecimal number. You consult a table of ASCII values, you'll see that these ASCII values represent the number $1A00, but how do we get from the four ASCII byte values to decode our number $1A00? We need a conversion algorithm.

We'll return to this problem in a moment and we'll study an algorithm that solves it. For now, let's look at what shifting and rotating means. Consider Figure 9.1.

The **ASL** instruction moves each bit of the byte, word or long word one or more bit positions to the left. Each time the bits are shifted, a 0 is inserted in the least significant bit position, DB0. The bit occupying the most significant bit position, DB7, DB15 or DB31 is moved into the carry bit and extended bit position of the condition code register, CCR. The bit that was in the CCR is discarded. Thus, if the instruction is **ASL.B #3,D0** is executed and <D0> = $AB005501. After the byte portion of D0 is shifted three times <D0> = $AB005508. The **ASL** instruction also has the effect of multiplying the data by 2 each time the bits are shifted to the left.

The other shift and rotate instructions operate in a similar manner as **ASL** and **ASR**. Figure 9.2 summarizes the behavior of these instructions. Comparing Figures 9.1 and 9.2, we see that some instructions seem to exhibit identical behavior, such as the **ASL** and **LSL** instructions, but the **ASR** and **LSR** instructions are slightly different.

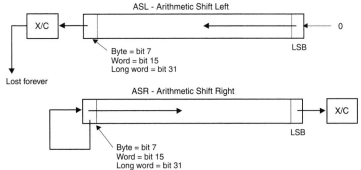

Figure 9.1: Operation of the Arithmetic Shift Left and the Arithmetic Shift Right Instructions.

Let's summarize some of the finer points about using the shift and rotate instructions.

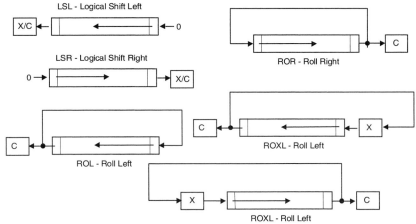

Figure 9.2: Logical shift left/right and roll left/right instructions.

- Instructions in this group may be byte, word or long word operations.
- Operands can only be the contents of Data registers D0-D7 or the contents of memory locations.
- When a data register is involved, the instruction must supply the number of times the bits are shifted.
- If the number of shifts is less than 8, use the immediate form.
- If the number of shifts is greater than 8, then the number must be in another data register.
- When a memory location is involved, only one bit is shifted at a time and the only word operands are allowed

Finally, we'll summarize this discussion by returning to the sample program that we discussed earlier. For the purpose of this example, we'll assume that there are four ASCII digits located in a memory buffer called "ascii_val". The program will convert the ASCII characters to a 4-digit hexadecimal number. The number will be stored back to a memory location called "hex_val". Pay particular attention to the way that the ASL instruction is used to move the bits.

The example program is called *Get Value*. It converts 4 ASCII digits stored in memory into a 4-digit long hexadecimal number.

```
**************************************************
* Get_value
* Converts 4 ASCII values to a 4-digit
* Input Parameters: None
* Assumptions: The buffer, ascii_val contains 4 valid ascii * characters
* in the range of 0...9, A...F, or a...f
**************************************************

*System equates
mask         EQU          $00FF          * Isolates the byte value
stack        EQU          $B000          * Location of stack pointer


* Program starts here

             ORG          $0400          * Program runs here
start        LEA          stack,SP       * Set-up the stack pointer

             CLR.W        D7             * We'll need this register
             LEA          ascii_val,A1   * A1 points to memory buffer

             MOVE.B       (A1)+,D0       * Get the first byte
             ANDI.W       #mask,D0       * Isolate the byte
             jsr          strip_ascii    * Get rid of the ascii code
             ASL.W        #8,D0          * Move left 8 bits
             ASL.W        #4,D0          * Move left 4 bits
             OR.W         D0,D7          * Load the bits into D7
             MOVE.B       (A1)+,D0       * Get the next byte
             ANDI.W       #mask,D0       * Isolate the byte
             jsr          strip_ascii    * Get rid of the ASCII code
             ASL.W        #8,D0          * Move left 8 bits
             OR.W         D0,D7          * Add the next hex digit
             MOVE.B       (A1)+,D0       * Get the next byte
             ANDI.W       #mask,D0       * Isolate the byte
             jsr          strip_ascii    * Get rid of the ASCII code
             ASL.W        #4,D0          * Move left 8 bits
             OR.W         D0,D7          * Add the next hex digit
             MOVE.B       (A1)+,D0       * Get next byte, point to con_val
             ANDI.W       #mask,D0       * Isolate the byte
             jsr          strip_ascii    * Get rid of the ASCII code
             OR.W         D0,D7          * Add the last hex digit
             MOVE.W       D7,(A1)+       * Save the converted value
```

```
            STOP           #$2700           * Return to the simulator

*******************************************************************
* SUBROUTINE: strip_ascii
* remove the ascii code from the digits 0-9,a-f, or A-F
* Input Parameters: <D0> = ascii code
*
* Return parameters: D0.B = number 0...F, returned as 00...0F
* Registers used internally: D0
* Assumptions: D0 contains $30-$39, $41-$46 or $61-66
*
*******************************************************************

  strip_ascii  CMP.B    #$39,D0        * Is it in range of 0-9?
               BMI      sub30          * It's a number
               CMP.B    #$46,D0        * Is is A...F?
               BMI      sub37          * It's A...F
               SUB.B    #$57,D0        * It's a...f
               BRA      ret_sa         * Go back
  sub37        SUB.B    #$37,D0        * Strip 37
               BRA      ret_sa         * Go back
  sub30        SUB.B    #$30,D0        * Strip 30
  ret_sa       RTS                     * Go back

  * Data

  ascii_val    DC.B     $31,$41,$30,$30  * Test value $1A00
  con_val      DS.W     1              * Save it here
               END      $400
```

Arithmetic Instructions

ADD (Add binary)

Add the source and destination operands and place the result in the destination operand. A data register must be the source or destination operand of an ADD instruction. In other words, you cannot directly add together the data contained in two memory locations. One of the operands must be contained in a data register. In fact, only the **MOVE** instruction is designed to operate with two effective addresses that could be memory locations.

Let's do an example. Assume that before the **ADD** instruction <D2> = $12345678, <D3> = $5F02C332

$$\text{ADD.B} \quad \text{D2,D3}$$

After the instruction is executed <D2> = $12345678, <D3> = $5F02C3AA

How are the condition codes affected?

- N = 1: NEGATIVE flag: Bit 7 is 1 so result is assumed to be negative.
- C = 0: CARRY flag: No carry out is generated from bit 7.
- X = 0: EXTEND flag: Used in certain operations, generally affected like CARRY.
- Z = 0: ZERO flag: Nonzero result.

237

- V = 1: OVERFLOW flag: This is tricky! The value in bit 7 of the destination register, D3, changed, so an overflow condition may have occurred.

ADDA (Add Address)

Adds data to an address register. Only word and longword operations are allowed and the condition codes are not affected.

ADDI (Add Immediate)

Adds number to data register or memory. Byte, word or longword operations are permitted. Consider this question. If <D2> = $25C30F7, what is the result of executing these two instructions? Note that they differ only in the size of the operation. To check your answer, write a test program and execute it in the simulator.

Code Example

```
ADDI.B    #$10,D2    * <D2> = ??, C = ??
ADDI.W    #$10,D2    * <D2> = ??, C = ??
```

CLR (Clear an operand)

The **CLR** instruction may be used on bytes, words and long words. All of the condition codes, except X, are affected.

CMP (Compare data with a data register)

The **CMP** instruction sets the condition codes accordingly. This instruction subtracts the source operand from the destination operand, but does not place the result back into the destination operand. Thus, neither operand is changed. Only the condition code flags are affected.

Summary of the 68K Instructions

Rather than continue to slog through every instruction in the instruction set of the 68K, let's briefly summarize the instructions by functional group.

Summary of Data Transfer Instructions

EXG	Exchange registers
LEA	Load effective address
LINK	Link and allocate
MOVE	Move data
MOVEA	Move address
MOVEM	Move multiple registers
MOVEP	Move peripheral data
MOVEQ	Move quick
PEA	Push effective address
SWAP	Swap register halves
UNLK	Unlink

Some of the data transfer instructions, such as **LINK** and **UNLK**, may seem rather strange to you. These are special purpose instructions that only exist to support high-level languages. We'll look at these instructions in more detail later on in this chapter.

Summary of Arithmetic Instructions

ADD	Add binary
ADDA	Add address
ADDI	Add immediate
ADDQ	Add quick
CLR	Clear operand
CMP	Compare
CMPI	Compare immediate
CPM	Compare memory
DIVS	Divide signed num.
DIVU	Divide unsigned
MULS	Multiply signed num.
NEG	Negate
NEGX	Negate with X
SUB	Subtract binary
SUBA	Subtract Address
SUBI	Subtract immediate
SUBQ	Subtract quick
SUBX	Subtract with X
TAS	Test and set
TST	Test
EXT	Extend sign

Summary of Privileged Instructions

ANDI SR	And immediate to Status Register
EORI SR	Exclusive OR immediate to Status Register
MOVE SR	Move to/from the Status Register
MOVE USP	Move to/from USP
RESET	Reset the processor
RTE	Return form exception
STOP	Stop the processor
CHK	Check register
ILLEGAL	Force an illegal instruction exception
TRAP	Trap call
TRAPV	Trap on overflow
ANDI CCR	AND immediate to condition code register
ORI CCR	OR immediate to condition code register
EORI CCR	Exclusive OR immediate to CCR
MOVE CCR	Move to/from CCR
NOP	No operation - Do nothing

Summary of Logical and Shift Instructions

AND	Logical AND
ANDI	AND immediate
OR	Logical OR
ORI	OR immediate
EOR	Exclusive OR
EORI	Exclusive OR immediate
NOT	Logical complement
ASL	Arithmetic shift left
ASR	Arithmetic shift right
LSL	Logical shift left
LSR	Logical shift right
ROL	Rotate left
ROR	Rotate right
ROXL	Rotate left with extend
ROXR	Rotate right with extend

Summary of Program Control Instructions

Bcc	Branch on state of conditional code (cc) flag
DBcc	Decrement and branch on cc
Scc	Set on cc
BRA	Branch always
BSR	Branch to subroutine
JMP	Unconditional jump
JSR	Jump to subroutine
RTR	Return and restore
RTS	Return from subroutine

Summary of Bit Manipulation Instructions

BCHG	Bit change
BCLR	Bit clear
BSET	Set bit
BTST	Test bit

Summary of Binary-Coded Decimal (BCD) Instructions

ABCD	Add BCD
NBCD	Negate BCD
SBCD	Subtract BCD

The BCD instructions deserve a word of explanation. They are a legacy from the early days of computers when many instruments were still designed using digital logic circuits that computed results and displayed them as decimal numbers, even though they were counting in binary. A BCD number is the same as the 4-bit hexadecimal number from 0 to 9. A BCD instruction will then generate a carry when the count goes from 9 to A, and the digit is returned to 0. In other words, base 10 counting instead of base 16. These instructions are rarely used today.

As you can see, we have quite a variety of 68K instructions available to us to use to solve a wide variety of real-world programming problems. The 68K instruction set architecture has stood the test of time and is still being designed into new products. Many are designed for managing applications running under operating systems, others to support high-level languages, and others for handling programming exceptions. As we've discussed several times during this introduction to assembly language processing, the important point is to first become comfortable with solving an algorithm using the instructions and addressing modes that you know well, and then move towards more efficient (and perhaps, more difficult) instructions and addressing modes. In fact, there is nothing sub-professional about consistently using a relatively small number of instructions. As you'll see in the next lesson the fact that most problems are solved using a fraction of the available instructions has led to the modern computer architecture, the *reduced instruction set computer,* or RISC.

Simulated I/O Using the **TRAP** **#15** Instruction

This section is included because we'll need to be able to write meaningful programs using the 68K simulator. So far, the programs that we've looked at and written have been totally self-contained. There has not been any interaction with a user. Computer programs like this are nice for studying architecture, but they are of very limited usefulness in the real world. For a computer to be useful, we have to be able to interact with it. In this short section we'll introduce you to a very useful feature of the simulator, the **TRAP** **#15** family of I/O routines.

In the previous section you saw an instruction called **TRAP**. The **TRAP** instruction is like a software-generated interrupt. When the processor executes a **TRAP** instruction, it will automatically pick up the TRAP address from a fixed location in the vector table, place the return address on the stack and begin processing at that new location. **TRAP** **#15** is just one of a list of **TRAP** locations in the Exception Vector Table that is maintained by the processor in the memory region from $080 to $0BC.

If you enter the Teesside version of the 68K simulator and type "HE" in response to the prompt you'll be taken into the help facilities. The **TRAP** **#15** instruction is a way that the designers of the simulator have developed to facilitate I/O between the user and the program. You can read about the **TRAP** **#15** instructions by studying the HELP facility that comes with the E68K program. Alternatively, you can read up on the **TRAP** **#15** facility in *Clements*[1].

Remember, the **TRAP** **#15** instruction is an artifact of the simulator. It was designed to allow I/O to take place between the simulator and the user. If you were really writing I/O routines, you probably would do some things differently, but lots of the things are the same.

Associated with the **TRAP #15** instruction are various tasks. Each task is numbered. Associated with each task is an *application programmer's interface,* or API, that explains how it does its work. For example, *task #0* prints a string to the display and adds a newline character so that the cursor advances to the beginning of the next line. In order to use *task #0* you must set up the following registers (this is the API):

- D0 holds the task number as a byte,
- A1 holds the memory address of the beginning of the string,
- D1 holds the length of the of the string to print,

Once you've set-up the three registers, you call **TRAP #15** as an instruction in your program and you're message will be printed to the display. Thus, the *TRAP* instructions are used as an interface between the 68K simulator and your PCs operating system. Here's a sample program that illustrates how it works. To output the string, "Hello world!" you could use the following code snippet.

"Hello World!" Example

```
************************************************
*
* Test program to print a string to the display
*
************************************************
           OPT         CRE
task0      EQU         00

           ORG         $400

start      MOVE.B      #task0,D0       * Load task number into D0
           LEA         string,A1       * Get address of string
           MOVE.W      str_len,D1      * Length of the string in D1
           TRAP        #15             * Do it
           STOP        #$2700          * Back to simulator

  * Data area

string     DC.B        'Hello world'   * Store the message here
str_len    DC.W        str_len-string  * Get the length of the string
           END         $400
```

Look at the line:

```
str_len    DC.W     str_len-string    * Get the length of the string
```

What's going on? We're letting the assembler do some work for us. The assembler is calculating the length of the string by subtracting the address of the start of the string (labeled "string") from the first address after the string, (labeled "str_len"). It is also a fairly common technique in assembly language to use this method to calculate the distance (in memory locations) between two places in the program. When you subtract pointers in C++ you're doing the same thing.

Task #1 is almost identical to *task #0* except that it does not print the "newline" character. This is handy when you want to prompt the user to enter information. So, you would issue a prompt using *task #1* rather than *task #0* .

Also, you might want to get information from the user. Such as where do they want to run the memory test and what test pattern do they want to use. You might also ask them if they want to run the test again with a different pattern.

Task #2 is another very useful function. Upon executing this instruction the simulator will wait for you to enter a string from the terminal, terminated by the ENTER key. The simulator will place the character string that you entered into a memory buffer pointed to by A1.

The **TRAP #15** has a number of facilities. The best way to understand how to use them is to read about them first, and then write several small test programs to test your knowledge.

Compilers and Assemblers

Up to now we've been keeping our distance from higher level languages, like C and C++. We did take a brief excursion to look at how we could mimic some typical C loop structures using assembly language, but for the most part, we've kept the two languages apart.

However, it should not come as a surprise that even though we might be writing a program in C++, we may be debugging in assembly language. Also, from the point of view of a computer's architecture, to try to understand why certain instructions and addressing modes exist at all. While a complete study of how a compiler does what it does and utilizes a computer's instruction set architecture is beyond the scope of this book, it can't hurt to take a peek under the covers and look at how the compiler does what it does. Therefore, we're going to see how a C language cross-compiler converts a C source file into assembly language.

Consider the following simple C program:

```
int funct(int,int *) ;
void main()
{
    int i = 0, aVar = 5555, j ;
    i++ ;
    j = funct(i, &aVar );
    j++ ;
}

int funct( int var, int * aPtr )
{
    return var + *aPtr ;
}
```

Here we declare 3 variables, i, j and aVar and do some simple manipulation. Of note, we pass two variables into the function, *funct.* One variable, i, is passed by value and the other variable, aVar, is passed by reference. This program was then compiled using a 68000 cross compiler manufactured by the Hewlett-Packard Company® (HP) several years ago for use with its embedded software development tools. This compiler generates an assembly language source file that is then assembled using a 68000 assembler, also manufactured by HP.

Here is the 68000 assembly language code created by the compiler for this program. The actual assembly language instruction or pseudo instructions are shown in gray colored font. I've added the gray colored blocks in order to explain why and what the compiler is doing. This assembly language source file would then be assembled by a 68000 assembler and an object file would be created. Modern C or C++ compilers generally skip the assembly language source file generation process and go right to the object code. However, for our purposes, this "old-fashioned" compiler allows to see exactly what is going on.

```
CHIP     68000 NAME      test3
```

CHIP: 68000 tells the assembler to use only 68000 conventions
NAME: tells assembler the name of the relocatable module to create

* Assembler options:
```
OPT      BRW,FRL,NOI,NOW
```

OPT BRW: All branches should use 16-bit displacements
OPT FRL: All forward references to absolute addresses should use 32-bit mode
OPT NOI: Do not list instructions that are not assembled due to conditional variables
OPT NOW: Do not print warnings during assembly

```
*
* Macro definition for calling run-time libraries:
* bytes per call = 6
*
```

CALL: A macro that takes as a dummy input parameter "routine" and generates the defined sequence of instructions each time it is used in place of an instruction. Macro definitions are a convenient way to group several assembly language instructions under the umbrella of a new instruction. In this case, it will be used to access a run-time subroutine that is located in a library module that will be pulled in when the program is linked together.

XREF is a pseudo op code that tells the assembler that routine is defined in another file.

```
CALL  MACRO    routine
      XREF     routine
      JSR      (routine,PC)
      ENDM
```

SECT: A pseudo-op that indicates that is a relocatable code block named "prog". The C indicates it is a code segment (instructions) and that it is section #2. The "P" is an internal HP type designator.

> The compiler often inserts NOP instructions in front of function calls to aid in debugging.

```
SECT    prog,2,C,P
        NOP
        NOP
        NOP
```

> Watch for interspersed C code and assembly code. The compiler will insert the C source instructions as comments in the assembly code. I added the next two instructions so that it would run in our simulation environment.

```
        ORG $400
        LEA $1000,SP
*    1              int funct(int,int *) ;
*    2              void main()
*    3              {
```

> XDEF: A pseudo op that tells the assembler that this function name, "_main", should be made available to other modules at link time. Thus, the completed program knows where to begin.

```
            XDEF    _main
```

> LINK: This instruction establishes the stack frame. Register A6 is the stack pointer for this function's (_main) stack. The instruction does several things:
> 1. The current value of A6 is pushed onto the stack,
> 2. The current value of the stack pointer, SP (A7), is transferred to A6,
> 3. The sign-extended, 16-bit displacement, -4, is added to the stack pointer in order to reserve 4 bytes on the stack. Thus, A6 is now a local stack pointer with a 4-byte stack. The system stack pointer, SP, sits below it. This operation reserves the space that the function main() needs to store its local variables.

```
_main
        LINK    A6,#-4
                MOVE.L  D3,-(A7)        * Save the current value of D3
                MOVE.L  D2,-(A7)        * Save the current value of D2
                MOVEQ   #$FF,D1         * Not exactly sure why
        MOVE.L  D1,(-4,A6)
```

```
* This puts FF at the bottom of the stack frame for _main
        MOVEQ   #$FF,D2         * Who knows?
        MOVE.L  D2,D3           * Ditto
```

> Register 'D2' is register variable 'S_i'. The compiler is smart
> enough to realize that it can save variables in registers instead of
> memory, thus running the program more efficiently. Below, it initial-
> izes D2 to 0.

```
        MOVEQ    #0,D2
```

> SET is a pseudo op code that gives a number a value that can be reas-
> signed. A variable given a name with the equate psuedo-op (EQU) can't
> be reassigned a new name or value.

```
S_aVar  SET      -4
```

> The value 5555 is placed in the stack frame. Register 'D3' will be
> assigned to the register variable 'S_j'. The ADDQ instruction imple-
> ments i++

```
        MOVE.L   #5555,(S_aVar+0,A6)
```

```
*    4              int i = 0, aVar = 5555, j ;
*    5              i++ ;


        ADDQ.L   #1,D2


*    6              j = funct(i, &aVar );
```

> The address of aVar will be placed in A0 and then the contents of
> the address that A0 is pointing to will be pushed onto the stack.
> This next block of code is called a prologue. Code like it is gener-
> ated whenever there is a function call made, or as we know it, a JSR
> instruction. Here:
> * A0 holds the address of aVar
> * The address of aVar is pushed onto the stack.
> * i is pushed onto the stack
> * The JSR instruction is implemented using PC relative addressing

```
        LEA      (S_aVar+0,A6),A0
        PEA      (A0)
        MOVEA.L  D2,A0
        PEA      (A0)
        JSR      (_funct+0,PC)

        ADDQ.L   #8,SP            * main goes out of scope
        MOVE.L   D0,D3            * D0 is the return register from JSR


*    7              j++ ;
  ADDQ.L   #1,D3            * This is j++
*    8              }
```

> The closing brace "}" causes the following function exit code to be generated.

```
functionExit1   NOP
                MOVE.L   (A7)+,D2      * Restore D2
                MOVE.L   (A7)+,D3      * Restore D3
                UNLK     A6      * De-allocate stack frame


returnLabel1      RTS
                NOP
                NOP
                NOP
*     9            int funct( int var, int * aPtr )
*     10           {
                XDEF     _funct
```

> Here is the subroutine entry point for "_funct". Notice how the compiler creates an assembly language label for the subroutine by placing an underscore character "_" in front of the function's name. Also notice that the function does not appear to need any variable storage space when the stack frame is allocated.

```
_funct
        LINK     A6,#-0
S_var   SET      8
S_aPtr  SET      12
*     11            return var + *aPtr ;
```

> This next block of code is quite interesting. Before the pointer is de-referenced to return the value of aVar, the compiler checks to see if the address of aVar is a NULL pointer. If it is, it calls a runtime error handler. This was the purpose of generating the MACRO code. The instructions do the following:
> - Get the value of aVar from the stack and put it in A0.
> - Move A0 into D1 to force the Z flag to be set if <A0> = 0
> - If A0 = 0 then there is a NULL pointer
> - If the pointer is bad it does a JSR to the library routine "ptrFault"

```
        MOVE.L   (S_aPtr+0,A6),A0
        MOVE.L   A0,D1
        BEQ.S    L0_APtrChk
        MOVEQ    #$FF,D0
        CMP.L    D0,D1
        BNE.S    L0_BPtrChk
L0_APtrChk
        CALL     ptrfault
```

> The next instruction was added so the program would run in the 68K
> simulator environment.

```
ptrFault   RTS
L0_PtrInfo
           DC.L    11
           DC.B    'test3.c'
           DC.B    0
```

> ALIGN: A pseudo op that forces instructions to be on word boundaries.
> The next3 instructions actually implements the C instruction:
>
> return var + *aPtr
>
> - The variable aVar is moved into register D0
> - i is added to D0 and the results put back into D0

```
           ALIGN   2
L0_BPtrChk
           MOVE.L  (S_var+0,A6),D0
           ADD.L   (A0),D0
           BRA     functionExit2
12              }
```

> "}" generates function exit code

```
functionExit2
           NOP              * This is the return for the function call
           UNLK    A6       * Deallocate stack frame returnLabel2
           RTS              * This is the return for the program

           END $400
```

The above code walk-through is a glimpse of how a compiler works. From an assembly language perspective some of the instruction sequence didn't make any sense. However, we must keep in mind the fact that the compiler is just a program and must be able to deal with all of the possibilities of the C language. This means that once a set of housekeeping rules are established for data handling, those rules are always observed, even if they are unnecessary or redundant in a particular situation. Thus, you can see why hand crafting assembly code can often result in performance improvements.

Now we can ask the $65,536 question. Does it work? Does the above assembly code correctly execute the original C code? Let's see. We'll actually run the code in the simulator. We'll also monitor the behavior of the stack because that's where the action is. The simulator code will be listed on the left and the state of the stack will be listed on the right. We'll indicate the current value of the stack frame pointer, A6 in light gray and the stack pointer, SP, in dark gray, as shown below:

```
A6   SP
```

```
PC=000400  SR=2000  SS=00A00000  US=00000000  X=0
A0=00000000  A1=00000000  A2=00000000  A3=00000000  N=0
A4=00000000  A5=00000000  A6=00000000  A7=00A00000  Z=0
D0=00000000  D1=00000000  D2=00000000  D3=00000000  V=0
D4=00000000  D5=00000000  D6=00000000  D7=00000000  C=0
---------->LEA.L $1000,SP
```

1000	00	00	00	00
OFFC	00	00	00	00
OFF8	00	00	00	00
OFF4	00	00	00	00

```
PC=000404  SR=2000  SS=00001000  US=00000000  X=0
A0=00000000  A1=00000000  A2=00000000  A3=00000000  N=0
A4=00000000  A5=00000000  A6=00000000  A7=00001000  Z=0
D0=00000000  D1=00000000  D2=00000000  D3=00000000  V=0
D4=00000000  D5=00000000  D6=00000000  D7=00000000  C=0
---------->LINK A6,#-4
```

```
PC=000408  SR=2000  SS=00000FF8  US=00000000  X=0
A0=00000000  A1=00000000  A2=00000000  A3=00000000  N=0
A4=00000000  A5=00000000  A6=00000FFC  A7=00000FF8  Z=0
D0=00000000  D1=00000000  D2=00000000  D3=00000000  V=0
D4=00000000  D5=00000000  D6=00000000  D7=00000000  C=0
---------->MOVE.L D3,-(SP)
```

1000	00	00	00	00
OFFC	00	00	00	00
OFF8	00	00	00	00
OFF4	00	00	00	00

```
PC=00040A  SR=2004  SS=00000FF4  US=00000000  X=0
A0=00000000  A1=00000000  A2=00000000  A3=00000000  N=0
A4=00000000  A5=00000000  A6=00000FFC  A7=00000FF4  Z=1
D0=00000000  D1=00000000  D2=00000000  D3=00000000  V=0
D4=00000000  D5=00000000  D6=00000000  D7=00000000  C=0
---------->MOVE.L D2,-(SP)
```

1000	00	00	00	00
OFFC	00	00	00	00
OFF8	00	00	00	00
OFF4	00	00	00	00

The above section of code shows the stack pointer being set to address $1000 and the LINK instruction being used to create a local stack frame. Here's what happened:

- Memory locations $0FFD through $1000 save the current value of register A6.
- 4 bytes are reserved on the stack and A6 becomes the stack pointer for those bytes. A6 is set to $0FFC.
- The system stack pointer, A7, is reset below the local stack to $0FF8

Next, D2 and D3 are pushed onto the stack because they're going to be needed.

```
PC=00040C SR=2004 SS=00000FF0 US=00000000          X=0  1000
A0=00000000 A1=00000000 A2=00000000 A3=00000000 N=0  OFFC
A4=00000000 A5=00000000 A6=00000FFC A7=00000FF0 Z=1  OFF8
D0=00000000 D1=00000000 D2=00000000 D3=00000000 V=0  OFF4
D4=00000000 D5=00000000 D6=00000000 D7=00000000 C=0  OFF0
---------->MOVEQ #-1,D1
```

1000	00	00	00	00
OFFC	00	00	00	00
OFF8	00	00	00	00
OFF4	00	00	00	00
OFF0	00	00	00	00

```
PC=00040E SR=2008 SS=00000FF0 US=00000000          X=0
A0=00000000 A1=00000000 A2=00000000 A3=00000000 N=1
A4=00000000 A5=00000000 A6=00000FFC A7=00000FF0 Z=0
D0=00000000 D1=FFFFFFFF D2=00000000 D3=00000000 V=0
D4=00000000 D5=00000000 D6=00000000 D7=00000000 C=0
---------->MOVE.L D1,-4(A6)
```

```
C=000412 SR=2008 SS=00000FF0 US=00000000           X=0  1000
A0=00000000 A1=00000000 A2=00000000 A3=00000000 N=1  OFFC
A4=00000000 A5=00000000 A6=00000FFC A7=00000FF0 Z=0  OFF8
D0=00000000 D1=FFFFFFFF D2=00000000 D3=00000000 V=0  OFF4
D4=00000000 D5=00000000 D6=00000000 D7=00000000 C=0  OFF0
---------->MOVEQ #-1,D2
```

1000	00	00	00	00
OFFC	00	00	00	00
OFF8	FF	FF	FF	FF
OFF4	00	00	00	00
OFF0	00	00	00	00

```
PC=000414 SR=2008 SS=00000FF0 US=00000000          X=0
A0=00000000 A1=00000000 A2=00000000 A3=00000000 N=1
A4=00000000 A5=00000000 A6=00000FFC A7=00000FF0 Z=0
D0=00000000 D1=FFFFFFFF D2=FFFFFFFF D3=00000000 V=0
D4=00000000 D5=00000000 D6=00000000 D7=00000000 C=0
---------->MOVE.L D2,D3
```

Note the highlighted instruction, MOVE.L D1, -4(A6). This is an example of address register indirect with offset addressing. The compiler is using A6 as a stack pointer to place data in its local stack frame.

The stack frame is simple the variable storage region that the compiler establishes on the stack for the variables of its function. The rest of the simulation will be presented without any further elaboration. We'll leave it as an exercise for you to work through the instructions and stack conditions and convince yourself that it really works as advertised.

```
PC=000416 SR=2008 SS=00000FF0 US=00000000          X=0  1000
A0=00000000 A1=00000000 A2=00000000 A3=00000000 N=1  OFFC
A4=00000000 A5=00000000 A6=00000FFC A7=00000FF0 Z=0  OFF8
D0=00000000 D1=FFFFFFFF D2=FFFFFFFF D3=FFFFFFFF V=0  OFF4
D4=00000000 D5=00000000 D6=00000000 D7=00000000 C=0  OFF0
---------->MOVEQ #0,D2
```

1000	00	00	00	00
OFFC	00	00	00	00
OFF8	FF	FF	FF	FF
OFF4	00	00	00	00
OFF0	00	00	00	00

```
PC=000418 SR=2004 SS=00000FF0 US=00000000        X=0
A0=00000000 A1=00000000 A2=00000000 A3=00000000 N=0
A4=00000000 A5=00000000 A6=00000FFC A7=00000FF0 Z=1
D0=00000000 D1=FFFFFFFF D2=00000000 D3=FFFFFFFF V=0
D4=00000000 D5=00000000 D6=00000000 D7=00000000 C=0
---------->MOVE.L #5555,-4(A6)
```

1000	00	00	00	00
OFFC	00	00	00	00
OFF8	00	00	15	B3
OFF4	00	00	00	00
OFF0	00	00	00	00

```
PC=000420 SR=2000 SS=00000FF0 US=00000000        X=0
A0=00000000 A1=00000000 A2=00000000 A3=00000000 N=0
A4=00000000 A5=00000000 A6=00000FFC A7=00000FF0 Z=0
D0=00000000 D1=FFFFFFFF D2=00000000 D3=FFFFFFFF V=0
D4=00000000 D5=00000000 D6=00000000 D7=00000000 C=0
---------->ADDQ.L #1,D2
PC=000422 SR=2000 SS=00000FF0 US=00000000        X=0
A0=00000000 A1=00000000 A2=00000000 A3=00000000 N=0
A4=00000000 A5=00000000 A6=00000FFC A7=00000FF0 Z=0
D0=00000000 D1=FFFFFFFF D2=00000001 D3=FFFFFFFF V=0
D4=00000000 D5=00000000 D6=00000000 D7=00000000 C=0
---------->LEA.L -4(A6),A0
PC=000426 SR=2000 SS=00000FF0 US=00000000        X=0
A0=00000FF8 A1=00000000 A2=00000000 A3=00000000 N=0
A4=00000000 A5=00000000 A6=00000FFC A7=00000FF0 Z=0
D0=00000000 D1=FFFFFFFF D2=00000001 D3=FFFFFFFF V=0
D4=00000000 D5=00000000 D6=00000000 D7=00000000 C=0
---------->PEA (A0)
```

1000	00	00	00	00
OFFC	00	00	00	00
OFF8	00	00	15	B3
OFF4	00	00	00	00
OFF0	00	00	00	00
OFEC	00	00	0F	F8

```
PC=000428 SR=2000 SS=00000FEC US=00000000        X=0
A0=00000FF8 A1=00000000 A2=00000000 A3=00000000 N=0
A4=00000000 A5=00000000 A6=00000FFC A7=00000FEC Z=0
D0=00000000 D1=FFFFFFFF D2=00000001 D3=FFFFFFFF V=0
D4=00000000 D5=00000000 D6=00000000 D7=00000000 C=0
---------->MOVEA.L D2,A0
PC=00042A SR=2000 SS=00000FEC US=00000000        X=0
A0=00000001 A1=00000000 A2=00000000 A3=00000000 N=0
A4=00000000 A5=00000000 A6=00000FFC A7=00000FEC Z=0
D0=00000000 D1=FFFFFFFF D2=00000001 D3=FFFFFFFF V=0
D4=00000000 D5=00000000 D6=00000000 D7=00000000 C=0
---------->PEA (A0)
```

1000	00	00	00	00
OFFC	00	00	00	00
OFF8	00	00	15	B3
OFF4	00	00	00	00
OFF0	00	00	00	00
OFEC	00	00	0F	F8
OFE8	00	00	00	01

```
PC=00042C SR=2000 SS=00000FE8 US=00000000        X=0
A0=00000001 A1=00000000 A2=00000000 A3=00000000 N=0
A4=00000000 A5=00000000 A6=00000FFC A7=00000FE8 Z=0
D0=00000000 D1=FFFFFFFF D2=00000001 D3=FFFFFFFF V=0
D4=00000000 D5=00000000 D6=00000000 D7=00000000 C=0
---------->JSR 28(PC)
```

```
C=00044A SR=2000 SS=00000FE4 US=00000000          X=0
A0=00000001 A1=00000000 A2=00000000 A3=00000000 N=0
A4=00000000 A5=00000000 A6=00000FFC A7=00000FE4 Z=0
D0=00000000 D1=FFFFFFFF D2=00000001 D3=FFFFFFFF V=0
D4=00000000 D5=00000000 D6=00000000 D7=00000000 C=0
---------->LINK A6,#0
PC=00044E SR=2000 SS=00000FE0 US=00000000          X=0
A0=00000001 A1=00000000 A2=00000000 A3=00000000 N=0
A4=00000000 A5=00000000 A6=00000FE0 A7=00000FE0 Z=0
D0=00000000 D1=FFFFFFFF D2=00000001 D3=FFFFFFFF V=0
D4=00000000 D5=00000000 D6=00000000 D7=00000000 C=0
```

1000	00	00	00	00
0FFC	00	00	00	00
0FF8	00	00	15	B3
0FF4	00	00	00	00
0FF0	00	00	00	00
0FEC	00	00	0F	F8
0FE8	00	00	00	01
0FE4	00	00	04	30
0FE0	00	00	00	00

```
---------->MOVEA.L 12(A6),A0
PC=000452 SR=2000 SS=00000FE0 US=00000000          X=0
A0=00000FF8 A1=00000000 A2=00000000 A3=00000000 N=0
A4=00000000 A5=00000000 A6=00000FE0 A7=00000FE0 Z=0
D0=00000000 D1=FFFFFFFF D2=00000001 D3=FFFFFFFF V=0
D4=00000000 D5=00000000 D6=00000000 D7=00000000 C=0
---------->MOVE.L A0,D1
PC=000454 SR=2000 SS=00000FE0 US=00000000          X=0
A0=00000FF8 A1=00000000 A2=00000000 A3=00000000 N=0
A4=00000000 A5=00000000 A6=00000FE0 A7=00000FE0 Z=0
D0=00000000 D1=00000FF8 D2=00000001 D3=FFFFFFFF V=0
D4=00000000 D5=00000000 D6=00000000 D7=00000000 C=0
```

1000	00	00	00	00
0FFC	00	00	00	00
0FF8	00	00	15	B3
0FF4	00	00	00	00
0FF0	00	00	00	00
0FEC	00	00	0F	F8
0FE8	00	00	00	01
0FE4	00	00	04	30
0FE0	00	00	00	00

```
---------->BEQ.S $0000045C
PC=000456 SR=2000 SS=00000FE0 US=00000000          X=0
A0=00000FF8 A1=00000000 A2=00000000 A3=00000000 N=0
A4=00000000 A5=00000000 A6=00000FE0 A7=00000FE0 Z=0
D0=00000000 D1=00000FF8 D2=00000001 D3=FFFFFFFF V=0
D4=00000000 D5=00000000 D6=00000000 D7=00000000 C=0
---------->MOVEQ #-1,D0
PC=000458 SR=2008 SS=00000FE0 US=00000000          X=0
A0=00000FF8 A1=00000000 A2=00000000 A3=00000000 N=1
A4=00000000 A5=00000000 A6=00000FE0 A7=00000FE0 Z=0
D0=FFFFFFFF D1=00000FF8 D2=00000001 D3=FFFFFFFF V=0
D4=00000000 D5=00000000 D6=00000000 D7=00000000 C=0
```

1000	00	00	00	00
0FFC	00	00	00	00
0FF8	00	00	15	B3
0FF4	00	00	00	00
0FF0	00	00	00	00
0FEC	00	00	0F	F8
0FE8	00	00	00	01
0FE4	00	00	04	30
0FE0	00	00	0F	FC

```
---------->CMP.L D0,D1
PC=00045A SR=2001 SS=00000FE0 US=00000000          X=0
A0=00000FF8 A1=00000000 A2=00000000 A3=00000000 N=0
A4=00000000 A5=00000000 A6=00000FE0 A7=00000FE0 Z=0
D0=FFFFFFFF D1=00000FF8 D2=00000001 D3=FFFFFFFF V=0
D4=00000000 D5=00000000 D6=00000000 D7=00000000 C=1
---------->BNE.S $00000470
PC=000470 SR=2001 SS=00000FE0 US=00000000          X=0
A0=00000FF8 A1=00000000 A2=00000000 A3=00000000 N=0
A4=00000000 A5=00000000 A6=00000FE0 A7=00000FE0 Z=0
D0=FFFFFFFF D1=00000FF8 D2=00000001 D3=FFFFFFFF V=0
```

```
D4=00000000 D5=00000000 D6=00000000 D7=00000000 C=1
---------->MOVE.L 8(A6),D0
PC=000474 SR=2000 SS=00000FE0 US=00000000          X=0
A0=00000FF8 A1=00000000 A2=00000000 A3=00000000 N=0
A4=00000000 A5=00000000 A6=00000FE0 A7=00000FE0 Z=0
D0=00000001 D1=00000FF8 D2=00000001 D3=FFFFFFFF V=0
D4=00000000 D5=00000000 D6=00000000 D7=00000000 C=0
---------->ADD.L (A0),D0
PC=000476 SR=2000 SS=00000FE0 US=00000000          X=0
A0=00000FF8 A1=00000000 A2=00000000 A3=00000000 N=0
A4=00000000 A5=00000000 A6=00000FE0 A7=00000FE0 Z=0
D0=000015B4 D1=00000FF8 D2=00000001 D3=FFFFFFFF V=0
D4=00000000 D5=00000000 D6=00000000 D7=00000000 C=0
---------->BRA.L $0000047A
PC=00047A SR=2000 SS=00000FE0 US=00000000          X=0
A0=00000FF8 A1=00000000 A2=00000000 A3=00000000 N=0
A4=00000000 A5=00000000 A6=00000FE0 A7=00000FE0 Z=0
D0=000015B4 D1=00000FF8 D2=00000001 D3=FFFFFFFF V=0
D4=00000000 D5=00000000 D6=00000000 D7=00000000 C=0
---------->NOP
PC=00047C SR=2000 SS=00000FE0 US=00000000          X=0
A0=00000FF8 A1=00000000 A2=00000000 A3=00000000 N=0
A4=00000000 A5=00000000 A6=00000FE0 A7=00000FE0 Z=0
D0=000015B4 D1=00000FF8 D2=00000001 D3=FFFFFFFF V=0
D4=00000000 D5=00000000 D6=00000000 D7=00000000 C=0
---------->UNLK A6
PC=00047E SR=2000 SS=00000FE4 US=00000000          X=0
A0=00000FF8 A1=00000000 A2=00000000 A3=00000000 N=0
A4=00000000 A5=00000000 A6=00000FFC A7=00000FE4 Z=0
D0=000015B4 D1=00000FF8 D2=00000001 D3=FFFFFFFF V=0
D4=00000000 D5=00000000 D6=00000000 D7=00000000 C=0
---------->RTS
PC=000430 SR=2000 SS=00000FE8 US=00000000          X=0
A0=00000FF8 A1=00000000 A2=00000000 A3=00000000 N=0
A4=00000000 A5=00000000 A6=00000FFC A7=00000FE8 Z=0
D0=000015B4 D1=00000FF8 D2=00000001 D3=FFFFFFFF V=0
D4=00000000 D5=00000000 D6=00000000 D7=00000000 C=0
---------->ADDQ.L #8,SP
PC=000432 SR=2000 SS=00000FF0 US=00000000          X=0
A0=00000FF8 A1=00000000 A2=00000000 A3=00000000 N=0
A4=00000000 A5=00000000 A6=00000FFC A7=00000FF0 Z=0
D0=000015B4 D1=00000FF8 D2=00000001 D3=FFFFFFFF V=0
D4=00000000 D5=00000000 D6=00000000 D7=00000000 C=0
---------->MOVE.L D0,D3
PC=000434 SR=2000 SS=00000FF0 US=00000000          X=0
```

1000	00	00	00	00
0FFC	00	00	00	00
0FF8	00	00	15	B3
0FF4	00	00	00	00
0FF0	00	00	00	00
0FEC	00	00	0F	F8
0FE8	00	00	00	01
0FE4	00	00	04	30
0FE0	00	00	0F	FC

1000	00	00	00	00
0FFC	00	00	00	00
0FF8	00	00	15	B3
0FF4	00	00	00	00
0FF0	00	00	00	00
0FEC	00	00	0F	F8
0FE8	00	00	00	01
0FE4	00	00	04	30
0FE0	00	00	0F	FC

1000	00	00	00	00
0FFC	00	00	00	00
0FF8	00	00	15	B3
0FF4	00	00	00	00
0FF0	00	00	00	00
0FEC	00	00	0F	F8
0FE8	00	00	00	01
0FE4	00	00	04	30
0FE0	00	00	0F	FC

```
A0=00000FF8 A1=00000000 A2=00000000 A3=00000000 N=0
A4=00000000 A5=00000000 A6=00000FFC A7=00000FF0 Z=0
D0=000015B4 D1=00000FF8 D2=00000001 D3=000015B4 V=0
D4=00000000 D5=00000000 D6=00000000 D7=00000000 C=0
---------->ADDQ.L #1,D3

PC=000436 SR=2000 SS=00000FF0 US=00000000          X=0
A0=00000FF8 A1=00000000 A2=00000000 A3=00000000 N=0
A4=00000000 A5=00000000 A6=00000FFC A7=00000FF0 Z=0
D0=000015B4 D1=00000FF8 D2=00000001 D3=000015B5 V=0
D4=00000000 D5=00000000 D6=00000000 D7=00000000 C=0
---------->NOP
PC=000438 SR=2000 SS=00000FF0 US=00000000          X=0
A0=00000FF8 A1=00000000 A2=00000000 A3=00000000 N=0
A4=00000000 A5=00000000 A6=00000FFC A7=00000FF0 Z=0
D0=000015B4 D1=00000FF8 D2=00000001 D3=000015B5 V=0
D4=00000000 D5=00000000 D6=00000000 D7=00000000 C=0
---------->MOVE.L (SP)+,D2

PC=00043A SR=2004 SS=00000FF4 US=00000000          X=0
A0=00000FF8 A1=00000000 A2=00000000 A3=00000000 N=0
A4=00000000 A5=00000000 A6=00000FFC A7=00000FF4 Z=1
D0=000015B4 D1=00000FF8 D2=00000000 D3=000015B5 V=0
D4=00000000 D5=00000000 D6=00000000 D7=00000000 C=0
---------->MOVE.L (SP)+,D3

PC=00043C SR=2004 SS=00000FF8 US=00000000          X=0
A0=00000FF8 A1=00000000 A2=00000000 A3=00000000 N=0
A4=00000000 A5=00000000 A6=00000FFC A7=00000FF8 Z=1
D0=000015B4 D1=00000FF8 D2=00000000 D3=00000000 V=0
D4=00000000 D5=00000000 D6=00000000 D7=00000000 C=0
---------->UNLK A6
PC=00043E SR=2004 SS=00001000 US=00000000          X=0
A0=00000FF8 A1=00000000 A2=00000000 A3=00000000 N=0
A4=00000000 A5=00000000 A6=00000000 A7=00001000 Z=1
D0=000015B4 D1=00000FF8 D2=00000000 D3=00000000 V=0
D4=00000000 D5=00000000 D6=00000000 D7=00000000 C=0
---------->RTS
```

1000	00	00	00	00
0FFC	00	00	00	00
0FF8	00	00	15	B3
0FF4	00	00	00	00
0FF0	00	00	00	00
0FEC	00	00	0F	F8
0FE8	00	00	00	01
0FE4	00	00	04	30
0FE0	00	00	0F	FC

1000	00	00	00	00
0FFC	00	00	00	00
0FF8	00	00	15	B3
0FF4	00	00	00	00
0FF0	00	00	00	00
0FEC	00	00	0F	F8
0FE8	00	00	00	01
0FE4	00	00	04	30
0FE0	00	00	0F	FC

1000	00	00	00	00
0FFC	00	00	00	00
0FF8	00	00	15	B3
0FF4	00	00	00	00
0FF0	00	00	00	00
0FEC	00	00	0F	F8
0FE8	00	00	00	01
0FE4	00	00	04	30
0FE0	00	00	0F	FC

You may be curious why the last instruction is an RTS. If this program were running under an operating system, and the C compiler expects that you are, this instruction would return the control to the operating system.

1000	00	00	00	00
0FFC	00	00	00	00
0FF8	00	00	15	B3
0FF4	00	00	00	00
0FF0	00	00	00	00
0FEC	00	00	0F	F8
0FE8	00	00	00	01
0FE4	00	00	04	30
0FE0	00	00	0F	FC

There's one last matter to discuss before we leave 68000 assembly language and move on to other architectures. This is the subject of instruction set decomposition. We've already looked at the architecture from the point of view of the instruction set, the addressing modes and the internal register resources. Now, we'll try to relate that back to our discussion of state machines to see how the actual encoding of the instructions takes place.

The process of converting machine language back to assembly language is called *disassembly.* A disassembler is a program that examines the machine code in memory and attempts to convert it back to machine language. This is an extremely useful tool for debugging and, in fact, most debuggers have a built-in disassembly feature.

You may recall from our introduction to assembly language that the first word of an instruction is called the *opcode word.* The op-code word contains an opcode, which tells the computer (what to do), and it also contains zero, one or two *effective address fields (EA).* The effective address fields contain the encoded information that tell the processor how to retrieve the operands from memory. In other words, what is the effective address of the operand(s).

As you already know, an operand might be an address register or a data register. The operand might be located in memory, but pointed to by the address register. Consider the form of the instructions shown below:

OP Code		Destination EA		Source EA	
DB15	DB12	DB11	DB6	DB5	DB0

The op code field consists of 4 bits, DB12 through DB15. This tells the computer that the instruction is a MOVE instruction. That is, move the contents of the memory location specified by the *source EA* to the memory location specified by the *destination EA.* Furthermore, the effective address field is further decomposed into a 3-bit Mode field and a 3-bit Subclass (register) field.

This information is sufficient to tell the computer everything it needs to know in order to complete the instruction, *but the op-code word itself may or may not contain all of the information necessary to complete the instruction.* The computer may have to go out to memory one or more additional times to fetch additional information about the source EA or the destination EA in order to complete the instruction. This begins to make sense if we recall that the op-code word is the part of the instruction that must be decoded by the microcode-driven state machine. Once the proper state machine sequence has been established by the decoding of the op-code word, the additional fetching of operands from memory, if necessary, can proceed.

For example, suppose that the source EA and destination EA are both data registers. In other words, the instruction is:

`MOVE.W  D0,D1`

Since both the source and the destination are internal registers, there is no additional information needed by the processor to execute the instruction. For this example, the instruction is the same length as the op code, 16-bits long, or one word in length.

However, suppose that the instruction is:

`MOVE.W  #$00D0,D1`

Now the # sign tells us that the source EA is an immediate operand (the hexadecimal number $00D0) and the destination EA is still register D1. In this case, the op code word would tell us that the source is an immediate operand, but it can't tell us what is the actual value of the immediate operand. The computer must go out to memory again and retrieve (fetch) the next word in memory after the op code word. This is the data value $00D0. If this instruction resided in memory at memory location $1000, the op code word would take up word address $1000 and the operand would be at memory location $1002.

The effective address field is a 6-bit wide field that is further subdivided into two, 3-bit fields called *mode* and *register*. The 3-bit wide mode field can specify one of 8 possible addressing modes and the register filed can specify one of 8 possible registers, or a *subclass* for the mode field.

The MOVE instruction is unique in that it has two possible effective address fields. All of the other instructions that may contain an effective address field have only one possible effective address. This might seem strange to you at first, but as you'll see, almost all of the other instructions that use two operands must involve a register and a single effective address.

Let's return our attention to the MOVE instruction. If we completely break it down to its constituent parts it would look like this:

0 0	Size	Destination Register	Destination Mode	Source Mode	Source Register
DB15 DB14	DB13 DB12	DB11 DB10 DB9	DB8 DB7 DB6	DB5 DB4 DB3	DB2 DB1 DB0

The subclasses are only used with mode 7 addressing modes. These modes are:

You may have noticed the Move to Address Register (MOVEA) instruction and wondered why we needed to create a special mnemonic for an ordinary MOVE instruction. Well first of all, address registers are extremely useful and important internal resources. Having an address register

Mode 7 addressing	Subclass
Absolute word	000
Absolute long word	001
Immediate	100
PC relative with displacement	010
PC relative with index	011

means that we can do register arithmetic and manipulate pointers. We can also compare the values in registers and make decisions based upon the address of data.

The MOVEA instruction is a more standard type of instruction because there is only one effective address for the source operand. The destination operand is one of the 7 address registers. Thus, the destination mode is hard coded to be an address register. However, we needed to create a special instruction for the MOVEA operation because a word fetch on an odd byte boundary is an illegal access and the MOVEA instruction prevents this from occuring.

The majority of the instructions take the form shown below.

Op Code		Register		Op Mode		Source or Destination EA	
DB15	DB12	DB11	DB9	DB8	DB6	DB5	DB0

The op code field would specify whether the instruction is an ADD, AND, CMP, etc. The register field specifies which of the internal registers is the source or destination of the operation. The *op mode field* is a new term. This specifies several one of 8 subdivisions for the instruction. For example, this field would specify if the internal register was the source or the destination of the operation and if the size was byte, word or long word. We'll see this a bit later.

Single operand instructions, such as CLR (set the contents of the destination EA to zero) have a slightly different form.

Op Code		Size		Destination EA	
DB15	DB8	DB7 DB6	DB5		DB0

The JMP (Jump) and JSR (Jump to Subroutine) are also single operand instructions. Both place the destination EA into the program counter so that the next instruction is fetched from the new location, rather than the next instruction in the sequence. The JSR instruction will also automatically place the current value of the program counter onto the stack before the contents of the PC are replaced with the destination EA. Thus, the return location is saved on the stack.

The branch instructions, Bcc (branch on condition code), is similar to the jump instruction in that it modifies the contents of the PC so that the next instruction may be fetched out of sequence. However, the branch instructions differ in two fundamental ways:

1. There is no effective address. A *displacement* value is *added to the current contents of the PC* so that the new value is determined as a positive or negative shift from the current value. Thus, a displacement is a relative jump, rather than an absolute jump.
2. The branch is taken only if the condition code being tested evaluates to true.

0	1	1	0	Condition		Displacement	
DB15			DB12	DB11	DB8	DB7	DB0

The displacement field is an 8-bit value. This means that the branch can move to another locations either +127 bytes or –128 bytes away from the present location. If a greater branch is desired, then the displacement field is set to all zeroes and the next memory word is used as an immediate value for the displacement. This gives a range of +16,383 to –16,384 bytes.

Thus, if the computer is executing a tight loop, then it will operate more efficiently if an 8-bit displacement is used. The 16-bit displacement allows it go further on a branch, but at a cost of an additional memory fetch operation.

Earlier, I said that the branch is executed if the condition code evaluates to true. This means that we can test more than just the state of one of the flags in the CCR. The table below shows us the possible ways that a branch condition may be evaluated.

CC	carry clear	0100	$\overline{C}$		LS	low or same	0011	$C + Z$
CS	carry set	0101	C		LT	less than	1101	$N*\overline{V} + \overline{N}*V$
EQ	equal	0111	Z		MI	minus	1011	N
GE	greater or equal	1100	$N*V + \overline{N}*\overline{V}$		NE	not equal	0110	$\overline{Z}$
GT	greater than	1110	$N*V*\overline{Z} + \overline{N}*\overline{V}*\overline{Z}$		PL	plus	1010	$\overline{N}$
HI	high	0010	$\overline{C}*\overline{Z}$		VC	overflow clear	1000	$\overline{V}$
LE	less or equal	1111	$Z + N*\overline{V} + \overline{N}*V$		VS	overflow set	1001	V

The reason for so many branch test conditions is that the branch instructions *BGT, BGE, BLT and BLE* are designed to be used with signed arithmetic operations and the branch instructions *BHI, BCC, BLS and BCS* are designed to be used with unsigned arithmetic operations.

Some instructions, such as **NOP** and **RTS** (Return from Subroutine) take no operands and are completely specified by their op code word.

Earlier we discussed the role of the op mode field on how a general instruction operation may be further modified. A good example to illustrate how the op mode field works is to examine the ADD instruction and all of its variations in more detail. Consider Figure 9.3

Notice how the op mode field, contained in bits 6, 7 and 8, define the format of the instruction. The **ADD** instruction is representative of most of the "normal" instructions. Other classes of instructions require special consideration. For example, the **MOVE** instruction contains two effective addresses. Also, all immediate addressing mode instructions have the source operand addressing mode hard-coded into the instruction and thus, it is not really an effective address.

	15	14	13	12	11	10	9	8	7	6	5	4	3	2	1	0
	1	1	0	1	n	n	n	OP	OP	OP	EA	EA	EA	EA	EA	EA
ADD.B <ea>,Dn	1	1	0	1	n	n	n	0	0	0	EA	EA	EA	EA	EA	EA
ADD.W <ea>,Dn	1	1	0	1	n	n	n	0	0	1	EA	EA	EA	EA	EA	EA
ADD.L <ea>,Dn	1	1	0	1	n	n	n	0	1	0	EA	EA	EA	EA	EA	EA
ADD.B Dn,<ea>	1	1	0	1	n	n	n	1	0	0	EA	EA	EA	EA	EA	EA
ADD.W Dn,<ea>	1	1	0	1	n	n	n	1	0	1	EA	EA	EA	EA	EA	EA
ADD.L Dn,<ea>	1	1	0	1	n	n	n	1	1	0	EA	EA	EA	EA	EA	EA
ADDA.W <ea>,An	1	1	0	1	n	n	n	0	1	1	EA	EA	EA	EA	EA	EA
ADDA.L <ea>,An	1	1	0	1	n	n	n	1	1	1	EA	EA	EA	EA	EA	EA

D   0..7   0..7

Figure 9.3 Op mode decomposition for the ADD instruction

This would include instructions such as add immediate (**ADDI**) and subtract immediate (**SUBI**).

Example of a Real Machine

Before we move on to consider other architectures let's do something a bit different, but still relevant to our discussion of assembly code and computer architecture. Up to now, we've really ignored the fact that ultimately the code we write will have to run on a real machine. Granted, memory test programs aren't that exciting, but at least we can imagine how they might be used with actual hardware.

In order to conclude our discussion of assembly language coding, let's look at the design of an actual 68000-based system.

Figures 9.4 and 9.5 are simplified schematic diagrams of the processor and memory portions of a 68K computer system. Not all the signals are shown, but that won't take anything away from the discussion. Also, we don't want the EE's getting too upset with us.

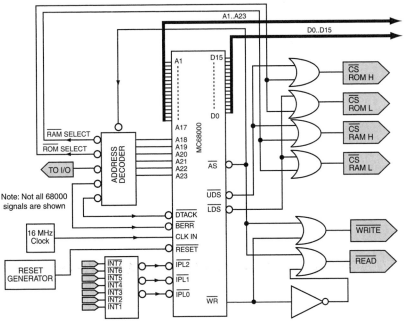

Figure 9.4 Simplified schematic diagram of a 68000-based computer system. Only the 68000 CPU is shown in this figure.

Referring to the address bus of Figure 9.4, we see that that there is no address bit labeled A0. The A0 is synthesized by the state machine by activating $\overline{UDS}$ or $\overline{LDS}$ as we saw in Figure 7.6. The OR gates on the right side of the figure are actually being used as negative logic AND gates, since all of the relevant control signals are asserted LOW.

The *Valid Address* signal in the 68K environment is called *Address Strobe* and is labeled $\overline{AS}$ in the diagram. This signal is routed to two places. It is used as a qualifying signal for the READ and WRITE operations and is gated through the two lower OR gates in the figure (keep thinking, " negative logic AND gates") and to the ADDRESS DECODER block. You should have no trouble identifying this functional block and how it works. In a pinch, you could probably design it yourself! Thus, we will not decode an address range until the $\overline{AS}$ signal is asserted LOW, guaranteeing a valid address. The I/O decoder also decodes our I/O devices, although they aren't shown in this figure.

Since the 68K does not have separate READ and WRITE signals, we synthesize the $\overline{READ}$ signal with the NOT gate and the OR gate. We also use the OR gate and the $\overline{AS}$ signal to qualify the READ and WRITE signals, with the OR gate functioning as a negative logic AND gate. Strictly speaking this isn't necessary because we are already qualifying the ADDRESS DECODER in the same way.

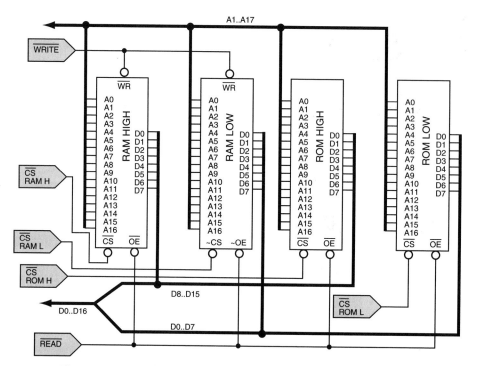

Figure 9.5 Simplified schematic diagram of the memory system for the 68K computer of Figure 9.4.

The last block of note is the Interrupt Controller block at the bottom left hand side of the diagram. The 68K uses 3 active LOW interrupt inputs labeled $\overline{IP0}$, $\overline{IP1}$ and $\overline{IP2}$. All three lines going low would signal a level 7 interrupt. We'll study interrupts in the next chapter. The inputs to the Interrupt Controller are 7 active low inputs labels INT1 – INT7. If the INT7 line was brought LOW, all of the output lines would also go LOW, indicating that INT7 is a level 7 interrupt. Now, if INT6 is low ($\overline{IP0} = 1$, $\overline{IP1} = 0$, $\overline{IP2} = 0$) and INT5 also went LOW, the outputs wouldn't change because the Interrupt controller both decodes the interrupt inputs and prioritizes them.

Figure 9.5 is the memory side of the computer. It is made up of two 128K × 8 RAM chips and two 128K x 8 ROM chips. Notice how the pairing of the devices and the two chip-select signals from the processor gives us byte writing control. The rest of this circuit should look very familiar to you.

The memory system in this computer design consists of 256K bytes of ROM, located at address $000000 through $3FFFF. The RAM is located at address $100000 through address $13FFFF. This address mapping is not obvious from the simplified schematic diagrams because you do not know the details of the circuit block labeled 'Address Decoder" in Figure 9.4.

Summary of Chapter 9

Chapter 9 covered:

- The advanced addressing modes of the 68K architecture and their relationship to high-level languages.
- An overview of the classes of instructions of the 68K architecture.
- Using the TRAP #15 instruction to simulated I/O.
- How a program, written in C executes in assembly language and how the C compiler makes use of the advanced addressing modes of the 68K architecture.
- How program disassembly is implemented and its relation to the architecture.
- The functional blocks of a 68K-based computer system.

Chapter 9: *Endnotes*

[1] Alan Clements, *68000 Family Assembly Language,* ISBN 0-534-93275-4, PWS Publishing Company, Boston, 1994, p. 704

Exercises for Chapter 9

1. Examine the block of assembly language code shown below.

   a. What is the address of the memory location where the byte, FF, is stored in the indicated instruction?

   b. Is this code segment relocatable? Why?

   ```
           org        $400
   start   movea.w    #$2000,A0
           move.w     #$0400,D0
           move.b     #$ff,($84,A0,D0)  *Where does ff go?
        ⇨  jmp        start
           end        $400
   ```

   Hint: Remember that displacements involve 2's complement numbers. Also, this instruction may also be written as:

   ```
   move.b #$FF,$84(A0,D0)
   ```

2. Examine the following code segment and then answer the question about the operation performed by the code segment. You may assume that the stack has been properly initialized. Briefly describe the effect of the highlighted instruction. What value is moved to what destination?

   ```
   00001000  41F9 00001018     10  START:   LEA       DATA+2,A0
   00001006  2C50              11           MOVEA.L   (A0),A6
   00001008  203C 00001B00     12           MOVE.L    #$00001B00,D0
   0000100E  2C00              13           MOVE.L    D0,D6
   00001010  2D80 6846         14           MOVE.L    D0,(70,A6,D6.L)
   00001014  60FE              15  STOP_IT: BRA       STOP_IT
   00001016  00AA0040 C8300000 16           DATA:     DC.L
   $00AA0040,$C8300000
   ```

3. Examine the following code segment and then answer the question about the operation performed by the code segment. You may assume that the stack has been properly initialized. Briefly describe the effect of the highlighted instruction. What is the value in register D0 after the highlighted instruction has completed?

   ```
   00000400 4FF84000    START:    LEA       $4000,SP
   00000404 3F3C1CAA              MOVE.W    #$1CAA,-(SP)
   00000408 3F3C8000              MOVE.W    #$8000,-(SP)
   ```

```
0000040C 223C00000010              MOVE.L    #16,D1
00000412 203C216E0000              MOVE.L    #$216E0000,D0
00000418 E2A0                      ASR.L     D1,D0
0000041A 383C1000                  MOVE.W    #$1000,D4
0000041E 2C1F                      MOVE.L    (SP)+,D6
00000420 C086                      AND.L     D6,D0
00000422 60FE          STOP_HERE:  BRA       STOP_HERE
```

4. Answer the following questions in a sentence or two.
   a. Why do local variables go out of scope when a C++ function is exited?
   b. Give two reasons why variables in high level languages, such as C, must be declared before they can be used?
   c. The 68000 assembly language instructions, LINK and UNLK would be representative of instructions that are created in order to support a high-level language. Why?

5. The following is a display of 32 bytes of memory that you might see from a "display memory" command in a debugger. What are the 3 instructions shown in this display?

```
00000400 06 79 55 55 00 00 AA AA 06 B9 AA AA 55 55 00 00
00000410 FF FE 06 40 AA AA 00 00 FF 00 00 00 00 00 00 00
```

6. Assume that you are trying to write a disassembler program for 68K instructions in memory. Register A6 points to the opcode word of the next instruction that you are trying to disassemble. Examine the following algorithm. Describe in words how it works. For this example, you may assume that <A6> = $00001A00 and <$00001A00> = %1101111001100001

```
shift         EQU     12              * Shift 12 bits


start         LEA     jmp_table,A0    *Index into the table
              CLR.L   D0              *Zero it
              MOVE.W  (A6),D0         *We'll play with it here
              MOVE.B  #shift,D1       *Shift 12 bits to the right
              LSR.W   D1,D0           *Move the bits
              MULU    #6,D0           *Form offset
              JSR     00(A0,D0)       *Jump indirect with index
              { Other instructions }

jmp_table     JMP     code0000
              JMP     code0001
              JMP     code0010
              JMP     code0011
              JMP     code0100
              JMP     code0101
              JMP     code0110
              JMP     code0111
              JMP     code1000
              JMP     code1001
              JMP     code1010
```

```
            JMP     code1011
            JMP     code1100
            JMP     code1101
            JMP     code1110
            JMP     code1111
```

7. Convert the memory test program from Chapter 9, exercise #9, to be relocatable. In order to see if you've succeeded write the program as follows:
   a. ORG the program at $400.
   b. Add some code at the beginning so when it begins to execute, it relocates itself to the memory region beginning at $000A0000.
   c. Test the memory region from $00000400 to $0009FFF0
   d. Make sure that you locate your stack so that it is not overwritten.

8. This exercise will extend your mastery of the 68K with the introduction of several new concepts to the memory test exercise from chapter 8. It will also be a good exercise for structuring your assembly code with appropriate subroutines. The exercise introduces the concept of user I/O through the use of the TRAP #15 instruction. You should read up on the TRAP #15 instructions by studying the HELP facility that comes with the E68K program. Alternatively, you can read up on the details of the TRAP #15 instruction on page 704 of the Clements[1] text.

   The TRAP #15 instruction is an artifact of the simulator. It was designed to allow I/O to take place between the simulator and the user. If you were really writing I/O routines, you probably would do some things differently, but lots of the things are the same.

   Associated with the TRAP #15 instruction are various tasks. Each task is numbered and associated with each task is an API that explains how it does its work. For example, task #0 prints a string to the display and adds a newline character so that the cursor advances to the beginning of the next line. In order to use task #0 you must set up the following registers (this is the API):

   - D0 holds the task number as a byte
   - The address of the beginning of the string is held in A1
   - The length of the string to print is stored as a word in D1

   Once you've set-up the three registers you then call TRAP #15 as an instruction and you're message is printed to the display. Here's a sample program that illustrates how it works. To output the string, "Hello world!", you might use the following code snippet:

```
*********************************************
*    Test program to print a string to the display
*
*********************************************
            OPT     CRE
task0       EQU     00

            ORG     $400

start       MOVE.B  #task0,D0           *Load task number into D0
            LEA     string,A1           *Get address of string
```

```
        MOVE.W     str_len,D1          *Get the length of the string
        TRAP       #15                 *Do it
        STOP       #$2700              * Back to simulator

* Data area

string     DC.B     'Hello world'       * Store the message here
str_len    DC.W      str_len-string     *Get the length of the string
        END        $400
```

Task #1 is almost identical to task #0 except that it does not print the newline character. This is handy when you want to prompt the user to enter information. So, you would issue a prompt using task #1 rather than task #0.

Also, you might want to get information from the user. Such as, "Where do they want to run the memory test and what test pattern do they want to use?" Also, you might ask them if they want to run the test again with a different pattern.

Problem Statement

1. Extend the memory test program (Chapter 9, exercise #9) to enable a user to enter the starting address of the memory test, the ending address of the memory test, and the word pattern to use for the test. Only the input test pattern will be used. There is no need to complement the bits or do any other tests.

2. The region of memory that may be tested will be between $00001000 and $000A0000.

3. The program first prints out a test summary heading with the following information:
   ADDRESS   DATA WRITTEN   DATA READ

4. Every time an error occurs you will print the address of the failed location, the data pattern that you wrote and the data pattern that you read back.

5. Locate your stack above $A0000.

6. When the program start running it prompts the user to enter the starting address (above $00001000) of the test . The user enters the starting address for the test. The program then prompts the user for the ending address of the test (below $FFFF). The user enters the address. Finally, the program prompts the user for the word pattern to use for the test.

You must check to see that the ending address is at least 1 word lengths away from the starting address and is greater than the starting address. You do not have to test for valid numeric entries. You may assume for this problem that only valid addresses and data values will be entered. Valid entries are the numbers 0 through 9 the lower case letters a through f and the upper case letters A through F.

Once this information is entered, the test summary heading line is printed to the display and the testing beings. Any errors that are encountered are printed to the display as described above.

Discussion

Once you obtain the string from the user you must realize that it is in ASCII format. ASCII is the code used to print characters and read characters. The ASCII codes for the numbers 0 thru 9 are $30..$39. The ASCII codes for the letters a through f are $61..$66 and the ASCII codes for the letters A through F are $41..$46.

So, if I prompt the user for an address and the user types in 9B56; when the TRAP #15 instruction completes there will be the following four bytes in memory: $39,$42,$35,$36. So, how do I go from this to the address $9B56? We'll you will have to write an algorithm. In other words, you must convert the ASCII values to their 16-bit word equivalent. Here is a good opportunity to practice your shifting and masking techniques. In order to get a number that is represented as ASCII 0-9, you must subtract $30 from the ASCII value. This leaves you with the numeric value. If the number is "B", then you subtract a different number to get the hex value $B. Likewise, if the number is "b", then you must subtract yet a different value.

Here's what the address $9B56 looks like in binary.

DB15				*DB12*				*DB7*				*DB3*			*DB0*
1	*0*	*0*	*1*	*1*	*0*	*1*	*1*	*0*	*1*	*0*	*1*	*0*	*1*	*1*	*0*

Now, if I have the value 9 decoded from the ASCII $39 and it is sitting in bit positions 0 through 3, how do I get it to end up in bit position 13 through 15? It makes good sense to work out the algorithm with a flow chart.

CHAPTER 10

The Intel x86 Architecture

Objectives

When you are finished with this lesson, you will be able to:
▶ *Describe the processor architecture of the 8086 and 8088 microprocessors;*
▶ *Describe the basic instruction set architecture of the 8086 and 8088 processors;*
▶ *Address memory using segment:offset addressing techniques;*
▶ *Describe the differences and similarities between the 8086 architecture and the 68000 architecture;*
▶ *Write simple program in 8086 assembly language using all addressing modes and instructions of the architecture.*

Introduction

Intel is the largest semiconductor manufacturer in the world. It rose to this position because of its supplier partnership with IBM. IBM needed a processor for its new PC-XT personal computer, under development in Boca Raton, FL. IBM's first choice was the Z-800 from Zilog, Inc., a rival of Intel's. Zilog was formed by Intel employees who worked on the original Intel 8080 processor. Zilog produced a code-compatible enhancement of the 8080, the Z80. The Z80 executed all of the 8080 instructions plus some additional instructions. It was also somewhat easier to interface to then the 8080.

Hobbyists embraced the Z80 and it became the processor of choice for almost all of the early PCs built upon the CP/M operating system developed by Gary Kidall at Digital Research Corporation. Zilog was rumored to be working on a 16-bit version of the Z80, the Z800, which promised dramatic performance improvement over the 8-bit Z80. IBM initially approached Zilog as the possible CPU supplier for the PC-XT. Today, it could be agued that Intel rose to that position of prominence because Zilog could not deliver on its promised delivery date to IBM.

In 1978 Intel had a 16-bit successor to the 8080, the 8086/8088. The 8088 was internally identical to the 8086 with the exception of an 8-bit eternal data bus, rather than 16-bit. Certainly this would limit the performance of the device, but to IBM it had the attractive feature of significantly lowering the system cost. Intel was able to meet IBM's schedule and a 4.077 MHz 8088 CPU deputed in the original IBM PC-XT computer. For an operating system, IBM chose a 16-bit CP/M look-alike from a small Seattle software company called, Microsoft. Although Zilog survived to this day, they never recovered from their inability to deliver the Z800. This should be a lesson that all software developers who miss their delivery targets should take this classic example of a missed schedule delivery to heart. As a post-script, the original Z80 also survives to this date and continues to be used as an embedded controller. It is hard to estimate its impact, but there must be billions of lines of Z80 code still being used in the world.

The growth of the PC industry essentially tracked Intel's continued development and ongoing refinement of the x86 architecture. Intel introduced the 80286 as a follow-on CPU and IBM introduced the PC-AT computer which used it. The 80286 introduced the concept of *protected mode*, which enabled the operating system to begin to employ task management, so that MS-DOS, which was a single tasking operating system, could now be extended to allow crude forms of multitasking to take place. The 80286 was still a 16-bit machine. During this period Intel also introduced the 80186/80188 integrated microcontroller family. These were CPUs with 8086/8088 cores and additional on-chip peripheral devices which made the device extremely attractive as a one-chip solution for many embedded computer products. Among the on-chip peripherals are:

- 2 direct memory access controllers (DMA)
- Three 16-bit programmable timers
- Clock generator
- Chip select unit
- Programmable Control Registers

The 80186/188 family is still extremely popular to this day, with other semiconductor companies building even more highly integrated variants, such as the E86™ family from Advanced Micro Devices (AMD) and the V-series from NEC. In particular, the combination of peripheral devices made the 186 families very attractive to disk drive manufacturers for controllers.

With the introduction of the 80386 the x86 architecture was finally extended to 32 bits. The PC world responded with hardware and software (Windows 3.0) which rallied around this architecture. The follow on processors to the 80386, the 80486 and Pentium families, continued to evolve the basic architecture defined in the i386.

Our interest in this family leads us to step back from the i386 architecture and to focus on the original 8086 family architecture. This architecture will provide us with a good point of reference for trying to gain a comparative understanding of the 68000 versus the 8086. While the 68000 is usually specified as a 16/32-bit architecture, the 8086 is also capable of doing many 32-bit operations as well. Also, the default operand size for the 68K is 16-bits, so the relative comparison of the two architectures is reasonably valid.

Therefore, we will approach our look at the x86 family from the perspective of the 8086. There are many other references available on the follow-on architectures to the 8086 and the interested reader is encouraged to seek them out.

As we begin to study the 8086 architecture and instruction set architecture it will become obvious that so much of focus is on the DOS operating system and the PC run time environment. From that perspective, we have an interesting counter example to the 68K architecture. When we studied the 68K, we were running in an "emulation" environment because the native instruction set of your PC and that of the 68K are incompatible. So, a virtual 68K is created programmatically and that program creates the virtual 68K machine for your code to execute on.

In the case of the 8086 ISA, you are executing in a "native" environment where the instruction set of your assembly language programs is compatible with the machine.

Thus, you can easily execute (well-behaved) 8086 programs that you create inside of DOS windows on your PC. In fact, the I/O from the keyboard and to the screen can easily be implemented with calls to the DOS operating through the *Basic Input Output System* (BIOS) of your PC. This is both a blessing and a curse because in trying to write and execute 8086 programs we will be forced to have to deal with interfacing to the DOS environment itself. This means learning some assembler directives that are new and are only applicable to a DOS environment.

Finally, mastering the basic instruction set architecture of the i86 requires a steeper learning curve then the 68K architecture that we started with. I'm sure this will start several minor religious wars, but please excuse me for allowing my personal bias to unveil itself.

In order to deal with these issues we'll opt for the "KISS" approach (*Keep It Simple, Stupid*) and keep our eyes on the ultimate objective. Since what we are trying to accomplish is a comparative understanding of modern computer architectures we will leave the issues of mastering the art of programming the i86 architecture in assembly language for another time. Therefore, we will place out emphasis on learning the addressing modes and instructions of the 8086 as our primary objective and make the housekeeping rules of DOS assembly language programming a secondary objective that we'll deal with on an as-needed basis. However, you should be able to obtain any additional information that I've omitted for the sake of clarity from any one of dozens of resources on i86 or DOS assembly language programming. Onward!

The Architecture of the 8086 CPU

Figure 10.1 is a simplified block diagram of the 8086 CPU. Depending upon your perspective, the block diagram may seem more complex to you than the block diagram of the 68000 processor in Figure 7.11. Although it seems quite different, there are similarities between the two. The general registers roughly correspond in purpose to the data registers of the 68K. The registers on the right side of the diagram, which is labeled the *bus interface unit*, or *BIU*, are called the *segment registers,* and roughly correspond to the address registers of the 68K.

The 8086 is strictly organized as two separate and autonomous functional blocks. The *execution* unit, or EU, handles the arithmetic and logical operations on the data and has a 6 byte *first-in, first-out (FIFO)* instruction queue (4 bytes on the 8088). The segment registers of the BIU are responsible for access instructions and operands from memory. The main linkage between the two functional blocks is the instruction queue, with the BIU looking ahead of the current instruction being executed in order to keep the queue filled with instructions for the EU to decode and operate on.

The symbol on the BIU side that looks like a carpenter's saw horse is called a *multiplexer,* or *MUX,* and its function is to combine the address and data information into a single, 20-bit external bus. The multiplexed (or shared) bus allows the 8086 CPU to have only 40 pins on the package, while the 68000 has 64 pins, primarily due to the extra pins required for the 23 address and 16 data pins. However, nothing is free. The multiplexed bus requires that systems using the 8086 must have external logic on the board to latch the address into holding registers during the first part of the bus cycle in order to have a stable address to present to memory during the second half of the

Figure 10.1: Simplified block diagram of the Intel 8086 CPU. From Tabak[1].

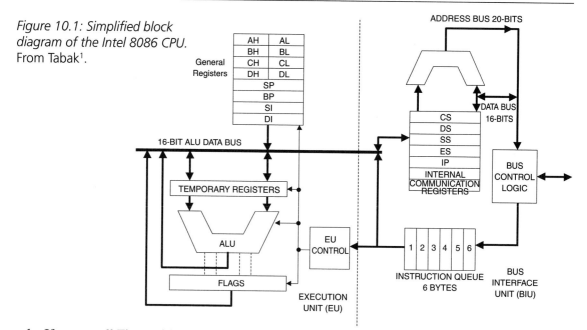

cycle. If you recall Figure 6.23, the timing diagram for a generic microprocessor, then we can describe the 8086 bus cycles as having four "T" states, labeled T1 to T4. If a wait state is going to be included in the cycle, then it comes as an extension of the T3 state.

During the falling edge of T1 the processor presents the 20-bit address to the external logic circuitry and issues a latching signal, *address latch enable,* or $\overline{\text{ALE}}$. $\overline{\text{ALE}}$ is used to latch the address portion of the bus cycle. Data is output on the rising edge of and is read in on the falling edge of T3. The 20-bit wide address bus gives the 8086 1 MByte address range. Finally, 8086 does not place any restrictions on the word alignment of addresses. A word can exist on an odd or even boundary. The BIU manages the extra bus cycle required to fetch both bytes of the word and aside from the performance penalty, the action is transparent to the software developer.

The programmer's model of the 8086 is shown in Figure 10.2.

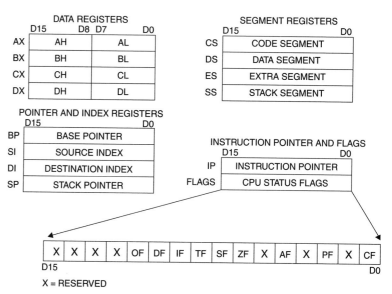

Figure 10.2: Programmer's model of the 8086 register set.

Data, Index and Pointer Registers

The eight, 16-bit general purpose registers are used for arithmetic and logical operations. In addition, the four data registers labeled AX, BX, CX and DX may be further subdivided for 8-bit operations into a high-byte or low-byte register, depending where the byte is to be stored in the register. Thus, for byte operations, the registers may be individually addressed. For example, **MOV AL,6D** would place the immediate hexadecimal value 6D into register AL. Yes, I know. It's backwards.

Also, when data stored in 16-bit registers are stored in memory they are stored in the reverse order from how they appear in the register. Thus, if <AX> = 109C and this data word is then written to memory at address 1000 and 1001, the byte order in memory would be:

$$<1000> = 9C, <1001> = 10 \qquad \{ \text{ Little Endian! Remember?} \}$$

Also, these data registers are not completely general purpose in the same way that D0 through D7 of the 68K are general purpose. The AX register, as well as its two half registers, AH and AL, are also known as an *accumulator*. An accumulator is a register that is used in arithmetic and logical operations. The AX register must be used when you are using the multiply and divide, **MUL and DIV**, instructions. For example, the code snippet:

```
MOV  AL,10
MOV  DH,25
MUL  DH
```

would do an 8-bit by 8-bit multiplication of the contents of the AL register with the contents of the DH register and store the resultant 16-bit value in the AX register. Notice that it was not necessary to specify the AL register in the multiplication instruction since it had to be used for a byte multiplication. For 16-bit operands, the result is stored in the DX:AX register pair with the high order word stored in the DX register and the low order word in the AX register. For example,

```
MOV  AX,0300
MOV  DX,0400
MUL  DX
```

Would result in <DX:AX> = 0001:D4C0 = 120,000 in decimal.

There are several interesting points here.

- The type of instruction to be executed (byte or word) is implied by the registers used and the size of the operands.
- Numbers are considered to be literal values without any special symbol, such as the '#' sign used in the 68000 assembler.
- Under certain circumstances, 16-bit registers may be ganged together to form 32-bit wide registers.
- Although not shown is this example, hexadecimal numbers are indicated with a following 'h'. Also, hexadecimal numbers beginning with the letters A through F should have a leading zero appended so that they won't be interpreted as labels.
  - 0AC55h = AC55 hex
  - AC55h = label 'AC55h'

The BX register can be used as a 16-bit offset memory pointer. The following code snippet loads absolute memory location 1000Ah with the value 0AAh.

```
mov AX,1000h
mov DS,AX
mov BX,000Ah
mov [BX],0AAh
```

In this example we see that:

- The segment register, DS, must be loaded from a register, rather than with an immediate value.

- Placing the BX register in parentheses changes the effective addressing mode to register indirect. The complete memory load address is [DS:BX]. Notice how the DS register is implied, it is not specified.

- The DS register is the default register used with data movement operations, just as the CS register would be used for referencing instructions. However, it is possible to override the default register by explicitly specifying the segment register to use. For example, this code snippet overrides the DS segment register and forces the instruction to use the [ES:BX] register for the operation.

  ```
  mov AX,1000h
  mov CX,2000h
  mov DS,AX
  mov ES,CX
  mov BX,000Ah
  es:mov w.[BX],055h
  ```

Also, notice how the 'w.' was used. This explicitly told the instruction to interpret the literal as a word value, 0055h. Otherwise, the assembler would have interpreted it as a byte value because of its representation as '055h'.

The CX register is used as the counter register. It is used for looping, shifting, repeating and counting operations. The following code snippet illustrated the CX register's *raison d'etre*.

```
        mov CX,5
myLoop: nop
        loop myLoop
```

The **LOOP** instruction functions like the **DBcc** instruction in 68K language. However, while the **DBcc** instruction can use any of the data registers as the loop counter, the **LOOP** instruction can only use the CX register as the loop counter. In the above snippet of code, the NOP instruction will be executed 5 times. Each time through the loop the CX register will be automatically decremented and the loop instruction will stop executing when <CX> = 0. Also notice how the label, 'myLoop', is terminated with a colon ':'. The 8086 assemblers require a colon to indicate a label. The label can also be placed on the line above the instruction or data:

```
myLoop: nop          myLoop:
                             nop
```

These are equivalent.

The DX register is the only register than may be used to specify I/O addresses. There is no I/O space segmentation required because the 8086 can only address 64K of I/O locations. The following code snippet reads the I/O port at memory location A43Eh and places the data into the AX register.

```
MOV  DX,0A43Eh
IN   AX,DX
```

Unlike the 68K, the i86 family handles I/O as a separate memory space with its own set of bus signals and timing.

The DX register is also used for 16-bit and 32-bit multiplication and division operations. As you've seen in a previous example, the DX register is ganged with the AX register when two 16-bit numbers are multiplied to give a 32-bit result. Similarly, the two registers are ganged together to form a 32-bit dividend when the divisor is 16-bits.

```
MOV DX,0200
MOV AX,0000
MOV CX,4000
DIV CX
```

This places the 32-bit number 00C80000h into DX:AX and this number is divided by 0FA0h in the CX register. The quotient, 0CCCh, is stored in AX and the remainder, 0C80h, is stored in DX.

The Destination Index and Source Index registers, DI and SI, are used with data movement and string operations. Each register has a specific purpose when indexing the source and destination operands. The DI and SI pointers also differ in that during string operations, the SI register is paired with the DS segment register and the SI register is paired with the ES segment register. During nonstring operations both are paired with the DS segment register.

This may seem strange, but it is a reasonable thing to do when, for example, you are trying to copy a string between two memory regions that are greater than 64K apart. Using two different segment registers automatically gives you the greatest possible span in memory. The following code snippet copies 5 bytes from the memory location initially pointed to by DS:SI to the memory location pointed to by ES:DI.

```
          MOV CX,0005
          MOV AX,1000h
          MOV BX,2000h
          MOV DS,AX
          MOV ES,BX
          MOV DI,200h
          MOV SI,100h
myLoop:   MOVSB
          LOOP myLoop
```

The program initializes the DS segment register to 1000h and the SI register to 100h. Thus the string to be copied is located at address <1000:0100>, or at the physical memory location 10100h. The ES segment register is initialized to 2000h and the DI register is initialized to 0200h, so the physical address in memory for the destination is 20200h. The CX register is initialized for a loop count of 5, and the LOOP instruction causes the **MOVSB** (*MOVStringByte*) instruction to execute 5 times. Each time the **MOVSB** instruction executes, the contents of the DI and SI registers are automatically incremented by 1 byte.

Like it or not, these are powerful and compact instructions.

The Base Pointer and Stack Pointer (BP and SP) general-purpose registers are used in conjunction with the Stack Segment (SS) register in the BIU and point to the bottom and top of the stack, respectively. Since the system stack is managed by a different segment register, we need additional offset registers to point to addresses located in the region pointed to by SS. Think of the SP register as "the" Stack Pointer, while the BP register is a general purpose memory pointer into the memory region pointed to by the SS segment register. The BP is used by high-level languages to provide support for the stack-based operations such as parameter passing and the manipulation of local variables. In that sense, the combination of the SS and BP take the place of the local *frame pointer* (often register A6) when high-level languages are compiled for the 68K.

All of the stack-based instructions (**POP**, **POPA**, **POPF**, **PUSH**, **PUSHA** and **PUSHF**) use the Stack Pointer (SP) register. The SP register is always used as an offset value from the Stack Segment (SS) register to point to the current stack location.

These pointer and index registers have one important difference with their 68K analogs. As noted in the above register definitions, these registers are used in conjunction with the segment registers in the BIU in order to form the physical memory address of the operand. We'll look at this in a moment.

Flag Registers

Some of the bits in the flag register have similar definitions to the bits in the 68K status register. Others do not. Also, the setting and resetting of the flags is more restrictive then in the 68K architecture. After the execution of an instruction the flags may be set (1), cleared or reset (0), unchanged or undefined. Undefined means that the value of the flag prior to the execution of an instruction may not be retained and its value after the instruction is executed can not be predicted[2].

- *Bit 0: Carry Flag (CF)* Set on a high-order bit carry for an addition operation or a borrow operation for a subtraction operation; cleared otherwise.
- *Bit 1:* Reserved.
- *Bit 2: Parity Flag (PF)* Set if the low-order 8 bits of a result contain an even number of 1 bits (even parity); cleared otherwise (odd parity).
- *Bit 3:* Reserved.
- *Bit 4: Auxiliary Carry (AF)* Set on carry or borrow from the low-order 4 bits of the AL general-purpose register; cleared otherwise.
- *Bit 5:* Reserved.
- *Bit 6: Zero Flag (ZF)* Set if the result is zero; cleared otherwise.
- *Bit 7: Sign Flag (SF)* Set equal to the value of the high-order bit of result. Set to 0 if the MSB = 0 (positive result). Set to 1 if the MSB is 1 (negative result).
- *Bit 8: Trace Flag (TF)* When the TF flag is set to 1, a trace interrupt occurs after the execution of each instruction. The TF flag is automatically cleared by the trace interrupt after the processor status flags are pushed onto the stack. The trace service routine can continue tracing by popping the flags back with a return from interrupt (IRET) instruction. Thus, this flag implements a single-step mechanism for debugging.

- *Bit 9: Interrupt-Enable Flag (IF)* When set to 1, maskable, or lower priority interrupts are enabled and may interrupt the processor. When interrupted, the CPU transfers control to the memory location specified by an interrupt vector (pointer).
- *Bit 10: Direction Flag (DF)* Setting the DF flag causes string instructions to auto-increment the appropriate index register. Clearing the flag causes the instructions to auto-decrement the register.
- *Bit 11: Overflow Flag (OF)* Set if the signed result cannot be expressed within number of bits in the destination operand; cleared otherwise.
- Bits 12-15: Reserved.

Segment Registers

The four 16-bit segment registers are part of the BIU. These registers store the segment (page) value of the address of the memory operand. The registers (CS, DS, ES and SS) define the segments of memory that are immediately addressable for code, or instruction fetches (CS), data reads and writes (DS and ES) and stack-based (SS) operations.

Instruction Pointer (IP)

The Instruction Pointer register contains the offset address of the next sequential instruction to be executed. Thus, it functions like the Program Counter register in the 68K. The IP register cannot be directly modified. Like the PC, nonsequential instructions which cause branches, jumps and subroutine calls will modify the value of the IP register.

These register descriptions have slowly been introducing us to a new way of addressing memory, called segment-offset addressing. Segment-offset addressing is similar to paging in many ways, but it is not quite the same as the paging method we've discussed earlier in the text. The segment register is used to point to the beginning of any one of the 64K sixteen-byte boundaries (called *paragraphs)* that can exist in a 20-bit address space. Figure 10.3 illustrates how the address is formed.

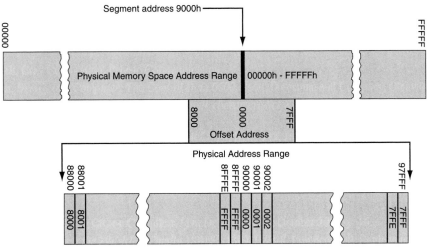

Figure 10.3: Memory address based upon segment and offset model.

It is obvious from this model that the same memory address can be specified in many different ways. This, in some respects, isn't very different from the aliasing problems that can arise from the 68K's addressing method, although the 68K addressing probably seems somewhat more straight-forward to you at this point. In any case, the 20-bit address is calculated by taking the 16-bit address value in the appropriate segment register and then doing an arithmetic shift left by 4 bit positions. Since each left shift has the effect of a multiplication by 2, four shifts multiply the address by 16 and result in the base addresses for the paragraph boundaries as shown in Figure 10.3. The offset value is a true negative or positive displacement centered about the paragraph boundary set by the segment register. Figure 10.3 shows that the physical address range that is accessible from a segment register value of 9000h extends from 88000h to 97FFFh. Offset addresses from 0000h to 7FFFh represent a positive displacement (towards higher addresses) and addresses from 0FFFFh to 8000h represent negative displacements.

Once the paragraph boundary is established by the segment register, the offset value (either a literal or register content) is sign extended and added to the shifted value from the segment register to form the full 20-bit address. This is shown in Figure 10.4.

While the physical address of memory operands is 20-bits, or 1 MByte, the I/O space address range is 16-bits, or 64K. The four, higher order address bits are not used when addressing I/O devices.

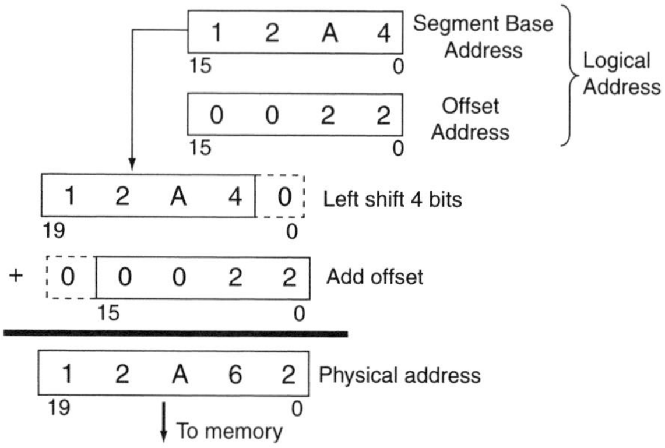

Figure 10.4: Converting the logical address to the physical address in the 8086 architecture.

One interesting point that should begin to become clear to you is that for most purposes, it doesn't matter what the physical address in memory actually works out to be. Most debugging tools that you will be working with present addresses to you in segment:offset fashion, so it isn't necessary for you to have to try to calculate the physical address in memory. Thus, if you are presented with the address 1000:0055, you can focus on the offset value of 55h in the segment paragraph starting at 1000h. If you have to go to physical memory, then you would convert this to 10055h, but this would be the exception rather than the rule.

Segment Registers

As you've seen in the previous examples, there are 4 segment registers located in the BIU. They are:

- Code Segment (CS) Register
- Data Segment (DS) Register
- Stack Segment (SS) Register
- Extra Segment (ES) Register

The CS register is the base pointer register for fetching instructions from memory. The DS register is the default segment pointer for all data operations. However, as you've seen the default registers can be overridden with an assembler register prefix on the instruction op-code. It is obvious that unlike the 68K, the architecture wants to have a clear division between what is to be fetched as instructions and what is to be manipulated as data.

The Stack Segment register functions like the stack pointer register in the 68K architecture when it is used with the SP register. The last register in the BIU is the Extra Segment Register, or ES. This register provides the second segment pointer for string-based operations.

We can see from these discussions that the behavior of the segment registers is very strictly defined. Since they are part of the BIU they cannot be used for arithmetic operations and their contents can be modified only with register-to-register transfers. This makes sense in light of the obvious division of the CPU into the EU and BIU.

In some respect we've been jumping ahead of ourselves with these code examples. The intent was to give you a feel for the way the architecture works before we dive in and try to systematically work through the instructions and effective addressing modes of the 8086 architecture.

Memory Addressing Modes

The 8086 architecture provides for 8 addressing modes. The first two modes operate on values that are stored in internal registers or are part of the instruction (immediate values). These are:

1. *Register Operand Mode:* The operand is located in one of the 8 or 16 bit registers. Example of Register Operand Mode instructions would be:

```
MOV     AX,DX
MOV     AL,BH
INC     AX
```

2. *Immediate Operand Mode:* The operand is part of the instruction. Examples of the Immediate Operand Mode are:

```
MOV     AX,0AC10h
ADD     AL,0AAh
```

There is no prefix, such as the '#' sign, to indicate that the operand is an immediate.

The next six addressing modes are used with memory operands, or data values that are stored in memory. The effective addressing modes are used to calculate and construct the offset that is combined with the segment register to create the actual physical address used to retrieve or write the memory data. The physical memory address is synthesized from the logical address contained in the segment register and an offset value, which may or may not be contained in a register. The segment registers are usually implicitly chosen by the type operation being performed. Thus, instructions are fetched relative to the value in the CS register; data is read or written relative to the DS register; stack operations are relative to the SP register and string operations are relative to the ES register. Registers may be overridden by prefacing the instruction with the desired register. For example:

```
ES:MOV AX,[0005h]
```

will fetch the data from ES:00005, rather than from DS:0005. The effective address (offset address) can be constructed by adding together any of the following three address elements:

    a. An 8-bit or 16-bit immediate displacement value that is part of the instruction,

    b. A base register value, contained in either the BX or BP register,

c. An index value stored in the DI or SI registers.

3. *Direct Mode:* The memory offset value is contained as an 8-bit or 16-bit positive displacement value.

Unlike the 68K, the offset portion of the address is always a positive number since the displacement is referenced from the beginning of the segment. The Direct Mode is closest to the 68K's absolute addressing mode, however, the difference is that there is always the implied presence of the segment register need to complete the physical address. For example:

```
MOV    AX,[00AAh]
```

copies the data from DS:00AA into the AX register, while,

```
CS:MOV    DX,[0FCh]
```

copies the data from CS:00FC into the DX register.

Notice how the square brackets are used to symbolize a memory instruction and the absence of brackets means an immediate value. Also, two memory operands are not permitted for the MOV instruction. The instruction,

```
MOV    [00AAh],[1000h]
```

is illegal. Unlike the 68K MOVE instruction, only one memory operand is permitted.

4. *Register Indirect Mode:* The operand offset is contained in one of the following registers:

- BP
- BX
- DI
- SI

The square brackets are used to indicate indirection. The following code snippet writes the value 55h to memory address DS:0100.

```
MOV    BX,100h
MOV    AL,55h
MOV    [BX],AL
```

5. *Based Mode:* The memory operand is the sum of the contents of the base register, BX or BP and the 8-bit or 16-bit displacement value. The following code snippet writes the value 0AAh to memory address DS:0104.

```
MOV    BX,100h
MOV    AL,0AAh
MOV    [BX]4,AL
```

The Based Mode instruction can also be written: **MOV [BX-4],AL.**

Since the displacement can be a positive or negative number, the above instruction is equivalent to: **MOV [BX + 0FFFCh],AL**

6. *Indexed Mode:* The memory operand is the sum of the contents of an index register, DI or SI, and the 8-bit or 16-bit displacement value. The following code snippet writes the value 055h to memory address ES:0104.

```
MOV    DI,100h
MOV    AL,0AAh
MOV    [DI]4,AL
```

The code snippet above illustrates another difference between using the index registers and the base registers. The DI register defaults to ES register as its address segment source, although the ES register may be overridden with a segment override prefix on the instruction. Thus, **DS:MOV [DI+4],AL** would force the instruction to use the DS register rather than the ES register.

7. *Based Indexed Mode:* The memory operand offset is the sum of the contents of one of the base registers, BP or BX and one of the displacement registers, DI or SI. The DI register is normally paired with the ES register in the Indexed Mode, but when the DI register is used in this mode, the DS register is the default segment register. The following code snippet writes the value 0AAh to memory location DS:0200h

```
MOV    DX,1000h
MOV    DS,DX
MOV    DI,100h
MOV    BX,DI
MOV    AL,0AAh
MOV    [DI+BX],AL
```

The instruction **MOV [DI+BX],AL** may also be written: **MOV [DI][BX],AL**

8. *Based Indexed Mode with Displacement:* The memory operand offset is the sum of the contents of one of the base registers, BP or BX, one of the displacement registers, DI or SI and an 8-bit or 16-bit displacement. The DI register is normally paired with the ES register in the Indexed Mode, but when the DI register is used in this mode, the DS register is the default segment register. The following code snippet writes the value 0AAh to memory location DS:0204h:

```
MOV    DX,1000h
MOV    DS,DX
MOV    DI,100h
MOV    BX,DI
MOV    AL,0AAh
MOV    [DI+BX]4,AL
```

The instruction **MOV [DI+BX]4,AL** may also be written: **MOV [DI][BX]4,AL or MOV [DI+BX+4],AL**.

Offset calculations can lead to very interesting and subtle code defects. Examine the code segment shown below:

```
MOV    DX,1000h
MOV    DS,DX
MOV    BX,07000h
MOV    DI,0FF0h
MOV    AL,0AAh
MOV    [DI+BX+0Fh],AL
```

This code assembles without errors and writes the value 0AAh to DS:7FFF. The physical address is 17FFFh. Now, examine the next code segment:

```
MOV    DX,1000h
MOV    DS,DX
MOV    BX,07000h
MOV    DI,0FF0h
MOV    AL,0AAh
MOV    [DI+BX+10h],AL
```

This code assembles without errors and writes the value 0AAh to DS:8000. However, the physical address is not 18000h, the next physical location in memory, it is actually 08000h. We have actually wrapped around the offset and gone from memory location 17FFFh to 8000h. If we had been in a loop, writing successive values to memory, we would have had an interesting problem to debug.

We can summarize the memory addressing modes as shown below. Consider that each of the items #1, #2, or #3 is optional, with the caveat that you must have at least one present to be able to specify an offset value. Wow! This almost makes sense. However, we need to keep in mind that unless the default segment register is overridden, the DI register will be paired with the ES register to determine the physical address while the other three registers will use the DS segment register. However, if the DI register is used in conjunction with either the BX or BP registers, then the DS segment register is the default.

#1		#2		#3
BX or BP	+	DI or SI	+	displacement

X86 Instruction Format

An 8086 instruction may be as short as 1 byte in length up through several bytes long. All assembly language instructions follow the format shown in Figure 10.5. Each field is interpreted as follows:

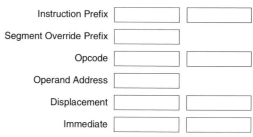

Figure 10.5: 8086 Instruction format.

1. *Instruction Prefixes:* Certain string operation instructions can be repeatedly executed with the REP, REPE, REPZ, REPNE and REPNZ prefixes. Also, there is a prefix term for asserting the BUS LOCK signal to the external interface.

2. *Segment Override Prefix:* In order to override the default segment register used for the memory operand address calculation, the segment override byte prefix is appended to the beginning of the instruction. Only one segment override prefix may be used. The format of the segment override prefix is shown in Figure 10.6.

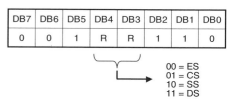

Figure 10.6: Format of the Segment Override Prefix.

3. *Opcode:* The op-code field specifies the machine language op-code for the instruction. Most op-codes are one byte in length, although there are some instructions which require a second op-code byte.

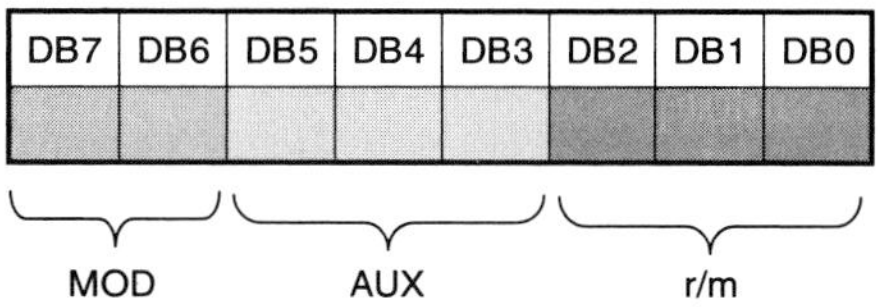

DB7	DB6	DB5	DB4	DB3	DB2	DB1	DB0

MOD   AUX   r/m

Figure 10.7: Format of the Operand Address Byte.

4. *Operand Address:* The operand address byte is somewhat involved. The form of the byte is shown in Figure 10.7. There are three fields, labeled *mod, aux* and *r/m*, respectively. The operand address byte controls the form of the addressing of the instruction.

The definition of the Mod Fields is as follows:

DB7	DB6	Description
1	1	r/m is treated as a register field
0	0	DISP Field = 0; disp-low and disp-high are absent
0	1	DISP Field = disp-low sign extended to 16-bits, disp-high is absent
1	0	DISP Field = disp-high:disp-low

The *mod (modifier) field* is used in conjunction with the r/m field to determine how the r/m field is interpreted. If *mod = 11*, then the *r/m* field is interpreted as a register operand. For a memory operand, the *mod* field determines if the memory operand is addressed directly or indirectly. For indirectly addressed operands, the mod field specifies the number of bytes of displacement that appears in the instruction.

The *register/memory (r/m) field* specifies if a register or a memory address is the operand. The type of operand is specified by the *mod field.* If the *mod* bits are 11, then the *r/m* field is treated as a register identifier. If the *mod* field has the values 00, 10 or 01, then the *r/m* field specifies one of the effective addressing modes that we've previously discussed. The *r/m* field values may be summarized as follows:

DB2	DB1	DB0	Effective address
0	0	0	[BX] + [SI] + DISP
0	0	1	[BX] + [DI] + DISP
0	1	0	[BP] + [SI] + DISP
0	1	1	[BP] + [DI] + DISP
1	0	0	[SI] + DISP
1	0	1	[DI] + DISP
1	1	0	[BP] + DISP (if mod = 00 then EA = disp-high:disp-low)
1	1	1	[BX] + DISP

When the *r/m* is used to specify a register then the bit code is as follows:

DB2	DB1	DB0	Word	Byte
0	0	0	AX	AL
0	0	1	CX	CL
0	1	0	DX	DL
0	1	1	BX	BL
1	0	0	SP	AH
1	0	1	BP	CH
1	1	0	SI	DH
1	1	1	DI	BH

Referring to the above table, you can see how the bit code for the register changes depending upon the size of operands in the instruction. The operand size is determined by the opcode portion of the instruction. We'll see how this works in a moment.

The *auxiliary field (Aux)* is used for two purposes. Certain instructions may require an opcode extension field and when needed, the extension is included in the *aux* field. Also, when the instruction requires a second register operand to be specified, the Auxiliary Field is used to specify the second register according to the above table.

5. *Displacement:* The displacement value is an 8-bit or 16-bit number to be added to the offset portion of the effective address calculation.
6. *Immediate:* The immediate field contains up to 2 bytes of immediate data.

As you can see, the 8086 instruction format is quite a bit more involved then the 68K instruction format. Much of this complexity is traceable to the segment:offset addressing format, but that isn't the whole story. We should consider the 8086 architecture as a whole to see how the bit patterns are used to create the path through the micro code that determines the state machine path for the instruction. The last piece of the puzzle that we need to consider is the opcode portion of instruction.

Recall that in the 68K architecture, the opcode portion of the instruction could readily be dissected into a fixed pattern of bits, a register field, a mode field, and an effective address field. In the 8086 architecture, the distinctions are less apparent and the control is more distributed. Let's consider the possible opcodes for the **MOV** instruction. The table below lists the possible instruction codes. The value in the Opcode column preceded by a forward slash '/' is the value in the Auxiliary Field of the Operand Address Byte. The value in the Opcode column following the plus sign '+' is the register code which is added to the base value of the opcode.

Format	Opcode	Description
MOV r/m8,r8	88 /r	Copy the byte value stored in register /r to the register or memory byte r/m8
MOV r/m16,r16	89 /r	Copy the word value stored in register /r to the register or memory word r/m16
MOV r8,r/m8	8A /r	Copy the byte value stored in r/m8 to the byte register /r
MOV r16,r/m16	8B /r	Copy the value stored in the register or memory word r/m16 to the word register /r
MOV r/m16,sreg	8C /sr	Copy the segment register /sr to the register or memory word r/m16
MOV sreg,r/m16	8E /sr	Copy the register or memory word r/m16 to the segment register
MOV AL,moffs8	A0	Copy the byte value stored in memory at segment:moffs8 to the AL register
MOV AX,moffs16	A1	Copy the word value stored in memory at segment:moffs16 to the AX register
MOV moffs8,AL	A2	Copy the contents of the AL register to the memory byte address at segment:moffs8
MOV moffs16,AX	A3	Copy the contents of the AX register to the memory word address at segment:moffs16
MOV r8,imm8	B0 + rb	Copy the immediate byte, imm8, to the register rb
MOV r16,imm16	B8 + rw	Copy the immediate word, imm16, to the register rw

Format	Opcode	Description
`MOV r/m8,imm8`	C6 /0	Copy the immediate byte, imm8 to the register or memory byte r/m8
`MOV r/m16,imm16`	C7 /0	Copy the immediate word, imm16, to the register or memory word r/m16

Referring to the opcode map, above, we can see that whenever a byte operation is specified then the least significant bit of the opcode is 0. Conversely, whenever a word operation is specified, the least significant bit is a 1. When the source or destination is a segment register, then the code for the appropriate segment register, DS, SS or ES appears in the *aux* field of the operand address byte as 011, 010 or 000, respectively. These are the same codes that are used for the segment override bits. The CS register cannot be directly modified by the MOV instruction.

This is a lot to try to absorb in a few pages of light reading, so let's try to pull it all together by looking at the bit patterns for some real instructions and see if they actually agree with what we would predict.

Mnemonic	Code	mod	aux	r/m
MOV AX,BX	8B C3	11	000	011

The opcode, 8Bh, tells us that this is a MOV instruction of the form MOV r/16,r/m16. The destination register is the word register AX. Also, the register is specified in the AUX field and its value is 000, which corresponds to the AX register. The MOD field is 11, which indicates that the r/m16 field is a register value. The r/m field vale is 011, which matches the BX register. Thus, our dissection of the instruction starts with the appropriate form of the opcode which leads us to the appropriate forms of the operand byte.

That was pretty straight forward. Let's ratchet it up a notch. Let's add a segment override and a memory offset.

Mnemonic	Code	mod	aux	r/m
ES:MOV [100h],CX	26 89 0E 00 01	00	001	110

The segment override byte, 26h, identifies the ES register and the opcode, 89h, tells us that the instruction is of the form **MOV r/m16,r16.** The *aux* field tells us that the source register is CX (001) and the *mod* field tells us that the register/memory address is a memory location but there is no displacement value needed to calculate the address offset. The *r/m* field value 110 indicates that the memory offset is provided as disp-high:disp-low, which is stored in memory as 00 01 (little Endian).

For the final example we'll put it all together.

Mnemonic	Code	mod	aux	r/m
CS:MOV [BX + DI + 0AAh],55h	2E C6 81 AA 00 55	10	000	001

Can we make sense of this instruction? The override byte indicates that the override segment is CS (01). The opcode, C6h, tells us that the instruction is of the form:

MOV r/m8,imm8

The Operand Address Byte gives us the mod value 10, which tells us that there is a displacement field present that must be used to calculate the offset address and it is off the form disp-high:disp-low.

We see this in the next two bytes of the instruction, AA 00. The *r/m* field value 001 then gives us the final form of the address calculation, [BX + DI + DISP]. The final byte of the instruction, 55h, is the 8-bit immediate value, imm8, as shown in the format of the opcode.

If we modify the instruction as follows:

```
CS:MOV        [BX+DI+0AAh],5555h
```

The instruction code becomes 2E C7 81 AA 00 55 55. The opcode changes from C6 to C7, indicating a word operation and the immediate field is now two bytes long. The entire instruction is now 7 bytes long, so we can see why the 8086 architecture must allow for nonaligned accesses from memory since the next instruction would be starting on an odd boundary if this instruction began on a word boundary.

8086 Instruction Set Summary

While it is beyond the scope of this text to cover the 8086 family instructions in detail we can look at representative instructions in each class and use these instructions as to gain an understanding of all of the instructions. Also, some of the instructions will also extend our insight into the overall x86 family architecture.

We can classify the original x86 instruction set into the following instruction families:

- Data transfer instructions
- Arithmetic instructions
- Logic instructions
- String manipulation instructions
- Control transfer instructions

Most of these instruction classes should look familiar to you. Others are new. The 68K family instruction set did not have dedicated instructions for loop control or string manipulation. Whether or not this represents a shortcoming of the architecture is difficult to assess. Clearly you are able to manipulate strings and form software loops in the 68K architecture as well. Let's now look at some representative instructions in each of these classes.

Data Transfer Instructions

Just as with the MOVE instruction of the 68K family, the MOV instruction is probably the most oft-used instruction in the instruction set. We've looked at the MOV instruction in some detail in this chapter, using it as a prototype instruction for understanding how the x86 instruction set architecture functions. We may summarize the instructions in the data transfer class in the following table[6]. Note that not all of the instruction mnemonics are listed and some f the instructions have additional variations with unique mnemonics. This table is meant to summarize the most general classification.

Mnemonic (© Intel)	Instruction	Description
MOV	MOVE	Copies data from a source operand to a destination operand
PUSH	PUSH	Creates storage space for an operand on the stack and then copies the operand onto the stack
POP	POP	Copies component from the top of the stack to memory or register

Mnemonic (© Intel)	Instruction	Description
XCHG	EXCHANGE	Exchanges two operands.
IN	INPUT	Read data from address in I/O space
OUT	OUTPUT	Write data to address in I/O space
XLAT	TRANSLATE	Translates the offset address of byte in memory to the value of the byte.
LEA	Load Effective Address	Calculates the effective address offset of a data element and loads the value into a register
LDS	Load DS with Segment and Register with Offset	Reads a full address pointer stored in memory as a 32-bit double word and stores the segment portion in DS and the offset portion in a register
LES	Load ES with Segment and Register with Offset	Same as LDS instruction with the exception that the ES register is loaded with the segment portion of the memory pointer rather than DS
LAHF	Load AH with Flags	Copies the Flag portion of the Processor Status Register to the AH register
SAHF	Store AH in Flags	Copies the contents of the AH register to the Flag portion of the Processor Status Register
PUSHF	Push Flags onto stack	Creates storage space on the stack and copies the flag portion of the Processor Status Register onto the stack.
POPF	Pop Flags from stack	Copies the data from the top of the stack to the Flag portion of the Processor Status Register and removes the storage space from the stack.

Reviewing the set of data transfer instructions, we see that all have analogs in the 68K instruction set except the XLAT, IN and OUT instructions.

Arithmetic Instructions

The following table summarizes the 8086 arithmetic instructions,

Mnemonic (© Intel)	Instruction	Description
ADD	Add	Add two integers or unsigned numbers
ADC	Add with carry	Add two integers or unsigned numbers and the contents of the Carry Flag
INC	Increment	Increment the contents of a register or memory value by 1
AAA	ASCII adjust AL after addition	Converts an unsigned binary number that is the sum of two unpacked binary coded decimal numbers to the unpack decimal equivalent.
DAA	Decimal Add Adjust	Converts an 8-bit unsigned value that is the result of addition to the correct binary coded decimal equivalent
SUB	Subtract	Subtract two integers or unsigned numbers
SBB	Subtract with borrow	Subtracts an integer or an unsigned number and the contents of the Carry Flag from a number of the same type
DEC	Decrement	Decrement the contents of a register or memory location by 1
NEG	Negate	Replaces the contents of a register or memory with its two's complement.

Mnemonic (© Intel)	Instruction	Description
CMP	Compare	Subtracts two integers or unsigned numbers and sets the flags accordingly but does not save the result of the subtraction. Neither the source nor the destination operand is changed
AAS	ASCII adjust after subtraction	Converts an unsigned binary number that is the difference of two unpacked binary coded decimal numbers to the unpacked decimal equivalent.
DAS	Decimal adjust after subtraction	Converts an 8-bit unsigned value that is the result of subtraction to the correct binary coded decimal equivalent
MUL	Multiply	Multiplies two unsigned numbers
IMUL	Integer Multiply	Multiplies two signed numbers
AAM	ASCII adjust for Multiply	Converts an unsigned binary number that is the product of two unpacked binary coded decimal numbers to the unpacked decimal equivalent.
DIV	Divide	Divides two unsigned numbers
IDIV	Integer divide	Divides two signed numbers
AAD	ASCII adjust after division	Converts an unsigned binary number that is the division of two unpacked binary coded decimal numbers to the unpacked decimal equivalent.
CWD	Convert word to double	Converts a 16-bit integer to a sign extended 32-bit integer
CBW	Convert byte to word	Converts an 8-bit integer to sign extended 16-bit integer

All of the above instructions have analogs in the 68K architecture with the exception of the 4 ASCII adjust instructions which are used to convert BCD numbers to a form that may be readily converted to ASCII equivalents.

Logic Instructions

The following table summarizes the 8086 logic instructions,

Mnemonic (© Intel)	Instruction	Description
NOT	Invert	One's complement negation of register or memory operand
SHL	Logical shift left	SHL and SAL shift the operand bits to the left, filling the shifted bit positions with 0's. The high order bits are shifted into the CF bit position
SAL	Arithmetic shift left	
SHR	Logical shift right	Shifts bits of an operand to the right, filling the vacant bit positions with 0's the low order bit is shifted into the CF position
SAR	Arithmetic shift right	Shifts bits of an operand to the right, filling the vacant bit positions with the original value of the highest bit. The low order bit is shifted into the CF position
ROL	Rotate left	Rotates the bits of the operand to the left and placing the high order bit that is shifted out into the CF position and low order bit position.

Mnemonic (© Intel)	Instruction	Description
ROR	Rotate right	Rotates the bits of the operand to the right and placing the low order bit that is shifted out into the CF position and the high order position.
RCL	Rotate through Carry Flag to the left	Rotates the bits of the operand to the left and placing the high order bit that is shifted out into the CF position and the contents of the CF into the low order bit position
RCR	Rotate through Carry Flag to the right	Rotates the bits of the operand to the right and placing the low order bit that is shifted out into the CF position and the contents of the CF into the high order bit position
AND	Bitwise AND	Computes the bitwise logical AND of the two operands.
TEST	Logical compare	Determines whether particular bits of an operand are set to 1. Results are not saved and only flags are affected.
OR	Bitwise OR	Computes the bitwise logical OR of the two operands.
XOR	Exclusive OR	Computes the bitwise logical exclusive OR of the two operands.

As you can see, the family of logical instructions is pretty close to those of the 68K family.

String Manipulation

This family of instructions has no direct analog in the 68K architecture and provides a powerful group of compact string manipulation operations. The following table summarizes the 8086 string manipulation instructions:

Mnemonic (© Intel)	Instruction	Description
REP	Repeat	Repeatedly executes a single string instruction
MOVS	Move a string component	Copy a byte element of a string (MOVSB) or word element of a string (MOVESW) from one location to another.
CMPS	Compare a string component	Compares the byte (CMPSB) or word (CMPSW) element of one string to the byte or word element of a second string
SCAS	Scan a string for component	Compares the byte (SCASB) or word (SCASW) element of a string to a value in a register
LODS	Load string component	Copies the byte (LODSB) or word (LODSW) element of a string to the AL (byte) or AX (word) register.
STOS	Store the string component	Copies the byte (STOSB) or word (STOSW) in the AL (byte) or AX (word) register to the element of a string

The **REPEAT** instruction has several variant forms that are not listed in the above table. Like the 68K **DBcc** instruction the repeating of the instruction is contingent upon a value of one of the flag register bits. It is clear how this set of instructions can easily be used to implement the string manipulation instructions of the C and C++ libraries.

The **REPEAT** instruction is unique in that it is not used as a separate opcode, but rather it is placed in front of the other string instruction that you wish to repeat. Thus,

REPMOVSB

would cause the MOVSB instruction to be repeated the number of times stored in the CX register.

Also, the MOVS instruction automatically advances or decrements the DI and SI registers' contents, so, in order to do a string copy operation, you would do the following steps:

1. Initialize the Direction Flag, DF,
2. Initial the counting register, CX,
3. Initialize the source index register, SI,
4. Initialize the destination index register, DI,
5. Execute the **REPMOVSB** instruction.

Compare this with the equivalent operation for the 68K instruction set. There is no direction flag to set, but steps 2 through 4 would need to be executed in an analogous manner. The auto incrementing or auto decrementing address mode would be used with the MOVE instruction, so you would mimic the MOVSB instruction with:

MOVE.B   (A0)+,(A1)+

The difference is that you would need to have an additional instruction, such as DBcc, to be your loop controller. Does this mean that the 8086 would outperform the 68K in such a string copy operation? That's hard to say without some in-depth analysis. Because the REPMOVSB instruction is a rather complex instruction, it is reasonable to assume that it might take more clock cycles to execute then a simpler instruction. The 186EM[4] processor from AMD takes $8 + 8*n$ clock cycles to execute the instruction. Here 'n' is the number of times the instruction is repeated. Thus, the instruction could take a minimum 808 clock cycles to copy 100 bytes between two memory locations. However, in order to execute the 68K MOVE and DBcc instruction pair, we would also have to fetch each instruction from memory over an over again, so the overhead of the memory fetch operation would be a significant part of the comparison.

Control Transfer

Mnemonic (© Intel)	Instruction	Description
CALL	Call procedure	*Suspends execution of the current instruction sequence, saves the segment (if necessary) and offset of the next instruction and transfers execution to the instruction pointed to by the operand.*
JMP	Unconditional jump	*Stops execution of the current sequence of instructions and transfer control to the instruction pointed to by the operand.*
RET	Return from procedure	*Used in conjunction with the CALL instruction the RET instruction restores the contents of the IP register and may also restore the contents of the CS register.*
JE	Jump if equal	*If the Zero Flag (ZF) is set control will be transferred to the address of the instruction pointed to by the operand. If ZF is cleared, the instruction is ignored.*
JZ	Jump if zero	
JL	Jump on less than	*If the Sign Flag (SF) and Overflow Flag (OF) are not the same, then control will be transferred to the address of the instruction pointed to by the operand. If they are the same the instruction is ignored.*
JNGE	Jump on not greater of equal	

Mnemonic (© Intel)	Instruction	Description
JB	Jump on below	If the Carry Flag (CF) is set control will be transferred to the address of the instruction pointed to by the operand. If CF is cleared, the instruction is ignored.
JNAE	Jump on not above or equal	
JC	Jump on carry	
JBE	Jump on below or equal	If the Carry Flag (CF) or the Zero Flag (ZF) is set control will be transferred to the address of the instruction pointed to by the operand. If CF and the CF are both cleared, the instruction is ignored.
JNA	Jump on not above	
JP	Jump on parity	If the Parity Flag (PF) is set control will be transferred to the address of the instruction pointed to by the operand. If PF is cleared, the instruction is ignored.
JPE	Jump on parity even	
JO	Jump on overflow	If the Overflow Flag (OF) is set control will be transferred to the address of the instruction pointed to by the operand. If OF is cleared, the instruction is ignored.
JS	Jump on sign	If the Sign Flag (SF) is set control will be transferred to the address of the instruction pointed to by the operand. If SF is cleared, the instruction is ignored.
JNE	Jump on not equal	If the Zero Flag (ZF) is cleared control will be transferred to the address of the instruction pointed to by the operand. If ZF is set, the instruction is ignored.
JNZ	Jump on not zero	
JNL	Jump on not less	If the Sign Flag (SF) and Overflow Flag (OF) are the same, then control will be transferred to the address of the instruction pointed to by the operand. If they are not the same, the instruction is ignored.
JGE	Jump on greater or equal	
JNLE	Jump on not less than or equal	If the logical expression ZF * (SF XOR OF) evaluates to TRUE then control will be transferred to the address of the instruction pointed to by the operand. If the expression evaluates to FALSE, the instruction is ignored.
JG	Jump on greater than	
JNB	Jump on not below	If the Carry Flag (CF) is cleared control will be transferred to the address of the instruction pointed to by the operand. If CF is set, the instruction is ignored
JAE	Jump on above or equal	
JNC		
JNBE	Jump on not below or equal	If the Carry Flag (CF) or the Zero Flag (ZF) are both cleared control will be transferred to the address of the instruction pointed to by the operand. If either flag is set, the instruction is ignored.
JA	Jump on above	
JNP	Jump on not parity	If the Parity Flag (PF) is cleared, control will be transferred to the address of the instruction pointed to by the operand. If PF is set, the instruction is ignored.
JO	Jump on odd parity	
JNO	Jump on not overflow	If the Overflow Flag (OF) is cleared control will be transferred to the address of the instruction pointed to by the operand. If OF is set, the instruction is ignored.
JNS	Jump on not sign	If the Sign Flag (SF) is cleared control will be transferred to the address of the instruction pointed to by the operand. If SF is set, the instruction is ignored.

Mnemonic (© Intel)	Instruction	Description
LOOP	Loop while the CX register is not zero	Repeatedly execute a sequence of instructions. The number of times the loop is repeated is stored in the CX register.
LOOPZ	Loop while zero	Repeatedly execute a sequence of instructions. The maximum number of times the loop is repeated is stored in the CX register. The loop is terminated before the count in CX reaches zero if the Zero Flag (ZF) is set.
LOOPE	Loop while equal	
LOOPNZ	Loop while not zero	Repeatedly execute a sequence of instructions. The maximum number of times the loop is repeated is stored in the CX register. The loop is terminated before the count in CX reaches zero if the Zero Flag (ZF) is cleared.
LOOPNE	Loop while not equal	
JCXZ	Jump on CX zero	If the previous instruction leaves 0 in the CX register, then control is transferred to the address of the instruction pointed to by the operand.
INT	Generate interrupt	The current instruction sequence is suspended and the Processor Status Flags, the Instruction Pointer (IP) register and the CS register are pushed onto the stack. Instruction continues at the memory address stored in appropriate interrupt vector location.
IRET	Return from interrupt	Restores the contents of the Flags register, the IP and the CS register.

Although the list of possible conditional jumps is long and impressive, you should note that most of the mnemonics are synonyms and test the same status flag conditions. Also, the set of JUMP-type instructions needs further explanation because of the segmentation method of memory addressing.

Jumps can be of two types, depending upon how far away the destination of the jump resides from the present location of the jump instruction. If you are jumping to another location in the same region of memory pointed to by the current value of the CS register then you are executing an *intrasegment jump*. Conversely, if the destination of the jump is beyond the span of the CS pointer, then you are executing an *intersegment jump*. Intersegment jumps require that the CS register is also modified to enable the jump to cover the entire range of physical memory.

The operand of the jump may take several forms. The following are operands of the jump instruction:

- *Short-label:* An 8-bit displacement value. The address of the instruction identified by the label is within the span of a signed 8-bit displacement from the address of the jump instruction itself.
- *Near label:* A 16-bit displacement value. The address of the instruction identified by the label is within the span of the current code segment. The value of
- *Memptr16 or Regptr16:* A 16-bit offset value stored in a memory location or a register. The value stored in the memory location or the register is copied into the IP register and forms the offset portion of the next instruction to be fetched from memory. The value in the CS register is unchanged, so this type of an operand can only lead to a jump within the current code segment.
- *Far-label or Memptr32:* The address of the jump operand is a 32-bit immediate value. The first 16-bits are loaded into the offset portion of the IP register. The second 16-bits are loaded into the CS register. The memptr32 operand may also be used to specify a double

word length indirect jump. That is, the two successive 16-bit memory locations specified by memptr32 contain the IP and CS values for the jump address. Also, certain register pairs, such as DS and DX may be paired to provide the CS and IP values for the jump.

The type of jump instruction that you need to use is normally handled by the assembler, unless you override the default values with assembler directives. Well discuss this point later on this chapter.

Assembly Language Programming the 8086 Architecture

The principles of assembly language programming that we've covered in the previous chapters are no different then those of the 68K family. However, while the principles may be the same, the implementation methods are somewhat different because:

1. the 8086 is so deeply linked to the architecture of the PC and its operating systems,
2. the segmented memory architecture requires that we declare the type of program that we intend to write and specify a *memory model* for the opcodes that the assembler is going to generate.

In order to write an executable assembly language program for the 8086 processor that will run natively on your PC you must, at a minimum, follow the rules of MSDOS®. This requires that you do not preset the values of the code segment register because it will be up to the operating system to initialize this register value when it loads the program into memory. Thus, in the 68K environment when we want to write relocatable code, we would use the PC or address register relative addressing modes. Here, we allow the operating system to specify the initial value of the CS register.

Assemblers such as Borland's Turbo Assembler (TASM®) and Microsoft's MASM® assembler handle many of these housekeeping tasks for you. So, as long as you follow the rules you may still be able to write assembly language programs that are well-behaved. Certainly these programs can run on any machine that is still running the 16-bit compatible versions of the various PC operating systems. The newer, 32-bit versions are more problematic because the run older DOS programs in an emulation mode which may or may not recognize the older BIOS calls. However, most simple assembly language programs which do simple console I/O should run without difficulty in a DOS window. Being a true Luddite, I still have my trusty 486 machine running good old MS-DOS, even though I'm writing this text on a system with Windows XP.

Let's first look at the issue of the segment directives and memory models. In general, it is necessary to explicitly identify the portions of your program that will deal with the code, the data and the stack. This is similar to what you've already seen. We use the directives:

- `.code`
- `.stack`
- `.data`

to denote the locations of these segments in your code (note that the directives are preceded by a period). For example, if you use the directive:

`.stack  100h`

you are reserving 256 bytes of stack space for this program. You do not have to specify where the stack itself is located because the operating system is managing that for you and the operating system is already up and running when it is loading this program.

The `.data` directive identifies the data space of your program. For example, you might have the following variables in your program:

```
        .data
var16       dw      0AAAAh
var8        db      55h
initMsg     db      'Hello World',0Ah,0Dh
```

This data space declares three variables, *var16, var8 and initMsg* and initializes them. In order for you to use this data space in your program you must initialize the DS segment register to address of the data segment. But since you don't know where this is, you do it indirectly:

```
MOV     AX,@data        ;Address of data segment
MOV     DS,AX
```

Here, @data is a reserved word that causes the assembler to calculate the correct DS segment value.

The `.code` directive identifies the beginning of your code segment. The CS register will initialized to point to the beginning of this segment whenever the program is loaded into memory.

In addition to identifying where in memory the various program segments will reside you need to provide the assembler (and the operating system with some idea of the type of addressing that will be required and the amount of memory resources that your program will need. You do this with the `.model` directive. Just as we've seen with the different types of pointers needed to execute an intrasegment jump and an intersegment jump, specifying the model indicates the size of your program and data space requirements. The available memory models are[3]:

- *Tiny:* Both program code and data fit within the same 64K segment. Also, both code and data are defined as *near*, which means that they are branched to by reloading the IP register.
- *Small:* Program code fits entirely within a single 64K segment and the data fits entirely within a separate 64K segment. Both code and data are near.
- *Medium:* Program code may be larger than 64K but program data must be small enough to fit within a single 64K segment. Code is defined as *far*, which means that both segment and offset must be specified while data accesses are all near.
- *Compact:* Program code fits within a single 64K segment but the size of the data may exceed 64K, with no single data element, such as an array, being larger than 64K. Code accesses are near and data accesses are far.
- *Large:* Both code and data spaces may be larger than 64K. However, no single data array may be larger than 64K. All data and code accesses are far.
- *Huge:* Both code and data spaces may be larger than 64K and data arrays may be larger than 64K. Far addressing modes are used for all code, data and array pointers.

The use of memory models is important because they are consistent with the memory models used by compilers for the PC. It guarantees that an assembly language module that will be linked in with modules written in a high level language will be compatible with each other.

Let's examine a simple program that could run on in a DOS emulation window on your PC.

```
        .MODEL      small
        .STACK      100h
```

```
        .DATA
PrnStrg             db      'Hello World$'              ;String to print
        .CODE
Start:
        mov     ax,@data            ;set data segment
        mov     ds,ax               ;initialize data segment register
        mov     dx,OFFSET   PrnStrg ;Load dx with offset to data
        mov     ah,09               ;DOS call to print string
        int     21h                 ;call DOS to print string
        mov     ah,4Ch              ;prepare to exit
        int     21h                 ;quit and return to DOS
        END     Start
```

As you might guess, this program represents the first baby steps of 8086 assembly language programming. You should be all teary-eyed, recalling the very first C++ program that you actually got to compile and run.

We are using the 'small' memory model, although the 'tiny' model would work just as well. We've reserved 256 bytes for the stack space, but it is difficult to say if we've used any stack space at all, since we didn't make any subroutine calls.

The data space is defined with the .data directive and we define a byte string, "Hello World$". The '$' is used to tell DOS to terminate the string printing. Borland[7] suggests that instruction labels be on lines by themselves because it is easier to identify a label and if an instruction needs to be added after the label it is marginally easier to do. However, the label may appear on the same line as the instruction that it references. Labels which reference instructions must be terminated with a colon and labels which reference data objects do not have colons. Colons are not used when the label is the target in a program, such as a for a loop or jump instruction.

The reserved word, *offset*, is used to instruct the assembler to calculate the offset from the instruction to the label, 'PrnStrg' and place the value in the DX register. This completes the code that is necessary to completely specify the segment and offset of the data string to print. Once we have established the pointer to the string, we can load the AH register with the DOS function call to print a string, 09. The call is made via a software interrupt, INT 21h, which has the same function as the TRAP #15 instruction did for the 68K simulator.

The program is terminated by a DOS termination call (INT 21h with AH = 4Ch) and the END reserved word tells the assembler to stop assembling. The label following the END directive tells the assembler where program execution is to begin. This can be different from the beginning of the code segment and is useful if you want to enter the program at some place other than the beginning of the code segment.

System Vectors

Like the 68K, the first 1K memory addresses are reserved for the system interrupt vectors and exceptions. In the 8086 architecture, the interrupt number is an 8-bit unsigned value from 0 to 255. The interrupt operand is shifted left 2 times (multiplied by 4) to obtain the address of the pointer to the interrupt handler code. Thus, the INT 21h instruction would cause the processor to vector through memory location 00084h to pick-up the 4 bytes of the segment and offset for the operating system entry point. In this case, the IP offset would be located at word address 00084h and the CS pointer would be located at word address 00086h. The function code, 09 in the AH register cause DOS to print the string pointed to by DS:DX.

System Startup

An 8086-based system comes out of RESET with all the registers set equal to zero with the exception of the CS register, which is set equal to 0FFFFh. Thus, the physical address of the first instruction fetch would be 0FFFFh:0000, or 0FFFF0h. This is an address located 16-bytes from the top of physical memory. Thus, an 8086 system usually has nonvolatile memory located in high memory so that it contains the boot code when the processor comes out of RESET. Once, out of RESET, the 16 bytes is enough to execute a few instructions, including a jump to the beginning of the actual initialization code. Once into the beginning of the ROM code, the system will usually initialize the interrupt vectors in low memory by writing their values to RAM, which occupies the low memory portion of the address space.

Wrap-Up

You may either be overjoyed or disappointed that this chapter is coming to an end. After all, we dissected the 68K instruction set and looked at numerous assembly language programming examples. In this chapter we looked at numerous code fragments and only one, rather trivial, program. What gives?

Earlier in the text we were both learning the fundamentals of assembly language programming and learning a computer's architecture at the same time. The architecture of the 68K family itself is fairly straight-forward and allows us to focus on basic principles of addressing and algorithms. The 8086 architecture is a more challenging architecture to absorb, so we delayed its introduction until later in the text. Now that you've been exposed to the general methods of assembly language programming, we could focus our efforts on mastering the intricacies of the 8086 architecture itself. Anyway, that's the theory.

In the next chapter we'll examine a third architecture that, once again, you may find either very refreshing or very frustrating, to work with. Frustrating because you don't have all the powerful instructions and addressing modes to work with that you have with the 8086 architecture; and refreshing because you don't have all of the powerful and complex instructions and addressing modes to master.

Summary of Chapter 10

Chapter 10 covered:

- The basic architecture of the 8086 and 8088 microprocessors
- 8086 memory models and addressing
- The instruction set architecture of the 8086 family
- The basics of 8086 assembly language programming

Chapter 10: *Endnotes*

[1] Daniel Tabak, *Advanced Microprocessors, Second Edition*, ISBN 0-07-062843-2, McGraw-Hill, NY, 1995, p. 186.

[2] Advanced Micro Devices, Inc, *Am186™ES and Am188™ES User's Manual*, 1997, p. 2-2.

[3] Borland,*Turbo Assembler 2.0 User's Guide,* Borland International, Inc. Scotts Valley, 1988.

[4] Advanced Micro Devices, Inc, *Am186™ES and Am188™ES Instruction Set Manual*, 1997.

[5] Walter A. Triebel and Avatar Singh, *The 8088 and 8086 Microprocessors*, Third Edition, ISBN 0-13-010560-0, Prentice-Hall, Upper Saddle River, NJ, 2000. Chapters 5 and 6.

[6] Intel Corporation, *8086 16-Bit HMOS Microprocessor*, Data Sheet Number 231455-005, September 1990, pp. 25–29.

[7] Borland, *op cit,* p. 83.

Exercises for Chapter 10

1. The contents of memory location 0C0020h = 0C7h and the contents of memory location 0C0021h = 15h. What is the word stored at 0C0020h? Is it aligned or nonaligned?

2. Assume that you have a pointer (segment:offset) stored in memory at byte addresses 0A3004h through 0A3007h as follows:

   <0A3004h> = 00
   <0A3005h> = 10h
   <0A3006h> = 0C3h
   <0A3007h> = 50h

   Express this pointer in terms of segment:offset value.

3. What would the offset value be for the physical memory address 0A257Ch if the contents of the segment register is 0A300h?

4. Convert the following assembly language instructions to their object code equivalents:
   ```
   MOV   AX,DX
   MOV   BX[SI],BX
   MOV   DX,0A34h
   ```

5. Write a simple code snippet that:
   a. Loads the value 10 into the BX register and the value 4 into the CX register,
   b. executes a loop that increments BX by 1 and decrements CX until the <CX> = 00

6. Load register AX with the value 0AA55h and then swap the bytes in the register.

7. What is are the contents of the AX register after the following two instructions are executed?
   ```
   MOV   AX,0AFF6h
   ADD   AL,47h
   ```

8. Suppose you want to perform the mathematical operation $X = Y*Z$, where:
   X is a 32-bit unsigned variable located at offset address 200h,
   Y is a 16-bit unsigned variable located at address 204h,
   Z is a 16-bit unsigned variable located at address 206h,

   write an 8086 assembly language code snippet that performs this operation.

9. Write a program snippet that moves 1000 bytes of data beginning at address 82000H to address 82200H.

10. Modify the program of problem 9 so that the program moves 1000 bytes of data from 82000H to C4000H.

CHAPTER 11

The ARM Architecture

● ●

Objectives

When you are finished with this lesson, you will be able to:
▶ *Describe the processor architecture of the ARM family;*
▶ *Describe the basic instruction set architecture of the ARM7 processors;*
▶ *Describe the differences and similarities between the ARM architecture and the 68000 architecture;*
▶ *Write simple code snippets in ARM assembly language using all addressing modes and instructions of the architecture.*

● ●

Introduction

We're going to turn our attention away from the 68K and 8086 architectures and head in a new direction. You may find this change of direction to be quite refreshing because we're going to look at an architecture that may be characterized by how it removed all but the most essential instructions and addressing modes. We call a computer that is built around this architecture a *RISC* computer, where RISC is an acronym for *Reduced Instruction Set Computer*. The 68K and the 8086 processors are characteristic of an architecture called *Complex Instruction Set Computer*, or *CISC*. We'll compare the two architectures in a later chapter. For now, let's just march onward and learn the ARM architecture as if we never heard of CISC and RISC.

In 1999 the ARM 32-bit architecture finally overtook the Motorola 68K architecture in terms of popularity[1]. The 68K architecture had dominated the embedded systems world since it was first invented, but the ARM architecture has emerged as today's most popular, 32-bit embedded processor. Also, ARM processors today outsell the Intel Pentium family by a 3 to 1 margin[2]. Thus, you've just seen my rationale for teaching these 3 microprocessor architectures.

If you happen to be in Austin, Texas you could head to the south side of town and visit AMD's impressive silicon foundry, FAB 25. In this modern, multibillion dollar factory, silicon wafers are converted to Athlon microprocessors. A short distance away, Freescale's (Motorola) FAB (fabrication facility) cranks out PowerPC® processors. Intel builds its processors in FABs in Chandler, Arizona and San Jose, CA, as well as other sites worldwide. Where's ARM's FAB located? Don't fret, this is a trick question. ARM doesn't have a FAB. ARM is a FABless manufacturer of microprocessors.

ARM Holdings plc was founded in 1990 as Advanced RISC Machines Ltd.[3] It was based in the United Kingdom as a joint venture between Acorn Computer Group, Apple and VLSI Technology.

ARM does not manufacture chips in its own right. It licenses its chip designs to partners such as VLSI Technology, Texas Instruments, Sharp, GEC Plessey and Cirrus logic who incorporate the ARM processors in custom devices that they manufacture and sell. It was ARM that created this model of selling *Intellectual Property, or IP*, rather than a silicon chip mounted in a package. In that sense it is no different then buying software, which, in fact, it is. A customer who wants to build a *system-on-silicon*, such as a PDA/Cell phone/Camera/MP3 player would contract with VLSI technology to build the physical part. VLSI, as an ARM licensee, offers the customer an encrypted library in a hardware description language, such as Verilog. Together with other IP that the customer may license, and IP that they design themselves, a Verilog description of the chip is created that VLSI can use to create the physical part. Thus, you can see with the emergence of companies like ARM, the integrated circuit design model predicted by Mead and Conway has come true.

Today, ARM offers a range of processor designs for a wide range of applications. Just as we've done with the 68K and the 8086, we're going to focus our efforts on the basic 32-bit ARM architecture that is common to most of the products in the family.

When we talk about the ARM processor, we'll often discuss it in terms of a *core*. The core is the naked, most basic portion of a microprocessor. When systems-on-silicon (or systems-on-chip) are designed, one or more microprocessor cores are combined with peripheral components to create an entire system design on a single silicon die. Simpler forms of SOCs have been around for quite a while. Today we call these commercially available parts *microcontrollers*. Both Intel and Motorola pioneered the creation of microcontrollers. Historically, a semiconductor company, such as Motorola, would develop a new microprocessor, such as the 68K family and sell the new part at a price premium to those customers who were doing the leading edge designs and were willing to pay a price premium to get the latest processor with the best performance.

As the processor gained wider acceptance and the semiconductor company refined their fabrication processes, they (the companies) would often lift the CPU core and place it in another device that included other peripheral devices such as timers, memory controllers, serial ports and parallel ports. These parts became extremely popular in more cost- conscious applications because of their higher level of integration. However, a potential customer was limited to buying the particular parts from the vendor's inventory. If the customer wanted a variant part, they either bought the closest part they could find and then placed the additional circuitry on a printed circuit board, or worked with the vendor to design a custom microcontroller for their products.

Thus, we can talk about Motorola Microcontrollers that use the original 68K core, called *CPU16*, such as the 68302, or more advanced microcontrollers that use the full 32-bit 68K core (*CPU32*) in devices such as the 68360. The 80186 processor from Intel uses the 8086 core and the SC520 (Elan) from AMD uses the 486 core to build an entire PC on a single chip. The original PalmPilot® used a Motorola 68328 microcontroller (code name Dragonball) as its engine. Thus, since all of the ARM processors are themselves cores, to be designed into systems-on-chip, we'll continue to use that terminology.

ARM Architecture

Figure 11.1 is a simplified schematic diagram of the ARM core architecture. Data and instructions come in and are routed to either the instruction decoder or to one of general purpose registers, labeled r0 – r15. Unlike the single bi-directional data bus to memory of the 68K and 8086 processors, the various ARM implementations may be designed with separate data busses and address busses going to instruction memory (code space) and data memory. This type of implementation, with separate data and instruction memories is called a *Harvard Architecture*.

All data manipulations take place in the *register file*, a group of 16, 32-bit wide, general- purpose registers. This is a very deferent concept from the dedicated address, data and index registers that we've previously dealt with. Although some of the registers do have specific functions, the rest of the registers may be used as source operands, destination

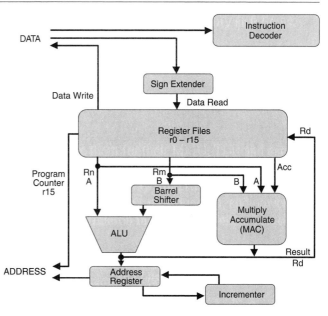

Figure 11.1: The ARM architecture.

operands or memory pointers. 8-bit and 16-bit wide data coming from memory to the registers is automatically converted to a sign extended 32-bit number before being stored in a register. These registers are said to be *orthogonal*, because they are completely interchangeable with respect to address or data storage.

All arithmetic and logical operations take place between registers. Instructions like the 68K's mixed memory operand and register addition, shown below, are not permitted.

```
        ADD  D5,$10AA    *Add D5 to $10AA and store result in $10AA
```

Also, most arithmetic and logical operations involve 3 operands. Thus, two operands, Rn and Rm, are manipulated and the result, Rd, is returned to the destination register. For example, the instruction:

```
                         ADD  r7,r1,r0
```

adds together the 32-bit signed contents of registers r0 and r1 and places the result in r7. In general, 3 operand instructions are of the form:

```
            opcode          Rd,Rn,Rm
```

with the Rm operand also passing through a barrel shifter unit before entering the ALU. This means that bit shifts may be performed on the Rm operand in one assembly language instruction. In addition to the standard ALU the ARM architecture also includes a dedicated *multiply-accumulate (MAC) unit* which can either do a standard multiplication of two registers, or accumulate the result with another register. MAC-based instructions are very important in signal processing applications because the MAC operation is fundamental to numerical integration. Consider the example numerical integration shown in Figure 11.2.

In order to calculate the area under the curve, or solve an integral equation, we can use a numerical approximation method. Each successive calculation of the area under the curve involves calculating the area of a small rectangular prism and summing all of the areas. Each area calculation is a multiplication and the summation of the prisms is the total area. The MAC unit does the multi-

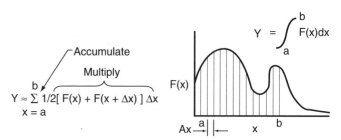

Figure 11.2: A multiply-accumulate (MAC) unit can accelerate numerical integration.

plication and keeps the running summation in one operation.

The ARM processors use a *load/store* architecture. *Loads* are data transfers from memory to a register in the register file and *stores* are data transfers from the register file to the memory. The load and store instructions use the ALU to compute address values that are stored in the address register for transfer to the address bus, or busses. The incrementer is used to advance the address register value for sequential memory load or store operations.

At any point in time, an ARM processor may be in one of seven operational modes. The most basic mode is called the *user mode.* The user mode has the lowest privilege level of the seven modes. When the processor is in user mode it is executing user code. In this mode there are 18 active registers; the 16 32-bit data registers and 2, 32-bit wide status registers. Of the 16 data registers, r13, r14, r15 are assigned to specific tasks.

- r13: Stack pointer. This register points to the top of the stack in the current operating mode. Under certain circumstances, this register may also be used as another general purpose register, however, when running under an operating system this register is usually assumed to be pointing to a valid stack frame.
- r14: Link register. Holds the return address when the processor takes a subroutine branch. Under certain conditions, this register may also be used as a general purpose register
- r15: Program counter: Holds the address of the next instruction to be fetched from memory.

During operation in the user mode the *current program status register (cpsr)* functions as the standard repository for program flags and processor status. The cpsr is also part of the register file and is 32-bits wide (although many of the bits are not used in the basic ARM architecture).

There are actually two program status registers. A second program status register, called the *saved program status register (spsr)* is used to store the state of the cpsr when a mode change occurs. Thus, the ARM architecture saves the status register in a special location rather than pushing it onto the stack when a context switch occurs. The program status register is shown in Figure 11.3.

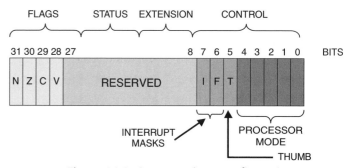

Figure 11.3: Status register configuration.

The processor status register is divided up into 4 fields; *Flags, Status, Extension and Control*. The Status and Extension fields are not implemented in the basic ARM architecture and are reserved for future expansion. The Flags field contains the four status flags:

- *N bit:* Negative flag. Set if bit 31 of the result is negative.
- *Z bit:* Zero flag. Set if the result is zero or equal.
- *C bit:* Carry flag. Set if the result causes an unsigned carry.
- *V bit:* Overflow flag. Set if the result causes a signed overflow.

The Interrupt Mask bits are used to enable or disable either of the two types of interrupt requests to the processor. When enabled, the processor may accept normal interrupt requests (IRQ) or Fast Interrupt Requests (FIQ). When either bit is set to 1 the corresponding type of interrupt is masked, or blocked from allowing an interrupt source from stopping the processor's current execution thread and servicing the interrupt.

The *Thumb Mode Bit* has nothing to do with your hand. It is a special mode designed to improve the code density of ARM instructions by compressing the original 32-bit ARM instruction set into a 16-bit form; thus achieving a 2:1 reduction in the program memory space needed. Special on-chip hardware does an on-the-fly decompression of the Thumb instructions back to the standard 32-bit instruction width. However, nothing comes for free, and there are some restrictions inherent in using Thumb mode. For example, only the general purpose registers r0–r7 are available when the processor is in Thumb mode.

Bit 0 through bit 4 define the current processor operating mode. The ARM processor may be in one of seven modes as defined in the following table:

Mode	Abbreviation	Privileged	Mode bits[4:0]
Abort	abt	yes	1 0 1 1 1
Fast Interrupt Request	fiq	yes	1 0 0 0 1
Interrupt Request	irq	yes	1 0 0 1 0
Supervisor	svc	yes	1 0 0 1 1
System	sys	yes	1 1 1 1 1
Undefined	und	yes	1 1 0 1 1
User	usr	no	1 0 0 0 0

The user mode has the lowest privilege level, which means it cannot alter the contents of the processor status register. In other words, it cannot enable or disable interrupts, enter Thumb mode or change the processor's operating mode. Before we discuss each of the modes we need to look at how each mode uses the register files. There are a total of 37 registers in the basic ARM architecture. We discussed the 18 that are accessible in user mode. The remaining 19 registers come into play when the other operating modes become active. Figure 11.4 shows the register configuration for each of the processor's operating modes.

When the processor is running in User Mode or System Mode the 13 general-purpose registers, sp ,lr, pc and cpsr registers are active. If the processor enters the Fast Interrupt Request Mode, the registers labeled r8_fiq through r14_fiq are automatically exchanged with the corresponding registers, r8 through r14. The contents of the current program status register, cpsr, are also automatically transferred to the saved program status register, spsr.

Thus, when a fast interrupt request comes into the processor, and the FIQ mask bit in the CPSR is enabled, the processor can quickly and automatically make an entire new group of registers available to service the fast interrupt request. Also, since the contents of the cpsr are needed to restore the context of the user's program when the interrupt request is serviced, the cpsr is automatically saved to the spsr_fiq. The spsr register automatically saves the context of the cpsr whenever any of the modes other than user and system become active.

Whenever the processor changes modes it must have be able to eventually return to the previous mode and correctly start again

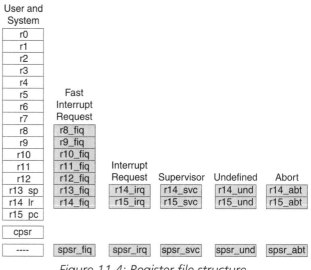

Figure 11.4: Register file structure.

from the exact point where it left off. Thus, when it enters a new mode, the new registers point to the memory stack for the new mode in r13_xxx and also the return address of the prior mode in r14_xxx. Thus, when the processor does switch modes the new context is automatically established and the old context is automatically saved. Of course, we could do all of this in software as well, but having these additional hardware resources offers better processor performance for time-critical applications.

Since the registers used are swapped in as a group (called a *bank switch*) their contents may be preloaded with the appropriate values necessary to service the fast interrupt request. Also, since the FIQ mode is a higher privilege mode, the program code used to service the fast interrupt request can also change the operational mode back to user or system when the interrupt is over. We'll discuss the general concepts of interrupts and servicing interrupts in more detail in a later chapter.

The other processor modes; Interrupt Request, Supervisor, Undefined and Abort all behave in much the same way as the Fast Interrupt Request mode. When the new mode is entered, the particular registers appropriate for that mode are bank switched with the registers of the user mode. The current contents of the base registers are not changed, so that when the user mode is restored, the old working registers are reactivated with their value prior to the bank switch taking place.

Our last remaining task is to look at the seven processor modes.

- *User mode:* This is the normal program execution mode. The processor has general use of registers r0 through r12 and the flag bits in the cpsr are changed according to the results of the execution of the assembly language instructions that can modify the flags.
- *System mode:* System mode is user mode with the higher privilege level. This means that in system mode the processor may alter the values in the cpsr. You will recall the 68K also had a user and supervisor mode, which gave access to additional instructions which could modify the bits in the status register. System mode would be the appropriate mode to use if your application was not so complex that it required the addition of an operating system.

For simpler applications, the accessibility of the processor modes would be an advantage, so the system mode would be an obvious choice.

- *Fast Interrupt Request Mode:* The FIQ and IRQ modes are designed for handling processor interrupts. The Fast Interrupt Request Mode provides more bank registers than the standard interrupt request mode so that the processor has less overhead to deal with if additional registers must be saved during the interrupt service routine. Also, the seven banked registers used in the FIQ mode are sufficient for creating a software emulation of a *Direct Memory Access (DMA)* controller.
- *Interrupt Request Mode:* Like the FIQ mode, the IRQ mode is designed for the servicing of processor interrupts. Two banked registers are available when the processor acknowledges the interrupt and switches context to the interrupt service routine.
- *Supervisor mode:* Supervisor mode is designed to be used with an operating system kernel. When in supervisor mode all the resources of the CPU are available. Supervisor mode is the active mode when the processor first comes out of reset.
- *Undefined mode:* This mode is reserved for illegal instructions or for instructions that are not supported by the particular version of the ARM architecture that is in use.
- *Abort mode:* Under certain conditions the processor may attempt to make an illegal memory access. The abort mode is reserved for dealing with attempts to access restricted memory. For example, special hardware might be designed to detect illegal write attempts to memory that is supposed to be read-only.

Conditional Execution

The ARM architecture offers a rather unique feature that we've not previously considered. That is, the ability to conditionally execute most instructions based upon the states, or logical combination of states, of the condition flags. The condition codes and their logical definitions are shown in the following table:

Code	Description	Flags	OP-Code[31:28]
EQ	Equal to zero	$Z = 1$	0 0 0 0
NE	Not equal to zero	$Z = 0$	0 0 0 1
CS HS	Carry set / unsigned higher or the same	$C = 1$	0 0 1 0
CC LO	Carry cleared / unsigned lower	$C = 0$	0 0 1 1
MI	Negative or minus	$N = 1$	0 1 0 0
PL	Positive or plus	$N = 0$	0 1 0 1
VS	Overflow set	$V = 1$	0 1 1 0
VC	Overflow cleared	$V = 0$	0 1 1 1
HI	unsigned higher	$\overline{Z} * C$	1 0 0 0
LS	unsigned lower or the same	$Z + \overline{C}$	1 0 0 1
GE	signed greater than or equal	$(N*V) + (\overline{N}*\overline{V})$	1 0 1 0
LT	signed less than	$N\ xor\ V$	1 0 1 1
GT	signed greater than	$(N*\overline{Z}*V) + (\overline{N}*\overline{Z}*\overline{V})$	1 1 0 0
LE	signed less than or equal	$Z + (N\ xor\ V)$	1 1 0 1
AL	always (unconditional)	not used	1 1 1 0
NV	never (unconditional)	not used	1 1 1 1

The condition code may be appended to an instruction mnemonic to cause the instruction to be conditionally executed, based upon the current state of the flags. For example:

$$\text{SUB} \quad \text{r3,r10,r5}$$

subtracts the contents of register r5 from register r10 and places the result in register r3. Writing the instruction as:

$$\text{SUBNE r3,r10,r5}$$

would perform the subtraction operation only if the zero flag = 0.

Thus, conditional execution of instructions could be implemented without the necessity of requiring the insertion of a special branch or jump instruction. Since many assembly language constructs involves jumping over the next instruction, the addition of a conditional execution mechanism greatly improves the code density of the instruction set.

Another unique feature of the ARM architecture is that the classes of instructions know as data processing instructions; including move instructions, arithmetic and logical instructions, comparison instructions and multiplication instructions, do not automatically change the state of the condition code flags. Just as in the previous example of the option to conditionally execute an instruction based upon the state of the condition code flag, it is possible to control whether or not the result of a data processing instruction will change the state of the appropriate flags. For example:

$$\text{SUB} \quad \text{r3,r10,r5}$$

performs the subtraction operation but does not change the state of the flags.

$$\text{SUBS} \quad \text{r3,r10,r5}$$

performs the same subtraction operation but does change the state of the flags.

$$\text{SUBNES} \quad \text{r3,r10,r5}$$

conditionally performs the same operation and changes the state of the flags. The conditional execution flag, 'NE', is appended first, followed by the conditional flag update, 'S'. Thus,

$$\text{SUBNES} \quad \text{r3,r10,r5}$$

is a legal opcode, but

$$\text{SUBSNE} \quad \text{r3,r10,r5}$$

is not legal and will cause an *unknown opcode* error.

Barrel Shifter

Referring to Figure 11.1 we see that the Rm operand passes through a barrel shifter hardware block before it enters the ALU. Thus, it is possible to execute both an arithmetic operation and a shift operation in one instruction. The barrel shifter can shift the 32-bit Rm operand any number of bit positions (up to 32 bits, left or right) before entering the ALU. The bit shift is entirely an asynchronous operation, so no additional clock cycles are required to facilitate the shift operation. We'll leave it to you as an exercise to design the gate implementation of a bit shifter for various degrees of shifts.

For example, the **MOV** instruction is used for register to register data transfers and for loading registers with immediate values. Consider the following instruction:

```
MOV  r6,r1           ;Copy r1 to r6
```

Now, consider the same instruction using with the logical shift left operation added:

```
MOV  r6,r1, LSL #3  ;Copies r1*8 to r6
```

In the second example, the contents of register r1 are shifted left by 3 bit positions (multiplied by 8) and then copied to register r6. The number of bits positions that are shifted may be specified as an immediate value or the contents of a register. This code snippet performs the same shift and load as in the previous example:

```
MOV  r9,#3           ;Initialize r9
MOV  r6,r1, LSL R9   ;Copies r1*8 to r6
```

The following table summarizes the barrel shifter operations:

Mnemonic	Operation	Shift Amount
LSL	Logical Shift Left	#0-31, or register
LSR	Logical Shift Right	#1-32, or register
ASR	Arithmetic Shift Right	#1-32, or register
ROR	Rotate Right	#1-32, or register
RRX	Rotate Right Extended	33

The Rotate Right Extended effectively shifts the bits right by 1 bit position and copies bit 31 into the carry flag position. Note that the flag bits are conditionally updated based upon the state of the S bit to the instruction mnemonic.

Operand Size

The ARM architecture permits operations on bytes (8-bit), half-words (16-bits) and words (32-bits). Once again, you are confronted with a slightly different set of definitions. However, being the most modern of the three architectures, the ARM definitions are probably the most natural and in closer agreement with the C language standards.

Bytes may be addressed on any boundary, while half-words must be aligned on even address boundaries (A0 = 0) and words must be aligned on quad byte boundaries (A0 and A1 = 0). This is even more restrictive than the 68K architecture, but is understandable because most ARM implementations would have a full 32-bit wide data path to memory. Thus, fetching a 32-bit wide memory value from any address where A0=0 and A0=1 would require two access operations, or the equivalent of a nonaligned access.

The ARM architecture supports both big endian and little endian data packing through the proper configuration of the core. However, the architecture does default to little endian as the native mode. Instructions for loading or storing data may optionally be modified by appending the operand type and size to the load or store op-code. The following table lists the type of load/store operations in the ARM architecture:

Mnemonic	Description	Operation
LDR	Load a word from memory into a register	Register < mem32
STR	Store the word contents of a register in memory	mem32 < Register
LDRB	Load a byte from memory into a register	Register < mem8
STRB	Store the byte contents of a register in memory	mem8 < Register
LDRH	Load a half-word from memory into a register	Register < mem16
STRH	Store the half-word contents of a register in memory	mem16 < Register
LDRSB	Load a signed byte into a register	Register < Sign extended mem8
LDRSH	Load a signed half-word into a register	Register < Sign extended mem16

Addressing Modes

The discussion of available addressing modes in the ARM architecture is somewhat more restrictive than the 8086 or the 68K architectures that we've previously studied. The reason for this is that the addressing modes which address external memory are limited to data transfer operations between the register files and memory. Between registers and with data processing instructions there are two addressing modes available, *register and immediate.*

```
SUB  r3,r2,r1    ; r3 < R2 - r1
SUB  r7,r7,#1    ; decrement r7
```

In the above example r7 is also the destination register for the subtraction operation. By subtracting 1 from r7 and storing the result back into r7, we've effectively executed a decrement instruction.

This may seem very strange, but not every possible immediate numbers between 0 and $2^{32} - 1$ may be specified. The reason for this is that the immediate field of the 32-bit instruction field is only 12 bits in length. Furthermore, this 12-bit field is further subdivided into an 8-bit constant operand and a 4-bit shift field. This is shown in Figure 11.5.

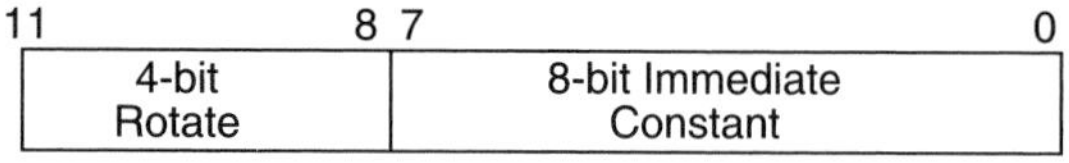

Figure 11.5: Bits 0-11 of the instruction word for defining an immediate operand.

The 32-bit immediate value is created by a roll right of the 8-bit immediate value an even number of bit positions. So a value of 0001 in bit field 11:8 would result in the 8-bit immediate value to be rolled right 2 bit positions. Thus, all valid immediate operands would consist of groups of up to 8 adjacent binary 1's located on even bit boundaries.

This may seem like an unreasonable constraint, but it might not be as restrictive as it first appears[6]. The assembler will try to convert out-of-range constant values by converting the sense of the operand, when it can. For example, the move (MOV) instruction will be converted to a *move not (MVN)* instruction, which negates all the bits of the word. Also, add instructions would be converted to subtract instructions to try to get the operand in range. In any case, if the assembler cannot bring the operand in range it will report it as an assembly error. The programmer may then add additional instructions, such as one or more logical OR instructions, to position additional bits into the immediate operand. In any case, most of the numbers that arise in real programming problems, such as byte values and pointer addresses, tend to follow these rules.

When memory-to-register and register-to-memory transfers are concerned, we have several addressing modes available to us. Also, we must differentiate between the addressing modes used for single register transfers to memory, and those instructions which result in block transfers between memory and the registers. Also note that there is no absolute addressing allowed in the ARM architecture. All memory addressing modes must involve one of the registers as the memory pointer, either by itself, or as a base register for an indexed operation. You might be wondering why this is. Since all ARM instructions (except the Thumb subset) are a single 32-bit word in length, we cannot use an absolute address as an operand, because exactly specifying a 32-bit address would require a second instruction word.

Consider the following example:

```
LDR    r12,[r7]
```

This instruction loads register r12 with the word contents of memory pointed to by register r7. The format of the load instruction is destination ← source. The store instruction has the same format:

```
STR    r12,[r7]
```

This instruction stores the word contents of r12 in the memory location pointed to by r7. The format is source → destination. Confusing? Yes it is.

In both cases, the value stored in r7 is the base address for the memory access. If we were using an indexed addressing mode, this register would provide the starting value for the address calculation. We can summarize the index modes for words or unsigned bytes in the following table:

Index mode	Description	Register offset	Immediate offset	Scaled register
Preindex without write back	*Address calculated prior to being used*	`[Rn,±Rm]`	`[Rn,#±Imm12]`	`[Rn,±Rm,shift #imm]`
Preindex with write back	*Address calculated prior to being used; base <- base + offset*	`[Rn,±Rm]!`	`[Rn,#±Imm12]!`	`[Rn,±Rm,shift #imm]!`
Postindex with write back	*Address calculated after being used; base <- base + offset*	`[Rn],±Rm`	`[Rn],#±Imm12`	`[Rn],#±Rm,shift #imm`

If the operand size is halfword, signed halfword, signed byte or doubleword, then the available indexing modes change slightly:

Index mode	Description	register offset	Immediate offset	scaled register
Preindex without write back	*Address calculated prior to being used*	`[Rn,±Rm]`	`[Rn,#±Imm8]`	N/A
Preindex with write back	*Address calculated prior to being used; base <- base + offset*	`[Rn,±Rm]!`	`[Rn,#±Imm8]!`	N/A
Postindex with write back	*Address calculated after being used; base <- base + offset*	`[Rn],±Rm`	`[Rn],#±Imm8`	N/A

Here we have used the following nomenclature:

Rn : Base register   Rm: Index Register   Imm8: 8-bit immediate offset
Imm12: 12-bit offset   !: With write back   shift: Shift operator

The following set of instructions give examples of several of the index addressing modes in the above tables:

Example #1   `LDR  r12,[r0,+r7]`

The contents of the base register, r0 and the index register, r7, are added together to form a memory address pointer, *addr* that lies on a quad-byte boundary. The word memory contents at *addr* are copied into register r12. The contents of r0 are not changed.

Example #2   `LDRB  r12,[r0,-#0x6A0]`

The immediate value 0x6A0 is subtracted from the contents of the base register, r0, to form a memory address pointer, *addr*. There is no restriction on the value of *addr*. The byte contents at *addr* are copied into register r12. The contents of r0 are not changed.

Example #3   `STR  r12,[r0],+#0x6A0`

The word contents of register r12 are written to the memory location pointed to by r0. After the data transfer takes place the immediate value 0x6A0 is added to the contents of r0 and the sum is stored back into r0.

Example #4   `STRH  r12,[r0,+r7]!`

The contents of the base register, r0, and the index register, r7, are added together to form a memory pointer, *addr*. The resultant address must be an on an even boundary. The halfword contents of register r12 are written to the memory location, *addr*. The contents of the base register are updated to contain the value *addr*.

Example #5   `LDR  r12,[r0,-r7, LSL #3]!`

Logical shift left of 3 bit positions is performed on the contents of the index register r7.

Thus, if <r7> = 0x00000AC4 before the scaling is performed, the value that is subtracted from r0 will be (0x00000AC4)x8 = 0x00005620.

The resulting value is subtracted from the contents of the base register r0 to form a memory pointer, *addr*. The resultant address must lie on a quad-byte boundary. The word value stored in memory at *addr* is copied into register r12 and register r0 is updated with the value, *addr*.

Example #6   `LDREQ  r12,[r0,-r7, LSL #3]!`
Execute Example #5 if the zero flag, Z, is set.

You may have missed something in all of this minute attention to detail so I'll mention it here. If the base register used in the address calculation is r15, the program counter register, then all of the indexed addressing operations are pc relative and position independent. Recall that with the 68K we had to explicitly choose a pc-relative addressing mode. With the ARM architecture pc-relative is simply a byproduct of choosing r15 as the base register.

Stack Operations

The ARM architecture directly addresses the need to have an efficient mechanism to support stack-based, high-level languages, with the addition of two variants of the load and store instruction, as well as additional addressing modes that support the *push* and *pop* behavior that we associate with a memory stack. We should also mention that these instructions are not solely limited to stack operations and are also very useful when multiple registers need to be saved.

The two instructions are:

LDM: Load multiple registers
STM: Store multiple registers

The syntax used for these is quite similar to the syntax of the 68K's **MOVEM** instruction. The registers to be saved or restored are placed in brackets and either a hyphen is used to indicate a contiguous range of registers or a comma-separated list of registers is used. The data transfers occur from the address stored in a base register, Rn, pointing into memory. After the transfer takes place the pointer register, Rn, can optionally be updated by adding the '!' symbol to the instruction, just as is done the indexing address modes that we've just studied. What is different about this addressing mode from the indexed modes is that the type of addressing mode to be used is appended to the instruction op-code, rather than placed in the operand field. For example, in order to load registers r0 through r3 from memory so that the memory pointer, r10, is incremented after each data transfer, we use the form of the instruction:

```
(1) LDMIA    r10,  {r0-r3}       or
(2) LDMIA    r10!, {r0-r3}
```

Assume that <r10> = 0xAABBCC00 before the instruction is executed. The first register, r0, is restored from memory address 0xAABBCC00. Register r1 is restored from address 0xAABBCC04, register r2 is restored from address 0xAABBCC08, r3 from address 0xAABBCC0C. After the instruction is executed:

Case (1) <r10> = 0xAABBCC00
Case (2) <r10> = 0xAABBCC0C

There are 4 addressing conditions for the multiple register transfer instructions:

Mnemonic	Description	Comments
IA	Increment After	*First data transfer occurs at memory location pointed to in base register, Rn. Subsequent transfers are from successively higher memory locations.*
IB	Increment Before	*First data transfer occurs at memory location 4 bytes higher in memory than initial value in base register, Rn. Subsequent transfers are from successively higher memory locations.*
DA	Decrement After	*First data transfer occurs at memory location pointed to in base register, Rn. Subsequent transfers are from successively lower memory locations.*
DB	Decrement Before	*First data transfer occurs at memory location 4 bytes lower in memory than initial value in base register, Rn. Subsequent transfers are from successively lower memory locations.*

The syntax of the multiple register data transfer operations may be summarized below:

LDM(optional conditional execution)(addressing mode) Rn(optional update), {reg. list}
STM(optional conditional execution)(addressing mode) Rn(optional update), {reg. list}

Thus, the instruction:

```
LDMNEIB     r11!, {r3,r5,r8}
```

Would load registers r3, r5 and r8 from the memory address pointed to by register r11+4 *if* the Zero Flag is cleared. Assuming the instruction was executed, the value in r11 would be incremented by 12 bytes.

In order to specifically implement the standard *push and pop* stack operations, ARM has defined additional mnemonics aliases for the 4 addressing modes that we've just described. Before we discuss them, we should spend a moment reviewing what we already know about stack operations.

Stacks are hardware implementations of *last in, first out (LIFO)* data structures. As we've seen in the 68K and 8086 architectures, the stack grows from higher memory addresses towards lower memory addresses. The stack pointer points to the data currently stored on the top of the stack. When data is fetched from the stack during a *pop* operation, the current value of the stack pointer is used and then the stack pointer is incremented by the size of the memory operand. When data is stored on the stack during a *push* operation, the stack pointer is first decremented by the appropriate amount and then the new data is added to the stack. By ARM's definition, this kind of a stack would use the *decrement before* addressing mode for stack pushes and the *increment after* addressing mode for pop operations.

However, we don't have to create a stack using that model. We could just as easily build a stack that grows towards higher memory and shrinks towards lower memory. This would certainly give more flexibility to hardware and software designers using the ARM processor. If we decided to have the stack grow upwards into higher memory it might better fit the image in our mind's eye of a "growing" stack. Let's consider this model for a stack. We would have to decide where we want the stack pointer to be and how we wanted it to behave. Suppose that we want to push a data item onto the stack. We have two choices:

1. The stack pointer points to the next available empty space on the stack. When we add an item the current value of the stack pointer provides the data pointer, then the stack pointer is incremented to prepare for the next data item.
2. The stack pointer points to the address of the last item placed on the stack. Before we place another item onto the stack we must increment the pointer in order to point to the next available memory address before we store the data.

Similarly, when we do a pop operation we also have two choices. Our choices for pop operations are not independent of the push operation. When we chose one strategy for the push, the pop strategy is then determined, and vice versa. For a pop operation:

1. If the stack pointer is pointing to the next available free memory space on the stack we must first decrement the stack pointer to point to the last item placed on the stack. Then we can load the data into the registers.
2. If the stack pointer is pointing to the last item placed onto the stack, then we can load memory immediately and then decrement the pointer value.

From the above discussion you can see that for the push operation, case #1 is an example of *increment after* and case #2 is an example of *increment before*. For the pop operations, case #1 is an example of *decrement before* and case #2 is an example of *decrement after.*

Now back to the stacks. These four stack implementation options are given the following names:

1. *Full Ascending:* This is a *full stack* model. The stack pointer points an address that was the last filled memory location on the stack. If the stack is growing up, then we need to use the increment before approach. The stack mnemonic that is an alias for the IB model is FA.

2. *Full Descending:* Still based upon the full model, we must decrement the stack pointer before storing the data in memory. If the stack is growing down, then we need to use the decrement before approach. The stack mnemonic that is an alias for the DB model is FD.

3. *Empty Ascending:* The *empty stack* model has the stack pointer pointing to the next available (empty) location on the stack. If the stack is growing up, then we need to use the increment after approach. The stack mnemonic that is an alias for the IA model is EA.

4. *Empty Descending:* If the stack is growing down, then we need to use the decrement after approach. The stack mnemonic that is an alias for the DA model is ED.

Thus, to push the 4 registers, r7 through r10, onto the stack using the full descending model, we would use the instruction:

```
STMFD    SP!, {r7-r10}
```

Note that 'SP' is an acceptable alias for register r13.

In some applications we actually might choose not to take advantage of the multiple data transfer instructions, even though they improve code density. Suppose that an interrupt comes into the processor for an external device that is extremely time-critical. It must be serviced as soon as possible, and it must be highly predictable when it will get serviced, or alternately, what is the longest amount of time that it would have to wait (*latency*) before it can be service. Most processors, including the ARM, have to finish the instruction that they are processing before an interrupt can be accepted. Using the multiple transfer instructions, we are increasing the latency for every register-to-memory data transfer operation taking place. Even though it may be more efficient to use the multiple data transfer instructions, we would use a group of discreet single data transfer operations so that the longest amount of time that the interrupt would have to wait is the executing time of a simple instruction, which may be one or two clock cycles, rather than 20 or more clock cycles.

ARM Instruction Set

The ARM instruction set architecture has been continuously evolving since the introduction of the ARM1 core and the ARMv1 instruction set architecture. The ARM family has been continuously evolving and today the ARM11 core supports the ARMv6 ISA. For our purposes, we've been focusing on, and will continue to focus on the ARMv4 architecture. This architecture also includes the ARMv4T architecture, which includes the 16-bit compressed Thumb instructions. We won't consider the Thumb instruction set in this overview, but the interested student is encouraged to consult the references at the end of the chapter in order to follow-up on the Thumb instruction set.

All ARM instructions are one 32-bit word long. Most, but not all, instructions execute in one clock cycle. Exceptions to this are the load and store multiple registers (LDM and STM) instructions and the loads and stores to slower memories when wait states must be added.

We can group the ARM instruction set into four general categories of instructions:

1. Data Processing Instructions
2. Load/Store Instructions
3. Branch Instructions
4. Control Instructions

We'll look at a representative sample of instructions in each of these categories.

1. Data Processing Instructions

The data processing instructions include the following two register data transfer instructions:

Mnemonic	Definition	Op Mode bits [25:21]
MOV	*Move a 32-bit value into a register*	1 1 0 1
MVN	*Move the complement of the 32-bit value into a register*	1 1 1 1

It may strike you as odd that a register move instruction is classified as a data processing instruction. Strictly speaking, it isn't. However, the ARM architecture provides that all data processing operations are between registers, while loads and stores are between memory and registers. So, loading a value into a register, whether that value is an immediate, or the contents of another register, classifies it as a data processing instruction.

The next group of data processing instructions contains the arithmetic instructions:

Mnemonic	Definition	Op Mode bits [25:21]
ADD	*Add two 32-bit numbers*	0 1 0 0
ADC	*Add two 32-bit numbers with carry*	0 1 0 1
SUB	*Subtract two 32-bit numbers*	0 0 1 0
SBC	*Subtract two 32-bit numbers with carry*	0 1 1 0
RSB	*Reverse subtract two 32-bit numbers*	0 0 1 1
RSC	*Reverse subtract two 32-bit numbers with carry*	0 1 1 1

The reverse subtract allow the two register operands used in the subtraction to be reversed, so that if Rd = Rn – Rm for the SUB instruction the instruction RSB would perform the operation Rn = Rm – Rn. Recall that the operand represented by Rm may also be a literal or a register shifted value, so the reverse subtract instructions provide further flexibility in the subtraction operation.

The logical operations are shown below:

Mnemonic	Definition	Op Mode bits [25:21]
AND	*Bitwise AND of two 32-bit operands*	0 0 0 0
ORR	*Bitwise ORR of two 32-bit operands*	1 1 0 0
EOR	*Bitwise Exclusive OR of two 32-bit operands*	0 0 0 1
BIC	*Bitwise logical clear (AND NOT)*	1 1 1 0

The BIC instruction is unfamiliar to us. Logically, it is Rd = Rn * ($\overline{Rm}$). Since the bits of Rm are inverted, any bit position in Rm that contains a 1, will cause a corresponding 1 bit in Rn to be cleared. Thus, if Rn = 0xAB and Rm = 0x01, then the BIC operation will result in Rd = 0xAA.

Mnemonic	Definition	Op Mode bits [25:21]
CMP	*Compare two 32-bit values*	1 0 1 0
CMN	*Compare negated*	1 0 1 1
TEQ	*Test two 32-bit numbers for equality*	1 0 0 1
TST	*Tests the bits of a 32-bit number (Logical AND)*	1 0 0 0

There are a total of 16 data processing instructions. The form of the instruction is as shown in one of the three possible forms shown in Figure 11.6. There are three forms of the instruction, depending if a shift is involved, or if the third operand is an immediate. Bit 25 is called the immediate bit

Figure 11.6: Format for the ARM data processing instructions.

Immediate operand

31 30 29 28	27 26 25	24 23 22 21	20	19 18 17 16	15 14 13 12	11 10 9 8	7 6 5 4 3 2 1 0
COND	0 0 1	OPCODE	S	Rn	Rd	Rotate	Immediate

Immediate shift operand

31 30 29 28	27 26 25	24 23 22 21	20	19 18 17 16	15 14 13 12	11 10 9 8 7	6 5	4	3 2 1 0
COND	0 0 0	OPCODE	S	Rn	Rd	Shift Immediate	Shift	0	Rm

Register operand shift

31 30 29 28	27 26 25	24 23 22 21	20	19 18 17 16	15 14 13 12	11 10 9 8	7	6 5	4	3 2 1 0
COND	0 0 0	OPCODE	S	Rn	Rd	Rs	0	Shift	1	Rm

and is 1 if the third operand is an immediate and 0 if it is a register. Let's look at some examples. If the instruction is of the form:

```
ADD    r0,r1,#0x00AC0000        ;r0 = r1 + Immediate
```

The instruction takes the form of an immediate operand. The instruction code is 0xE28108AB. We can see the breakdown as follows:

- *Bits 31:28* = 1 1 1 0. Always execute.
- *Bits 27:25* = 0 0 1. Fixed
- *Bits 24:21* = 0 1 0 0. ADD opcode.
- *Bit 20* = 0. Result of instruction sets flags. 1 = set flag, 0 = don't set flag.
- *Bits 19:16* = 0 0 0 1. Second operand register = r1.
- *Bits 15:12* = 0 0 0 0. Result register = r0.
- *Bits 11:8* = 1 0 0 0. Immediate operand 0xAC is rolled 16 times (2*<Rotate>) to the right.
- *Bits 7:0* = 1 0 1 0 1 1 0 0, Immediate operand.

When a register is specified as the second operand, and a shift value is also specified as an immediate operand, then the Immediate Shift Operand form is used. The instruction:

```
ADD    r3,r7,r4          ;r3 = r7 + r4
```

has the instruction code 0xE0873004. The least significant 3 hex digits mean that the shifted register is r7, the type of shift is logical shift left and the number of shifts performed is 0.

If we change the instruction to:

```
ADD    r3,r7,r4, LSL #4        ;r3 = r7 + (r4*16)
```

the instruction code becomes 0xe0873204, which decodes to an LSL of 4 bits.

The last example uses a fourth register to provide the bit shift count.

```
ADD    r5,r6,r7, LSR r8       ;r5 = r6 + shifted by r8
```

The instruction code is 0xe0865837. Let's decode this one.

- *Bits 3:0* = 0 1 1 1. Register r7 provides the bit shift count.
- *Bits 6:4* = 0 1 1 . Logical shift right.
- *Bit 7* = 0. Fixed value.
- All other bits are decoded in the same manner as the previous examples.

The compare group of instructions does not produce a result other than to set the flags. Therefore, the result field is not used. In ARM jargon, that field is defined as *SBZ* (should be zero). Any other value in that field may produce unpredictable results.[1] Another feature of the compare instructions is that they will always set the flags, so the state of the S bit is ignored.

The next group of data processing instructions is the very powerful set of multiplication instructions. There are size multiplication instructions, as shown in the following table:

Mnemonic	Description	Syntax
MUL	Multiply two 32-bit numbers, produce a 32-bit result: Rd = Rm * Rs	`MUL{cond}{S}  Rd, Rm, Rn`
MLA	Multiply two 32-bit numbers, and add 3rd number for a 32-bit result: Rd = Rn + (Rm * Rs)	`MLA{cond}{S}  Rd, Rm, Rn, Rs`
UMULL	Multiply two unsigned 32-bit numbers, produce an unsigned 64-bit resulted in two registers: [RdHi][RdLo] = Rm * Rs	`UMULL{cond}{S} RdLo,RdHi,Rm,Rs`
UMLAL	Multiply two unsigned 32-bit numbers and add an unsigned 64-bit number in two registers to produce an unsigned 64-bit resulted in two registers: [RdHi][RdLo] = [RdHi][RdLo] + Rm * Rs	`UMLAL{cond}{S} RdLo,RdHi,Rm,Rs`
SMULL	Multiply two signed 32-bit numbers, produce a signed 64-bit result in two registers	`SMULL{cond}{S} RdLo,RdHi,Rm,Rs`
SMLAL	Multiply two signed 32-bit numbers and add a signed 64-bit number in two registers to produce a signed 64-bit resulted in two registers: [RdHi][RdLo] = [RdHi][RdLo] + Rm * Rs	`SMLAL{cond}{S} RdLo,RdHi,Rm,Rs`

As a class of instructions, the multiple instructions also take longer than one cycle to execute.

Finally, it may surprise you that the ARM instruction set does not contain any division instructions. Sloss et al[7] describe approximation methods that may be used to convert division operations to multiplications.

2. Load/Store Instructions

All data transfers between registers and memory use the load and store class of instructions. All memory addresses are generated using a base register pointer, summed with an additional immediate offset value, register values or scaled register values. In addition, the calculated memory address pointer may be used without updating the base register pointer with the new address value. Finally, the address calculation may take place before or after the address is used in the instruction.

The load/store operations must also deal with the size and type of the operands, since bytes and half-words are also permitted.

---

[1] There's a wonderful story about the intrepid hobbyists/pioneers of the PC industry. It became sort of a cottage industry to try to figure out what the unimplemented op-codes did. In other words, "What would happen if the SBZ field was set to 011?" Sometimes some very interesting undocumented instructions were discovered and were actually designed into commercial products. Unfortunately, when the CPU manufacturer revised the chip, they often changed the codes for unsupported instruction codes, figuring, "Who would use them?" You can imagine the uproar when products started failing when a new batch of processors was plugged in.

Since we've already discussed much of the operation of the load/store instructions as part of our discussion of the addressing modes that they use, we'll just take a brief look at the format of the load/store instruction word.

The load register instruction may take any of the following forms:

Mnemonic	Description	Syntax
LDR	Load a register from a 32-bit memory word	LDR{cond} Rn,<address mode>
LDRB	Load a register from an 8-bit memory byte	LDRB{cond} Rn,<address mode>
LDRH	Load a register from an 16-bit memory half-word	LDRH{cond} Rn,<address mode>
LDRSB	Load a register from an 8-bit signed memory byte	LDRSB{cond} Rn,<address mode>
LDRSH	Load a register from a 16-bit signed memory half-word	LDRSH{cond} Rn,<address mode>

Special mnemonics must be used for the signed byte and half-word data types because these values are sign extended to 32-bits when the register is loaded from memory. No special instruction is necessary for a 32-bit value in memory because it is the native data size of the ARM architecture.

Figure 11.7 shows the format for the load/store instructions for word or unsigned bytes. Note that the load and stores are almost identical, the difference being the state of the L bit in bit position 20. The instruction format is slightly different for half-words and signed bytes.

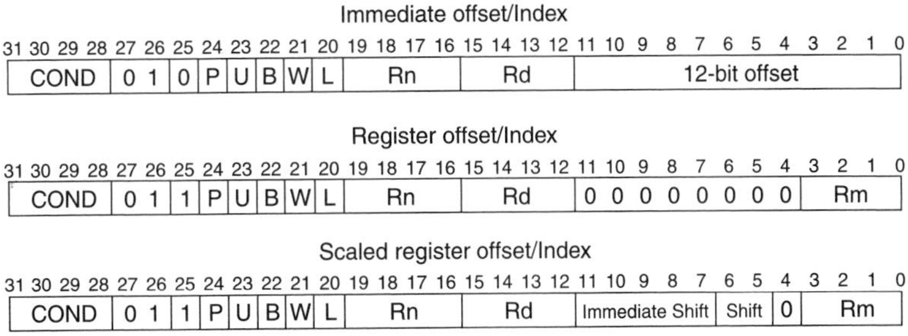

Figure 11.7: Format for the ARM word or unsigned byte load/store instructions.

The meanings of the common fields are as follows:

- *Bits 31:28:* Conditional execution fields.
- *Bits 27:25:* Fixed.
- *Bit 24: For* pre-index mode, P = 1. The offset is applied to the base register and the sum of the base register and the offset is used as the memory load/store address. For post-index mode P = 0. The base register is used as the memory pointer and then the sum of the offset and base register value is written back to the base register.
- *Bit 23: The When* U = 1 the offset is added to the base register value to form the memory address. If U = 0 the offset is subtracted from the base register value.
- *Bit 22:* When B = 1 the memory access is an unsigned byte. If B = 0 the access is a 32-bit word.
- *Bit 21:* If the P bit = 1, then the W bit determines if the calculated memory address value is written back to update the base register. If W = 0 the base register is not updated. When the P bit = 0 and W = 1 the current access is treated as a user mode access. When the P bit = 0 and W = 0 then it is treated as a normal memory access.

- *Bit 20:* If L = 1 then the operation is a memory load. If L = 0 then it is a memory store operation.
- *Bits 19:16:* Base register pointer.
- *Bits 15:12:* Destination register for load operation or source register for store operation.
- *Bits 11:0:* Addressing mode dependent.

Let's look at some examples of memory load and store operations.

Instruction	Description	Instruction code
LDR r5, [r8]	Load r5 with the word pointed to by r8	0xE5985000
LDRSH r6, [r0, -r2]!	Load register r5 with the signed half-word pointed to by r0 – r2. Update r0 with computed address value after the memory load operation.	0xE13060F2
LDRNE r0, [r9,#-12]	Conditionally execute if the Zero Flag = 0. Load register r0 with the word value pointed to by register r3 minus 12 bytes. The value in r9 is not changed.	0x1519000C
LDRVCB r11, [r4],r2,LSL #4	Conditionally execute if the Overflow Flag = 0. Load register r11 with the unsigned byte pointed to by r4. Then update r4 so that r4 = r4 + r2*16	0x76D4B202
LDR R8, [PC,r5]	Load register r8 with the word value pointed to by the sum of the current value of the program counter (r15) and r5.	0xE79F8005
STRB r7, [r3,#0xAA]!	Store an unsigned byte from register r7 to the memory location pointed to by the sum of r3 + 0xAA. Write back the sum to register r3.	0xE5E370AA
STRCCH r11, [r2,#-&A]	Conditionally execute if the Carry Flag = 0. Store the half-word in register r11 in the memory location pointed to by r2 – 10. r2 is unchanged.	0x3142B0BA
STREQ r0, [r4,r5,lsr #7]	Conditionally execute if the Zero Flag = 1. Store the word in r0 to the memory address pointed to by r4 + the result of r5 shifted right 7 bit positions. r5 is unchanged.	0x078403A5
STRB r6, [r4],r3	Store the byte in register r6 to the memory address pointed to by r4. Then add the contents of r3 to r4 and update r4 with the sum.	0xE6C46003
STRPLH r11, [r9,#-2]!	Conditionally execute if the Negative Flag = 0. Store the half word contents of r11 to the memory address pointed to by r9 – 2. Update r9 with the new address.	0x5169B0B2

The above table should give you a sense of the syntax for the various forms of the single item data transfer instructions work and how they are coded in a single 32-bit instruction.

Let's now look at several forms of the load and store operations for multiple data items.

The general form of the load multiple registers and store multiple registers is shown in Figure 11.8. Each bit in the bit

31 30 29 28	27 26 25	24	23	22	21	20	19 18 17 16	15 14 13 12 11 10 9 8 7 6 5 4 3 2 1 0
COND	1 0 0	P	U	S	W	L	Rn	Register List

Figure 11.8: Format for the ARM multiple register load/store operation.

field 15:0 corresponds to a register to be loaded or stored. A 1 in the bit position indicates that the corresponding register is to be loaded or stored. The lowest numbered register is stored at the lowest memory address and the highest numbered register is stored at the highest memory address.

The definition of the bit fields is as follows:

- *Bits 31:28:* Conditional execution fields.
- *Bits 27:25:* Fixed
- *Bit 24:* When P = 1 the address is incremented or decremented prior to the memory access (pre-indexing). When P = 0 the current memory pointer address is used first, and then the memory pointer is changed (post-indexing).
- *Bit 23:* When U = 1 the memory addresses are incremented with each transfer. When U = 0 the memory addresses are decremented.
- *Bit 22:* When S = 1 and the LDM instruction is loading the program counter (r15), then the current program status register (CPSR) will be loaded from the saved program status register (SPSR). If the load operation does not involve r15 and for all STM instructions, the S bit indicates that when the processor is in privileged mode, the standard user mode registers are transferred and not the registers of the current mode. The state of the S bit is set by appending the up-carat symbol, '^', to the end of the instruction.
- *Bit 21:* If W=1, the pointer register will be permanently updated after the multiple register transfer occurs. Since each data transfer is 4 bytes long, the memory pointer will be updates by 4 times the number of registers transferred. If W = 0, the register will not be updated.
- *Bit 20:* If L = 1 then a memory to register (load) operation will take place. If L = 0 then a register to memory (store) operation will occur.
- *Bits 19:16:* Denotes the pointer register.
- *Bits 15:0:* Register list.

The general syntax of the load multiple or store multiple instructions is shown below. The terms in braces are optional.

```
LDM or STM{Condition}XY     Rn{!}, <register list>{^}
```

Here, XY represents:

- IA: Increment After
- IB: Increment Before
- DA: Decrement After
- DB: Decrement Before

The following are two representative forms of the load and store multiple instructions.

Instruction	Description	Instruction code
`LDMDB    SP!,{r0-r3,r5,r7-r9}`	*Load the registers r0,r1,r2,r3,r5,r7,r8 and r9 from the block of memory pointed to by stack pointer (SP) register r13. Load register r8 first from the address SP-4 and continue to decrement SP until r0 is loaded. Update the SP with the address of the last memory word loaded into register r0.*	*0xE93D03AF*
`STMNEIA  r0,{r2-r9}`	*Conditionally execute this instruction if the Zero Flag = 0. Store the contents of registers r2 through r9 in the block of memory pointed to by r0. Store register r2 and then increment r0 for the next store operation. After the multiple data transfer is completed the value of r0 is restored to its previous value.*	*0x188003FC*

The swap instruction (SWP) is a special type of load store operation. It is designed to swap the contents of memory location with the contents of a register. Now, you might argue that this is a nice instruction to have, but it doesn't quite fit into our streamlined model of a computer's instruction set architecture. For example, couldn't you use a traditional algorithm to exchange the contents of memory and a register? For example, suppose we want to exchange the contents of r0 with the contents of the memory location pointed to by r10:

```
MOV  r8,r0        ;Move r0 to a temporary register
LDR  r0,[r10]     ;Get memory, ½ of the swap done
STR  r8,[r10]     ;Save r8, swap completed
```

The corresponding form of the swap instruction is:

```
SWP  r0, r0, [r10]   ;Exchange <r10> with r0
```

The general form of the swap instruction is:

```
SWP{B}{Condition}        Rd,Rm,[Rn]
```

Where register Rd is loaded from the memory location pointed to by Rn and the contents of the memory is overwritten by the value in Rm. Thus, in the general case, the exchange can be between two registers and a single memory location.

The question still remains, "Why have the swap instruction at all?" The answer is that the swap instruction is an *atomic* operation. An atomic operation cannot be interrupted. Most instructions are atomic. That is, once an instruction starts and the processor receives an external interrupt, the instruction must complete before the interrupt can be taken care of. In the above example of the memory to register exchange operation, we need to use 3 instructions to complete the data transfer. These 3 instructions are not atomic because an interrupt could cause a gap to occur in the exchange of data. If the interrupt also changed the data in these registers or memory, then the data exchange might become corrupted. The swap instruction is a way to lock the bus so that it must complete before another event can take control.

3. Branch Instructions

There are two forms of the branch instruction; *branch (B)* and *branch with link (BL)*. The instructions are similar with the exception that the branch with link instruction automatically saves the address of the next instruction after the BL instruction in the link register, r14. This is just a subroutine call. To return from the subroutine, you just copy the link register to the program counter: **MOV PC, LR**.

The range of the branch instruction is +/- 32 megabytes. Just like the 68K, the branch instruction in the ARM architecture is a pc-relative displacement. The displacement is added or subtracted from the current

31 30 29 28	27 26 25 24	23 22 21 20 19 18 17 16 15 14 13 12 11 10 9 8 7 6 5 4 3 2 1 0
COND	1 0 1 L	24-bit offset

Figure 11.9: Format for the ARM branch and branch with link instructions

value of the pc and the pc is reloaded with this new value, causing a program branch to occur. The form of the branch instruction is shown in Figure 11.9.

The branch address is calculated as follows:
1. The 24 bit offset value is sign-extended to 32 bits.
2. The result is shifted left by 2 bit positions (multiplied by 4) to provide a word-aligned displacement value, or effectively, a 26-bit word address.
3. The displacement is added to the program counter and the result is stored back into the pc.

4. Software Interrupt Instructions

The software interrupt instruction is designed to allow application code to change the program execution context through a vector stored in memory. This instruction is similar to the TRAP instruction of the 68K and the INT instruction of the 8086. In general, the software interrupt (SWI) is used by an application to make a call to operating system services.

Since the SWI instruction is used to change context, it must also save the current processor context so that it can return after the interrupt. The action of the SWI is as follows:
1. Save the address of the instruction after the SWI instruction in register r14_svc.
2. Save the CPSR in SPSR_svc.
   Enter supervisor mode and disable the normal interrupts, but not the fast interrupt request.
3. Load the PC with address 0x00000008 and execute the instruction there.

Rather than use the exception vector table as an indirect address to the start of the software interrupt service routine, the vector table location contains space for one instruction, which is then used as a branch to the start of the code. This may seem strange if you think about the 68K's vector table organization, but with the ARM architecture it really doesn't matter. Since all instructions are one word long, you don't need to use an indirect pointer to get to the start of the ISR code. Motorola must use a vector because an unconditional jump instruction would take up too much space. However, since an ARM instruction fits into the same space as an address, either method would work.

The software interrupt instruction also contains a 24-bit immediate operand field that may be used to pass parameters to the interrupt service routine. Thus, instead of using multiple software interrupt vectors, a single vector is used, but information about the type of interrupt service being requested can be passed in the operand field of the instruction.

5. Program Status Register Instructions

The last ARM instruction category that will look at contains two instructions that implement a load or store operation between the CPSR or SPSR registers and the general purpose registers. The syntax for the instructions is as follows:

```
MRS{condition} Rd, <cpsr or spsr>
MSR{condition} <cpsr or spsr>_<fields>, Rm
MSR{condition} <cpsr or spsr>_<fields>, #Immediate
```

The MRS instruction moves the current value of the CPSR or SPSR to a general purpose register. The MSR instruction moves the contents of a general purpose register or an immediate value into the CPSR or SPSR.

Some comments on the status register instructions. This instruction will be ignored if the processor is in user mode and the instruction attempts to modify any other field besides the Flag Field. It must be in one of the privileged modes for this instruction to be executable because the program status registers may only be modified when the processor is in one of the privileged modes.

The values for the field variables are as follows:

- _C: The Control Field represents bits 0 through 7 of the program status register. This is further subdivided as:
  - Bits 0:4: Processor mode
  - Bit 5: Enable Thumb Mode
  - Bit 6: Enable Fast Interrupt Request Mode
  - Bit 7: Enable Interrupt Request Mode
- _X: The Extension Field represents bits 8:15. Currently this field is not used, but is reserved by ARM for future expansion. These bits should not be modified.
- _S: The Status Field represents bits 16:23. Currently this field is not used, but is reserved by ARM for future expansion. These bits should not be modified.
- _F: The flag field represents bits 24:31. This field is further subdivided as:
  - Bit 28: V bit- Overflow Flag
  - Bit 29: C bit- Carry Flag
  - Bit 30: Z bit- Zero Flag
  - Bit 31: N bit- Negative Flag

The immediate operand can only modify the bits in the Flag Field. Also, in order not to inadvertently modify bits in the program status register that should not be modified, the program status register should be modified using the following three steps:

1. Copy the contents of the PSR into a general purpose register using the MRS instruction,
2. Modify the appropriate bits in the general purpose register,
3. Copy the general purpose register back into the PSR using the MSR instruction.

The following instruction sequence enables the FIR mode.

```
MRS    r6, c_spsr    ;Copy the spsr to r6
MOV    r7, #&40      ;Set bit 6 to 1
ORR    r6,r7,r6      ;Set the bit
MSR    c_spsr, r6    ;Reload the register
```

The field bits are logically OR'ed together, so that, for example, you may use cxsf_cpsr to modify all the fields of the cpsr.

MSR

31 30 29 28	27 26 25 24 23	22	21 20	19 18 17 16	15 14 13 12	11 10 9 8 7 6 5 4 3 2 1 0
COND	0 0 0 1 0	R	0 0	1 1 1 1	Rd	0 0 0 0 0 0 0 0 0 0 0 0

MSR Immediate Form

31 30 29 28	27 26 25 24 23	22	21 20	19 18 17 16	15 14 13 12	11 10 9 8	7 6 5 4 3 2 1 0
COND	0 0 1 1 0	R	1 0	field_mask	1 1 1 1	Rotate	Immediate

The formats of the MSR and MRS instructions are shown in Figure 11.10. If the R bit = 1 the program status register used is the

MSR Register Form

31 30 29 28	27 26 25 24 23	22	21 20	19 18 17 16	15 14 13 12	11 10 9 8 7 6 5 4	3 2 1 0
COND	0 0 0 1 0	R	1 0	field_mask	1 1 1 1	0 0 0 0 0 0 0 0 0	Rm

Figure 11.10: Format for the ARM modify status register instructions.

SPSR, if R = 0 the program status register is the CPSR register. The immediate filed is rotated by the rotate field value to move the bits to the Flag bits position of the PSR.

ARM System Vectors

The system vectors for the ARM architecture are rather sparse, compared to the 68K and 8086 architectures. There are a total of 8 system vectors, shown in the following table:

Exception Vector	Address
Reset	0x00000000
Undefined Instructions	0x00000004
Software Interrupt	0x00000008
Prefetch Abort	0x0000000C
Data Abort	0x00000010
Reserved	0x00000014
Interrupt Request	0x00000018
Fast Interrupt Request	0x0000001C

The Fast Interrupt Request vector is the last vector in the table for a reason that may not be so obvious. Recall that each vector is a 32-bit word, capable of holding just one instruction. That instruction will generally be a branch instruction to the starting point of the user's service routine. The FIR vector sits at the top of the table so that the FIR service routine can begin at address 0x0000001C and continue on from there, without the need to add a branch instruction to get to the real code. If you want to be fast, every clock cycle counts!

The Prefetch Abort vector is used when the processor attempts to fetch an instruction from an address without having the correct permissions to access that instruction. It is called a pre-fetch abort because the actual instruction decoding takes place after the instruction is fetched, but the exception actually occurs during the prefetching of the instruction. We'll look into this more deeply when we study pipelined processors in a later chapter.

The Data Abort vector is like the Prefetch Abort vector, except for data. Thus, a Data Abort Exception will occur when the processor attempts to fetch data from a memory region without the correct access permissions.

The Reset vector is also unique because when it is asserted the processor will immediately stop execution and begin the reset sequence. With other exceptions, the processor will complete the current instruction before accepting the exception sequence. Of course, this makes good sense, since a reset has no need to restore the system context, so you might as well get on with it as soon as possible.

Summary and Conclusions

The ARM instruction set is a thoroughly modern, 32-bit RISC instruction set. Unlike the 8086 and 68K processors, all instructions are the same length and, with few, exceptions, execute in one clock cycle. The register set is almost completely general-purpose. Only three of the 16 registers user mode registers have dedicated uses. All data processing instructions take place between registers and all memory operations are restricted to memory-to-register load operations and register-to-memory store operations. All memory accesses use one of the general purpose registers as a base address memory pointer. Additional effective addressing modes enhance this model by adding incrementing, decrementing, index register, immediate offset values and scaled register modes.

While you might disagree with this, the ARM instruction set architecture is quite a bit simpler and more restrictive than the architectures that we've previously examined. This simplicity places more dependence upon the compiler to be able to generate the most optimal code flow, and hence, the most efficient code.

With this chapter's overview of the ARM architecture we will be leaving the study of common architectures and move on to other topics. We'll return to the study of architecture once again when we consider pipelines in detail in a later chapter. At that time we'll return to the ARM architecture once again, but hopefully, we'll stay at a higher level the next time around. While a certain percentage of those of you reading these chapters may have found this as exciting as watching paint dry, these is a method to the madness. In order to understand a computer's architecture from a software perspective, we must look at examples of how the various bit patterns are used to form the instruction words.

Dr. Science, a performer on National Public Radio, once said,

> "I like to read columns of random numbers, looking for patterns."

Summary of Chapter 11

Chapter 11 covered:

- A brief history of the evolution of the ARM architecture
- An overview of the ARM7TDMI processor architecture
- An introduction to the ARM instruction set and addressing modes

Chapter 11: *Endnotes*

[1] Jim Turley, *RISCy Business*, Embedded Systems Programming, March, 2003, p. 37.

[2] *Ibid.*

[3] ARM Corporate Backgrounder, http://www.arm.com/miscPDFs/3822.pdf, p. 1.

[4] Andrew N. Sloss, Dominic Symes and Chris Wright, *ARM System Developer's Guide*, ISBN 1-55860-874-5, Morgan-Kaufmann, San Francisco, CA.

[5] Dave Jagger, Editor, *Advanced RISC Machines Architectural Reference Manual*, ISBN 0-13-736299-4, Prentice-Hall, London.

[6] Steve Furber, *ARM System-on-chip Architecture*, ISBN 0-201-67519-6, Addison-Wesley, Harlow, England.

[7] Andrew N. Sloss, Dominic Symes and Chris Wright, *ibid,* pp. 143–149.

Exercises for Chapter 11

1. What are the operating modes of the ARM system? How do they compare with the 68K?

2. Why is there a Fast Interrupt Request Mode and how is it implemented?

3. Compare the 16 base registers of the ARM architecture with the 16 registers of the 68K architecture.

4. Is the instruction,

   MOV   r4,#&103

   a legal or illegal instruction? Why? Note: &103 is the ARM notation for a hexadecimal number.

5. Write a code snippet that loads register r4 with the immediate value &103.

6. Initialize register r7 with the value &06AA4C01.

7. Suppose that the contents of register r8 = &0010AA00 and the contents of register r6 = &0000CFD3. What will be the value stored in register r11 after the instruction:

   ADD r11,r8,r6 LSL #2

8. Rewrite the following 68K instruction as an equivalent ARM operation. Hint: Don't forget the flags.

   ADD.L   D3,$00001000

9. Assume that <r1> = &DEF02340. Describe as completely as you can the operation performed by the instruction:

   `LDRNEH   r4,[r1,#4]!`

CHAPTER **12**

Interfacing with the Real World

. . . .

Objectives

When you are finished with this lesson, you will be able to:
- ▶ *Describe why interrupts are inherent in computer/real-world interaction;*
- ▶ *Explain why interrupts are prioritized;*
- ▶ *Understand the concept of I/O ports;*
- ▶ *Explain how analog signals are converted to the digital domain and vice versa;*
- ▶ *Understand the tradeoffs associated with speed versus accuracy in the analog to digital conversion process.*

. . . .

Introduction

In the previous lesson we saw that a computer that operated only within its own environment, and couldn't interact with the real world is a rather useless computer. It is a nice environment for studying architecture, but that's about all its good for. It was somewhat refreshing (I hope) when you were able to add input and output activity (I/O) to your programs using the TRAP #15 instructions. Now, let's begin our discussion of computers and the real-world by consider Figure 12.1. The drawing inside the dotted lines in the figure represents the minimum number of components necessary to have an operating computer. Outside of the dotted lines is everything else that we need to make it do useful work.

As you can see, a processor, memory array, glue logic (memory decoding and such) and clocks form the basic computer, but this computer is relatively worthless in human terms. We somehow need to be able to interact (*interface)* with external stimuli.

The external world *(real world)* has its own sets of constraints that we must be able to deal with. The real world is a very messy place compared to the motherboard of your PC. Some of these constraints are:

- real world events are generally not digital in nature;
- real world events happen at much different rates than the fundamental clock cycle in the computer;
- real world events are transitory and may be lost if not serviced within the appropriate timeframe;
- real world events often take place in environments that are dirty, wet, extremely hot or cold, or have large amounts of background noise (electrical interference);

322

- failures in computers that service real world events have real world consequences. Critical systems may fail (Y2K) or human life may be lost.

If we are going to accept the fact that we need to make some order out of the chaos of the real-world, we first need to understand how the real-world and the computer can communicate with each other. Since events happen at such different rates, we need methods that will allow us to synchronize the worlds inside and outside of our computer environment. The first of these methods that we need to discuss is the concept of interrupts.

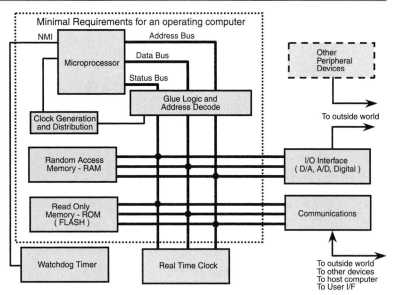

Figure 12.1: A typical computer system. The functional blocks shown inside of the dotted lines are the minimal requirements for a computer to actually run.

Interrupts

So far we've examined the computer system in terms of the processor and memory. Add a clock and this is a functional, but useless, computer. On occasion, throughout the previous lessons, you might have noticed the word *interrupt* sprinkled here and there. Now let's take the time to understand just what the interrupt is all about. Recall that we actually alluded to interrupts when we studied the computer architectures in the previous chapters. In particular, we looked at the ARM architecture and its interrupt and fast interrupt request modes. We also looked at software interrupts and how they were used to change the context of the processor and access the operating system. Now, let's step back and look at the interrupt process itself.

In order for a computer to be worth the cost of the electricity that you feed into it, it must be able to interact with you and its environment. You manipulate a keyboard and a mouse. The computer responds with actions on the screen, sound, disk access, etc. Sometimes unexpected events, called *exceptions*, occur and the computer has to be able to deal with them. A typical exception might be the result of an errant pointer causing the program to attempt a memory fetch from a region of memory where no physical memory is present. Or, you try to divide by zero. Duh!

When asynchronous events, both internal and external to the processor, need to grab the attention of the processor, they do it by generating an interrupt. The interrupt forces the processor to stop its normal program execution and start executing another block of code called the *interrupt service routine (ISR)*. After the ISR program code completes and the interrupt is taken care of, the processor returns to where it left off and resumes execution of your application code.

Suppose we didn't have interrupts. In order for a processor to take care of these events it would have to periodically check each event that might require servicing to see if the event is ready for service. A good analogy is the ringer on the telephone. Here you are, eating dinner (the application), when the phone rings (the interrupt). You put down your fork and answer the phone (ISR). You tell the telemarketer that you don't want a free vacation at the exciting Riviera Resort in Newfoundland and go back to your dinner (return from the interrupt).

Now, suppose that the ringer on your phone is broken. The only way for you to tell if someone is calling you is to pick up the phone every few seconds and say, "Hello, hello, is anyone there?" This gets very old, very fast. The analogous process in a computer is called *polling*. In a polled system, the computer periodically checks each event that might require servicing as part of its regular program code. Polling is still a very acceptable way to program a computer when the application lends itself to a polled structure. A burglar alarm controller is a perfect example of a polled system. The program checks every sensor in turn to see if it has been tripped by a burglar or by the family cat. If a sensor has tripped, the program turns on the alarm. The program runs in a continuous loop, called a *polling loop*, checking the sensors.

As you might imagine, when a computer is polling its peripheral devices to see if they need to be serviced, it isn't doing much of anything else. That's why we have interrupts. The interrupts, because they are asynchronous, happen more or less randomly, and the processor deals with them on an "as needed" basis. However, we should not lose sight of the fact that interrupts could also happen at very precise intervals, as well as randomly. For example, your Windows Operating System has external timers providing clock ticks every few milliseconds, and other systems may have interrupts appearing regularly every few microseconds. The key point is that the interrupts are not synchronized to our program's execution. Now let's examine some of the common types of interrupts that you might encounter in a computer system.

The most common interrupt is the RESET ($\overline{\text{RST}}$). RESET is a very dramatic interrupt. It starts the processor from the beginning. It does not, as do other interrupts, return to a point in the application where it left off. The RESET interrupt assumes that everything is suspect and you truly want to start from the beginning. When you assert the RESET by pressing the RESET button on your computer you cause the following sequence of events to occur:

1. clear the contents of the internal registers;
2. establish the processor in a known state; and
3. begin program execution from a known memory location.

Please note that the above process is a general list of actions for the RESET interrupt. As we previously saw, different processors start in different ways. Modern Pentium and Athlon CPUs have very complex start-up sequences when RESET is asserted.

If you examine the RESET interrupt at the hardware level you might be surprised to see that you have to hold the RESET input asserted for quite a number of clock pulses in order for the RESET to work properly. This is another clue that our algorithmic state machine is busy behind the scenes. It is typical that the RESET input may have to be asserted for 50 to several hundred clock cycles in order to bring the state machine to the correct state.

Interrupts, because they are asynchronous, can interrupt each other. A processor can be in an ISR when another interrupt comes in. What does the processor do? In order to deal with this situation interrupts may be prioritized. A more important (higher priority) interrupt may always preempt a lower priority interrupt. What are some examples of high and low priority interrupts? Answering this question is not always as straight-forward as it seems. Sometimes we are concerned with the window of time that is available to us to service the interrupt. If we are trying to capture and process a fast data stream, such as a digital video camcorder, and we don't want to drop any frames, then we might give that interrupt a higher priority.

Another factor might be criticality of the interrupt. Most laptop computers have a high priority interrupt driven by the circuitry that monitors the battery's energy level. When the battery has almost lost its ability to power the computer, a high-priority ISR automatically takes over and saves the state of the computer so you can shut down and recover when the battery is recharged. Speaking of criticality of the interrupt, the highest priority interrupt mechanisms are usually reserved for protecting human life.

A lower priority interrupt will have to wait if a higher priority interrupt is being serviced because the higher priority interrupt can *mask* the interrupt signal from the lower priority interrupt. This is why you get the hourglass in Windows when your PC is busy with the disk drive. It's Windows way of telling you that the mouse is waiting its turn while the disk data is being read.

In many situations, particularly with operating systems such as Windows and Linux, the time it takes for the computer and operating system to respond to an interrupt, is unpredictable, and may not be fast enough, to reliably service all of the interrupts in the allotted amount of time. In order to deal with computer-based systems that must function reliably while dealing with real-world events, a different type of operating system was designed. Operating systems that must be able to handle external events occurring in real-world time frames are called *real-time operating systems,* or RTOSs. Unlike Windows, an RTOS is not an egalitarian operating system. The operating system on your PC schedules tasks in a round robin fashion. Each task that is executing receives a slice of time from the operating system (usually abbreviated O/S), independent of how "important" that task may be. Sometimes the O/S will, in the background, give tasks that appear to be doing input and output more time than tasks that are not. However, the key is that we cannot predict with high confidence that under all conditions, that all tasks will be executed in order of their criticality and that all interrupts will be serviced in the required time windows. Of course, getting a faster computer always helps, but sometimes economic reality rears its ugly head and that is not a viable option.

RTOSs use a very different scheduling mechanism than PC O/Ss. An RTOS task of higher priority that is ready to run will always preempt a lower priority task that is currently running. Also, the kernel software of the O/S is carefully designed to minimize the time spent switching between tasks. This is the reason why the ARM architecture has a fast interrupt request mode and banked registers. Both features are architectural enhancements for speeding up the response time of the processor to the stringent requirements of servicing interrupts in a timely manner.

Figure 12.2 is a graph of the behavior of a computer system running under the control of an RTOS. The x-axis is measured in seconds with each time tick shown in 100 microseconds. The y-axis shows the various tasks and interrupts of the system. The interrupts and tasks are arranged in order of descending priority. The highest priority interrupt, Int_1 has a higher priority than Int_2, which has a higher priority than any task. The lowest priority task is the system in its idle state.

Notice how each task of higher priority may preempt the lower priority task until the higher priority task either runs to completion or must wait for information before it can proceed. Consider the task labeled Task_3. When Task_3 becomes active it preempts Task_5 and continues to run until it stops at about Time = 10.4155 seconds. At that time Task_5 starts again until it is preempted once again, this time by Task_1, the highest priority task.

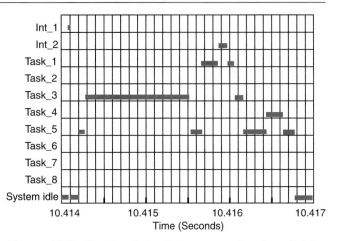

Figure 12.2: Graph of the interrupt and task switching response of a typical real-time operating system. The X axis spans approximately 3 milliseconds of execution time.

Task_1 runs but is preempted by Int_2. When the ISR for Int_2 completes, Task_1 starts up again and then stops, allowing Task_3 to once again take over. When Task_3 finally stops executing Task_5 can begin to execute. However, Task_4 suddenly comes alive and preempts Task_5. Finally, Task_4 completes and Task_5 can finish execution. When Task_5 completes the system returns to the idle state.

If all this seems very complicated, you're right. Even in this "relatively simple" example, you can begin to see the potential problems that might emerge. What if Task_5 keeps getting preempted by the higher priority tasks and interrupts above it and never runs to completion? That is certainly a possibility. We can also, have certain sequence of events cause the system to lock up. One such situation is called *priority inversion*. Priority inversion occurs when higher priority tasks are suspended from running because lower priority tasks were themselves preempted, but the lower priority task retains control of a system resource, such as a memory buffer, that the higher priority task needs to use. A very interesting priority inversion problem occurred with the *Mars Rover* project[1] when the rover had just halted on Mars. The mission control team back in Pasadena ran simulations with an identical rover running in the Jet Propulsion Laboratory's back yard and they discovered that the rover's RTOS had locked up due to a priority inversion. They were able to correct the problem and upload new code to the rover's 8086-based computer system.

You've seen from the above example that interrupts, because they are asynchronous, can interrupt each other. A processor can be in an ISR when another interrupt comes in. What does the processor do? In order to deal with this situation interrupts are generally prioritized. A more important (higher priority) interrupt may always preempt a lower priority interrupt, just as a higher priority task may preempt a lower priority task. What are some examples of high and low priority interrupts? A lower priority interrupt will have to wait if a higher priority interrupt is being serviced because the higher priority interrupt can *mask* the interrupt signal from the lower priority interrupt.

This is why you get the hourglass in Windows when your PC is busy with the disk drive. It's Windows way of telling you that the mouse is waiting its turn while the disk data is being read. Recall that the ARM processor did not have prioritized interrupts. The two interrupts, FIR and IRQ were of equal priority and they could be always be asserted if they were enabled in the PSR.

However, in most computers there is always a highest priority interrupt. This is the interrupt that cannot be ignored and must always be serviced immediately. This is the *nonmaskable interrupt (NMI)*. Generally we reserve the NMI for catastrophic events, like a power loss to the system, or the detection of a memory failure. In an aircraft, the NMI might be reserved for an impending life-threatening situation. The NMI is a true interrupt in the sense that once it is serviced, execution can return to the point in your code where the interrupt first took place.

The remaining interrupts are prioritized and can be assigned their priority level through various external and internal methods. We won't be concerned with how that is accomplished in this text, other than to realize that a lower priority interrupt must have to wait if a higher priority interrupt is being service. In the Motorola 68K architecture, the lowest priority interrupt is a priority-1 interrupt. The NMI interrupt is a priority-7 interrupt. When an interrupt is being serviced, only an interrupt with a higher priority level can take over. Also note that we don't associate an interrupt priority level with a nonmaskable interrupt because it is hardwired into the state machine of the processor.

Exceptions

Exceptions are similar to interrupts but they are generated by program-related events, such as a memory access error (no memory around), an illegal instruction (pointer error), divide by zero error, or other program faults that require special handling. From a structural point of view they are handled like an interrupt, except they cannot be masked out. If the exception-generating situation occurs, the exception handling process begins immediately.

Motorola 68K Interrupts

The Motorola 68K handles interrupts in a fairly standard manner, so we'll use this architecture as our prototype and take a few moments to discuss it. In the address space of the 68K processor, the first 1024 bytes of memory are reserved for dealing with exceptions and interrupts. Thus, byte addresses from 0x000000 to 0x0003FF are reserved for interrupts and exceptions and a user program should not start below address 0x000400. We call each of these addresses *vectors* because they point to the program code that is designed to service the exception pointed to by the appropriate vector. In other words, the contents of the memory locations associated with these interrupt vectors are themselves addresses, the address of the location in memory where the ISR is located.

In each long word memory location the programmer places the address of the first instruction for the corresponding ISR or exception. For example, the ISR for a NMI is stored at address 0x00007C. The processor executes the following sequence in response to the NMI.

- completes the current instruction
- saves a copy of the status register on the stack
- saves the address of the next instruction to be executed on the stack
- switches to supervisor mode

- fetches the 32-bit data located at memory location 0x00007C
- begins execution of the ISR at the address of the memory location stored at 0x00007C

This type of addressing is also called *indirect addressing*. The data stored at a location in memory is actually the address of the real data. In C and C++ we call this a pointer. Thus, the first 256 long words of memory is reserved for *system vectors*. These system vectors are pointers to other regions of memory where the actual exception code or ISR's are stored.

The first two long word addresses in memory, 0x000000 and 0x000004, have special significance. After a RESET is asserted, the 68K will fetch the vector at 0x000000 and place it in the stack pointer register. It will then fetch the vector at 0x000004 and place it in the *program counter register*. It will then begin program execution at the address in the program counter register.

Interrupts tell us when an outside world task needs to be serviced, but how do we actually exchange real data with the outside world? One of the most basic tasks that we have in interfacing to the outside world is the task of synchronizing events in the computer, which may be changing hundreds of millions of times a second, with events in the outside world, which may or may not change over the course of hours. One of the simplest ways to synchronize these two timescales is to use the "D-type" flip-flop as a storage register. D registers or latches are typically used to synchronize external world events to the processor bus. We call them *I/O ports* rather than registers when they are used to interface to the outside world.

You've already seen how the Intel X86 family, treat I/O ports as a separate address space, while the ARM and 68K architectures treat I/O devices as part of the processor's memory space. Thus, the 8086 has a separate assembly language instruction for reading and writing to the I/O space than for reading and writing to memory. Sometimes the I/O space has less stringent timing and easier address decoding than the memory space. Having the I/O devices mapped into the memory space of the processor results in a simpler instruction set because I/O transfers are the same as memory transfers. In both systems we use interrupts and status register are used to signal when data is available.

Let's examine the operation of a simple 8-bit I/O port to see how we might interface the computer to the outside world. Figure 12.3, is a simple schematic diagram of an *I/O port*.

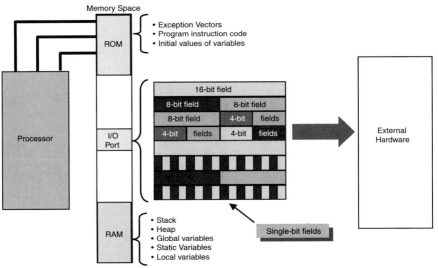

Figure 12.3: Generalized view of an I/O port within the memory space of a microprocessor.

I/O ports can be single ports, or entire consecutive blocks of I/O. For example, the graphics chip inside of your PC may have a hundred or more I/O ports (called *register maps*) associated with data transfer and control of the graphics environment. In Figure 12.3 we see that this I/O device is divided into separate ports that are single bit fields, 4-bit wide fields and larger. Each port is configured as needed for the I/O function that it will perform.

Let's consider a more specific circuit arrangement as shown in Figure 12.4. The I/O port in Figure 12.4 appears to the computer as two consecutive memory locations. The actual device that forms the I/O port has two component parts associated with it. The portion

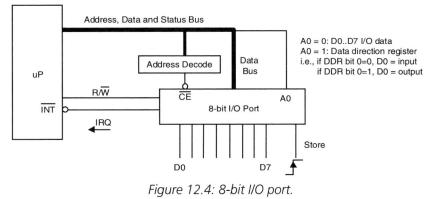

Figure 12.4: 8-bit I/O port.

located at the even address (A0 = 0) is the I/O port itself and the portion located at the odd address (A0 = 1) determines how the I/O port will be used. It is a constraint of most I/O ports that they cannot simultaneously be both in input and an output. We must program the individual bits of the device to be either inputs or outputs.

The address decoding block determines where in the address space of the processor, the I/O ports will become active. For example, let's assume that the port occupies byte addresses $00A600 and $00A601, respectively. The actual I/O port appears at the even address and a control register appears at the odd address. We call this control register the *data direction register,* or DDR.

When we write to the DDR we are programming the configuration of the I/O port on a bit-by-bit basis. Any bit position that has a "0" written to it makes the corresponding bit of the I/O port an input, any bit position that has a 0 written to it becomes an output. Yes, it would make more sense to make a "1" correspond to an input and "0" correspond to an output, but hardware designers are such a bunch of kidders. Thus, writing $FF to the DDR makes all of the bits of the I/O port output bits and writing a $00, makes all of the bits input bits. Writing $AA to the DDR port makes the odd bits inputs and the even bits outputs, and so on.

Assuming that we program the DDR to $FF, we now have an 8-bit output port. From the computer side, we can write a value to this register as if it was any other memory location, but we can then see the data on the output side of the I/O port. Also, the data is permanent. It can remain on the output side of the port for days without stopping the computer. We now have a digital control signal that we can use for whatever tasks we want. Now let's assume that we want to use the port as an input port.

When outside data is presented to the port it must be written into the port with a positive going clock edge. The data is now stored in the input portion of the port, but in general, the computer has no way of knowing that the data is there, or that the data stored there may have changed. This is a subtle, but very important difference between an I/O port and a memory location. Also, if I write a data value to memory and then immediately read it back, I expect to see the data that I just wrote. However, with an I/O port, the data that I write to the port (output) will generally not be the data that I read from the port (input) because the output and input portions of the port are different.

With memory, we assume that the data stored there will not change unless we somehow change it. You are all familiar, I assume, with the dangers of creating global variables. Global variables can be so easily changed in unexpected ways that programmers are especially vigilant when dealing with them. What about I/O port variables? We have the same problem, only worse. With an I/O port we must assume that the data stored in the port (on the input side) can change spontaneously, and all of the rules of memory integrity that we are accustomed to may no longer hold true.

Since I/O ports are often handled as if they are memory, but are not, compilers have to have special instructions on how to deal with I/O ports. The most general way to tell a compiler not to assume a memory location (variable) is simple memory is to use the key word *volatile*. For example:

```
volatile unsigned short int * foo;
```

would tell the compiler that foo is a pointer to a positive, 16-bit memory variable that may spontaneously change value without warning. This means that the compiler should not make any assumptions about optimizing the code with that pointer. The value dereference by the pointer should not be assigned to registers or any other form of optimization. The pointer is always used to change the memory value or to read it, that's all.

As an example, let's consider a real-life I/O port. The example we'll look at is a universal asynchronous receiver/transmitter, or *UART*. This name may not mean much to you because you most likely know it as a *com port* on your PC, such as COM 1 or COM 2. A com port is a serial transmitting and receiving device. Suppose that we have a UART in our computer and the hardware designer has designed it so that it is memory mapped as a 16-bit wide I/O port at address $006000. Here are the operational specifications:

- 8 bits of serial data (one byte) is sent and received at bit positions DB8 – DB15.
- DB0 – DB7 provide status information about the device.
- DB0 = DATA READY (DR) status bit.
- DR = 1 means that data is available and may be read from DB8 – DB15.
- DR = 0 means that no data is currently waiting to be read.
- DB1 = TRANSMITTER BUFFER (TB) status bit.
- TB = 0 means that the transmitter buffer is currently idle and data may be transferred to it to be sent.
- TB = 1 means that the transmitter buffer is currently sending data and new data should not be written to it.

Looking at the above specification we know that:

- If we read DB8 – DB15 when DR = 0, we have no guarantee if the data is valid or garbage.
- If we read DB8 – DB15 when DR = 1, we will read valid data and the UART will automatically reset DR to 0.
- If we write to DB8 – DB15 when TB = 1 then we may overwrite the data being sent and corrupt it.

Figure 12.5 shows the UART as it resides inside of the computer.

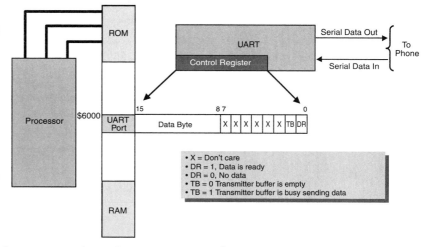

Let's look at two small snippets of 68K assembly language code that illustrates how we would read and write to the UART. Note that in this example we'll use a polling loop rather than an interrupt service routine. This makes the coding a bit easier to illustrate, although you will see immediately how inef-

Figure 12.5: Schematic representation of a UART device. In this example, the UART sends serial data from the computer to a dial-up modem.

ficient it is. A 50K baud modem, which is typical for a dial-up connection, sends and receives approximately 5000 characters per second. How did I know this? Well BAUD is the telephone jargon for bits per second. With the transmission overhead, it takes 10 bits to transmit an 8-bit character, so 50K baud is really 5000 bytes per second of actual data flow.

If my computer is running at a 1GHz clock rate, then 200,000 computer clock pulses will transpire in the time it takes for the UART to send one character. If the computer can average about 1 instruction every 2 clock pulses, then it could have executed about 100,000 instructions while it waited for the UART to send one character. At least with interrupts, it could be doing something else while it waits, but with a polling loop, it will run around the loop several tens of thousands of times waiting for the UART to finish. Sigh....

```
* A short program to test for serial data ready to be read:
START      LEA      $00006000,A5      * Load address
LOOP1      MOVE.W   (A5),D0           * Get UART status
           ANDI.W   #$0001,D0         * Test DR
           BEQ      LOOP1             * Keep waiting
```

```
A short program to see if data can be sent
START      LEA      $00006000,A5      * Load address
LOOP2      MOVE.W   (A5),D0           * Get UART status
           ANDI.W   #$0002,D0         * Test DR
           BNE LOOP2                  * Keep waiting
```

We've been sticking pretty much with assembly language to illustrate various aspects of computer architecture. However, the majority of programmer program in the higher level languages. So, it is fair to ask how a C++ programmer might deal with the hardware restrictions imposed by an I/O port. Remember, an I/O port is at a fixed address in memory. A C++ compiler generally wants to be able to manage the allocation of space for memory variables (or call upon the operating system for assistance). Therefore, we sort of have to beat the compiler into submission in order to assign variables to a specific memory address.

Let's look at the same assembly code example, but this time in C++.

```
char *p_status; // Pointer to the status port
char *p_data;   // Pointer to the data port

p_status = (char*) 0x6001; // Assign pointer to status port

p_data =   (char*) 0x6000; // Assign pointer to data port

do { } while (( *p_status & 0x01) == 0 ); // Wait

char inData = *p_data;
```

In order to assign a pointer to the I/O device addresses we had to explicitly *cast* the pointer to 0x6000 and 0x6001. This method of addressing hardware is fairly typical of how C++ is used in embedded systems programming.

Analog-to-Digital (A/D) and Digital-to-Analog (D/A) Conversion

The real world is an analog world. What does that mean? It means that the physical events that occur in the real world can take on an infinite number of possible values. The speed of the wind, the temperature outside, the intensity of sunlight are all physical quantities that can vary smoothly, and infinitely over a range of values. For example, the outdoor temperature range over the earth might vary from –100 degrees Fahrenheit at the South Pole in the winter to +140 degrees Fahrenheit in Death Valley at noon on a summer day. Although we usually report the temperature in whole degrees, there is nothing preventing the temperature from being 72.56435791 degrees outside. Our problem is how accurately, quickly and cheaply do we want to measure it.

In order to measure these physical quantities we first need an electrical circuit or device that provides an electrical "analogy" of the physical quantity. Analog electronics is just the circuitry that can be used to represent a smoothly varying electrical signal that is the analog of a physical quantity. For example, I might want to be able to control the temperature inside of a furnace. In order to accurately maintain a fixed temperature, or a temperature profile that varies over time, I first need to be able to measure the temperature in the furnace. Typically, a thermocouple or platinum resistance thermometer (PRT) would be used to measure the temperature in the furnace. Other circuits would then be used to take the measured temperature as input signal and drive to power control to the furnace as an output signal.

The PRT might have a range of resistance values that vary smoothly from 100 ohms at room temperature to 2000 ohms at 800 degrees Fahrenheit. We would say that the PRT has a *transfer function* of 2.6 ohms per degree. The important point is that for a smoothly varying physical quantity, temperature, analog electronics gives us a smoothly varying electrical signal. This is the essence of analog electronics. Most of the time, although not all of the time, we'll want to convert our physical quantity to an analog voltage. This may take several steps because the sensor that we use might first convert the physical signal to an electrical current or an electrical resistance. Also, the magnitude of these signals may be too small to measure without using an amplifier of some kind.

Our computer is a digital device. It deals with numbers in discrete quanta called bits. An 8-bit number gives us a range of values from 0 to 255, or –128 to +127. A 16-bit number gives us a range of values from 0 to 65,535 or from –32,768 to 32,767, and so on. The problem then becomes, "How do we convert from an analog value to a digital equivalent and from a digital value to its analog equivalent?" The former process is called *analog-to-digital conversion*, or A/D conversion, and the latter is called *digital-to-analog conversion*, or D/A conversion.

There are A/D and D/A converters in your PC. Where are they? Your sound card can digitize your voice (A/D) and store it as a *.wav file. You can play the *.wav file through your speakers (D/A). The video output to your monitor consists of rapidly changing analog voltages that drive the red, green and blue "guns" inside of your monitor. Here are some examples of physical quantities that we convert to voltages.

- Thermocouple: Measurement of temperature
- Resistance thermometer: Definition of temperature
- Strain gauge: Measurement of displacement or pressure
- Microphone: Measurement of sound level
- Magnetic pickup: Measurement of disk data
- Photocell: Measurement of light intensity

As you might expect, the more accurately we want to measure an analog voltage, the more time consuming and expensive it becomes. Thus, when we talk about A/D conversion we tend to focus our attention on the resolution and the speed of doing a conversion from analog to digital. An A/D converter with 12-bit resolution might take 10 microseconds to convert an analog signal to a digital representation. What does that mean? Suppose that the physical quantity that we are interested in measuring can be converted to an analog voltage over a range of 0 volts to 10 volts, with 0 volts being the lowest possible value that the physical quantity can measure to and 10 volts is the highest voltage that we can use to measure the physical quantity. Note that this is over a range of values. So, let's assume that the physical quantity to be measured is temperature, our analog circuitry might be designed so that:

1. If the temperature is anything less than 100 degrees Fahrenheit, the analog voltage is 0 volts.
2. If the temperature is between 100 degrees and 500 degrees Fahrenheit, the analog voltage varies linearly with temperature from 0 volts to 10 volts. The transfer function in this case is 40 degrees per volt.
3. At all temperatures above 500 degrees Fahrenheit the analog voltage remains at 10 volts.

Now, our 12-bit converter has 2^{12} possible values from 0 to 4095, so our 10 volt analog range can be divided into 4095 slices. Each increment of the digital signal represents a change of 10/4095 or about 2.44×10^{-3} volts per bit. Thus, the temperature could vary by (40 degrees per volt) x $(2.44 \times 10^{-3}$ volts per bit) or about 0.01 degrees before we noticed a change in the temperature.

If we had used an 8-bit A/D then the temperature could vary by about 1.5 degrees before we could detect any change. Thus, the resolution of an A/D converter is a function of the number of bits of resolution in the digital output and the range of analog voltages representing the input span.

As our entry point into this discussion of real-world interfacing, let's do a case study of a real-world situation[2]. It happens to be a situation that I was involved in during the Y2K changeover.

Although there were only a few, isolated Y2K problems that actually occurred, that doesn't mean that the Y2K problem wasn't a real one. In fact, the problem, and its potential for life-threatening failures to occur, was very real indeed.

The situation we'll study is part of the monitoring and control system for a nuclear power plant. The plant's Y2K Program Management Team is reviewing all of their computer systems, looking for potential Y2K-related weaknesses in the systems. Let's suppose that we were hired as very high-priced consultants and asked to help them determine if they have a Y2K problem. We'll do an analysis and make a presentation to the Y2K Program Manager and his team.

Figure 12.6 is the most important slide in the presentation. Let's have a look. Referring to Figure 12.6, we see that the data concentrator is an embedded computer system that is designed to monitor a number of remote sensing devices (approximately 256 of them) and then periodically update the central computer system with the status of these remote sensors. It does that by sending an ASCII character stream over an RS-232 data loop.

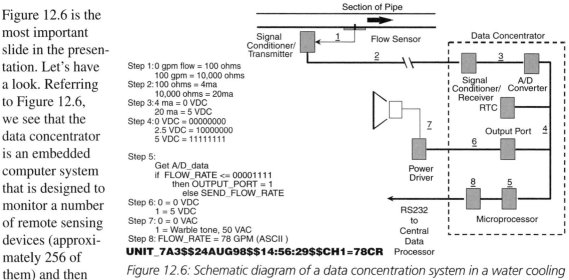

Figure 12.6: Schematic diagram of a data concentration system in a water cooling pipe of a nuclear reactor power plant. The characters in bold are the serial data stream sent by the local data concentrator to the power plant's central computer.

The particular sensor that we're interested in is shown at the top of the figure. We want to monitor the flow rate of water in a pipe. The water is designed to keep the reactor cool. If water stops flowing through the pipe, we would then want to be able to shut down the reactor and warn the workers before there is any damage or threat to human life.

Let's follow the process as it actually works in the power plant:

1. A sensor in the water pipe measures the flow rate of water in the pipe. The transfer function of the sensor converts the pressure of the water in the pipe to an electrical resistance. For this sensor, we already know that a resistance in the sensor of 100 ohms means that there is no water flowing in the pipe. A resistance of 10,000 ohms means that the water flow rate is 100 gallons per minute (G/M). The sensor's resistance is linear over the range of 100 ohms to 10,000 ohms. We can show that the flow rate can be expressed in terms of the sensor's resistance by the equation:

$$\text{FLOW (G/M)} = 0.01(\text{ RESISTANCE}) - 1$$

2. Since the sensor is located several hundred meters away from the data concentrator it is necessary to convert the resistance value to an electrical current value. The exact reason for this is part of the EE's secret oath, so my lips are sealed. Just take my word for it. A standard transmission method in industrial environments is the *4 to 20 milliampere current loop*. This means that we use an electrical current that can vary from a low value of 4 milliamperes to a high value of 20 milliamperes (mA), with everything in between. This conversion is done by a signal conditioner/transmitter device that is located close by the sensor. The 4–20 mA transmitter converts the resistance with yet another transfer function:

$$100 \text{ ohms resistance} \rightarrow 4 \text{ mA}$$
$$10000 \text{ ohms resistance} \rightarrow 20 \text{ mA}$$

3. Back at the data concentrator, the signal from the water pipe is received as a 4–20 mA current loop and once again goes through a conversion process to an electrical voltage in another signal conditioner/receiver. Now, our transfer function is:

$$4 \text{ mA} \rightarrow 0 \text{ volts}$$
$$20 \text{ mA} \rightarrow 5 \text{ volts}$$

Therefore after 3 conversion processes we can say that within the data concentrator 0 volts means no flow of water and 5 volts means 100 gallons per minute of flow.

4. Inside the data concentrator is an analog to digital converter with 8-bits of resolution. This means that we have 256 unique digital codes, $00 through $FF, to represent an analog voltage range of 0 volts to 5 volts with an infinite number of possible values within that range. What does this mean? Assume that the analog voltage is 0 volts. Its pretty obvious the digital code should be $00. Now let's slowly raise the analog voltage. When will the digital code flip from $00 to $01? That's almost impossible to answer exactly, but we can get some idea if we divide the analog range of 5 volts by the number of possible digital ranges we can have. Thus, 5 volts / 255 = 0.0196 volts per digital range. So, very roughly, we should expect that when the analog voltage rises to about 0.020 volts or so, out A/D output should be $01. Thus, we have an uncertainty of around 0.020 volts in the A/D conversion process, which also leads to an uncertainty in the actual flow rate in the pipe. Want more accuracy? Use an A/D converter with more bits of resolution. Also, we know that 2.5 volts should give us an A/D code of $80.

5. Now we're ready to use the computer. The computer reads the output of the A/D converter. If the digital code is $0F or lower, we have a serious problem. Go to step 6, else go to step 8 because if it is greater than $0F then there is no need to sound a warning siren.

6. The data concentrator has an output port that can send a signal to a siren. Sending a digital 0 or digital 1 to the output port places a voltage of 0 volts or 5 volts, respectively, on the output of the port and that becomes an input to the warning siren.

7. An input voltage of 5 volts on the siren turns it on and it does what loud sirens do. The nuclear plant immediately goes into emergency shutdown mode.

8. The data concentrator converts the output of the A/D converter to a flow rate and bundles it with some identifying information, time stamps it, and sends it on to the control room computer. The data stream from the concentrator looks like:

UNIT_7A3$$24AUG98$$14:56:29$$CH1=78CR

The message could be deciphered as follows:
1. This is data concentrator 7A3 reporting.
2. $$ is a field delimiter
3. The date is August 24, 1998
4. The time is 14:56:29
5. The value read on channel 1 is 78 gpm.
6. CR (carriage return) is the end of message delimiter

Where is the Y2K problem here? Look at field value #3. The date is represented as a 2 digit number. Will it be interpreted as 1998 or 2098? We can't say, but UNIT 7A3 is introducing an ambiguity in its data stream that must be investigated further. End of presentation. Where do we send the bill?

Let's get back to A/D conversion. The typical methods of A/D conversion in use today are

- flash conversion
- successive approximation
- single slope
- dual slope
- voltage to frequency

Flash conversion is the fastest, most expensive and least accurate, typically only 6 to 12 bits of resolution. Successive approximation is best balance of cost, accuracy and speed, with anywhere from 12 to 20 bits of resolution available in commercial devices. Single slope is the simplest type of A/D and has modest resolution, 12 to 14 bits. Dual slope (and triple slope) is the most accurate with up to 22 to 24 bits of resolution. Voltage to frequency conversion, or V/F, is the slowest but has the ability to average out noise in the signal better than any other method.

Since the design of A/D and D/A circuitry is usually the responsibility of the Electrical Engineers, we'll have to be careful here, so as not to disturb the sleeping dragon with our brazenness. We'll take a compromise position so that you'll be able to speak to an EE about A/D conversion issues in your computer system, but you won't know enough to threaten their job security.

We need to start our study of A/D and D/A conversion by learning about an import circuit element called a *comparator.* The comparator lives in the nether world between analog and digital. Its inputs are analog and its output is digital. Consider the circuit in Figure 12.7.

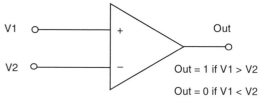

Figure 12.7: Analog comparator.

The circuit has two inputs, a + input and a – input and a single output. The inputs accept analog voltages over some range of values. Assume that the input voltage range is from – 10 volts to +10 volts, or +/– 10 volts. Any analog voltage from –10 volts to +10 volts may be applied to either the + or the – inputs without doing any damage to the circuit.

Over the range of –10 volts to +10 volts, if the voltage at the + input (V1) is *more positive* than the voltage at the – input (V2), the digital output of the comparator is TRUE (1). If the voltage at the – input is more positive than the voltage at the + input then the output is FALSE (0). Figure 12.8 illustrates this behavior. Assume that the gray line labeled "Comparator threshold voltage" in Figure 12.8 is connected to the – input (V2) and that input is held at a steady value of 0 volts. Assume further that a sine wave voltage with positive and negative peaks of +/– 5 volts is connected to the + input (V1) of the comparator.

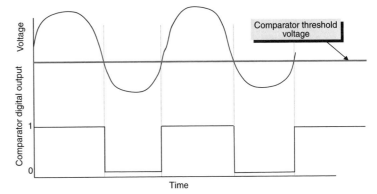

Figure 12.8: Transfer function for the analog comparator of Figure 12.7.

Each time the sine wave rises about the reference voltage (0 volts) the comparator output goes to 1. Each time it falls below the reference voltage, the output goes to zero. In a way, the comparator is a 1-bit A/D converter. We know if the unknown voltage is either greater than or less than the reference voltage, but that's all we know. In effect, the comparator is answering the logical question,

True or false, is the voltage at input V1 more positive than the voltage at input V2?

Let's increase the complexity just a bit. Assume now that our sine wave voltage varies over the range of 0 volts, minimum value, to 4 volts (Vmax), the maximum value. Now, let's take three comparators and label them A, B, and C. We'll connect the + input of all the comparators to our sine wave voltage and we'll connect the – input of each comparator as follows:

1. Comparator A to 75% of Vmax, or 3 volts
2. Comparator B to 50% of Vmax, or 2 volts
3. Comparator C to 25% of Vmax, or 1 volt

Now consider the output of Figure 12.9. As you can see, the same unknown voltage is simultaneously applied to all 3 of the + inputs of 3 identical comparators. However,

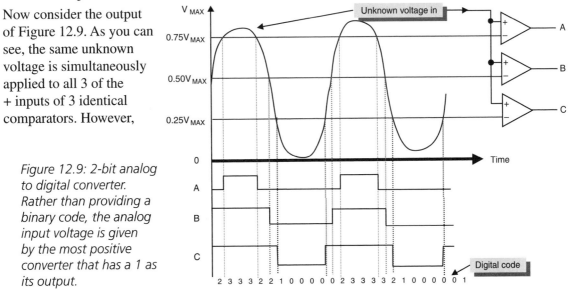

Figure 12.9: 2-bit analog to digital converter. Rather than providing a binary code, the analog input voltage is given by the most positive converter that has a 1 as its output.

each – input is connected to an equally spaced, but different, reference voltage. This circuit tells us more, but it is still very crude. For example, if the output of comparator A is 1, then we know that the analog voltage is greater than 3 volts, but less than 4 volts, the maximum value of the sine wave. If the output of comparator B is a 1 but comparator A is a 0, then the voltage is greater than 2 volts, but less than 3 volts.

On the bottom of Figure 12.9 we can see the digital code that represents the instantaneous value of the analog voltage. You can begin to see what would happen as we added more and more comparators. This is, in fact, exactly what we'll do. With 3 comparators we can "digitize" our unknown voltage by dividing it into 4 possible buckets:

- Less than 1 volt
- Between 1 volt and 2 volts
- Between 2 volts and 3 volts
- Greater than 3 volts

If we had more comparators, we could have even more buckets with better accuracy. However, you'll begin to see the problem. This is a binary progression. With 15 comparators we get 16 ranges, but this is only 4 bits of resolution. In order to get 8 bits of resolution, we'll need 255 comparators and additional circuitry to convert the value of the ladder of comparators to a digital code. But wait, there's more. Somehow, we've got to also supply each comparator with its proper value of the threshold voltage. This means we need to be able to generate 255 different analog threshold voltages, one for each comparator. Since the solution to this part of the problem is so interesting, we need to take a bit of a sidetrack for a moment and deal with it.

At the risk of incurring the wrath of the Union of Electrical Engineers, we're going to learn the fundamental equation of electrical engineering. Strictly speaking, this is bonus material. We could probably learn as much as we need to about computer architecture without learning *Ohm's Law*, but it's a good equation to know, even if you never have to use it in a professional situation. Also, it will help us to understand some of the circuitry to come, so it is worth doing just for that reason. Ohm's Law relates current, voltage and electrical resistance through a simple equation.

Ohm's law states that the voltage across a circuit element, V = current through that circuit element, I, multiplied by the electrical resistance of the circuit element, R. Simply stated,

$$V = I \times R.$$

Stated in terms of units, the voltage (measured in volts) equals the current (measured in amperes) times the resistance (measured in ohms). Thus, a current of 1 Ampere flowing through a resistance of 1 ohm ($R = 1\Omega$) will produce a voltage across the resistance of 1 volt. Another common example: a light bulb in your house is connected to the 120V AC line. If there is 1A of current flowing through the bulb, its resistance is 120 ohms.

Typical ranges of values are:

- Voltages: 1mV to 1KV, or 10^{-3} volt to 10^3 volts
- Current: 1uA to 10A, or 10^{-6} amps to 10amps
- Resistance: 1Ω to 1M Ω, or 1 ohm to 10^6 ohms

Before we move on and use Ohm's Law, let's think about what it means. A good analogy is a garden hose connected to an outdoor faucet. Let's assume that we're dealing with water pressure instead of voltage and the amount of water flowing through the hose instead of current. Also, lets assume we have a bunch of hoses at our disposal, in different diameters and different lengths, starting from a tiny hose with a 1/8" diameter inside diameter up to one with a 4" diameter bore (fire hose), and lengths from a few feet to several hundred feet.

Suppose that we have lots of water pressure at the faucet (high voltage). For most of the hoses, we get plenty of water (current) so the resistance of the hose isn't too bad. However, if we use the longest hose with the smallest diameter, the flow out the end is barely a trickle.

Now suppose that we have low water pressure (low voltage). Even with a short, fat hose (low resistance), we still have very low water flow. That's Ohm's Law in action. Now consider Figure 12.10. Case 1 shows Ohm's Law with a single resistance, a circuit element called a *resistor*. The resistor could be the filament of a light bulb, a resistor on a circuit board, or the resistor inside of an integrated circuit. Notice how the current, I, through the resistance element, R, produces a voltage, V, across the resistance element.

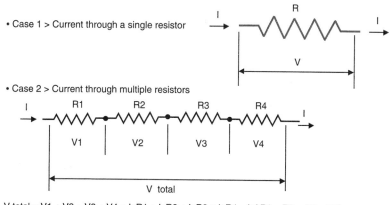

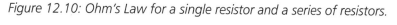

$$V \text{ total} = V1 + V2 + V3 + V4 = IxR1 + IxR2 + IxR3 + IxR4 = Ix(R1 + R2 + R3 + R4)$$
$$R \text{ total} = R1 + R2 + R3 + R4$$
$$V \text{ total} = I \times R \text{ total}$$

Figure 12.10: Ohm's Law for a single resistor and a series of resistors.

Case 2 is similar to Case 1, except that we now have a series of resistors. The current flows from one resistor to the next. Since the current is always the same through each resistor, we can simplify the equations as shown. As you can see, the total voltage measured across the entire circuit is simply the sum of the individual voltages developed across each individual resistor. However, the important point for our A/D discussion is that if the resistance values are all the same, the voltage measured across each resistor is the same.

Suppose that in Figure 12.10 we had 10 resistors, each resistor measured exactly 1000 ohms. Now, we connect one end of the resistor chain to 10 volts and the other end to 0 volts. The 10 resistors in series add up to 10,000 ohms and the total voltage across all 10 resistors is 10 volts. By Ohm's Law, the current flowing in the circuit, $I = 10V/10,000\Omega$, or 1 milliamp (1 mA). Now 1 mA flowing through each resistor produces a voltage of exactly one volt, so the junction of each resistor is exactly 1 volt higher than the previous one. This type of circuit is called a *voltage divider*, because it divides the total voltage into smaller pieces.

Now we have enough information to understand how to really build the A/D converter (but don't tell the International Brotherhood of Electronic Circuit Designers). Figure 12.11 is a simplified picture of a *flash A/D converter*. It gets its name because it is extremely fast. The flash converter consists of a stack of comparators.

Commercially available circuits have as many as 4,096 of them. The minus input of each comparator is connected to one of the junctions (taps) of the voltage divider. The voltage divider has one more resistor in it than the number of comparators in the circuit. Each resistor is the same value, so each tap increments the voltage by the same step. The end of the resistance chain is connected to a stable reference voltage, such as 10.000V.

The key point is that this chain of resistors, each one with the same value, enables us to divide a reference voltage into as many arbitrary reference points as we need to have in order to produce a flash A/D converter of sufficient accuracy. However, we can't just go on indefinitely. Each time we ask for another bit of resolution, the number of comparators and resistors that we need goes up by a factor of two.

Now you might be thinking that they can build integrated circuits with 50 million transistors on it, why can't they build an A/D converter with a paltry 8 thousand resistors and comparators on it? The problem is that analog circuits can't be compressed the way strictly digital circuitry can and resistors are particularly big circuit elements. Also, the very feature that makes digital circuits attractive to us, their insensitivity to voltage levels over a fairly wide range is exactly opposite of what we need for analog electronics. Thus, making 8,192 equal resistors and 8,192 comparators with exactly the same switching characteristics is a Herculean (read expensive) feat of circuit design.

The circuit of Figure 12.11 works exactly the same as the simple model that we looked at in Figure 12.9, with one exception. We added a logic circuit that converts the digital code coming out from the stack of comparators to a real binary code. Thus, if we had a flash converter with 255 comparators and the unknown voltage was exactly 1/2 of the reference voltage, we would expect that the lowest 128 comparators would all have an output of 1 and the next 127 comparators would have an output of 0. The digital circuit would then convert the input code to a true binary output code. What would the algorithm look like? You know enough to do the design, although it might be a bit tedious. You've got 255 inputs and 8 outputs. It just might take more than a few logic gates to do the circuit implementation.

The logic circuit that we need to translate the comparator number with the highest logical '1' output to a binary code is called a *priority encoder*. Thus, 255 comparators would result in a true 8-bit binary code for the output.

The flash A/D converter is fast because the amount of time it takes to digitize the

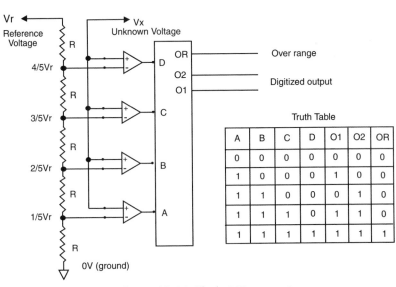

Figure 12.11: Flash A/D converter.

A	B	C	D	O1	O2	OR
0	0	0	0	0	0	0
1	0	0	0	1	0	0
1	1	0	0	0	1	0
1	1	1	0	1	1	0
1	1	1	1	1	1	1

signal is simply a function of how fast the signal can ripple through the comparator and the priority encoder circuit. It is common that flash A/D converters can digitize a sample every 10 nanoseconds, or less, in some extreme cases. The flash converter is the favorite device for people doing research on shock waves because you can obtain a lot of bomb blast data before the blast destroys all of your sensors.

Let's summarize what we know about the flash converter:

- The range of allowable input voltages is 0V to 4/5Vr
- The resolution of our A/D converter is 2 bits
- Assuming that reference voltage, V_{ref} = 5 Volts, we can only digitize the unknown voltage, Vx, to an accuracy of 1 Volt
- In order to achieve 8-bit accuracy we require 256 equally matched resistors and 255 equally match comparators.

The flash A/D converter is the simplest to understand from a conceptual point of view, but it is not the most commonly used type of A/D converter, due primarily to the exponential complexity that arises when we try to get accuracy beyond 10-bits. Remember, each extra bit of resolution requires double the amount of matched comparators and resistors.

In order to understand how the common A/D converters work, we need to understand the circuitry of a D/A converter. The digital-to-analog converter has the opposite functionality of an A/D converter because it produces an analog output voltage that is based upon a digital input code. The circuit shown below in Figure 12.12 is a simple 4-bit D/A converter.

The symbols that look like the number 8 are circuit elements called *current sources*. Each one generates constant and stable flow of current as long as there is a complete circuit. The magnitude of the current from each current source increases in a binary progression, 1, 2, 4,

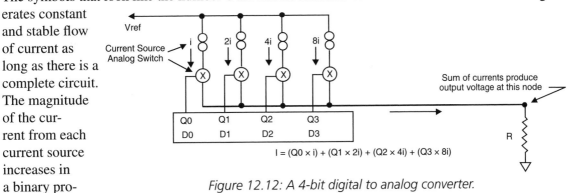

Figure 12.12: A 4-bit digital to analog converter.

$$I = (Q0 \times i) + (Q1 \times 2i) + (Q2 \times 4i) + (Q3 \times 8i)$$

8, 16 and so on. Suppose that the value of 'i' in Figure 12.12 is 0.1 mA. The current being output towards ground by the first current source is 0.1 mA. The current from the second current source is 0.2 mA, and so on for each of the other current sources. Each current source also has a switch (the switch is represented by the symbol of a circle with an "x" in the center) that is controlled by a digital input signal, represented as a 4-bit output port. If the digital signal is 0, no current flows from that current source. If the digital signal is 1, the current flows down through the common conductor wire and through the resistor to ground.

All the current coming from each current source comes together and flows through a resistor, R. In this circuit, if R = 1000 ohms and all of the current sources are turned on, then we have 15 × 0.1mA, or 1.5 mA flowing through a 1000 ohm resistor. This gives us a voltage of 1.5 volts. Thus, a digital code from 0 to $F, will give us an analog voltage out from 0 to 1.5 volts in steps of 0.1 volts. Now we're ready to understand how real A/D converters actually work.

Figure 12.13, is a simplified diagram of a 16-bit analog to digital converter. At it's heart is a 16-bit D/A converter and a comparator. The operation of the circuit is very straightforward. We start by applying the digital code $0000 to the D/A converter.

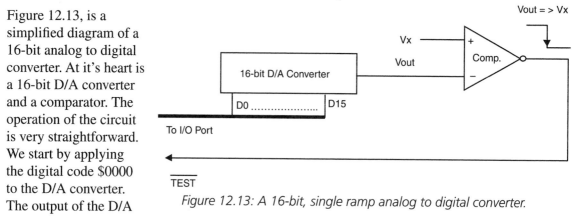

Figure 12.13: A 16-bit, single ramp analog to digital converter.

The output of the D/A converter is 0 volts. The output is applied to the minus input of the comparator. The voltage that we want to digitize is applied to the positive input of the comparator. We then add a count of 1 to the digital code and apply the 16-bit code to the D/A converter. The output voltage will increase slightly because we have 65,534 codes to go. However, we also check the output of the comparator to see if it changed from 1 to 0. When the comparator's output changes state, then we know that the output voltage of the D/A converter is just slightly greater than the unknown voltage.

When it changes state we stop counting up and the digital code at the time the comparator changes state is the digital value of the unknown analog voltage. We call this a *single-ramp* A/D converter because we increase the test voltage in a linear ramp until the test voltage and the unknown voltage are equal.

Imagine that you were building a single ramp A/D converter as part of a computer-based data logging system. You would have a 16-bit I/O port as your digital output port and a single bit ($\overline{TEST}$) input to sample the state of the comparator output. Starting from an initialized state you would keep incrementing the digital code and sampling the $\overline{TEST}$ input until you saw the $\overline{TEST}$ input go low. The flow chart of this algorithm for the single ramp A/D is shown in Figure 12.14.

The single ramp has the problem that the digitizing time is

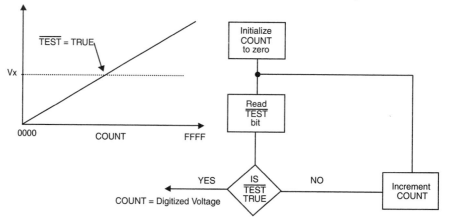

Figure 12.14: Algorithm for the single ramp A/D converter.

variable. A low voltage will digitize quickly, a high voltage will take longer. Also, the algorithm of the single ramp is analogous to a linear search algorithm. We already known that a binary search is more efficient than a linear search, so as you might imagine, we could also use this circuit to do a binary progression to zero in on the unknown voltage. This is called the *successive approximation* A/D converter and it is the most commonly used design today.

The algorithm for the successive approximation A/D converter is just as you would expect of a binary search. Instead of starting at the digital code of 0x0000, we start at 0x8000. We check to see if the comparator output is 1 or 0, and we either set the next most significant bit to 1 or to 0. Thus, the 16-bit A/D can determine the unknown voltage in 16 tests, rather than as many as 65,535.

The last type of A/D converter is the voltage to frequency converter, or V/F converter. This converter converts the input voltage into a stream of digital pulses. The frequency of this pulse stream is proportional to the analog voltage. For example, a V/F converter can have a transfer function of 10 KHz per volt. So at 1 volt in, it has a frequency output of 10,000 Hertz. At 10 volts input, the output is 100,000 Hertz, and so on. Since we know how to accurately measure quantities related to time, it is possible to very accurately measure frequency and count pulses, we are effectively doing a voltage to time conversion.

The V/F converter has one very attractive feature. It is extremely effective in filtering out noise in an input signal. Suppose that the output of the V/F converter is around 50,000 Hz. Every second, the V/F emits approximately 50,000 pulses. If we keep counting and accumulating the count, in 10 seconds we count 500,000 pulses, in 100 seconds we count 5,000,000 pulses, and so forth.

On a finer scale, perhaps each second the count is sometimes slightly greater than 50,000, sometimes slightly less. The longer we keep counting, the more we are averaging out the noise in our unknown voltage. Thus, if we are willing to wait long enough, and our input voltage is stable for that period of time, we can average it to a very high accuracy.

Now that we understand how an analog to digital converter actually works, let's look at a complete data logging system that we might use to measure several analog inputs.

Figure 12.15 is a simplified schematic diagram of such a data logger. There are several circuit elements in Figure 12.15 that we haven't discussed before. For the purposes of this example it isn't necessary to go into a detailed analysis of how they work. We'll just look at their overall operation in the context of understanding how the process of data logging takes place.

The block marked 'Signal Conditioning' is usually a set of amplifiers or other form of signal converters. The purpose is to convert the analog signal from the sensor to a voltage that is in the range of the A/D converter. For example, suppose that we are trying to measure the output signal from a sensor whose output voltage range is 0 to 1 mV. If we were to feed this signal directly into an A/D converter with an input range of 0–10 volts, we would never see the output of the sensor. Thus, it is likely that we would use an analog amplifier to amplify the sensor's signal from the range of 0 to 0.001 volts to a range of 0 to 10 volts.

Presumably each analog channel has different amplification requirements, so each channel is handled individually with its own amplifier or other type of signal conditioner. The point is that we want each channel's sensor range to be optimally matched to the input range of the A/D converter.

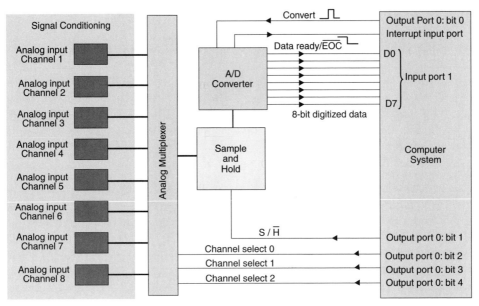

Figure 12.15: Simplified schematic diagram of a computer-based data logger.

Notice that the data logging system is designed to monitor 8 input channels. We could connect an A/D converter to each channel, but usually that is not the most economical solution. Another analog circuit element, called an *analog multiplexer*, is used to sequentially connect each of the analog channels to the A/D converter. In a very real sense, the analog multiplexer is like a set of tri-state output devices connected to a common bus. Only one output at a time is allowed to be connect to the bus. The difference here is that the analog multiplexer is capable of preserving the analog voltage of its input signal.

The next device is called a *sample and hold module, or S/H*. This takes a bit more explaining to make sense of. The S/H module allows us to digitize an analog signal that is changing with time. Previously we saw that it can take a significant amount of time to digitize an analog voltage. A single-ramp A/D might have to count up several thousand counts before it matched the unknown voltage. Through all of these examples we always assumed that the unknown analog voltage was nice and constant. Suppose for a moment that it is the sound of a violin that we are trying to faithfully digitize. At some instant of time we want to know a voltage point on the violin's waveform, but what is it? If the unknown voltage of the violin changes significantly during the time it takes the A/D converter to digitize it, then we may have a very large error to deal with. The S/H module solves this problem.

The S/H module is like a video freeze-frame. When the digital input is in the sample position ($S/\overline{H} = 1$) the analog output follows the digital input. When the $S/\overline{H}$ input goes low, the analog voltage is frozen in time, and the A/D converter can have a reasonable chance of accurately digitizing it. To see why this is, let's consider a simple example. Suppose that we are trying to digitize a sine wave that is oscillating at a frequency of 10 KHz. Assume that the amplitude of the sine wave is ±5 volts. Thus,

$$V(t) = 5\sin(\omega t)$$

where ω is the angular frequency of the sine wave, measured in radians per second. If this is new to you, just trust me and go with the flow. The angular frequency is just $2\pi f$, where f is the actual frequency of the sine wave in Hertz (cycles per second). The rate of change of the voltage is just the first derivative of $V(t)$:

$$dV/dt = -5\omega\cos(\omega t) = -10\pi f\cos(\omega t).$$

The maximum rate of change of the voltage with time occurs when $\cos(\omega t) = 1$, so

$$dV/dt(\text{maximum}) = -10\pi f \text{ or } -31.4 \times 10\text{x}10^3.$$

Thus, the maximum rate of change of the voltage with time is 0.314 volts per microsecond. Now, if our A/D converter requires 5 microseconds to do a single conversion, then the unknown voltage may change as much as ~1.5 volts during the time the conversion is taking place. Since this is usually an unacceptable large error source, we need the S/H module to provide a stable signal to the A/D converter during the time that the conversion is taking place.

We now know enough about the system to see how it functions. Let's do a step-by-step analysis:

1. Bits 2:4 of output port 0 select the desired analog channel to connect to the S/H module.
2. The conditioned analog voltage appears at the input of the S/H module.
3. Bit 1 of output port 0 goes low and places the S/H module in hold mode. The analog input voltage to be digitized is now locked at its value the instant of time when S/H went low.
4. Bit 0 of output port 0 issues a positive pulse to the A/D converter to trigger a convert cycle to take place.
5. After the required conversion interval, the *end-of-conversion* signal ($\overline{\text{EOC}}$) goes low, causing an interrupt to the computer.
6. The computer goes into its ISR for the A/D converter and reads in the digital data.
7. Depending on its algorithm, it may select another channel and read another input value, or continue digitizing the same channel as before.

Figure 12.16 summarizes the degree of difficulty required to build an A/D converter of arbitrary speed and accuracy. The areas labeled "SK", although theoretically rather straightforward to do, often require application-specific knowledge. For example, a heart monitor may be relatively slow and medium accuracy, but the requirements for electrically protecting the patient from any shock hazards may impose addition requirement for a designer.

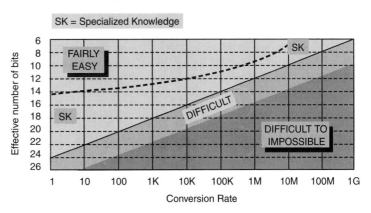

Figure 12.16: Graph summarizing degree of difficulty producing an A/D converter of a given accuracy and conversion rate. From Horn[3].

The Resolution of A/D and D/A Converters

Before we leave the topic of analog-to-digital and digital-to-analog converters we should try to summarize our discussion of what we mean by the *resolution* of a converter. The discussion applies equally to the D/A converter, but is somewhat easier to explain from the perspective of the A/D converter, so that's what we'll do. When we try to convert an analog voltage, or current or resistance (remember Ohm's Law?) to a corresponding digital value, we're faced with a fundamental problem. The analog voltage is a continuously variable quantity while the digital value can only be represented in discrete steps.

You're already familiar with this problem from your C++ programming classes. You know, or should know, that certain operations are potentially dangerous because they could result in erroneous results. In programming, we call this a "round-off error". Consider the following example:

```
float A = 3.1415906732678;
float B = 3.1415906732566;
if ( A == B )
      {do something}
else
      {do something else}
```

What will it do? Unless you knew how many digits of precision you can represent with a float on your computer, you may or may not get the result you expect. We have the same problem with A/D converters. Suppose I have a precision voltage source. This is an electronic device that can provide a very stable voltage for long periods of time. Typically, special batteries, called *standard cells,* are used for this. Let's say that we just spent $500 and sent our standard cell back to the National Institute for Standards and Testing in Gaithersburg, MD.

After a few weeks we get the standard cell back from NIST with a calibration certificate stating that the voltage on our standard cell is +1.542324567 volts at 23 degrees, Celsius (there is a slight voltage versus temperature shift, but we can account for it). Now we hook this cell up to our A/D converter and take a reading. What will we measure?

Right now you don't have enough information to answer that so let's be a bit more specific:

A/D range: 0 volts – +2.00 volts
A/D resolution: 10 bits
A/D accuracy: +/– 1/2 least significant bit (LSB)

This means that over the analog input range of 0.00 to +2.00 volts, there are 1024 digital codes available to us to represent the analog voltage. We know that 0.00 volts should give us a digital value of 00 0000 0000 and that +2.00 volts should give us a digital value of 11 1111 1111, but what about everything in between? At what point does the digital code change from 0x000 to 0x001? In other words, how sensitive is our A/D converter to changes, or fluctuations, in the analog input voltage?

Let's try to figure this out. Since there are 1023 intervals between 0x000 and 0x3FF we can calculate what interval in the analog voltage corresponds 1 change of the digital voltage.

Therefore $2.00 / 1023 = 1.9550 \times 10^{-3}$ volts. Thus, every time the analog voltage changes by about 2 millivolts (mV) we should see that the digital code also changes by 1 unit. This value of 2 mV is also what we would call the least significant bit because this amount of voltage change would cause the LSB to change by 1.

Consider Figure 12.17. The stair-step looking curve represents the transfer function for our A/D converter. It shows us how the digital code will change as a function of the analog input voltage. Notice how we get a digital code of 0x000 up until the analog voltage rises to almost 1 mV. Since the accuracy is 1/2 of the LSB, we have a range of analog voltage centered about each analog interval (vertical dotted lines). This is the region defined by the horizontal portion of the line. For example, the digital code will be $001 for an analog voltage in the range of just under 1 mV to just under 3 mV.

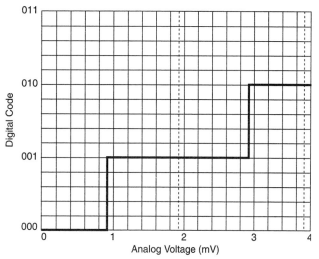

Figure 12.17: Transfer function for a 10-bit A/D converter over a range of 0 to 2.00 volts. Accuracy is 1/2 LSB.

What happens if our analog voltage is right at the switching point? Suppose it is just about 0.9775 mV? Will the digital code be $000 or $001? The answer is, "Who knows?" Sometimes it might digitize as $000 and other times it might digitize as $001.

Now back to our standard cell. Recall that the voltage on our standard cell is +1.542324567 volts. What would the digital code be? Well +1.542324567 / 1.9550×10^{-3} = 788.913, which is almost 789. In hexadecimal, 789_{10} equals 0x315, so that's the digital code that we'd probably see.

Is this resolution good enough? That's a hard question to answer unless we know the context of the question. Suppose that we're given the task of writing a software package that will be used to control a furnace in a manufacturing plant. The process that takes place in this furnace is quite sensitive to temperature fluctuations, so we must exhibit very tight control. That is, the temperature must be held at exactly 400 degrees Celsius, +/– 0.1 degree Celsius. Now the temperature in the furnace is being monitored by a thermocouple whose voltage output is measured as follows:

> Voltage output @ 400 degrees Celsius = 85.000 mV
> Transfer function = .02 mV / degree Celsius

so far, this doesn't look too promising. But we can do some things to improve the situation. The first thing we can do is amplify the very low level voltages output by the thermocouple and raise it to something more manageable. If we use an amplifier that can amplify the input voltage by a factor of 20 times (gain = 20) then our analog signal becomes:

> Voltage output @ 400 degrees Celsius (X 20) = 1.7000 V
> Transfer function (X 20) = .4 mV / degree Celsius

Now the analog voltage range is OK. Our signal is 1.7 volts at 400 degrees Celsius. This is less than the 2.00 maximum voltage of the A/D converter, so we're not in any danger of going off scale. What about our resolution? We know that our analog signal can vary over a range of almost 2 mV before the A/D converter will detect a change. Referring back to the specifications for our amplified thermocouple, this means that the temperature could shift by about 5 degrees Celsius before the A/D converter could detect a variation. Since we need to control the system to better than 0.1 degree, we need to use an A/D converter with better resolution. How much better? Well, we would predict that a change in the temperature of 0.1 degree Celsius would cause a voltage change of 0.04 mV. Therefore, we've got to improve our resolution by a factor of 2 mV / 0.04 mV or 50 times!

Is this possible? Let's see. Suppose we decided to sell our 10-bit A/D converter on eBay and use the proceeds to buy a new one. How about a 12-bit converter? That would give us 4096 digital codes. Going from 1024 codes to 4096 codes is only a 4× improvement in resolution. We need 50X. A 16-bit A/D converter gives us 65,536 codes. This is a 64× improvement. That should work just fine! Now, we have:

A/D range: 0 volts – +2.00 volts
A/D resolution: 16 bits
A/D accuracy: +/– 1/2 least significant bit (LSB)

Our analog resolution is now 2.00 volts / 65,535 or 0.03 mV per digital code step. Since we need to be able to detect a change of 0.04 mV, this new converter should do the job for us.

Summary of Chapter 12

Chapter 12 covered:

- The concepts of interrupts as a method of dealing with asynchronous events,
- How a computer system deals with the outside world through I/O ports
- How physical quantities in the real world events are converted to a computer-compatible format and vice versa through the processes of analog-to-digital conversion and digital-to-analog conversion.
- The need for an analog to digital interface device called a comparator.
- How Ohm's Law is used to establish fixed voltage points for A/D and D/A conversion.
- The different types of A/D converters and their advantages and disadvantages.
- How accuracy and resolution impact the A/D conversion process.

Chapter 12: *Endnotes*

[1] Glenn E. Reeves, "Priority Inversion: How We Found It, How We Fixed It," Dr. Dobb's Journal, November, 1999, p. 21.

[2] Arnold S. Berger, *A Brief Introduction to Embedded Systems with a Focus on Y2K Issues,* Presented at the Electric Power Research Institute Workshop on the Year 2000 Problem in Embedded Systems, August 24–27, 1998, San Diego, CA.

[3] Jerry Horn, *High-Performance Mixed-Signal Design,* http://www.chipcenter.com/eexpert/jhorn/jhorn015.html.

Exercises for Chapter 12

1. Write a *subroutine* in Motorola 68000 assembly language that will enable a serial UART device to transmit a string of ASCII characters according to the following specification:

   a. The UART is memory mapped at byte address locations $2000 and $2001.
   b. Writing a byte of data to address $2000 will automatically start the data transmission process and it will set the *Transmitter Buffer Empty Flag* (TBMT) in the STATUS register to 0.
   c. When the data byte has been sent, the TBMT flag automatically returns to 1, indicating that TBMT is TRUE.
   d. The STATUS register is memory mapped at byte address $2001. It is a READ ONLY register and the only bit of interest to you is DB0, the TBMT flag.
   e. The memory address of the string to be transmitted is passed into the subroutine in register A6.
   f. The subroutine does not return any values.
   g. All registers used inside the subroutine must be saved on entry and restored on return.
   h. All strings consist of the printable ASCII character set, 00 thru $7F, located in successive memory locations and the string is terminated by $FF.

   The UART is shown schematically in the figure shown below:

   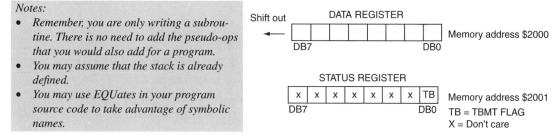

   Notes:
   - *Remember, you are only writing a subroutine. There is no need to add the pseudo-ops that you would also add for a program.*
   - *You may assume that the stack is already defined.*
   - *You may use EQUates in your program source code to take advantage of symbolic names.*

2. Examine the block of 68K assembly language code shown below. There is a serious error in the code. Also shown is the content of the first 32-bytes memory.

   a. What is the bug in the code?
   b. What will the processor do when the error occurs? Explain as completely as possible, given the information that you have.

```
                org         $400
       start     lea         $2000,A0
                move.l      #$00001000,D0
                move.l      #$0000010,D1
       loop      divu        D1,D0
                move.l      D0,(A0)+
                subq.b      #08,D1
                bpl         loop
                end         $400
```

Memory contents (Partial)
```
00000000     00 00 A0 00 00 00 04 00 00 AA 00 00 00 AA 00 00
00000010   00 AA 00 00 00 CC AA 00 00 AA 00 00 00 AA 00 00
```

Note: The first few vectors of the Exception Vector Table are listed below:

Vector #	Memory Address	Description
0	$00000000	RESET: supervisor stack pointer
1	$00000004	RESET: program counter
2	$00000008	Bus Error
3	$0000000C	Address Error
4	$00000010	Illegal instruction
5	$00000014	Zero Divide
6	$00000018	CHK Instruction
7	$0000001C	TRAPV instruction
8	$00000020	Privilege violation

3. Assume that you have two analog-to-digital converters as shown in the table, below:

Converter type	Resolution (bits)	Clock rate (MHz)	Range (volts)
Single Ramp	16	1.00	0 to +6.5535
Successive Approximation	16	1.00	0 to +6.5535

How long (time in microseconds) will it take each type of converter to digitize an analog voltage of +1.5001 volts?

4. Assume that you may assign a priority level from 0 (lowest) to 7 (highest, NMI) for each of the following processor interrupt events. For each of the following events, assign it a priority level and briefly describe your reason for assigning it that level.

a. Keyboard strike input.
b. Imminent Power failure.
c. Watchdog timer.
d. MODEM has data available for reading.
e. A/D converter has new data available.
f. 10 millisecond real time clock tick.
g. Mouse click.
h. Robot hand has touched solid surface.
i. Memory parity error.

5. Assume that you have an 11-bit A/D converter that can digitize an analog voltage over the range of –10.28V to + 10.27volts. The output of the A/D converter is formatted as a 2's complement positive or negative number, depending upon the polarity of the analog input signal.
   a. What is the minimum voltage that an analog input voltage could change and *be guaranteed to be detected* by a change in the digital output value?
   b. What is the binary number that represents an analog voltage of –5.11 volts?
   c. Suppose that the A/D converter is connected to a microprocessor with a 16-bit wide data bus. What would the *hexadecimal* number be for an analog voltage of +8.96V? Hint: It is not necessary to do any rescaling of the 11-bit number to 16-bits.
   d. Assume that the A/D converter is a successive approximation-type A/D converter. How many samples must it take before it finally digitizes the analog voltage?
   e. Suppose that the A/D converter is being controlled by a 1 MHz clock signal and a sample occurs on the rising edge of every clock. How long will it take to digitize an analog voltage?

6. Assume that you are the lead software designer for a medical electronics company. Your new project is to design the some of the key algorithms for a line of portable heart monitors. In order to test some of your algorithms you set up a simple experiment with some of the preliminary hardware. The monitor will use a 10-bit analog to digital converter (A/D) with an input range of 0 to 10 volts. An input voltage of 0 volts results in a binary output of 0000000000 and an input voltage of 10 volts results in a binary output of 1111111111. It digitizes the analog signal every 200 microseconds. You decide to take some data. Shown below is a list of the digitized data values (in hex).

   2C8, 33B, 398, 3DA, 3FC, 3FB, 3D7, 393, 334, 2BF, 23E, 1B8, 137, 0C4, 067, 025, 003, 004, 028, 06C, 0CB, 140, 1C1, 247

   Once you collect the data you want to write it out to a strip chart meter and display it so a doctor can read it. The strip chart meter has an input range of –2 volts to +2 volts. Fortunately, your hardware engineer has designed a 10-bit digital to analog (D/A) circuit such that a binary digital input value of 0000000000 cause an analog output of –2 volts and 1111111111 causes an output of +2 volts. You write a simple algorithm that sends the digitized data to the chart so you can see if everything is working properly.
   a. Show what the chart recorder would output by plotting the above data set on graph paper.
   b. Is there any periodicity to the waveform? If so, what is the period and frequency of the waveform?

7. Suppose that you have a 14-bit, successive approximation, A/D converter with a conversion time of 25 microseconds.
   a. What is the maximum frequency of an AC waveform that you can measure, assuming that you want to collect a minimum of 4 samples per cycle of the unknown waveform?
   b. Suppose that the converter can convert an input voltage over the range of –5V to +5V, what is the minimum voltage change that should be measurable by this converter?
   c. Suppose that you want to use this A/D converter with a particular sample and hold circuit (S/H) that has a droop rate of 1 volt per millisecond. Is this particular S/H circuit compatible with the A/D converter? If not, why?

8. Match the applications with the best A/D converter for the job. The converters are listed below:
   A. 28-bit successive approximation A/D converter, 2 samples per second
   B. 12-bit, successive approximation A/D, 20 microsecond conversion time.
   C. 0 – 10 KHz voltage to frequency converter, 0.005% accuracy.
   D. 8-bit flash converter, 20 nanosecond conversion time.

   a. Artillery shell shock wave measurements at an Army research lab. _____
   b. General purpose data logger for weather telemetry._____
   c. 7-digit laboratory quality digital voltmeter._____
   d. Molten steel temperature controller in a foundry._____

9. Below is a list of "C" function prototypes. Arrange them in the correct order to interface your embedded processor to an 8-channel 12-bit A/D converter system.

   a. boolean Wait(int)   /* True = done, int defines # of   */
                                        /* milliseconds to wait before timeout */
   b. int GetData (void)   /* Returns the digitized data value */
   c. int ConfidenceCheck(void)   /* Perform a confidence check on the */
                                        /* hardware */
   d. void Digitize(void)   /* Turn on A/D converter to digitize */
   e. void SelectChannel(int)   /* Select analog input channel to read */
   f. void InitializeHardware(void)   /* Initialize the state of the hardware to a */
                                        /* known condition */
   g. void SampleHold(boolean)   /* True = sample, False = hold */

10. Assume that you have 16-bit D/A converter, similar in design to the one shown in Figure 12.12. The current source for the least significant data bit, D0, produces a current of 0.1 microamperes. What is the value of the resistor needed so that the full scale output of the D/A converter is 10.00 volts?

CHAPTER 13

Introduction to Modern Computer Architectures

. . . .

Objectives

When you are finished with this lesson, you will be able to:
▶ *Describe the basic properties of CISC and RISC architectures;*
▶ *Explain why pipelines are used in modern computers;*
▶ *Explain the advantages of pipelines and the performance issues they create;*
▶ *Describe how processors can execute more than one instruction per clock cycle;*
▶ *Explain methods used by compilers to take advantage of a computer's architecture in order to improve overall performance.*

. . . .

Today, microprocessors span a wide range of speed, power, functionality and cost. You can pay less than 25 cents for a 4-bit microcontroller to over $10,000 for a space-qualified custom processor. There are over 300 different types of microprocessors in use today. How do we differentiate among such a variety of computing devices? Also, for the purposes of this text we will not consider mainframe computers (IBM, VAX, Cray, Thinking Machines, and so forth), but rather, we'll confine our discussion to the world of the microprocessor.

There are three main microprocessor architectures in general use today. These are: CISC, RISC, DSP. We'll discuss what the acronyms stand for in a little while, but for now, how do we differentiate among these multiple devices? What factors identify or differentiate the various families? Let's first try to identify the various ways that we can rack and stack the various configurations.

1. Clock speed: Processors today may run at clock speeds from essentially zero, to multiple gigahertz. With modern CMOS circuit design, the amount of power a device consumes is generally proportional to its clock frequency. If you want a microprocessor to last 2 years running on an AAA battery on the back of a whale, then don't run the clock very fast, or better yet, don't run it at all, but wake up the processor every so often to do something useful and then let it go back to sleep.

2. Bus width: We can also differentiate processors by their data path width: 4, 8, 16, 32, 64, VLIW (very long instruction word). In general, if you double the width of the bus, you can roughly speed-up the processing of an algorithm between 2 and 4 times.

3. Processors have varying amounts of addressable address space, from 1 Kbyte for a simple microcontroller to multi-gigabyte addressing capabilities in the Pentium, SPARC, Athlon and Itanium class machines. A PowerPC processor from Freescale has 64-bit memory addressing capabilities.

4. Microcontroller/Microprocessor/ASIC: Is the device strictly a CPU, such as a Pentium or Athlon? Is it an integrated CPU with peripheral devices, such as a 68360? Or is it a library of encrypted Verilog or VHDL code, such as an ARM7TDMI, that will ultimately be destined for a custom integrated circuit design?

As you've seen, we can also differentiate among processors by their *instruction set architectures (ISA)*. From a software developer's perspective, this is the architecture of a processor and the differences between the ISA's determine the usefulness of a particular architecture for the intended application. In this text we've studied the Motorola 68K ISA, the Intel x86 and the ARM v4 ISAs, but they are only three of many different ISA's in use today. Other examples are 29K, PPC, SH, MIPS and various DSP ISA's. Even within one ISA we can have over 100 unique microprocessors or integrated device. For example, Motorola's microprocessor family is designated by 680X0, where the X substitutes for the numbers of various family members. If we take the microprocessor core of the 68000 and add some peripheral devices to it, it becomes the 6830X family. Other companies have similar device strategies.

Modern processors also span a wide range of clock speeds, from 1MHz or less, to over 3 GHz (3000 MHz). Not too long ago, the CRAY supercomputer cost over $1M and could reach the unheard of clock speed of 1 GHz. In order to achieve those speeds the engineers at CRAY had to construct exotic, liquid cooled circuit boards and control signal timing by the length of the cables that carried them. Today, most of us have that kind of performance on our desktop. In fact, I'm writing this text on a PC with a 2.0 GHz AMD Athlon processor that is now consider to be third generation by AMD. Perhaps if this text is really successful, I can use my royalty checks to upgrade my PC to an Athlon™ 64. Sigh…

Processor Architectures, CISC, RISC and DSP

The 68K processor and its instruction set, the 8086 processor and its instruction set are examples of the *complex instruction set computer (CISC)*, architecture. CISC is characterized by having many instructions and many addressing modes. You've certainly seen for yourself many assembly language instructions and variations on those instructions we have. Also, these instructions could vary greatly in the number of clock cycles that one instruction might need to execute. Recall, the table shown below. The number of clock cycles to execute a single instruction varied from 8 to 28, depending upon the type of MOVE being executed.

Instruction	Clock Cycles	Instruction Time (usec)*
MOVE.B #$FF,$1000	28	1.75
MOVE.B D0,$1000	20	1.25
MOVE.B D0,(A0)	12	0.75
MOVE.B D0,(A0)+	8	0.50

Assuming a 16 MHz clock frequency

Having variable length instruction times is also characteristic of CISC architectures. The CISC instruction set can be very compact because these complex instructions can each do multiple operations. Recall the DBcc, or the test condition, decrement and branch on condition code instruction. This is a classic example of a CISC instruction. The CISC architecture is also called the *von Neumann architecture*, after John von Neumann, who is credited with first describing the design that bears his name. We'll look at an aspect of the Von Neumann architecture in a moment.

CISC processors have typically required a large amount of circuitry, or a large amount of area on the silicon integrated circuit die. This has created two problems for companies trying to advance the CISC technology: higher cost and slower clock speeds. Higher costs can result because the price of an integrated circuit is largely determined the fabrication yield. This is a measure of how many good chips (yield) can be harvested from each silicon wafer that goes through the IC fabrication process. Large chips, containing complex circuitry, have lower yields than smaller chips. Also, complex chips are difficult to speed up because distributing and synchronizing the clock over the entire area of the chip becomes a difficult engineering task.

A computer with a von Neumann architecture has a single memory space that contains both the instructions and the data, see Figure 13.1 The CISC computer has a single set of busses linking the CPU and memory. Instructions and data must share the same path to the CPU from memory, so if the CPU is writing a data value out to memory, it cannot fetch the next instruction to be executed. It must wait until the data has been written before proceeding. This is called the *von Neumann bottleneck* because it places a limitation on how fast the processor can run.

Howard Aiken of Harvard University invented the Harvard architecture (he must have been too modest to place his name on it). The Harvard architecture features a separate instruction memory and data memory. With this type of a design, both data

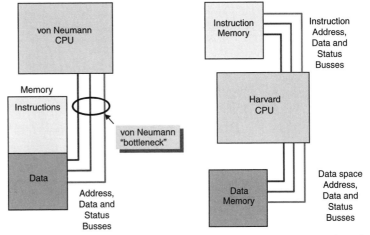

Figure 13.1: Memory architecture for the von Neumann (CISC) and Harvard (RISC) architectures.

and instructions could be operated on independently. Another subtle difference between the von Neumann and Harvard architectures is that the von Neumann architecture permits self-modifying programs, the Harvard architecture does not. Since the same memory space in the von Neumann architecture may hold data and program code, it is possible for an instruction to change the instruction in another portion of the code space. In the Harvard Architecture, loads and stores can only occur in the data memory, so self-modifying code is much harder to do.

The Harvard architecture is generally associated with the idea of *a reduced instruction set computer, or RISC,* architecture, but you can certainly design a CISC computer with the Harvard Architecture. In fact, it is quite common today to have CISC architectures with separate on-chip cache memories for instructions and data.

The Harvard architecture was used commercially on the Am29000 RISC microprocessor, from Advanced Micro Devices (AMD). While the Am29K processor was used commercially in the first

LaserJet series of printers from Hewlett-Packard, designers soon complained to AMD that 29K-based designs were too costly because of the need to design two completely independent memory spaces. In response, AMD's follow-on processors all used a single memory space for instructions and data, thus forgoing the advantages of the Harvard architecture. However, as we'll soon see, the Harvard architecture lives on in the inclusion of on-chip instruction and data caches in many modern microprocessors. Today, you can design ARM processor implementations with either a von Neumann or Harvard architecture.

In the early 1980's a number of researchers were investigating the possibility of advancing the state of the art by streamlining the microprocessor rather than continuing the spiral of more and more complexity[1,2]. According to Resnick[3], Thornton[4] explored aspects of certain aspects of the RISC architecture in the design of the CDC 6600 computer in the late 60's. Among the early research carried out by the computer scientists who were involved with the development of the RISC computer were studies concerned with what fraction of the instruction sets were actually being used by compiler designers and high-level languages. In one study[5] the researchers found that 10 instructions accounted for 80% of all the instructions executed and only 30 instructions accounted for 99% of all the executed instructions. Thus, what the researchers found was that most of the time, only a fraction of the instructions and addressing modes were actually being used. Until then, the breadth and complexity of the ISA was a point of pride among CPU designers; sort of a "My instruction set is bigger than your instruction set" rivalry developed.

In the introductory paragraph to their paper, Patterson and Ditzel note that,

> *Presumably this additional complexity has a positive tradeoff with regard to the cost-effectiveness of newer models. In this paper we propose that this trend is not always cost-effective, and in fact, may even do more harm than good. We shall examine the case for a Reduced Instruction Set Computer (RISC) being as cost-effective as a Complex Instruction Set Computer (CISC).*

In their quest to create more and more complex and elegant instructions and addressing modes, the CPU designers were creating more and more complex CPUs that were becoming choked by their own complexity. The scientists asked the question, "Suppose we do away with all but the most necessary instructions and addressing modes. Could the resultant simplicity outweigh the inevitable increase in program size?"

The answer was a resounding, "Yes!" Today RISC is the dominant architecture because the gains over CISC were so dramatic that even the growth in code size of 1.5 to 2 times was far outweighed by the speed improvement and overall streamlining of the design. Today, a modern RISC processor can execute more than one instruction per clock cycle. This is called a *superscalar architecture.* When we look at pipelines, we'll see how this dramatic improvement is possible. The original RISC designs used the Harvard architecture, but as caches grew in size, they all settled on a single external memory space.

However, everything isn't as simple as that. The ISA's of some modern RISC designs, like the PowerPC, has become every bit as complex as the CISC processor it was designed to improve upon. Also, aspects of the CISC and RISC architectures have been morphing together, so drawing distinctions between them is becoming more problematic. For example, the modern Pentium and

Athlon CPUs execute an ISA that has evolved from Intel's classic x86, CISC architecture. However, internally, the processors exhibit architectural features that would be characteristic of a RISC processor. Also, the drive for speed has been led by Intel and AMD, and today's 3+ gigahertz processors are the Athlons and Pentiums.

Both CISC and RISC can get the job done. Although RISC processors are very fast and efficient, the executable code images for RISC processors tend to be larger because there are fewer instructions available to the compiler. Although this distinction is fading fast, CISC computers still tend to be prevalent in control applications, such as industrial controllers, instrument controllers.

On the other hand, RISC computers tend to prevail in data processing applications where the focus of the algorithm is

```
Data in >>> Do some processing >>> Data out.
```

The RISC processor, because of its simplified instruction set and high speed, is well suited for algorithms that stress data movement, such as might be used in telecommunications or games.

The *digital signal processor (DSP)* is a specialized type of mathematical data processing computer. DSPs do math instead of control (CISC) or data manipulation (RISC). Traditionally, DSP were classic CISC processors with several architectural enhancements to speed-up the execution of special categories of mathematical operations. These additions were circuit elements such as barrel shifters and multiply/accumulate (MAC) instructions (See figure 11.2) that we looked at when we examined the ARM multiplier block. Recall, for example, the execution of an inner loop:

- fetch an X constant and a Y variable
- multiply them together and accumulate (SUM) the result
- check if loop is finished

The DSP accomplished in one instruction what a CISC processor took eight or more instructions to accomplish. Recall from integral calculus that calculating the integral of a function is the same as calculating the area under the curve of that function. We can solve for the area under the curve by multiplying the height at each point by the width of a small rectangular approximation under the curve and the sum the area of all of these individual rectangles. This is the MAC instruction in a DSP.

The solution of integrals is an important part of solving many mathematical equations and transforming real time data. The domain of the DSP is to accept a stream of input data from an A/D converter operate on it and output the result to a D/A converter. Figure 13.2 shows a continuously varying signal going into the DSP from the A/D converter and the output of the DSP going to a D/A converter. The DSP is processing the data stream in real time. The analog data is converted to its digital representation and then reconverted to analog after processing.

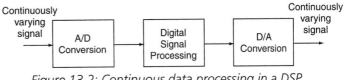

Figure 13.2: Continuous data processing in a DSP.

Several "killer apps" have emerged for the DSP. The first two were the PC modem and PC sound card. Prior to the arrival of the PC, DSP were special use devices, mostly confined to military and CIA types of applications. If you've ever participated in a conference call and spoken on a speakerphone, you've had your phone conversation processed by a DSP. Without the DSP, you would get the annoying echo and screech effect of feedback. The DSP implements an echo cancellation algorithm that removes the audio feedback from your conversation as you are speaking in real time. The newest application of the DSP in our daily lives is the digital camera. These devices contain highly sophisticated DSP processors that can process a 3 to 8 megapixel image, converting the raw pixel data to a compress jpeg image, in just a few seconds.

An Overview of Pipelining

We'll need to return to our discussion of CISC and RISC in a little while because it is an important element of the topic of pipelining. However, first we need to discuss what we mean by *pipelining*. In a sense, pipelining in a computer is a necessary evil. According to Turley[6]:

> *Processors have a lot to do and most of them can't get it all done in a single clock cycle. There's fetching instructions from memory, decoding them, loading or storing operands and results, and actually executing the shift, add, multiply, or whatever the program calls for. Slowing down the CPU clock until it can accomplish all this in one cycle is an option, but nobody likes slow clock rates. A pipeline is a compromise between the amount of work a CPU has to do and the amount of time it has to do it.*

Let's consider this a bit more deeply. Recall that our logic gates, AND, OR, NOT, and the larger circuit blocks built from them, are electronic circuits that take a finite amount of time for a signal to propagate through from input to output. The more gates a signal has to go through, the longer the propagation delay in the circuit.

Consider Figure 13.3. Here's a complex functional block with 8 inputs and 3 outputs. We can assume that it does some type of byte processing. Assume that each functional block in the circuit has a propagation delay of X nanoseconds. The blocks can be simple gates, or more complex functions, but for simplicity, each block has the same propagation delay through it.

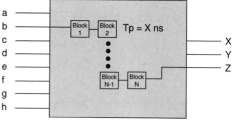

Figure 13.3: Propagation delay through a series of functional blocks.

Also, let's assume that an analysis of the circuit shows that the path from input *b* through to output *Z* is the longest path in the circuit. In other words, input *b* must pass through N gates on its way to output *Z*. Thus, whatever happens when a set of new inputs appear on *a* through *h* we have to wait until input *b* finally ripples through to output *Z* we can consider the circuit to have stabilized and that the output data on *X*, *Y* and *Z* to be correct.

If each functional block has a propagation delay of X ns, then the *worst case* propagation delay through this circuit is N*X nanoseconds. Let's put some real numbers in here. Assume that X = 300 picoseconds (300×10^{-12} seconds) and N = 6. The propagation delay is 1800 picoseconds (1800 ps). If this circuit is part of a synchronous digital system being driven by a clock, and we expect that this circuit will do its job within one clock cycle, then the maximum clock rate that we can have in this system is (1/1800 ps) = 556 MHz.

Keep in mind that the maximum speed that entire computer can run at will be determined by this one circuit path. How can we speed it up? We have several choices:

1. Reduce the propagation delay by going to a faster IC fabrication process,
2. Reduce the number of gates that the signal must propagate through,
3. Fire your hardware designers and hire a new batch with better design skills,
4. Pipeline the process.

All of the above options are usually considered, but the generally accepted solution by the engineering team is #4, while upper management usually favors option #3. Let's look at #4 for a moment. Consider Figure 13.4

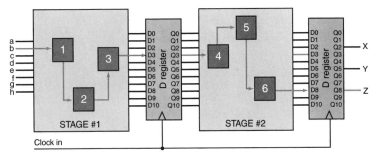

Now, each stage of the system only has a propagation delay of 3 blocks, or 900 ps, for the entire block. To be sure, we also have to add in the propagation delay of

Figure 13.4: A two stage implementation of the digital circuit. The propagation delay through each stage has been reduced from 6 to 3 gate delays.

the 'D' type register, but that's part of the overhead of the compromise. Assume that at time, t = 0, there is a rising clock edge and the data is presented to the inputs *a* through *h*. After, 900 ps, the intermediate results have stabilized and appear on the inputs to the first D register, on its inputs, D0 through D10. After any time greater than 900 ps, a clock may come along again and, after a suitable propagation delay through the register, the intermediate results will appear on the *Q0* through *Q10* outputs of the first D register. Now, the data can propagate through the second stage and after a total time delay of 900 ps + t_p (register), the data is stable on the inputs to the second D register. At the next clock rising edge the data is transferred to the second D register and the final output of the circuit appears on the outputs X,Y and Z.

Let's simplify the example a bit and assume that the propagation delay through the D register is zero, so we only need to consider the functional blocks in stages #1 and #2. It still takes a total of 1800 picoseconds for any new data to make it through both stages of the pipeline and appear on the outputs. However, there's a big difference. On each rising edge of the clock we can present new input data to the first stage and because we are using the D registers for intermediate storage and synchronization, the second stage can still be processing the original input variables while the first stage is processing new information.

Thus, even though it still takes the same amount of time to completely process the first data, through the pipeline, which in this example is two clock cycles, every subsequent result (X, Y and Z) will appear at intervals of 1 clock cycle, not two.

The ARM instruction set architecture that we studied in Chapter 11 is closely associated with the ARM7TDMI core. This CPU design has a 3-stage pipeline. The ARM9 has a 5-stage pipeline. This is shown in Figure 13.5.

In the *fetch stage* the instruction is retrieved from memory. In the *decode stage*, the 32-bit instruction word is decoded and the instruction sequence is determined. In the *execute stage* the instruction is carried out and any results are written back to the registers. The ARM9TDMI core uses a 5-stage pipeline. The two additional stages, Memory and Write allow the ARM9 architecture to have approximately 13% better instruction throughput than the ARM7 architecture[7]. The reason for this is illustrates the advantage of a multi-stage pipeline design. In the ARM7, the Execute Stage does up to three operations:

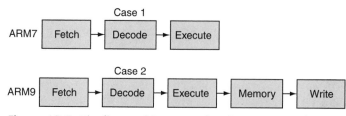

Figure 13.5: Pipeline architectures for the ARM7 and ARM9 CPU core designs.

1. Read the source registers,
2. Execute the instruction,
3. Write the result back to the registers.

In the ARM7 design, the registers are read during the decode stage. The execute stage only does instruction execution, and the Write Stage handles the write-back to the destination register. The memory stage is unique to the ARM9 and doesn't have an analog in the ARM7. The ARM7 supports a single memory space, holding instructions and data. When, it is fetching a new instruction, it cannot be loading or storing to memory. Thus, the pipeline must wait (stall) until either the load/store or instruction fetch is completed. This is the von Neumann bottleneck. The ARM9 uses a separate data and instruction memory model. During the Memory Stage, load/stores can take place simultaneously with the instruction fetch in stage 1.

Up to now, everything about the pipeline seemed made for speed. All we need to do to speed up the processor is to make each stage of the pipeline have finer granularity and we can rev up the clock rate. However, there is a dark side to this process. In fact, there are a number of potential *hazards* to making the pipeline work to its maximum efficiency. In the ARM7 architecture, when a branch instruction occurs, and the branch is taken, what do we do? There are two instructions stacked-up behind the branch instruction in the pipeline and suddenly, they are worthless. In other words, we must *flush* the pipe and start to refill it again from the memory location of the target of the branch instruction.

Recall Figure 13.4. It took two clock cycles for the first new data to start exiting the pipeline. Suppose that our pipeline is a 3-stage design, like the ARM7. It will take 3 clocks for the target of the branch instruction to make it down the pipe to completion. These additional clocks are extra cycles that diminish the throughput every time a branch is taken. Since most programs take a branch of some kind, on average, every 5 to 7 instructions, things can get very slow if we are doing a lot of branching. Now, consider the situation with a 7 or 9-stage pipeline. Every nonsequential instruction fetch is a potential roadblock to efficient code flow through the processor.

Later in this chapter we'll discuss some methods of mitigating this problem, but for now, let's just be aware that the pipeline architecture is not all that you might believe it to be. Finally, we need to cover few odds and ends before we move on. First, it is important to note that the pipeline does not have to be clocked with exactly the same clock frequency as the entire CPU. In other words, we could use a frequency divider circuit to create a clock that is ¼ of that of the system clock.

We might then use this slower clock to clock the pipeline, while the faster clock provides us with a mechanism to implement smaller state machines within each stage of the pipeline. Also, it is possible that certain stages might cause the pipeline to stall and just mark time during the regular course of program execution. Load or store to external memory, or instruction fetches will generally take longer than internal operations, so there could easily be a one or two clock cycle stall every time there is an external memory operation.

Let's return to our study of comparative architectures and look at a simpler, nonpipelined architecture. First we need to appreciate that a processor is an expensive resource, and, just like expensive machinery, want to always keep it busy. An idle processor is wasting space, energy, time, etc. What we need is a way to increase performance. In order to understand the problem, let's look at how a processor, like the 68K, might execute a simple memory-resident instruction, such as MOVE.W $XXXX,$YYYY.

According to the 68K Programmer's Manual, this MOVE.W instruction requires 40 clock cycles to execute. The minimum 68K instruction time requires seven clock cycles. Where is all this time being used up?

A new instruction fetch cycle begins when the contents of the PC are transferred to the address lines. Several cycles are required as the memory responds with the instruction op-code word.

1. The instruction op-code is decoded by the processor.
2. Time is required as the processor generates the first operand (source address) from memory.
3. Time is required to fetch the source data from memory.
4. Time is required to fetch the second operand (destination address) from memory.
5. Time is required to write the data to the destination address.

While all this activity is going on, most of the other functional blocks in the processor are sitting idle. Thus, another of the potential advantages of pipelining is to break up these execution tasks

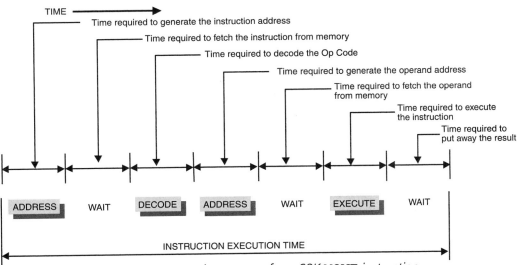

Figure 13.6: Execution process for a 68K MOVE instruction.

into smaller tasks (stages) so that all of the available resources in the CPU are being utilized to the greatest extent possible. This is exactly the same idea as building a car on an assembly line. The overall time to execute one instruction will generally take the same amount of time, but we don't need to wait the entire 40 clock cycles for the next instruction to come down the pipe. As we've seen with the ARM processor, and assuming that we didn't flush the pipe, the next instruction only needs to go through last stage to finish one stage behind. What we have improved is the *throughput* of the system.

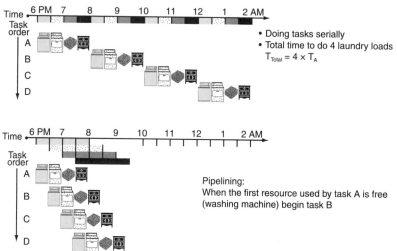

In Figure 13.7 we see that if it takes us two hours to do a load of laundry, which includes four tasks (washing, drying folding and putting away), then it would take eight hours to do four loads of laundry if they're done serially. However, if we overlap the tasks such that the next wash load is started as soon as the first wash is completed, then we can complete the entire four loads of laundry in three and a half hours instead of eight.

Figure 13.7: Doing the laundry: An example of serial task execution versus overlapping task. From Patterson and Hennessy[8].

The execution of an instruction is a good example of a physical process with logical steps that can be divided into a series of finer steps. Let's look at an example of a three-stage process:

1. Do the first thing. . . .
2. Do the second thing. . . .
3. Do the third thing. . . .
4. Release result. . . .
5. Go back to step #1.

With pipelining, this becomes:

1. Do the first thing, hold the result for step #2.
2. Get the result from step #1 and do the second thing, hold the result for step #3.
3. Get the result from step #2 and do the third thing.
4. Release result.
5. Go back to step #1.

Now, let's look at the pipelined execution of the 3-stage processes, shown in Figure 13.8. Presumably, we can create a pipeline because the hardware in Stage #1 becomes idle after executing its task. Thus, we can start another task in Stage #1 before the entire process is completed. The time for the first result to exit the pipe is called the *flowthrough time*. The time for the pipe to produce subsequent results is called the *clock cycle time*.

We see that for the pipeline process, it takes a total time of $5 \times T_S$ to process 3 instructions. Without a pipeline, the 3 instructions, executed serially, would take $9 \times T_S$ to process.

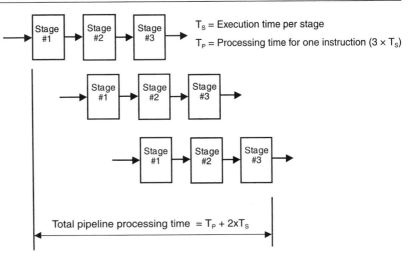

T_S = Execution time per stage

T_P = Processing time for one instruction $(3 \times T_S)$

Total pipeline processing time $= T_P + 2 \times T_S$

Figure 13.8: Decomposing a process into stages.

Let's stop for a moment and ask a reasonable question. As we've previously noted, programs execute out-of-sequence instructions every five to seven instructions. This means that a JSR, BRANCH, or JMP instruction occurs with such regularity that our pipelining process could be a lot more complex than we first assumed. For example, what happens in Figure 13.8 if the first instruction down the pipe is a BNE instruction and the branch is taken? This is not an easy question to answer, and in general, we won't attempt to solve the problem in this text. However, let's look at some of the factors that can influence pipelining and make some general comments about how such problems might be dealt with in real systems.

Let's consider this question. What architectural decisions could we make to facilitate the design of a pipeline-based computer?

- All instructions are the same length.
- There are just a few instruction formats.
- Memory operands appear only in loads and stores.
- ALU operations are between registers

What kinds of architectural decisions make CPU designers want to retire and become mimes in Boulder, Colorado?

- Variable length instructions
- Instructions with variable execution times
- Wide variety of instructions and addressing modes
- Instructions that intermix register and memory operands

In other words, RISC processors lend themselves to a pipelined architecture and CISC processors make it more difficult. That isn't to say that you can't do a CISC-based pipelined design, because you can. It is just that designing a traditional CISC architecture with a pipelined approach is extremely difficult. For example, AMD's 64-bit processor family uses a 12-stage pipeline and 5 of the 12 stages are used to break down the x86 CISC instructions into something that can continue to be processed in the pipeline.

Now we need to consider other hazards which have less to do with the type of architecture and everything to do with the type of algorithm. These are hazards that are dependent upon how well the architecture and code flow can cooperate. Pipelines are sensitive to three additional hazards:

- Structural hazards: Suppose we had only one memory and we need to fetch an instruction from one pipeline when we're trying to read an operand from another pipeline.
- Control hazards: We need to worry about branch instructions.
- Data hazards: An instruction depends on the result of a previous instruction that is still in the process of being executed.

We saw how the ARM architecture handled the structural hazard. By keeping two separate memory spaces and maintaining a separate data and instruction memory, data fetches and instruction fetches could be isolated from each other and be allowed to overlap. Branches are a fact of life, how can we mitigate the damage? Also, the bigger the pipeline, the bigger the problem caused by branches. Now with a complex pipeline, you've got a sequence of instructions in various stages of decoding and execution. It's not unusual for the processor to discover that a branch needs to be taken while the instruction after the branch is already executing, or nearly completed.

One method of solving this problem is with *branch prediction*. That is, build into the CPU some intelligence to make educated guesses about the upcoming branch. Turley[9] notes that studies have shown that most backward branches are taken and most forward branches are not. You might have already guessed this from some of the loops you've written in your assembly language programming exercises. Thus, when the CPU encounters a branch instruction with a target address less than the current value in the PC, it automatically begins filling the pipeline from the new address. When it sees a forward branch it ignores it and continues to load sequential instructions. There's no real down side to doing this. If the branch predictor fails, then the pipeline must be flushed and the correct instruction restarted. This certainly slows down execution, but does not create a fatal flaw.

Another technique is called *dynamic branch prediction*. Suppose that the processor maintained a large number of small binary counters, i.e., two-bit counters. The counters have a minimum binary value of 00 and a maximum binary value of 11. The minimum and maximum values are terminal states, or saturation values. When the counter is in state 00 and it is decremented, it remains in state 00. Likewise, when it is in state 11 and it is incremented, it remains in state 11.

Each of the 4 states is assigned a value:
- 00 = Strongly not taken
- 01 = Weakly not taken
- 10 = Weakly taken
- 11 = Strongly taken

Counters are associated with as many branch addresses as possible. Every time the corresponding branch is taken, the counter is incremented. Every time the branch is not taken, the counter is decremented. Now the branch predicting logic has some statistical evidence available to make decisions about how the branch instruction will be handled. Clearly this method may be scaled with more bits for the counters and more counters in the prediction circuitry.

Some processors, such as AMD's 29K family use special on-chip caches called *branch target caches*[9], to store the first four instructions of previously taken branches. In the next chapter we'll look at caches in more detail, but for now, let's just assume that there is a hardware-based algorithm that constantly updates the branch target cache memory with the most recently used branches and overwrites the least recently used branches in the cache. By combining the branch prediction mechanism with the branch target cache, the processor can mitigate the penalties caused by out-of-sequence instruction execution.

Modern RISC processors are every bit as complex as their CISC relatives, but it is the instruction set architecture that allows them to achieve the speed improvements of higher clock speeds and pipelining. Figure 13.9 is a schematic diagram of the Freescale (formerly Motorola) 603e PowerPC RISC processor.

You may be familiar with this processor because the 603e is used in Apple Computer's Macintosh G series. The 603e is a superscalar processor. It has 3 separate instruction processing systems that can concurrently execute instructions through separate pipelines.

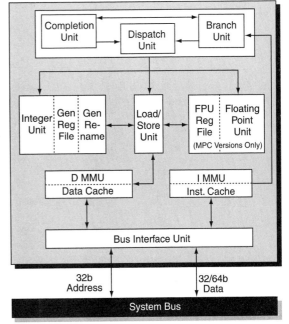

Figure 13.9: PowerPC 603e RISC processor. Courtesy of the Freescale Corporation.

The concept of a pipelining architecture also leads us to the idea of a computer as a loose assortment of internal resources, pipelines and control systems. Rather than a tightly-integrated architecture of the 68K, we can look at a system like that of the PowerPC 603e. Although it isn't apparent from Figure 13.9, we can picture the processor as shown in Figure 13.10.

As each instruction progresses down its pipe, the control system continuously monitors for available resources that can be used to execute the instruction. A good analogy is going to the Laundromat to

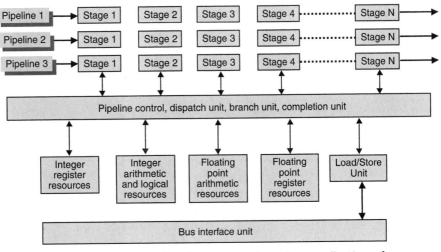

Figure 13.10: A pipelined processor viewed as a collection of resources and pipelines.

do your laundry instead of doing it at home. The Laundromat has multiple washing machines, dryers and folding tables. If one resource is busy, another one is available, unless, of course, you're in a hurry, in which case Murphy's Law dictates that all machines are either in use, or broken.

If our figure of merit is, "How many loads of laundry come in the door of the Laundromat and how many leave the Laundromat per unit of time?", it is clear that the Laundromat is a lot more efficient at processing laundry than the washer and dryer in someone's home. However, it is a lot more expensive to equip a Laundromat than it is to put a washer and a dryer in your basement.

We call this process of dividing the instruction execution into resources and then using these resources as they become available *dynamic scheduling*. Very complex hardware within the processor performs the scheduling tasks. The hardware tries to find instructions that it can execute. In some cases, it will try to execute instructions out of order, or it might go ahead and execute the instruction after a branch instruction under the assumption that the branch might not be taken. We call this *speculative execution*.

Another aspect of pipeline efficiency that is often overlooked is the efficiency of the compiler that is being used to support the processor. One of the original justifications of CISC processors is that the complex instructions would make the job of compilers easier to do. However, that proved not to be the case because compiler designers were not taking advantage of these specialized instructions.

RISC architectures, on the other hand, had a different strategy. Put more of the responsibility for improving the processor's throughput on the squarely onto the backs of the compilers. Many of the advantages of the RISC architecture can be negated by the use of an inefficient compiler. Without mentioning any names, there is a commercially available RISC microprocessor that is supported by 3 separate C++ compilers. The difference between the best and the worst compiler in terms of measured performance is a factor of 2. This is the same as running one processor at half the clock speed of the other, a very significant difference.

Figure 13.11 presents some interesting compiler data. While the data was compiled (so to speak) for the Intel x86 architecture executing on a PC, we can see how sensitive even this architecture is

Scenario	Compiler A	Compiler B	Compiler C	Compiler D	Compiler E	Compiler F	Compiler G	Compiler H	Compiler I
Dhrystones/s									
Dhrystone	1,847,675	1,605,696	**2,993,718**	2,496,510	2,606,319	2,490,718	2,263,404	2,450,608	484,132
time (ms)									
Int2string a (sprintf())	7642	5704	5808	7714	7933	9419	7802	7813	**5539**
Int2string b (STLSoft)	3140	1289	3207	1679	**1156**	1624	1808	1843	DNC
StringTok a (Boost)	4746	3272	DNC	6809	**1450**	2705	2641	2341	DNC
StringTok b (STLSoft)	636	809	**280**	385	382	579	383	406	DNC
RectArr (1 iteration)	1082	910	997	1590	859	915	**824**	887	DNC
RectArr (10 iterations)	6922	3168	5589	3853	1649	1995	*1533*	1828	DNC
zlib (small)	92	110	88	92	87	87	91	**78**	90
zlib (large)	**8412**	12,550	8847	11,310	9390	10,875	10,266	9117	15,984

Figure 13.11: Execution times (in milliseconds) of nine benchmark scenarios for a range of X86 compilers. The figures in **bold** *are the best results. From Wilson[10].*

to the quality of the code generated by each of the compilers in the test. For the *Dhrystone* benchmark, there is more than a 6X difference in performance between Compiler I and Compiler C. The interested reader is encouraged to read the original article in order to see the actual names of the compilers and manufacturers.

Let's look at some of the characteristics of the RISC architecture and see how it relates to our discussion of pipelining. Most RISC processors have many of the following characteristics.

- Instructions are conceptually simple.
- Memory/Register transfers are exclusively LOADS and STORES.
  - LOAD = memory to register
  - STORE = register to memory
- All arithmetic operations are between registers.
- Instructions are uniform in length.
- Instructions use one (or very few) instruction formats.
- Instruction sets are orthogonal (No special registers for addresses and data).
- Little or no overlapping functionality of instructions.
- One (or very few) addressing mode.
- Almost all instructions execute in 1 clock cycle (throughput).
- Optimized for speed.
- RISC processors tend to have many registers.
  - The AM29K has 256 general-purpose registers.
  - It is extremely useful for compiler optimizations not to have to store intermediate results back to external memory.
- Multiple instruction pipes mean that more than one instruction can be executed per clock cycle.

Today, the RISC architectures dominate most high-performance processors. Some examples of modern RISC processors are

- Motorola, IBM: PowerPC 8XX, 7XX, 6XX, 4XX
- Sun: SPARC
- MIPS: RXXXX
- ARM: ARM7,9
- HP: PA-RISC
- Hitachi: SHX

Until about 10 years ago, computer professionals were still debating the relative merits of the CISC and RISC architectures. Since then, however, the RISC architecture has won the performance war. The basic hardware architecture of RISC is simpler, cheaper and faster. The speed increases have more than compensated for the increased number of instructions. It is easier to compile for RISC than for CISC.

The compiler technology for the Intel X86 architecture is basically as good as it gets because a huge effort by compiler designers (Microsoft) finally made it as efficient as RISC. Also, as you've seen, it is easier to introduce parallelism into the RISC architecture. Pipelining can be more predictable because of the instruction uniformity. Also, today, modern applications are data intensive.

Desktop computers are used for multimedia and games. Workstations are used for data intensive applications such as design and simulation. Embedded systems are used for games (Nintendo) and telecommunications applications (routers, bridges, switches).

Motorola stopped CISC development at the 68060 and designed two new RISC architectures, ColdFire and PowerPC, to replace the 680X0 family. AMD stopped with the 586 and Intel stopped with the original Pentium. The modern processors from AMD and from Intel are a merging of the RISC and CISC technologies, and have been designed to be able to take in the original Intel X86 instruction set and translate it internally to a sequence of RISC-like instructions that the native architecture can handle very rapidly.

As we've seen, RISC processors use pipelining to accelerate instruction decoding an program execution. RISC's usually do not allow self-modifying code; that is, overwriting the instruction space to change an instruction. Pure RISC processors use Harvard architecture to eliminate the von Neumann bottleneck due to instructions and data sharing a single address and data bus. However, most modern processors still have one external memory bus, but substitute internal instruction and data caches for the Harvard architecture.

RISC processors use large register sets to reduce memory to register latencies. These registers tend to be general purpose, and any register could be used as a memory pointer for an indirect address. Also, some RISC processors have separate floating point register sets. All RISC processors have separate instruction decoding and instruction execution units and many have separate floating-point execution units.

RISC processors often use *delayed branches* in order to avoid the branch penalty. The CPU always executes the instruction in the *branch delay slot*. If branch succeeds, the instruction is annulled. RISC processors may also depend upon the compiler to avoid the *load-use penalty*. This penalty arises because operations involving memory operands inherently take longer than operations between internal registers. We must first load the memory operand before we can use it. Here's the problem. Suppose that we're doing an arithmetic operation in which two operands are added together to produce a result. The first instruction loads a memory operand into a register and the next instruction takes that register and adds it to a second register to produce a result. Since most memory operations take longer than operations between registers. The load operation will stall the pipeline for at least one clock cycle while the data is brought to the register. Here's where the focus is on the compiler. A good compiler that is architecturally aware will see the load instruction and move it up in the program so that both registers are loaded with the appropriate operands when the instruction is to be executed. This is a good example of out-of-order execution. Of course, if by moving an instruction up in the sequence we destroy the logical flow of the algorithm, we haven't saved anything at all.

We cannot discuss the performance improvements of the RISC architecture without discussing some of the improvements in compiler technology that takes advantage of the RISC architecture. RISC computers should use optimizing compilers to best take advantage of the speed-enhancing features of the processor. Looking at the assembly-language output of an optimizing RISC compiler is an exercise in futility because the compiler will typically rearrange instructions to take best advantage of the processor's parallelism. Some compiler techniques are:

- Wherever possible, fill load-delay and branch-delay slots with independent instructions.
- Aggressively use registers to avoid memory load/store penalties.
- Move LOAD instructions as early as possible in the instruction stream to avoid load-delay penalties.
- Move instructions that evaluate branch addresses as early as possible in the instruction stream to avoid branch-delay penalties.
- Move condition code testing instructions as early as possible so that branch-test sequences can be optimized.
- Fix branch tests so that the most common result of the test is not to take the branch.
- Unroll loops (inline the loop code) to avoid penalty of nonsequential instruction execution.
- Try to maximize the size of the *basic block* (a block of code with a single entry and exit point and no internal loops).

Summary of Chapter 13

- Pipelining improves performance by increasing instruction throughput, not decreasing instruction execution time.
- Complications arise from hazards.
- To make processors faster, we try to start more than one instruction at a time by utilizing separate instruction pipes. We call it a superscalar architecture.
- Compilers can rearrange the instruction code they generate to improve pipeline performance.
- Hardware scheduling is used by modern RISC microprocessors because, internally, the processor tries to keep its resources as busy as possible through dynamic scheduling.

Chapter 13: *Endnotes*

[1] David A. Patterson and David R. Ditzel, *The Case for the Reduced Instruction Set Computer*, ACM SIGARCH Computer Architecture News,8 (6), October 1980. pp. 25–33.

[2] Manolis G. H. Kavantis, *Reduced Instruction Set Computer Architectures for VLSI*, The MIT Press, Cambridge, MA, 1986.

[3] David Resnick, Cray Corporation, Private Communication.

[4] J.E. Thornton, *Design of a Computer: The Control Data 6600*, Scott, Foresman and Company, Glenview, Ill, 1970.

[5] W.C. Alexander and D.B. Wortman, *Static and Dynamic characteristics XPL Programs*, pp. 41–46, November 1975, Vol. 8, No. 11.

[6] Jim Turley, *Starting Down the Pipeline Part I*, Circuit Cellar, Issue 143, June, 2002, p. 44.

[7] See Sloss et al, Chapter 11.

[8] David A. Patterson and John L. Hennessy, *Computer Organization and Design, Second Edition*, ISBN 1-5586-0428-6, Morgan-Kaufmann Publishers, San Francisco, 1998, p. 437.

[9] Jim Turley, *op cit*, p. 46.

[10] Matthew Wilson, "Comparing C/C++ Compilers," Dr. Dobbs Journal, October 2003, p. 16.

[11] Daniel Mann, *Programming the 29K RISC Family*, ISBN 0-1309-1893-8, Prentice-Hall, Englewood Cliffs, NJ, 1994, p. 10.

[12] Mike Johnson, *Superscalar Microprocessor Design*, ISBN 0-1387-5634-1, Prentice-Hall, Englewood Cliffs, NJ, 1991.

Exercises for Chapter 13

1. Examine the 2 assembly language code segments. For simplicity, we'll use 68K assembly language instructions. One of the segments would lead to more efficient operation of a pipeline then the other segment. Which segment would result in the more efficient pipeline of the code and why?

Segment A	
MOVE.W	D1,D0
MOVE.W	#$3400,D2
ADDA.W	D7,A2

Segment B	
MOVE.W	D1,D0
ADD.W	D0,D3
MULU	D3,D1

2. Discuss in a few sentences what problems might arise by increasing the number of stages in a pipeline. Modern microprocessors, such as the Pentium from Intel and the Athlon from AMD have similar benchmark results, but the AMD processors generally run at clock speeds that are about 65% of the clock speed of the Pentium processor with a comparable benchmark score. Discuss in a few sentences why this is the case.

3. For each of the following assembly language instructions, indicate whether or not it is generally characteristic of a *RISC* instruction. For simplicity, we'll use 68000 mnemonics for the instructions.
   a. MOVE.L $0A0055E0,$C0000000
   b. ADD.L   D6,D7
   c. MOVE.L   D5,(A5)
   d. ANDI.W   #$AAAA,$10000000
   e. MOVE.L   #$55555555,D3

4. The instruction:

   MOVEM.L   D0-D3/D5/A0-A2,-(SP)

   is typical of a CISC instruction (although the ARM ISA has a similar instruction). Rewrite this instruction as a series of RISC instructions assuming that the only effective addressing modes available are:
   a. immediate
   b. data register direct
   c. address register direct
   d. address register indirect

   Remember: You must use only constructs that would be typical of a RISC instruction set architecture.

5. Assume that the mnemonic LD (LOAD) is used for a memory to register transfer and ST (STORE) is used for a register to memory transfer. Assume that the following sequence of instructions is part of the assembly language for an arbitrary RISC processor containing 64 general-purpose registers (R0 – R63), each register is capable of being used as an address register (memory pointer) or a data register for arithmetic operations. Discuss in a sentence or two why the instruction sequence is characteristic of a RISC processor.

    Assume the instruction format is: op-code source operand, destination operand

```
*****************************************
* This code sequence adds together two numbers stored in
* memory and writes the result back to a third memory
* location.
*****************************************
addr1    EQU        $00001000      * First operand
addr2    EQU        $00001004      * Second operand
result   EQU        $00001006      * Result

start    LD         #addr1,R0      * Set up first address
         LD         #addr2,R1      * Set up second address
         LD         #result,R3     * Pointer to result
         LD         (R0),R60       * Get first operand
         LD         (R1),R61       * Get second operand
         ADD        R60,R61        * Do the add
         ST         R61,(R3)       * Save the result
```

    `* End of code sequence.`

6. One way that a compiler will try to optimize a program is to transform certain loop structures, such as the *for* loop, into a longer succession of in-line instructions. Why would this be advantageous?

7. Suppose that a certain processor has a 7-stage pipeline and runs at 100 MHz. The pipeline is designed so that 2 clock cycles are required at each stage. Also, assume that there is a basic block of ten instructions about to enter the pipeline.
    a. Assuming that no pipeline stalls occur, how much time will elapse from the time that the first instruction enters the pipeline to when it is retired at the end of the pipeline?
    b. What is the total elapsed execution time for all ten instructions in the basic block, assuming that, at different times, two of the instructions each cause the pipeline to stall for 4 clock cycles?

8. Consider the following snippet of 68K assembly language instructions. Reorder the instructions so that they may execute more efficiently in a pipelined microprocessor architecture.

```
MOVE.W         D1,D0
ADD.W          D0,D3
MULU           D3,D1
LEA            (A4),A6
MOVE.W         #$3400,D2
ADDA.W         D7,A2
MOVE.W         #$F6AA,D4
```

Memory Revisited, Caches and Virtual Memory

● ●

Objectives

When you are finished with this lesson, you will be able to:
- ▶ *Explain the reason for caches and how caches are organized;*
- ▶ *Describe how various caches are organized;*
- ▶ *Design a typical cache organization;*
- ▶ *Discuss relative cache performance;*
- ▶ *Explain how virtual memory is organized; and*
- ▶ *Describe how computer architecture supports virtual memory management.*

● ●

Introduction to Caches

As an introduction to the topic of caches and cache-based systems, let's review the types of memories that we discussed before. The major types of memories are static random access memory (SRAM), dynamic random access memory (DRAM), and nonvolatile read-only memory (ROM). SRAM memory is based on the principle of the cross-coupled, inverting logic gates. The output value feeds back to the input to keep the gate locked in one state or the other. SRAM memory is very fast, but each memory cell required five or six transistors to implement the design, so it tends to be more expensive than DRAM memory.

DRAM memory stores the logical value as charge on a tiny charge-storage element called a capacitor. Since the charge can leak off the capacitor if it isn't refreshed periodically, this type of memory must be continually read from at regular intervals. This is why it is called dynamic RAM rather than static RAM. The memory access cycles for DRAM is also more complicated than for static RAM because these refresh cycles must be taken into account as well.

However, the big advantage of DRAM memory is its density and low cost. Today, you can buy a single in-line memory module, or SIMM for your PC with 512 Mbytes of DRAM for $60. At those prices, we can afford to put the complexity of managing the DRAM interface into specialized chips that sit between the CPU and the memory. If you're a computer hobbyist who likes to do your own PC upgrading, then you've no doubt purchased a new motherboard for your PC featuring the AMD, nVidia, Intel or VIA "chipsets." The chipsets have become as important a consideration as the CPU itself in determining the performance of your computer.

Our computer systems demand a growing amount of memory just to keep up with the growing complexity of our applications and operating systems. This chapter is being written on a PC with

1,024 Mbytes (1 Gbyte) of memory. Today this is considered to be more than an average amount of memory, but in three years it will probably be the minimal recommended amount. Not too long ago, 10 Mbytes of disk storage was considered a lot. Today, you can purchase a 200 Gbyte hard disk drive for around $100. That's a factor of 10,000 times improvement in storage capacity. Given our insatiable urge for ever-increasing amounts of storage, both volatile storage, such as RAM, and archival storage, such as a hard disk, it is appropriate that we also look at ways that we manage this complexity from an architectural point of view.

The Memory Hierarchy

There is a hierarchy of memory. In this case we don't mean a pecking order, with some memory being more important than others. In our hierarchy, the memory that is "closer" to the CPU is considered to be higher in the hierarchy then memory that is located further away from the CPU. Note that we are saying "closer" in a more general sense then just "physically closer" (although proximity to the CPU is an important factor as well). In order to maximize processor throughput, the fastest memory is located the closest to the processor. This fast memory is also the most expensive. Figure 14.1 is a qualitative representation of what is referred to as

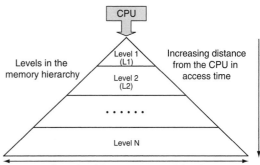

Figure 14.1: The memory hierarchy. As memory moves further away from the CPU both the size and access times increase.

the *memory hierarchy*. Starting at the pinnacle, each level of the pyramid contains different types of memory with increasingly longer access times.

Let's compare this to some real examples. Today, SRAM access times are in the 2–25ns range at cost of about $50 per Mbyte. DRAM access times are 30–120ns at cost of $0.06 per Mbyte. Disk access times are 10 to 100 million ns at cost of $0.001 to $0.01 per Mbyte. Notice the exponential rise in capacity with each layer and the corresponding exponential rise in access time with the transition to the next layer.

Figure 14.2, shows the memory hierarchy for a typical computer system that you might find in your own PC at home. Notice that there could be two separate caches in the system, an on-chip cache at level 1, often called an *L1 cache*, and an off-chip cache at level 2, or an *L2 cache*. It is easily apparent that the capacity increases and the speed of the memory decreases at each level of the hierarchy. We could also imagine that at a final level to this pyramid, is the Internet. Here the capacity is almost infinite, and it often seems like the access time takes forever as well.

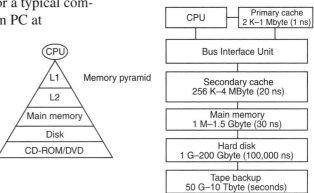

Figure 14.2: Memory hierarchy for a typical computer system.

Locality

Before we continue on about caches, let's be certain that we understand what a cache is. A cache is a nearby, local storage system. In a CPU we could call the register set the zero level cache. Also, on-chip, as we saw there is another, somewhat larger cache memory system. This memory typically runs at the speed of the CPU, although it is sometimes slower then regular access times. Processors will often have two separate L1 caches, one for instructions and one for data. As we've seen, this is an internal implementation of the Harvard architecture.

The usefulness of a cache stems from the general characteristics of programs that we call *locality*. There are two types of locality, although they are alternative ways to describe the same principle. *Locality of Reference* asserts that program *tend to* access data and instructions that were recently accessed before, or that are located in nearby memory locations. Programs tend to execute instructions in sequence from adjacent memory locations and programs tend to have loops in which a group of nearby instructions is executed repeatedly. In terms of data structures, compilers store arrays in blocks of adjacent memory locations and programs tend to access array elements in sequence. Also, compilers store unrelated variables together, such as local variables on a stack. *Temporal locality* says that once an item is referenced it will tend to be referenced again soon and *spatial locality* says that nearby items will tend to be referenced soon.

Let's examine the principle of locality in terms of a two-level memory hierarchy. This example will have an upper-level (cache memory) and a lower level (main memory).The two-level structure means that if the data we want isn't in the cache, we will go to the lower level and retrieve at least one block of data from main memory. We'll also define a *cache hit* as a data or instruction request by the CPU to the cache memory where the information requested is in cache and a *cache miss* as the reverse situation; the CPU requests data and the data is not in the cache.

We also need to define a *block* as a minimum unit of data transfer. A block could be as small as a byte, or several hundred bytes, but in practical terms, it will typically be in the range of 16 to 64 bytes of information. Now it is fair to ask the question, "Why load an entire block from main memory? Why not just get the instruction or data element that we need?" The answer is that locality tells us that if the first piece of information we need is not in the cache, the rest of the information that we'll need shortly is probably also not in the cache, so we might as well bring in an entire block of data while we're at it.

There is another practical reason for doing this. DRAM memory takes some time to set up the first memory access, but after the access is set up, the CPU can transfer successive bytes from memory with little additional overhead, essentially in a burst of data from the memory to the CPU. This called a *burst mode access*. The ability of modern SDRAM memories to support burst mode accesses is carefully matched to the capabilities of modern processors. Establishing the conditions for the burst mode access requires a number of clock cycles of overhead in order for the memory support chip sets to establish the initial addresses of the burst. However, after the addresses have been established, the SDRAM can output two memory read cycles for every clock period of the external bus clock. Today, with a bus clock of 200MHz and a memory width of 64-bits, that translates to a memory to processor data transfer rate of 3.2 GBytes per second during the actual burst transfer.

Let's make one more analogy about a memory hierarchy that is common in your everyday life. Imagine yourself, working away at your desk, solving another one of those interminable problem sets that engineering professors seem to assign with depressing regularity. You exploit locality keeping the books that you reference most often, say your required textbooks for your classes, on your desk or bookshelf. They're nearby, easily referenced when you need them, but there are only a few books around.

Suppose that your assignment calls for you to go to the engineering library and borrow another book. The engineering library certainly has a much greater selection than you do, but the retrieval costs are greater as well. If the book isn't in the engineering library, then the Library of Congress in Washington, D.C. might be your next stop. At each level, in order to gain access to a greater amount of stored material, we incur a greater penalty in our access time. Also, our unit of transfer in this case is a book. So in this analogy, one block equals one book.

Let's go back and redefine things in terms of this example:
- *block:* the unit of data transfer (one book),
- *hit rate:* the percentage of the data accesses that are in the cache (on your desk)
- *miss rate:* the percentage of accesses not in the cache (1 – hit rate)
- *hit time:* the time required to access data in the cache (grab the book on your desk)
- *miss penalty:* the time required to replace the block in the cache with the one you need (go to the library and get the other book)

We can derive a simple equation for the *effective execution time*. That is the actual time, on average, that it takes an instruction to execute, given the probability that the instruction will, or will not, be in the cache when you need it. There's a subtle point here that should be made. The miss penalty is the time delay imposed because the processor must execute all instructions out of the cache. Although most cached processors allow you to enable or disable the on-chip caches, we'll assume that you are running with the cache on.

Effective Execution Time = hit rate × hit time + miss rate × miss penalty

If the instruction or data is not in the cache, then the processor must reload the cache before it can fetch the next instruction. It cannot just go directly to memory to fetch the instruction. Thus, we have the block of time penalty that is incurred because it must wait while the cache is reloaded with a block from memory.

Let's do a real example. Suppose that we have a cached processor with a 100 MHz clock. Instructions in cache execute in two clock cycles. Instructions that are not in cache must be loaded from main memory in a 64-byte burst. Reading from main memory requires 10 clock cycles to set up the data transfer but once set-up, the processor can read a 32-bit wide word at one word per clock cycle. The cache hit rate is 90%.

1. The hard part of this exercise is calculating the miss penalty, so we'll do that one first.
    a. 100 MHz clock -> 10 ns clock period
    b. 10 cycles to set up the burst = 10 × 10 ns = 100 ns
    c. 32-bit wide word = 4 bytes -> 16 data transfers to load 64 bytes
    d. 16 × 10 ns = 160 ns
    e. Miss penalty = 100 ns + 160 ns = 260 ns

2. Each instruction takes 2 clocks, or 20 ns to execute.
3. Effective execution time = 0.9x20 + 0.1x260 = 18 + 26 = 44 ns

Even this simple example illustrates the sensitivity of the effective execution time to the parameters surrounding the behavior of the cache. The effective execution time is more than twice the in-cache execution time. So, whenever there are factors of 100% improvement floating around, designers get busy.

We can thus ask some fundamental questions:
1. How can we increase the cache hit ratio?
2. How can we decrease the cache miss penalty?

For #1, we could make the caches bigger. A bigger cache holds more of main memory, so that should increase the probability of a cache hit. We could change the design of the cache. Perhaps there are ways to organize the cache such that we can make better use of the cache we already have. Remember, memory takes up a lot of room on a silicon die, compared to random logic, so adding an algorithm with a few thousand gates might get a better return then adding another 100K to the cache.

We could look to the compiler designers for help. Perhaps they could better structure the code so that it would be able to have a higher proportion of cache hits. This isn't an easy one to attack, because cache behavior sometimes can become very counter-intuitive. Small changes in an algorithm can sometimes lead to big fluctuations in the effective execution time. For example, in my Embedded Systems Laboratory class the students do a lab experiment trying to fine tune an algorithm to maximize the difference in measured execution time between the algorithm running cache off and cache on. We turn it into a small contest. The best students can hand craft their code to get a 15:1 ratio.

Cache Organization

The first issue that we will have to deal with is pretty simple: "How do we know if an item (instruction or data) is in the cache?" If it is in the cache, "How do we find it?" This is a very important consideration. Remember that your program was written, compiled and linked to run in main memory, not in the cache. In general, the compiler will not know about the cache, although there are some optimizations that it can make to take advantage of cached processors. The addresses associated with references are main memory addresses, not cache addresses. Therefore, we need to devise a method that somehow *maps* the addresses in main memory to the addresses in the cache.

We also have another problem. What happens if we change a value such that we must now write a new value back out to main memory? Efficiency tells us to write it to the cache, but this could lead to a potentially disastrous situation where the data in the cache and the data in main memory are no longer *coherent* (in agreement with each other). Finally, how do we design a cache such that we can maximize our hit rate? We'll try to answer these questions in the discussion to follow.

In our first example our block size will be exactly one word of memory. The cache design that we'll use is called a *direct-mapped cache.* In a direct-mapped cache, every word of memory at the lower level has exactly one location in the cache where it might be found. Thus, there will be lots of memory locations at the lower level for every memory location in the cache. This is shown in Figure 14.3

Referring to Figure 14.3, suppose that our cache is 1,024 words (1K) and main memory contains 1,048,576 words (1M). Each cache location maps to 1,024 main memory locations. This is fine, but now we need to be able to tell which of the 1,024 possible main memory locations is in a particular cache location at a particular point in time. Therefore, every memory location in the cache needs to contain more information than just the corresponding data from main memory.

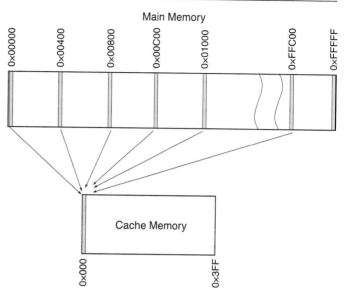

Figure 14.3: Mapping of a 1K direct mapped cache to a 1M main memory. Every memory location in the cache maps to 1024 memory locations in main memory.

Each cache memory location consists of a number of cache entries and each cache entry has several parts. We have some cache memory that contains the instructions or data that corresponds to one of the 1,024 main memory locations that map to it. Each cache location also contains an *address tag,* which identifies which of the 1,024 possible memory locations happens to be in the corresponding cache location. This point deserves some further discussion.

Address Tags

When we first began our discussion of memory organization several lessons ago, we were introduced to the concept of paging. In this particular case, you can think of main memory as being organized as 1,024 pages with each page containing exactly 1,024 words. One page of main memory maps to one page of the cache. Thus, the first word of main memory has the binary address 0000 0000 0000 0000 0000. The last word of main memory has the address 1111 1111 11 11 1111 1111. Let's split this up in terms of page an offset. The first word of main memory has the page address 00 0000 0000 and the offset address 00 0000 0000. The last page of main memory has the page address 11 1111 1111 and the offset address 11 1111 1111.

In terms of hexadecimal addresses, we could say that the last word of memory in page/offset addressing has the address $3FF/$3FF. Nothing has changed, we've just grouped the bits differently so that we can represent the memory address in a way that is more aligned with the organization of the direct-mapped cache. Thus, any memory position in the cache also has to have storage for the page address that the data actually occupies in main memory.

Now, data in a cache memory is either copies of the contents of main memory (instructions and/or data) or newly stored data that are not yet in main memory. The cache entry for that data, called a *tag*, contains the information about the block's location in main memory and validity (coherence) information. Therefore, every cache entry must contain the instruction or data contained in main memory, the page of main memory that the block comes from, and, finally, information about whether the data in the cache and the data in main memory are coherent. This is shown in Figure 14.4.

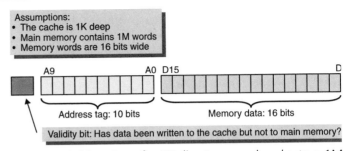

We can summarize the cache operation quite simply. We must maximize the probability that whenever the CPU does an instruction fetch or a data read, the instruction or data is available in the cache. For many CPU designs, the algorithmic state machine design that is used to manage the

Figure 14.4: Mapping of a 1K direct mapped cache to a 1M main memory. Every memory location in the cache maps to 1024 memory locations in main memory.

cache is one of the most jealously guarded secrets of the company. The design of this complex hardware block will dramatically impact the cache hit rate, and consequently, the overall performance of the processor.

Most caches are really divided into three basic parts. Since we've already discussed each one, let's just take a moment to summarize our discussion.

- *cache memory:* holds the memory image
- *tag memory:* holds the address information and validity bit. Determines if the data is in the cache and if the cache data and memory data are coherent.
- *algorithmic state machine:* the cache control mechanism. Its primary function is to guarantee that the data requested by the CPU is in the cache.

To this point, we've been using a model that the cache and memory transfer data in blocks and our block size has been one memory word. In reality, caches and main memory are divided into equally sized quantities called *refill lines*. A refill line is typically between four and 64 bytes long (power of 2) and is the minimum quantity that the cache will deal with in terms of its interaction with main memory. Missing a single byte from main memory will result in a full filling of the refill line containing that byte. This is why most cached processors have burst modes to access memory and usually never read a single byte from memory. The refill line is another name for the data block that we previously discussed.

Today, there are four common cache types in general use. We call these:

1. direct-mapped
2. associative
3. set-associative
4. sector mapped

The one used most is the four-way set-associative cache, because it seems to have the best performance with acceptable cost and complexity. We'll now look at each of these cache designs.

Direct-Mapped Cache

We've already studied the direct-mapped cache as our introduction to cache design. Let's re-examine it in terms of refill lines rather than single words of data. The direct-mapped cache partitions main memory into an XY matrix consisting of K columns of N refill lines per column.

The cache is one-column wide and N refill lines long. The Nth row of the cache can hold the Nth refill line of any one of the K columns of main memory. The tag address holds the address of the memory column. For example, suppose that we have a processor with a 32-bit byte-addressable address space and a 256K, direct-mapped cache. Finally, the cache reloads with 64 bytes long refill line. What does this system look like?

1. Repartition the cache and main memory in terms of refill lines.
   a. Main memory contains 2^{32} bytes / 2^6 bytes per refill line = 2^{26} refill lines
   b. Cache memory contains 2^{18} bytes / 2^6 bytes per refill line = 2^{12} refill lines
2. Represent cache memory as single column with 2^{12} rows and main memory as an XY matrix of 2^{12} rows by 2^{26} / 2^{12} = 2^{14} columns. See Figure 14.5.

In Figure 14.5 we've divided main memory into three distinct regions:

- offset address in a refill line;
- row address in a column; and
- column address.

We map the corresponding byte positions of a refill line of main memory to the byte position in the refill line of the cache. In other words, the offset addresses are the same in the cache and in main memory. Next, every row of the cache corresponds to every row of main memory. Finally, the same row (refill line) within each column of main memory maps to the same row, or refill line, of the cache memory and its column address is stored in the tag RAM of the cache.

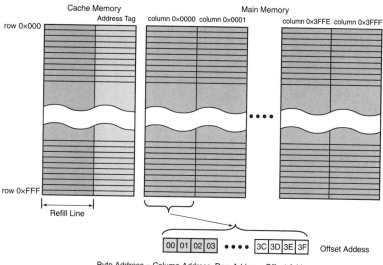

Figure 14.5: Example of a 256Kbyte direct-mapped cache with a 4Gbyte main memory. Refill line width is 64 bytes.

The address tag field must be able to hold a 14-bit wide column address, corresponding to column addresses from 0x0000 to 0x3FFF. The main memory and cache have 4096 rows, corresponding to row addresses 0x000 through 0x3FF.

As an example, let's take an arbitrary byte address and map it into this column/row/offset schema.

Byte address = 0xA7D304BE

Because not all of the boundaries of the column, row and offset address do not lie on the boundaries of hex digits (divisible by 4), it is will be easier to work the problem out in binary, rather than hexadecimal. First we'll write out the byte address 0xA7D304BE as a 32-bit wide number and then group it according to the column, row and offset organization of the direct mapped cache example.

```
1010 0111 1101 0011 0000 0100 1011 1110    * 8 hexadecimal digits
Offset: 11 1110 = 0x3E
Row: 1100 0001 0010 = 0xC12
Column: 10 1001 1111 0100 = 0x29F4
```

Therefore, the byte that resides in main memory at address 0xA7D304BE resides in main memory at address 0x29F4, 0xC12, 0x3E when we remap main memory as an XY matrix of 64-byte wide refill lines. Also, when the refill line containing this byte is in the cache, it resides at row 0xC12 and the address tag address is 0x29F4. Finally, the byte is located at offset 0x3E from the first byte of the refill line.

The direct mapped cache is a relatively simple design to implement but it is rather limited in its performance because of the restriction placed upon it that, at any point in time, only one refill line per row of main memory may be in the cache. In order to see how this restriction can affect the performance of a processor, consider the following example.

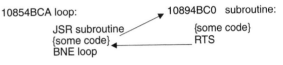

The two addresses for the loop and for the subroutine called by the loop look vaguely similar. If we break these down into their mappings in the cache example we see that for the loop, the address maps to:

- Offset = 0x0A
- Row = 0x52F
- Column = 0x0421

The subroutine maps to:

- Offset = 0x00
- Row = 0x52F
- Column = 0x0422

Thus, in this particular situation, which just might occur, either through an assembly language algorithm or as a result of how the compiler and linker organize the object code image, the loop, and the subroutine called by the loop, are on the same cache row but in adjacent columns.

Every time the subroutine is called, the cache controller must refill row 0x52F from column 0x422 before it can begin to execute the subroutine. Likewise, when the RTS instruction is encountered, the cache row must once again be refilled from the adjacent column. As we've previously seen in the calculation for the effective execution time, this piece of code could easily run 10 times slower then it might if the two code segments were in different rows.

The problem exists because of the limitations of the direct mapped cache. Since there is only one place for each of the refill lines from a given row, we have no choice when another refill from the same row needs to be accessed.

At the other end of the spectrum in terms of flexibility is the *associative cache*. We'll consider this cache organization next.

Associative Cache

As we've discussed, the direct-mapped cache is rather restrictive because of the strict limitations on where a refill line from main memory may reside in the cache. If one particular row refill line address in the cache is mapped to two refill lines that are both frequently used, the computer will be spending a lot of time swapping the two refill lines in and out of the cache. What would be an improvement is if we can map any refill line address in main memory to any available refill line position in the cache. We call a cache with this organization an *associative cache*. Figure 14.6 illustrates an associative cache.

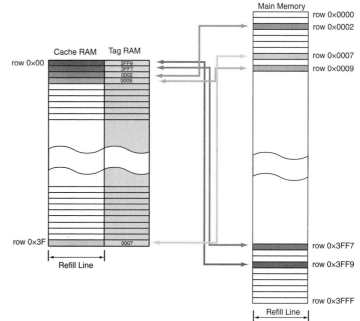

Figure 14.6: Example of a 4 Kbyte associative cache with a 1M main memory. Refill line width is 64 bytes. [NOTE: This figure is included in color on the DVD-ROM.]

In Figure 14.6, we've taken an example of a 1 Mbyte memory space, a 4 Kbyte associative cache, and a 64 byte refill line size. The cache contains 64 refill lines and main memory is organized as a single column of 2^{14} refill lines (16 Kbytes).

This example represents a *fully associative* cache. Any refill line of main memory may occupy any available refill position in the cache. This is as good as it gets. The associative cache has none of the limitations imposed by the direct-mapped cache architecture. Figure 14.6 attempts to show in a multicolor manner, the almost random mapping of rows in the cache to rows in main memory. However, the complexity of the associative cache grows exponentially with cache size and main memory size. Consider two problems:

1. When all the available rows in the cache contain valid rows from main memory, how does the cache control hardware decide where in the cache to place the next refill line from main memory?
2. Since any refill line from main memory can be located at any refill line position in the cache, how does the cache control hardware determine if a main memory refill line is currently in the cache?

We can deal with issue #1 by placing a binary counter next to each row of the cache. On every clock cycle we advance all of the counters. Whenever, we access the data in a particular row of the cache, we reset the counter associated with that row back to zero. When a counter reaches the maximum count, it remains at that value. It does not roll over to zero.

All of the counters feed their values into a form of priority circuit that outputs the row address of the counter with the highest count value. This row address of the counter with the highest count value is then the cache location where the next cache load will occur. In other words, we've implemented the hardware equivalent of a *least recently used (LRU)* algorithm.

The solution to issue #2 introduces a new type of memory design called a *contents addressable memory (CAM)*. A CAM memory can be thought of as a standard memory turned inside out. In a CAM memory, the input to the CAM is the data, and the output is the address of the memory location in the CAM where that data is stored. Each memory cell of the CAM memory also contains a data comparator circuit. When the tag address is sent to the CAM by the cache control unit all of the comparators do a parallel search of their contents. If the input tag address matches the address stored in a particular cache tag address location, the circuit indicates an address match (hit) and outputs the cache row address of the main memory tag address.

As the size of the cache and the size of main memory increases, the number of bits that must be handled by the cache control hardware grows rapidly in size and complexity. Thus, for real-life cache situations, the fully associative cache is not an economically viable solution.

Set-Associative Cache

Practically speaking, the best compromise for flexibility and performance is the *set-associative cache* design. The set-associative cache combines the properties of the direct-mapped and associative cache into one system. In fact, the four-way set-associative cache is the most commonly used design in modern processors. It is equivalent to a multiple column direct mapping. For example, a two-way set-associative cache has two direct mapped columns. Each column can hold any of the refill lines of the corresponding row of main memory.

Thus, in our previous example with the direct-mapped cache, we saw that a shortcoming of that design was the fact that only one refill line from the entire row of refill lines in main memory may be in the corresponding refill line position in the cache. With a two-way set-associative cache, there are two cache locations that are available at any point in time to hold two refill lines from the corresponding row of main memory.

Figure 14.7 shows a two-way set-associative cache design.

Within a row, any two of the refill lines in main memory may be mapped by the tag RAM to either of the two refill line locations in the cache. A one-way set-associative cache degenerates to a direct-mapped cache.

The four-way set associative cache has become the *de facto standard* cache design for mod-

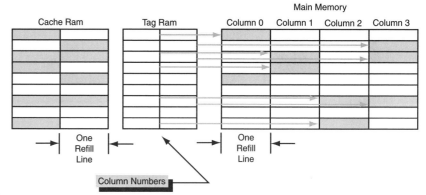

Figure 14.7: Two-way set-associative cache design. From Baron and Higbie[1]. [NOTE: This figure is included in color on the DVD-ROM.]

ern microprocessors. Most processors use this design, or variants of the design, for their on-chip instruction and data caches.

Figure 14.8 shows a 4-way set associative cache design with the following specifications:

- 4 Gbyte main memory
- 1 Mbyte cache memory
- 64 byte refill line size

The row addresses go from 0x000 to 0xFFF (4,096 row addresses) and the column addresses go from 0x0000 to 0x3FFF (16,384 column addresses). If we compare the direct mapped cache with a 4-way set associative cache of the same size, we see that the direct mapped cache has ¼ the number of columns and 4 times the number of rows. This is the consequence of redistributing the same number of refill lines from a single column into a 4 × N matrix.

This means that in the 4-way set associative cache design each row of the cache maps 4

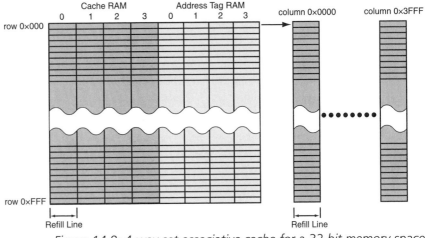

Figure 14.8: 4-way set associative cache for a 32-bit memory space and 1 Mbyte cache. The refill line size is 64 bytes.

times the number of main memory refill lines as with a direct-mapped cache. At first glance, this may not seem like much of an improvement. However, the key is the associativity of the 4-way design. Even though we have 4 times the number of columns, and 4 possible places in the cache to map these rows, we have a lot more flexibility in which of the 4 locations in the cache the newest refill line may be placed. Thus, we can apply a simple LRU algorithm on the 4 possible cache locations. This prevents the kind of thrashing situation that the direct mapped cache can create.

You can easily see the additional complexity that the 4-way set associative cache requires over the direct mapped cache. We now require similar additional circuitry as with the fully associative cache design to decide on cache replacement strategy and to detect address tag hits. However, the fact that the associativity extends to only 4 possible locations makes the design much simpler to implement. Finally, keep in mind that a direct-mapped cache is just a 1-way set associative cache.

The last cache design that we'll look at is the *sector-mapped cache*. The sector-mapped cache is a modified associative mapping. Main memory and refill lines are grouped into *sectors (rows)*. Any main memory sector can be mapped into a cache sector and the cache uses an associative memory to perform the mapping. The address in tag RAM is sector address. One additional complexity introduced by the sector mapping is the need for *validity bits* in the tag RAM. Validity bits keep track of the refill lines from main memory that are presently contained in the cache. Figure 14.8a illustrates the sector-mapped cache design.

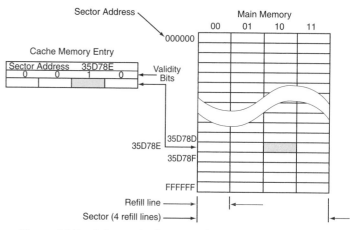

In this example we are mapping a memory system with a 32-bit address range into a cache of arbitrary size. There are four refill lines per sector and each refill line contains 64 bytes. In this particular example, we show that refill line 01 of sector 0x35D78E is valid, so the validity bit is set for that refill line.

It may not be obvious why we need the validity bits at all. This simple example should help to clarify the point. Remember, we map main

Figure 14.8a: Schematic diagram of a sector mapped cache for a memory system with a 32-bit address range. In this example there are 4 refills per sector and each refill line contains 64 bytes. Only the refill line at sector address 0x35D78E and position 10 is currently valid.

memory to the cache by sector address, refill lines within a sector maintain the same relative sector position in main memory or the cache, and we refill a cache sector one refill line at a time. Whew!

Since the cache is fully associative with respect to the sectors of main memory, we use an LRU algorithm of some kind to decide which refill line in the cache can be replaced. The first time that a refill line from a new sector is mapped into the cache and the sector address is updated, only the refill line that caused the cache entry to be updated is valid. The remaining three refill lines, in positions 00, 01 and 11, correspond to the previous sector, and are do not correspond to the refill lines of main memory at the new sector address. Thus, we need validity bits.

By grouping a row of refill lines together, we reduce some of the complexity of the purely associative cache design. In this case, we reduce the problem by a factor of four. Within the sector, each refill line from main memory must map to the corresponding position in the cache. However, we have another level of complexity because of the associative nature of the cache, when we load a refill line from a sector into the cache, the address in the tag RAM must correspond to the sector address of the refill line just added. The other refill lines will probably have data from other sectors. Thus, we need a validity bit to tell us which refill lines in a cache sector correspond to the correct refill lines in main memory.

Figure 14.9 is a graph of the miss rate versus cache associativity for different cache sizes.

Clearly the added dimension of an associative cache greatly improves the cache hit ratio. Also, as the cache size increases the sensitivity to the degree of associativity decreases, so there is apparently no improvement in the miss ratio by going from a 4-way cache to an 8-way cache.

Figure 14.10 shows how the cache miss rate decreases as we increase the cache size and the refill line size. Notice, however, that the curves are asymptotic and there is not much improvement, if any, if we increase the refill line size beyond 64 bytes. Also, the improvements in performance begin to decrease dramatically once the cache size is about 64 Kbytes in size. Of course, we know now that this is a manifestation of locality, and gives us some good data to know what kind of a cache will best suit our needs.

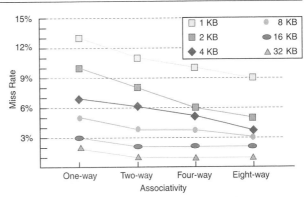

Let's look at performance more quantitatively. A simplified model of performance would be given by the following pair of equations.

1. *Execution time* = (execution cycles + stall cycles) × (cycle time)
2. *Stall cycles* = (# of instructions) × (miss ratio) × (miss penalty)

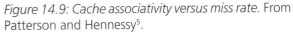

Figure 14.9: Cache associativity versus miss rate. From Patterson and Hennessy[5].

The execution time, or the time required to run an algorithm depends upon two factors. First, how many instructions are actually in the algorithm (execution cycles) and how many cycles were spent filling the cache (stall cycles). Remember, that the processor always executes from the cache, so if the data isn't in the cache, it must wait for the cache to be refilled before it can proceed.

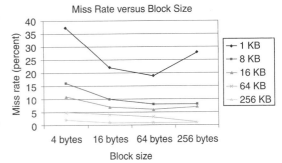

The stall cycles are a function of the cache-hit rate, so it depends upon the total number of

Figure 14.10: Miss rate versus block size. From Agarwal[2].

instructions being executed. Some fraction of those instructions will be cache misses, so for each cache miss, we incur a penalty. The penalty being the time required to refill the cache before execution can proceed.

Thus, we can see two strategies for improving performance:

1. decrease the miss ratio, or
2. decrease the miss penalty.

What happens if we increase block size? According to Figure 14.10, we might get some improvement as we approach 64 bytes, but the bigger the refill lines become, the bigger the miss penalty because we're refilling more of the cache with each miss, so the performance may get worse, not better.

We also saw why the four-way set-associative cache is so popular by considering the data in Figure 14.9. Notice that once the cache itself becomes large enough, there is no significant difference in the miss rate for different types of set-associative caches.

Continuing with our discussion on improving overall performance with caches, we could do something that we really haven't considered until now. We can improve our overall performance for a given miss rate by decreasing the miss penalty. Thus, if a miss occurs in the primary cache, we can add a second level cache in order to decrease the miss penalty.

Often, the primary cache (L1) is on the same chip as the processor. We can use very fast SRAM to add another cache (L2) above the main memory (DRAM). This way, the miss penalty goes down if data is in 2nd level cache. For example, suppose that we have a processor that executes one instruction per clock cycle (cycles per instruction, or CPI = 1.0) on a 500 MHz machine with a 5 percent miss rate and 200ns DRAM access times.

By adding an L2 cache with 20ns access time, we can decrease the overall miss rate to 2 percent for both caches. Thus, our strategy in using multilevel caches is to try and optimize the hit rate on the 1st level cache and try to optimize the miss rate on the 2nd level cache.

Cache Write Strategies

Cache behavior is relatively straightforward as long as you are reading instructions and data from memory and mapping them to caches for better performance. The complexity grows dramatically when newly generated data must be stored in memory. If it is stored back in the cache, the cache image and memory are no longer the same (coherent). This can be a big problem, potentially life-threatening in certain situations, so it deserves some attention. In general, cache activity with respect to data writes can be of two types:

Write-through cache: data is written to the cache and immediately written to main memory as well. The write-through cache accepts the performance hit that a write to external memory will cause, but the strategy is that the data in the cache and the data in main memory must always agree.

Write-back cache: data is held until bus activity allows the data to be written without interrupting other operations. In fact the write-back process may also wait until it has an entire block of data to be written. We call the write-back of the data a *post-write.* Also, we need to keep track of which cache cells contain incoherent data and a memory cell that has an updated value still in cache is called a *dirty cell.* The tag RAM of caches that implement a write-back strategy must also contain validity bits to track dirty cells.

If the data image is not in cache, then there isn't a problem because the data can be written directly to external memory, just as if the cache wasn't there. This is called a *write-around cache* because noncached data is immediately written to memory. Alternatively, if there is a cache block available with no corresponding dirty memory cells, the cache strategy may be to store the data in cache first, then do a write through, or post-write, depending upon the design of the cache.

Let's summarize and wrap-up our discussion of caches.

- There are two types of locality: spatial and temporal.
- Cache contents include data, tags, and validity bits.
- Spatial locality demands larger block sizes.
- The miss penalty is increasing because processors are getting faster than memory, so modern processors use set-associative caches.
- We use separate I and D caches.

In order to avoid the von Neumann bottleneck,

- multi-level caches used to reduce miss penalty (assuming that the L1 cache is on-chip); and
- memory system are designed to support caches with burst mode accesses.

Virtual Memory

Remember the memory hierarchy? Once we move below the main memory in our memory hierarchy we need the methods provided by virtual memory to map *lower memory* in the hierarchy to *upper memory* in the hierarchy. We use large quantities of slower memory (hard disk, tape, the Internet) to supply instructions and data to our main physical memory. The virtual memory manager, usually your operating system, allows programs to pretend that they have limitless memory resources to run in and store data.

Why does Win9X, Win2K and Win XP recommend at least 64–128 Mbytes of main memory? Because if the computer's physical memory resources are too limited, the PC ends up spending all of its time swapping memory images back and forth to disk. In our previous discussion of caches we considered the penalty we incurred if we missed the cache and had to go to main memory. Consider what the penalty is if we miss main memory and have to go to the disk. The average disk access time for a hard disk is about 1 ms and the average for noncached memory is 10 ns. Therefore the penalty for not having data in RAM is $10^{-3}/ 10^{-8} = 10,000$ X slower access time for the hard disk, compared to RAM.

With virtual memory, the main memory acts as a cache for the secondary storage, typically your hard disk. The advantage of this technique is that each program has the illusion of having unlimited amounts of physical memory. In this case, the address that the CPU emits is a *virtual address*. It doesn't have any knowledge of whether or not the virtual address really exists in physical memory, or is actually mapped to the hard disk. Nor does it know if the operating system has another program currently loaded into the physical memory space that the virtual address is expecting.

By having a system based on a virtual memory model, program relocation becomes the natural method of building executable object code images and the system is able to assign attributes to each program, typically in the form of protection schemes, that allow them to remain isolated from each other while both are running. The now defunct Digital Equipment Corporation (DEC) is generally credited with inventing the idea of virtual memory and literally built their company on top of it. The term, *DEC VAX* was synonymous with the company. VAX stood for *virtual address extension*, or virtual memory.

Figure 14.11, right, is a simplified schematic diagram of a virtual memory model. As these processors become more complex we need to partition the CPU apart from other hardware that might be managing memory. Thus, the CPU puts out an address for data or instructions.

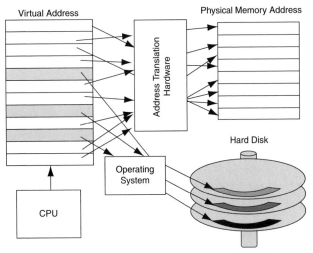

Figure 14.11: A virtual memory model. The CPU presents a virtual address, dedicated hardware and the operating system map the virtual address to physical memory or to secondary storage (hard disk).

The address is a *virtual address* because the CPU has no knowledge of where the program code or data actually is residing at any point in time. It might be in the I-cache, D-cache, main memory, or secondary storage. The CPU's function is to provide the virtual address request to the rest of the system.

If the data is available in the L1 cache, the CPU retrieves it and life is good. If there is a data miss, then the CPU is stalled while the refill line is fetched from main memory and the cache is replenished. If the data is not in main memory then the CPU must continue to wait until the data can be fetched from the hard disk. However, as you'll see, if the data is not in physical memory, the instruction is aborted and the operating system must take over to fetch the virtual address from the disk.

Since virtual memory is closely associated with operating systems, the more general term that we'll use to describe the memory space that the CPU thinks it has available is *logical memory*. Although we may use these terms in closely related ways, there is a real distinction between logical memory, physical memory and virtual memory, with virtual memory being the hard disk, as managed by the operating system. Let's look at the components of a virtual memory system in more detail. Refer to Figure 14.12.

The CPU executes an instruction and requests a memory operand, or the program counter issues the address of the next in-

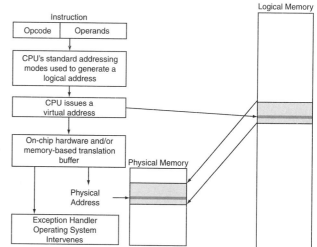

Figure 14.12: Components of a virtual memory system.

struction. In any case, standard addressing methods (addressing modes) are used and an address is generated by the CPU. This address is pointing to an address location in the *logical memory space* of the processor. Special hardware within the processor, called the *memory management unit (or MMU)* but outside of the CPU, maps the logical addresses to physical addresses in the computer's memory space. If the requested address is in physical memory at the time the instruction is issued, the memory request is fulfilled by the physical memory. However, in all probability, this will be a burst assess to refill the cache, not to fetch a single word.

However, it is possible that the memory access request cannot be fulfilled because the memory data is located in virtual memory on the disk. The memory management hardware in the processor detects that the address is not in physical memory and generates an exception. The exception is similar to an internally generated interrupt, but differs because a true interrupt will allow the present instruction to complete before the interrupt is accepted. With an instruction exception, the instruction must be aborted because the operating system must take over and handle the exception request.

When we were dealing with the on-chip caches we saw that the most effective block size to transfer between main memory and the cache is the refill line, typically 64 bytes in length. When we are dealing with virtual memory, or data held on the hard disk, we have to increase the size of the blocks because the miss penalty is thousands of times greater than a main memory miss penalty.

Remember, the hard disk is a mechanical device with moving and rotating elements. The fastest hard drives rotate at 10,000 RPM, or about 167 revolutions per second. If the data on a hard disk track just missed the disk's read/write head, then we'll have to wait about another 60 ms, for the data to come around. Also, it takes about 1 ms for the head to go from track to track. Therefore, if we do have to go out to virtual memory, then we should read back enough data to make it worth the wait.

If you have ever *defragmented* your hard disk, then you know the speed-up in performance that you can realize. As you constantly add and delete material from your hard disk fewer and fewer contiguous regions of the hard drive (sectors) remain available for storing programs and data files. This means that if you have to retrieve data located in 4 sectors of the hard drive (approximately 2000 bytes) the data might actually located in 4 sectors that are spread all over the drive. Let's do a simple calculation and see what happens.

Suppose that the time it takes the disk's read/write heads to move from one track to the next track is 5 milliseconds. Also, let's assume that in a heavily fragmented disk, measurements have shown that, on average, sectors are located about 10 tracks away from each other. This is an average value, sometimes we'll be on adjacent cylinders, sometimes we won't. Recall that a cylinder is just the track that we happen to be on, but considering that the disk has multiple platters and multiple heads, we might think of the multiple tracks forming a cylinder.

Finally, we can assume that every time we have to go to a new cylinder, we'll wait, on average, 1/2 of a rotation (rotational latency) of the disk before the data comes under the heads.

Therefore, for this case we'll need to wait:
1. 5 milliseconds per track X 10 tracks per access + 1/2*(1/167) = 53 milliseconds.
2. 4 access x 53 milliseconds per access = 212 milliseconds.

Now, suppose that the data is located in 4 consecutive sectors on the disk. In other words, you have defragmented your hard drive. It takes the same time to get to the sector of interest, and the same delay for the first sector to appear, but then we can read the sectors in rapid succession. My hard drive has 64 sectors per track. Thus, at 10,000 rpm, it takes (4/64)*(1/167) = 374 microseconds to read the 4 sectors.

Thus, the time to read the data is 50 milliseconds + 3 milliseconds + 374 microseconds = 53.00374 milliseconds to read the data; a savings of over 150 milliseconds. Now locality tells us that this is a wise thing to do because we'll probably need all 2000 bytes worth of data soon anyway.

Pages

In a virtual memory system we call the blocks of memory that can map as a unit between physical memory and secondary storage (disk, tape, FLASH card) *pages*. The virtual memory system loads into physical memory only those parts of the program

- that can be held by physical memory;
- that have been allocated by the operating system; *and/or*
- that are currently in use.

Once again, we see paging in use. However, while paging is an interesting property of binary memory addresses in physical memory, *it is most important in virtual memory* systems because there is a natural mapping between sectors on a hard disk and pages in a virtual memory. We can divide our logical address into two parts

1. *virtual page number* (or page number)
2. *word offset* (or byte offset) within the page

Recall the paging example for physical memory. If we assume a 16-bit address $F79D, then we can represent it as

- page # = F7,
- byte offset = 9D

Thus, in this example we have 256 pages with 256 bytes per page. Different operating systems utilize different size pages. When the data is not in memory, the exception handler causes a *page fault* to occur. The operating system (O/S) must take over to retrieve the page from the disk. As we've discussed, a page fault is huge miss penalty. Therefore, we want to minimize its impact by making our pages fairly large (for example, 1 to 8 KB).

The operating system (O/S) manages the page to memory translation through the MMU. The strategy is to reduce page faults as its highest priority. A page fault and subsequent reloading of main memory from the disk can take hundreds of thousands of clock cycles. Thus, LRU algorithms are worth the price of their complexity. With the O/S taking over, the page faults can be handled in software instead of hardware. The obvious tradeoff is that it will be much slower than managing it with a hardware algorithm, but much more flexible. Also, virtual memory systems use write-back strategies because the write-through strategy is too expensive. Did you ever wonder why you must shut down your computer by shutting down the O/S? The reason is that the O/S is using a write-back strategy to manage virtual memory. If you just turn off the computer before allowing the O/S to close the open files, you run the risk of losing data stored on the physical page but not yet written back to the virtual page, the hard disk.

Figure 14.13 is a simplified schematic diagram of the paging system. Since we can easily break the full 32-bit address into a page segment and an offset segment, we can then map the page segment into physical memory or virtual memory, the hard disk. Since the hard drive stores data according to a head, track and sector model, we need the services provided by the O/S to map the hard drive to virtual memory. We call this type of system a *demand-paged virtual memory* because the O/S loads the pages into physical memory on demand.

Since a page fault is such a big penalty, we may transfer and store more than one page at a time. We use the term *page frames* to refer to blocks of physical memory that are designed to hold the pages transferred from the hard disk. These may be anywhere from 1,024 to 8,192 bytes in size. Finally, just as the cache

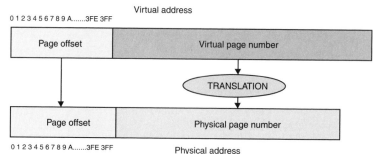

Figure 14.13: Schematic representation of a paging system. The full virtual address is separated into a page/offset representation. The page numbers are mapped to physical and virtual memory.

needs a tag memory to store the mapping information between the cache and main memory, the O/S maintains a *page map* to store the mapping information between virtual memory and physical memory.

The page map may contain other information that is of interest to the operating system and the applications that are currently running. One portion of the page map, the *page table*, is the portion of the page map that is owned by the operating system and keeps track of the pages that are currently in use and that are mapped to physical memory and virtual memory.

Some of the page table entries can be summarized as follows:

- *virtual page number*: the offset into the page table
- *validity bit*: whether or not the page is currently in memory
- *dirty bit*: whether or not the program has modified (written to) the page
- *protection bits*: which user (process, program) may access the page
- *page frame number*: the page frame address if the page is in physical memory

Translation Lookaside Buffer (TLB)

Most computer systems keep their page table in main memory. The processor may contain a special register, the *page-table base register,* which points to the beginning of the table. Only the O/S can modify the page table base register using supervisor mode instructions. In theory, main memory (noncached) accesses could take twice as long because the page table must be accessed first whenever a main memory access occurs.

Therefore, modern processors maintain a *translation look-aside buffer (TLB)* as part of the page map. The TLB is an on-chip cache, usually fully associative, designed for virtual memory management. The TLB holds the same information as part of the page table and maps virtual page numbers into page frame numbers. The TLB cache algorithm holds only most recently accessed pages and flushes the least recently used (LRU) entries from TLB. The TLB holds only the mapping of valid pages (not dirty).

Figure 14.14 shows the components of a paging system implemented as a combination of on-chip hardware, the TLB and a page table in main memory that is pointed to by the page table base register. The CPU generates the virtual address, which is first tested for a match in on-chip cache. If there is a cache miss, the pipeline for that instruction stalls (we're assuming that this is a superscalar processor and other instruction pipes are still

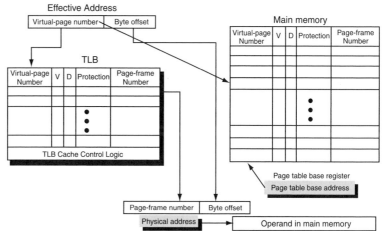

Figure 14.14: Components of a paging system.
From Baron and Higbie[3].

processing instructions) and the cache control logic signals the TLB to see if the page is in physical memory. If it's in the TLB then the refill line from the physical memory is retrieved by the bus logic and placed in the cache, processing can then proceed.

If the page is not listed in the TLB, then the processor uses the page table base address register to establish a pointer and begins searching the page table for the page to see if its in physical memory. If it is in

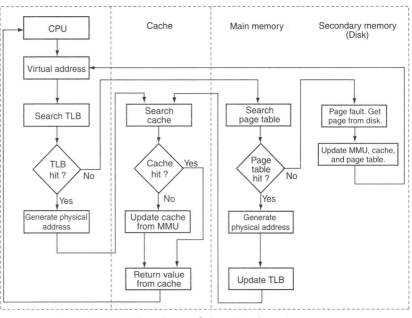

Figure 14.15: Flowchart of the virtual paging process.
From Heuring and Jordan[4].

physical memory, the TLB is updated and the refill line is retrieved and placed in the cache. If the page is in virtual memory, then a page fault must be generated and the O/S intervenes to retrieve the page from the disk and place it in the page frame. The virtual paging process is shown as a flow chart in Figure 14.15.

Protection

Since the efficient management of the virtual memory depends upon having specialized hardware to support the paging processes, we need to look at another important function that this hardware must perform—protecting various memory regions from inadvertently, or maliciously, accessing each other.

In the 68K we've already seen a simple protection scheme, using *supervisor and user modes*. When the processor is in user mode only a subset of the instruction set is available. If a program attempts to execute a supervisor mode instruction, an exception is generated. Presumably, the exception vector would be an entry point into the O/S and the O/S would then deal with the illegal instruction.

Also, we've seen that the 68K actually supports two separate stack pointers, a user stack pointer and a supervisor stack pointer. However, the 68K has no mechanism for protecting one region of memory from illegal accesses by other regions of memory. To do this we need the protection hardware of the MMU. MMU's use to be separate and distinct chips from the CPU, but as we have gone to higher level's of integration, the MMU circuitry has been incorporated into the processor and most modern high-performance processors have on-chip MMUs.

Therefore, we can summarize some of the hardware-based protection functionality as follows.

- user mode and kernel (supervisor) mode
- controlled user read-only access to user/kernel mode bit and a mechanism to switch from user to kernel mode, typically via an O/S system call
- TLB
- page table base address pointer
- MMU

Finally, let's look at two modern processors and compare them for their MMU, TLB and cache support. This is shown in the table below[5]. We're not going to discuss this table at all because there is one very, very key piece of information that you should take away from studying the information presented there. This table is a data summary of two of today's most modern, and powerful, commercially available microprocessors. There is no information in those two comparison charts that should now be foreign to you.

Characteristic	Intel Pentium Pro	Freescale PowerPC 604
Virtual address range	32 bits	52 bits
Physical address range	32 bits	32 bits
Page size	4 KB, 4 MB	4 KB, selectable, 256 MB
TLB organization	Separate TLB's for instruction and for data. Both are 4-way set associative using pseudo-LRU replacement strategy. Instruction TLB: 32 entries, Data TLB: 64 entries. TLB misses are handled in hardware	Separate TLB's for instruction and for data. Both are 2-way set associative using LRU replacement strategy Instruction TLB: 128 entries, Data TLB: 128 entries. TLB misses are handled in hardware
Cache organization	Split instruction and data caches	Split instruction and data caches
Cache size	8 KB each	16 KB each
Cache associativity	4-way set associative	4-way set associative
Replacement strategy	Approximate LRU replacement	LRU replacement
Block size	32 bytes	32 bytes
Write policy	Write-back	Write-back or write-through

Summary of Chapter 14

Chapter 14 covered:

- The concept of a memory hierarchy,
- The concept of locality and how locality justifies the use of caches,
- The major types of caches, direct mapped, associative, set-associative and sector mapped
- Measurements of cache performance,
- The architecture of virtual memory,
- The various components of virtual memory and how the operating system manages the virtual memory resources.

Chapter 14: *Endnotes*

[1] Robert J. Baron and Lee Higbie, *Computer Architecture,* ISBN 0-2015-0923-7, Addison-Wesley, Reading, MA, 1992, p. 201.

[2] A. Agarwal, *Analysis of Cache Performance for Operating Systems and Multiprogramming,* PhD Thesis, Stanford University, Tech. Report No. CSL-TR-87-332 (May 1987).

[3] Robert J. Baron and Lee Higbie, *op cit,* p. 207.

[4] Vincent P. Heuring and Harry F. Jordan, *Computer Systems Design and Architecture*, ISBN 0-8053-4330-X, Addison-Wesley Longman, Menlo Park, CA, 1997, p. 367.

[5] Patterson and Hennessy, *op cit,* p. 613.

[6] Daniel Mann, *op cit.*

Exercises for Chapter 14

1. a. What does the term "Memory Hierarchy mean"? Illustrate it and use some typical values to make your point.

   b. What is spatial and temporal "locality"? What is the significance of locality with respect to on-chip caches.

   c. Why does an on-chip cache get refilled with a "refill line" instead of just fetching just the data or instruction of interest directly from memory?

   d. Contrast the operation of a "write-back cache" architecture with the architecture of a "write-through" cache.

2. On-chip caches are very small compared to the size of main memory, yet cache hit rates are typically better than 90% for most programs. Explain why this is true.

3. Assume that the C++ program shown below has been compiled to run on a 68000-based computer. The instruction code resides in the memory range of $00001000 to $00001020. The variable, DataStream[10] is located in the heap memory and occupied the address range of $00008000 to $00008028. All other variables are stored on the stack frame starting at address $FFFFFF00. Explain as completely as you can how the following program would exhibit the principles of temporal and spacial locality.

```
int main(void)
{
        int count ;
        const int maxcount = 10 ;
        int DataStream[10] ;
        for ( count = 0 ; count <= maxcount ; count++ )
          *(DataStream + count) = count*count ;
        return 0;
}
```

4. Consider the following C++ declaration and diagram. The numbers represent the byte addresses in memory of the various array elements.

   Briefly discuss whether or not this declaration is an example of the principle of locality.

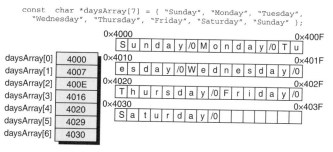

```
const char *daysArray[7] = { "Sunday", "Monday", "Tuesday",
  "Wednesday", "Thursday", "Friday", "Saturday", "Sunday" };
```

5. Assume that you have a microprocessor with a 20-bit byte addressing range and a 16-bit wide data bus. The processor has an on-chip cache with the following properties:
   - The cache is direct-mapped.
   - The cache is 4096 bytes in size, excluding the Tag RAM.
   - One refill line is 64 bytes long.
   a. How many refill lines are there in the main memory?
   b. How many refill lines are there in the cache memory?
   c. How many rows and columns are there in the organization of the main memory in terms of the organization of the direct-mapped cache?
   d. How many addressing bits are required in the TAG memory associated with each cache memory in order to address all the refill lines across one row of main memory? You may neglect any other bits, such as a dirty-write bit.
   e. Draw a sketch of this cache organization. Show the address components of the memory organization as completely as you can.

6. Suppose that you have a microprocessor with 1 megabyte addressing range and a 4 kilobyte direct mapped cache. The block size for a refill line in the cache is 64 bytes.
   a. How many refill lines are in the cache?
   b. How should main memory be organized (rows and columns) in order to be compatible with this cache design?
   c. How many bits are required for the tag memory?
   d. Suppose that you are trying to fetch an instruction from address $3FB0A. What are the row and column addresses of the refill line that contains this memory location?

7. Assume that the cache hit rate for the on-chip cache for a certain RISC processor is 98% and instructions that are located in-cache execute in 1 clock cycle. Instructions that are not found in the on-chip cache will cause the processor to stop program execution and refill a portion of the cache. This operation takes 100 clock cycles to execute. What is the effective execution time in nanoseconds for this RISC processor if the clock frequency is 100 MHz?

8. Assume that the cache hit rate for the on-chip cache for a certain 32-bit RISC processor is 98% and instructions that are located in-cache execute in 1 clock cycle. Instructions that are not found in the on-chip cache will cause the processor to stop program execution and do a burst memory read access of a single, 64-byte, refill line of the main memory and then write it in the cache. All external memory fetches require 2 clock cycles per memory fetch. What is the effective execution time in nanoseconds for an instruction on this processor if the clock frequency is 100 MHz?

9. Consider the components of a virtual memory system. In particular, consider the structure of the TLB. Why is it necessary to have a validity bit for each entry of the TLB since it is caching the page table from memory and would routinely get updated when a virtual page operation occurs?

CHAPTER 15

Performance Issues in Computer Architecture

Objectives

When you are finished with this lesson, you will be able to describe:
  ▶ *How various aspects of the hardware can impact the performance of a computer system;*
  ▶ *How hardware-software partitioning decisions affect performance;*
  ▶ *Methods use to incrementally improve the performance of computer systems;*
  ▶ *The use and misuse of benchmarks in measuring computer performance;*
  ▶ *Methods for measuring performance;*
  ▶ *How to achieve specified performance goals for a project.*

Introduction

One chapter in a textbook is a woefully inadequate amount of space to cover the topic of computer performance. However, that's never stopped me in the past, so let's consider the problem and its scope before we dive in. When we talk about performance, I'm sure that many of you assume that we are referring to how well the latest and greatest PC performs on a series of side-by-side tests, called *benchmarks*. The benchmark test might take the form of measuring the amount of time needed to recalculate a large spreadsheet, or the number of frames per second a particular graphics-intensive video game might play at. These two examples of performance benchmarks may or may not be valid to you if your criterion for selecting a PC is not spreadsheet manipulations or video games.

If you are tasked with designing software, as most of you who are reading this book right now will be doing some day, then you are faced with a different aspect of performance. Your issue becomes one of, "How do I guarantee that the software I write meets the performance requirements as per the design specification?" Thus, your issue is not one of choosing the hottest CPU *du jour*, it is trying to make sure that your software meets its performance goals as designed, rather than meeting its performance goals by costly tweaking of the code or re-designs.

In this chapter we'll look at performance from three points of view:

1. How does the hardware directly impact performance?
2. How can you measure performance?
3. How do you design with performance in mind?

Since this text is about hardware, we'll tackle item #1 first. Next, we'll look at methods used to measure the performance of a computer system. Finally, we'll look at performance from the perspective of you, the software developer.

Hardware and Performance

It may seem just slightly improbable that the same microprocessor that drives your DVD player at home could also run some of the hottest video games on the market today. Why choose video games? Well, we've come to associate video games with high-performance computers and video gaming magazines are always comparing this computer to that, or this video card to that one. Of course, just because I said we could run it, it doesn't mean that you would want to play the video game with it. Why? The action would be unacceptably slow. Compared to your 2 GHz Athlon or Pentium, the frame rate would likely be significantly less than 1 frame per second, so you could easily get a latte while the screen updates.

What are the factors that make the video game unacceptably slow in one situation, versus allowing you to play it in real time in another? In order to answer this question, let's actually go through the exercise of trying to port a typical pc-based video game to an 8-bit computer.

For the sake of our experiment, we'll assume that we have an 8-bit processor with a 10 MHz clock. The processor is capable of directly addressing 1 MB of memory and the processor is part of a single-board computer with 32 MB of RAM and a video chip that can drive a standard PC monitor (VGA) to 1,024 by 768 pixel resolution at 8-bit color depth. The video memory is capable of displaying one frame at a time and is memory mapped into a 3 MB block of memory. Special hardware allows the memory to be *dual-ported*, so the video chip can read the memory in order to display the current frame while the processor can update the video frame by writing to the memory block and also read the data at a memory address. External hardware is I/O mapped as a base memory pointer, so writing to this hardware register allows the processor to directly read and write to memory on any 1 MB boundary. Finally, the video chip only supports the process of mapping the video memory block to the screen. There is no graphics processing done in the video chip.

The first problem we face is that it is highly unlikely that the 8-bit microprocessor has the same instruction set architecture as the PC (unless we choose a 4.77 MHz Intel 8088 processor), so we'll need to call the video game manufacturer and talk them into giving us the C++ source code to the game. Of course, they'll immediately agree to our request, so that problem is taken care of. Next, we'll need to rewrite any assembly language routines that were written for the game and re-write them in our ISA.

All of the operating system calls and graphics routines will need to be re-written and recompiled as well. Since we only have a 1 MB memory model, we'll need to take the assembly language output of the compiler and the linker map and insert assembly language instructions in the code to form a bridge over the 1 MB boundaries. Each time that the code must traverse a boundary, we insert the I/O call to remap the base address pointer.

Finally, we'll need to write the graphics engine emulation drivers so that the video displays properly on screen; and assuming that we accomplished all of these tasks, and by now you're wondering what moron thought up this coding problem, we can load the game onto the board and let it rip. Actually, it doesn't rip, it is painfully slow. So slow, that we can't play it. But the point is, it runs. The code does what it is designed to do and the program doesn't crash, but the performance is abysmal.

Let's compare the performance specifications of our PC to our 8-bit, single-board computer. The table below lists the major differences between the two systems. We can see that the deck is really stacked against the 8-bit SBC. For starters, the PC clock is 200 times faster than the SBC clock. Now, we need to be a little cautious here because the PC clock rate is the internal clock rate of the CPU. The external memory typically clocks at 200 MHz for PC3200 DDR memory.

Parameter	PC	SBC	Comment
Clock speed	2,000 MHz	10 MHz	
Data bus width	32-bits	8-bits	
Addressable memory	4,096 MB	32 MB	
Clocks per instruction	<1	8-20	CPU on PC has multiple pipelines
Floating point calculations	In hardware	None	FPU on-chip
Cache	On-chip I cache and D-cache	None	
External memory	256 MB	32 MB	Not a factor
Graphics acceleration	On video card	None	

Also, external PC memory will be SDRAM, so there is a penalty of several clock cycles each time it does a burst memory load. However, the on-chip caches keep the external memory loads to a minimum. We can't say exactly what the cache hit ration is, but 90% is probably a fair guess.

Can we thus make the assumption that a faster clock equates to better performance? In this case it certainly is a significant factor, but can we make the general statement that faster clock speeds equate to better performance? An Intel Pentium 4 processor with a 3.2 GHz clock speed and an AMD 3000+ Athlon processor have similar performance results on comparative benchmarks, but the Athlon's actual clock frequency is approximately 2.2 GHz.

This created a perception problem for AMD. For most PC purchasers, the clock frequency equates to performance[1]. In order to remain competitive, AMD was forced to use a virtual clock frequency rating of 3000+. It is virtual because it means that the AMD part performs a bit better (the +) than the comparable Pentium processor running at a true clock frequency of 3,000 MHz.

Knowing what we know about pipelined processors we might surmise that AMD's Athlon architecture allows the processor to do "more" at each pipeline stage than the Intel Pentium architecture. However, because it does more, it has to run more slowly. Thus, there is a difference in the clock frequencies between the two parts.

With the exception of the desktop computing environment, most computer systems try to run as slowly as possible, not as fast as possible. For processors based upon the CMOS technology, speed equates to power dissipation. Figure 2.15 showed the relatively linear increase in power dissipation with clock frequency CMOS processors. With increased clock frequencies comes a corresponding rise in system costs. A faster processor requires faster support chips and faster memories. As speed goes up, so does the power requirements. Faster bus speeds also mean more potential for *radio frequency interference (RFI)* and more work to incorporate *electromagnetic shielding* into the product design to suppress RFI emissions. If you've ever wondered why all electronic equipment in the passenger cabin on commercial airliners must be turned off below an altitude of 10,000 feet, this is the reason. The airlines (and the FAA) are concerned about potential interference with their navigational equipment during takeoffs and landings so they require all electronic devices (which means anything with a clock oscillator) to be turned off.

Since most embedded systems are highly cost-sensitive, performance takes on an almost inverse relationship to CPU clock speed. The question that an embedded design team will ask is, "How slow can we run the processor in order to still get the job done?" For them, performance is a measure of what is adequate, not what is the fastest.

Even Intel, the leader of the "clock frequency equals performance" group, is now looking at other ways to address the future performance issues of their processors. Rattner[2] describes a "right-hand turn" at Intel. The company's Centrino® processor uses the Banias CPU technology which focused on limiting power consumption at every step of the process, rather than increasing the speed of the clock. Rather than continually scaling- up the clock speed, Intel's future processors will feature multiple CPU cores operating in parallel.

The next place we lose performance is due to the data bus width. Our 8-bit wide data bus can only deal with integers in the range of −128 to 127. Any numbers either more negative or more positive will require multiple memory accesses. Thus, the wider the data bus, the more efficient the system is at accessing memory and the more efficient the CPU is internally at moving the data values around and operating on them. For most processing algorithms, excluding ASCII string manipulation, doubling the width of the data bus can improve performance anywhere between two times and four times.

Having a wider memory bus means that more data can be moved between the CPU and memory at each access. Even though the Pentium and Athlon processors are 32-bit machines, their external memory bus is width is 64-bits, enabling it to achieve burst data rates of 3.2 GB per second. How did we get that? Using PC3200 SDRAM memory, we are able to transfer data from memory on each phase of the 200 MHz memory clock, or a 400 MHz data rate. Since we are sending or receiving 8 bytes in parallel during every memory cycle; which equates to 8 x 400 MHz, or 3200 MB/sec. Also, having the ability to directly address 4 GB of memory means that memory accesses won't require the overhead of paging instructions every time a 1 MB memory boundary is crossed.

Since our 8-bit processor has no on-chip caches, external memory is the only memory, so all memory accesses take place according to the memory fetch model of the processor. We can assume that the CPU requires a minimum of 4 clock cycles for each memory access. We'll also assume, for simplicity, that our 32 MB of memory is 0 wait state memory, so all memory accesses require just 4 clocks to complete a read or a write. Thus, reading data from memory takes place at 2.5 MB per second, a factor of 1280 times slower than the PC. But wait, things keep getting worse. Since the CPU on the PC has on-chip I-cache and D-caches, access to operands and instructions take place at the full internal clock speed of the processor, or 2 GHz.

Next we have an interesting difference to consider. The PC's CPU contains an on-chip floating point math accelerator and dedicated floating point registers. The 8-bitter must emulate all floating point instructions using algorithms that perform integer arithmetic operations on numbers represented in floating point. As we'll soon see in the next chapter, dedicated hardware can greatly accelerate an algorithm over general-purpose programming methods.

Finally, the off-chip support hardware, in the form of the graphics accelerator on the PC's video card allows it to off-load the graphics-intensive operations from the CPU. The 8-bit processor must calculate all of the graphical transformations on its own.

It should be clear from this discussion that the effect of the hardware architecture on performance cannot be minimized. A back-of-the envelope calculation could yield a video frame rate of

75 frames per second for the PC and less than 1 frame per hour for the 8-bit processor; and even though this is a rather ludicrous example, it does factor in the significant issues relating hardware and performance. Let's try to summarize them:

- Faster clock rates generally mean higher performance, but it isn't universally the case,
- Wider bus widths mean more data can be transferred in each operation and larger numbers can be manipulated in a single instruction,
- Larger addressable memory means more efficient memory accesses,
- Multiple pipelines means that CPUs can often retire (complete the execution of) more than one instruction per clock cycle,
- On-chip caches provide instructions and data at the speed of the clock,
- Dedicated on-chip hardware, such as floating point units greatly accelerate the execution of instructions designed to manipulate numbers represented in floating point format.
- Dedicated off-chip hardware accelerators, such as graphical processors, sound chips, Ethernet chips, etc. can off-load the processing of I/O data from the CPU.

Do any of these bullet points come to you as a great awakening? Well, probably not, but you can't be sure. At least we went through the process of enumerating them. However, buried in the above discussion there's an interesting concept that we need to consider in a bit more depth. In two instances, we considered how specialized hardware could be used to accelerate floating point and image manipulation calculations. The use of hardware to accelerate an algorithm is a fairly common way to realize increased performance. Figure 15.1 illustrates this principle.

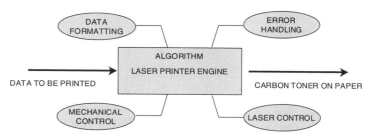

Figure 15.1: A generalize model of an algorithm to print an image on a laser printer. From Berger[3].

In this algorithm, data, in the form of image information, is received by the laser engine and this information is used to control the exposure of a photosensitive drum to a laser beam. Wherever the laser beam has written on the drum, fine particles of carbon black can be made to adhere to the photosensitized regions. This image can then be transferred and fused to ordinary paper to give us an output from the printer.

In describing this algorithm we did make some mention of the hardware involved, but the imagining process itself was intentionally left vague. How are the bits of data transformed into a modulated beam of laser light? We could do it in software, or we can imagine that some kind of a dedicated hardware engine is doing the transformation for us. In fact, both methods are widely used today and different manufacturers can achieve competitive results using different performance strategies. Company A might focus on fine tuning their software algorithm while company B might relax the software processing requirements and use a less-powerful processor, but accelerate the transformation process (called *banding*) by using a dedicated hardware accelerator.

Today, there's a third alternative. With so much processing power available on the PC, many printer manufacturers are significantly reducing the price of their laser printers by equipping the printer with the minimal intelligence necessary to operate the printer. All of the processing requirements have been placed back onto the PC in the printer drivers.

We call this phenomenon the *duality* of software and hardware since either, or both, can be used to solve an algorithm. It is up to the system architects and designers to decide upon the partitioning of the algorithm between software (slow, low-cost and flexible) and hardware

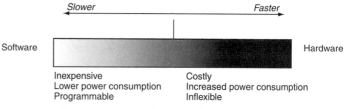

Figure 15.2: Hardware/software trade-off.

(fast, costly and rigidly defined). This duality is not black or white. It represents a spectrum of trade-offs and design decisions. Figure 15.2 illustrates this continuum from dedicated hardware acceleration to software only.

Thus, we can look at performance in a slightly different light. We can also ask, "What are the architectural trade-offs that must be made to achieve the desired performance objectives?

With the emergence of hardware description languages we can now develop hardware with the same methodological focus on the algorithm that we apply to software. We can use object oriented design methodology and UML-based tools to generate C++ or an HDL source file as the output of the design. With this amount of fine-tuning available to the hardware component of the design process, performance improvements can become incrementally achievable as the algorithm is smoothly partitioned between the software component and the hardware component.

Overclocking

A very interesting subculture has developed around the idea of improving performance by *overclocking* the processor, or memory, or both. Overclocking means that you deliberately run the clock at a higher speed then it is supposedly designed to run at. Modern PC motherboards are amazingly flexible in allowing a knowledgeable, or not-so-knowledgeable, user to tweak such things as clock frequency, bus frequency, CPU core voltage and I/O voltage.

Search the Web and you'll find many websites dedicated to this interesting bit of technology. Many of the students whom I teach have asked me about it each year, so I thought that this chapter would be an appropriate point to address it. Since overclocking is, by definition, violating the manufacturer's specifications, CPU manufacturers go out of their way to thwart the zealots, although the results are often mixed.

Modern CPUs generally *phase lock* the internal clock frequency to the external bus frequency. A circuit, called a *phase-locked loop (PLL)*, generates an internal clock frequency that is a multiple of the external clock frequency. If the external clock frequency is 200 MHz (PC3200 memory) and the multiplier is 11, the internal clock frequency would be 2.2 GHz. The PLL circuit then divides the internal clock frequency by 11 and uses the divided frequency to compare itself with the external frequency. The local frequency difference is used to speed-up or slow down the internal clock frequency.

You can overclock your processor by either:

1. Changing the internal multiplier of the CPU, or
2. Raising the external reference clock frequency.

CPU manufacturers deal with this issue by hard-wiring the multiplier to a fixed value, although enterprising hobbyists have figured out how to break this code. Changing the external clock frequency is relatively easy to do if the motherboard supports the feature, and may aftermarket motherboard manufacturers have added features to cater to the overclocking community. In general, when you change the external clock frequency you also change the frequency of the memory clock.

OK, so what's the down side? Well, the easy answer is that the CPU is not designed to run faster than it is specified to run at, so you are violating specifications when you run it faster than it is designed to run. Let's look at this a little deeper. An integrated circuit is designed to meet all of its performance parameters over a specified range of temperature. For example the Athlon processor from AMD is specified to meet its parametric specifications for temperatures less than 90 degrees Celsius. Generally, every timing parameter is specified with three parameters, minimum, typical and maximum (worst case) over the operating temperature range of the chip. Thus, if you took a large number of chips and placed them on an expensive parametric testing machine, you would discover a bell-shaped curve for most of the timing parameters of the chip. The peak of the curve would be centered about the typical values and the maximum and minimum ranges define either side of typical. Finally, the colder that you can maintain a chip, the faster it will go. Device physics tells us that electronic transport properties in integrated circuits get slower as the chip gets hotter.

If you were to look closely at an IC wafer fully of just-processed Athlons or Pentiums, you would also see a few different looking chips evenly distributed over the surface of the wafer. These chips are the chips that are actually used to characterize the parameters of each wafer manufacturing batch. Thus, if the manufacturing process happens to go really well, you get a batch of faster than typical CPUs. If the process is marginally acceptable, you might get a batch of slower than typical chips.

Suppose that, as a manufacturer, you have really fine-tuned the manufacturing process to the point that all of your chips are much better than average. What do you do? If you've ever purchased a personal computer, or built one from parts, you know that faster computers cost more because the CPU manufacturer charges more for the faster part. Thus, an Athlon XP processor that is rated at 3200+ is faster than an Athlon XP rated at 2800+ and should cost more. But suppose that all you have been producing are the really fast ones. Since you still need to offer a spectrum of parts at different price points, you mark the faster chips as slower ones.

Therefore, overclockers may use the following strategies:

1. Speed up the processor because it is likely to be either conservatively rated by the manu-facturer or is intentionally rated below its actual performance capabilities for marketing and sales reasons,
2. Speed up the processor and also increase the cooling capability of your system to keep the chip as cool as possible and to allow for the additional heat generated by a higher clock frequency.
3. Raise either or both the CPU core voltage and the I/O voltage to decrease the rise and fall times of the logic signals. This has the effect of raising the heat generated by the chip.

4. Keep raising the clock frequency until the computer becomes unstable, then back off a notch or two,

5. Raise the clock frequency, core voltage, I/O voltage until the chip self-destructs.

The dangers of overclocking should now be obvious:

1. A chip that runs hotter is more likely to fail,
2. Depending upon typical specs does not guarantee performance over all temperatures and parametric conditions,
3. Defeating the manufacturers thresholds will void your warranty,
4. Your computer may be marginally stable and have a higher sensitivity to failures and glitches.

That said should you overclock your computer to increase performance? Here's a guideline to help you answer that question:

If your PC is hobby activity, such as game box, then by all means, experiment with it. However, if you depend upon your PC to do real work, then don't tempt fate by overclocking it. If you really want to improve your PC's performance, add some more memory.

Measuring Performance

In the world of the personal computer and the workstation, performance measurements are generally left to others. For example, most people are familiar with the *SPEC* series of software benchmark suites. The SPECint and SPECfp benchmarks measured integer and floating point performance, respectively. SPEC is an acronym for the Standard Performance Evaluation Corporation, a nonprofit consortium of computer manufacturers, system integrators, universities and other research organizations. Their objective is to set, maintain and publish a set of relevant benchmarks and benchmark results for computer systems[4].

In response to the question, "Why use a benchmark?" The SPEC Frequently Asked Question page notes,

Ideally, the best comparison test for systems would be your own application with your own workload. Unfortunately, it is often very difficult to get a wide base of reliable, repeatable and comparable measurements for comparisons of different systems on your own application with your own workload. This might be due to time, money, confidentiality, or other constraints.

The key here is that best benchmark test is your actual computing environment. However, few people who are about to purchase a PC have the time or the inclination to load all of their software on several machines and spend a few days with each machine, running their own software applications in order to get a sense of relative strengths of each system. Therefore, we tend to let others, usually the computer's manufacturer, or a third-party reviewer, do the benchmarking for us. Even then, it is almost impossible to be able to compare several machines on an absolutely even playing field. Potential differences might include:

- Differences in the amount of memory in each machine,
- Differences in memory type in each machine, (PC2700 versus PC3200)

- Different CPU clock rates,
- Different revisions of hardware drivers,
- Differences in the video cards,
- Differences in the hard disk drives (serial ATA or parallel ATA, SCSI or RAID)

In general, we will put more credence in benchmarks that are similar to the applications that we are using, or intend to use. Thus, if you are interested in purchasing high-performance workstations for an animation studio you likely choose from the graphics suite of tests offered by SPEC.

In the embedded world, performance measurements and benchmarks are much more difficult to acquire and make sense of. The basic reason is that embedded systems are not standard platforms the way workstations and PCs are standard. Almost every embedded system is unique in terms of the CPU, clock speed, memory, support chips, programming language used, compiler used and operating system used.

Since most embedded systems are extremely cost sensitive, there is usually little or no margin available to design the system with more theoretical performance then it actually needs "just to be on the safe side". Also, embedded systems are typically used in real time control applications, rather than computational applications. Performance of the system is heavily impacted by the nature and frequency of the real time events that must be serviced within a well-defined window of time or the entire system could exhibit catastrophic failure.

Imagine that you are designing the flight control system for a new fly-by-wire jet fighter plane. The pilot does not control the plane in the classical sense. The pilot, through the control stick and rudder pedals, sends requests to the flight control computer (or computers) and the computer adjusts the wings and tail surfaces in response to the requests. What makes the plane so highly maneuverable in flight also makes it difficult to fly. Without the constant control changes to the flight surfaces, the aircraft will spin out of control. Thus, the computer must constantly monitor the state of the aircraft and the flight control surfaces and make constant adjustments to keep the fighter flying.

Unless the computer can read all of its input sensors and make all of the required corrections in the appropriate time window, the aircraft will not be stable in flight. We call this condition *time critical*. In other words, unless the system can respond within the allotted time, the system will fail.

Now, let's change employers. This time you are designing some of the software for a color photo printer. The Marketing Department has written a requirements document specifying a 4 page-per-minute output delivery rate. The first prototypes actually deliver 3.5 pages per minute. The printer keeps working, no one is injured, but it still fails to meet its design specifications. This is an example of a *time sensitive* application. The system works, but not as desired. Most embedded applications with real-time performance requirements fall into one or the other of these two categories.

The question still remains to be answered, "What benchmarks are relevant for embedded systems?" We could use the SPEC benchmark suites, but are they relevant to the application domain that we are concerned with. In other words, "How significant would a benchmark that does a prime number calculation be in comparing the potential use of one of three embedded processors in a furnace control system?"

For a very long time there were no benchmarks suitable for use by the embedded systems community. The available benchmarks were more marketing and sales devices then they were usable technical evaluation tools. The most notorious among them was the MIPS benchmark. The MIPS benchmark means *millions of instructions per second*. However, it came to mean,

Meaningless Indicator of Performance for Salesmen.

The MIPs benchmark is actually a relative measurement comparing the performance of your CPU to a VAX 11/780 computer. The 11/780 is a 1 MIPS machine that can execute 1757 loops of the Dhrystone[5] benchmark in 1 second. Thus, if your computer executes 2400 loops of the benchmark, it is a 2400/1757 = 1.36 MIPS machine. The Dhrystone benchmark is a small C, Pascal or Java program which compiles to approximately 2000 lines of assembly code. It is designed to test the integer performance of the processor and does not use any operating system services.

There is nothing inherently wrong with the Dhrystone benchmark, except that people started using it to make technical decisions which created economic impacts. For example, if we choose processor A over processor B because its better Dhrystone benchmark results, that could result in the customer using many thousands of A-type processors in their new design. How could you make your processor look really good in a Dhrystone benchmark? Since the benchmark is written in a high-level language, a compiler manufacturer could create specific optimizations for the Dhrystone benchmark. Of course, compiler vendors would never do something like that, but everyone constantly accused each other of similar shortcuts. According to Mann and Cobb[6],

> *Unfortunately, all too frequently benchmark programs used for processor evaluation are relatively small and can have high instruction cache hit ratios. Programs such as Dhrystone have this characteristic. They also do not exhibit the large data movement activities typical of many real applications.*

Mann and Cobb cite the following example,

> *Suppose you run Dhrystone on a processor and find that the µP (microprocessor) executes some number of iterations in P cycles with a cache hit ratio of nearly 100%. Now, suppose you lift a code sequence of similar length from your application firmware and run this code on the same µP. You would probably expect a similar execution time for this code.*

> *To your dismay, you find that the cache hit rate becomes only 80%. In the target system, each cache miss costs a penalty of 11 processor cycles while the system waits for the cache line to refill from slow memory; 11 cycles for a 50 MHz CPU is only 220 ns. Execution time increases from P cycles for Dhrystone to $(0.8 \times P) + (0.2 \times P \times 11) = 3P$. In other words, dropping the cache hit rate to 80% cuts overall performance to just 33% of the level you expected if you had based your projection purely on the Dhrystone result.*

In order to address the benchmarking needs of the embedded systems industry, a consortium or chip vendors and tool suppliers was formed in 1997 under the leadership of Marcus Levy, who was a Technical Editor at EDN magazine. The group sought to create, *meaningful performance benchmarks for the hardware and software used in embedded systems*[7]. The EDN Embedded Microprocessor Benchmark Consortium (EEMBC, pronounced "Embassy") uses real-world benchmarks from various industry sectors.

The sectors represented are:

- Automotive/Industrial
- Consumer
- Java
- Networking
- Office Automation
- Telecommunications
- 8 and 16-bit microcontrollers

For example, in the Telecommunications group there are five categories of tests; and within each category there are several different tests. The categories are:

- Autocorrelation
- Convolution encoder
- Fixed-point bit allocation
- Fixed-point complex FFT
- Viterbi GSM decoder

If these seem a bit arcane to you, they most certainly are. These are algorithms that are deeply ingrained into the technology of the Telecommunications industry. Let's look at an example result for the EEMBC Autocorrelation benchmark on a 750 MHz Texas Instruments TMS320C4X Digital Signal Processor (DSP) chip. The results are shown in Figure 15.3.

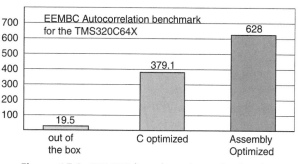

Figure 15.3: EEMBC benchmark results for the Telecommunications group Autocorrelation benchmark[8].

The bar chart shows the benchmark using a C compiler without optimizations turned on; with aggressive optimization; and with hand-crafted assembly language fine-tuning. The results are pretty impressive. There is a almost a 100% improvement in the benchmark results when the already optimized C code is further refined by hand crafting in assembly language. Also, both the optimized and assembly language benchmarks outperformed the nonoptimized version by factors of 19.5 and 32.2, respectively.

Let's put this in perspective. All other things being equal, we would need to increase the clock speed of the out-of-the-box result from 750 MHz to 24 GHz to equal the performance of the hand-tuned assembly language program benchmark.

Even though the EEMBC benchmark is vast improvement there are still factors that can render comparative results rather meaningless. For example, we just saw the effect of the compiler optimization on the benchmark result. Unless comparable compilers and optimizations are applied to the benchmarks, the results could be heavily skewed and erroneously interpreted.

Another problem that is rather unique to embedded systems is the issue of *hot boards*. Manufacturers build *evaluation boards* with their processors on them so that embedded system designers who don't yet have hardware available can execute benchmark code or other evaluation programs on the processor of interest. The evaluation board is often priced above what a hobbyist would be willing to spend, but below what a first-level manager can directly approve. Obviously, as a manufacturer, I want my processor to look its best during a potential design win test with my evaluation board. Therefore, I will maximize the performance characteristics of the evaluation board so that the benchmarks come out looking as good as possible. Such boards are called hot boards and they usually don't represent the performance characteristics of the real hardware. Figure 15.4 is an evaluation board for the AMD AM186EM microcontroller. Not surprising, it was priced at $186.

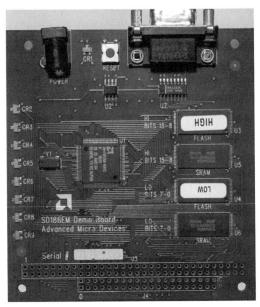

Figure 15.4: Evaluation board for the AM186EM-40 Microcontroller from AMD.

The evaluation board contained the fastest version of the processor then available (40 MHz), and RAM memory that is fast enough to keep up without any additional wait states. All that is necessary to begin to use the board is to add a 5 volt DC power supply and an RS232 cable to the COM port on your PC. The board comes with an on-board monitor program in ROM that initiates a communications session on power-up. All very convenient, but you must be sure that this reflects the actual operating conditions of your target hardware.

Another significant factor to consider is whether or not your application will be running under an operating system. An operating system introduces additional overhead and can decrease performance. Also, if your application is a low-priority task, it may become starved for CPU cycles as higher priority tasks keep interrupting.

Generally, all benchmarks are measured relative to a timeline. Either we measure the amount of time it takes for a benchmark to run, or we measure the number of iterations of the benchmark that can run in a unit of time, day a second or a minute. Sometimes we can easily time events that take enough time to execute that we can use a stopwatch to measure the time between writes to the console. You can easily do this by inserting a *printf() or cout* statement in your code. But what if the event that you're trying to time takes milliseconds or microseconds to execute? If you have operating system services available to you then you could use a high resolution timer to record your entry and exit points. However, every call to an O/S service or to a library routine is a potentially large perturbation on the system that you are trying to measure; a sort of computer science analog of Heisenberg's Uncertainty Principle.

In some instances, evaluation boards may contain I/O ports that you could toggle on and off. With an oscilloscope, or some other high-speed data recorder you could directly time the event or events with minimal perturbation on the system. Figure 15.5 shows a software timing measurement made using an oscilloscope to record the entry and exit points to a function. Referring to the figure, when the function is entered an I/O pin is turned on and then off, creating a short pulse. On exit, the pulse is recreated. The time difference between the two pulses measures the amount of time taken by the function to execute. The two vertical dotted lines are cursors that can be placed on the waveform to determine the timing reference marks. In this case, the time difference between the two cursors is 3.640 milliseconds.

Another method is to use the digital hardware designer's tool of choice, the logic analyzer. Figure 15.6 is photograph of a TLA7151 logic analyzer manufactured by Tektronix, Inc. In the photograph the logic analyzer has a multi-wire connected to the busses of the computer board through a dedicated cable. It is a common practice, and a good idea, for the circuit board designer to provide a dedicated port on the board to enable a logic analyzer to easily be connected to the board. The logic analyzer allows the designer to record the state of many digital bits at the same time. Imagine that you could simultaneously record and timestamp 1 million samples of a digital system that is 80 digital bits wide. You might use 32 bits for the data, 32-bits for the address bus, and the remaining 16-bits for various status signals. Also, the circuitry within the logic analyzer can be programmed to only record a specific pattern of bits. For example, suppose that we programmed the logic analyzer to record only data writes to memory address 0xAABB0000. The logic analyzer would monitor all of the bits, but only record the 32-bits on the data bus whenever the address matches 0xAABB00 AND the status bits indicate a data write is in process. Also, every time that the logic analyzer records a data write event, it time stamps the event and records the time along with the data.

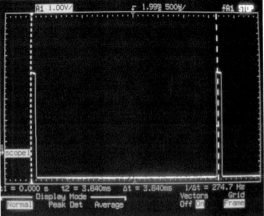

Figure 15.5: Software performance Measurement made using an oscilloscope to measure the time difference between a function entry and exit point.

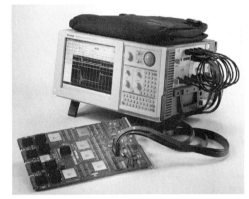

Figure 15.6: Photograph of the Tektronix TLA7151 logic analyzer. The cables from the logic analyzer probe the bus signals of the computer board. Photograph courtesy of Tektronix, Inc.

The last element of this example is for us to insert the appropriate reference elements into our code so that the logic analyzer can detect them and record when they occur. For example, let's say that we'll use the bit pattern 0xAAAAXXXX for the entry point to a function and 0x5555XXXX for the exit point. The 'X's' mean "don't care" and may be any value, however, we would probably want to use them to assign unique identifiers to each of the functions in the program.

Let's look at a typical function in the program. Here's the function:

```
int typFunct( int aVar, int bVar, int cVar)
{

    -----------------      /* Lines of code */
    -----------------
    -----------------
    -----------------

}
```

Now, let's add our measurement "tags." We call this process *instrumenting* the code. Here's the function with the instrumentation added:

```
int typFunct( int aVar, int bVar, int cVar)
{
    *(volatile unsigned int*) 0xAABB0000 = 0xAAAA03E7
    -----------------      /* Lines of code */
    -----------------
    -----------------
    -----------------
    *(volatile unsigned int*) 0xAABB0000 = 0x555503E7
}
```

This rather obscure C statement, *(unsigned int*) 0xAABB0000 = 0xAAAA03E7 creates a pointer to the address 0xAABB0000 and immediately writes the value 0xAAAA03E7 to that memory location. We can assume that 0x03E7 is the code we've assigned to the function, typFunct(). This statement is our tag generator. It creates the data write action that the logic analyzer can then capture and record. The keyword, volatile, tells the compiler that this write should not be cached. The process is shown schematically in Figure 15.7. Let's summarize the data shown in Figure 15.7 in a table.

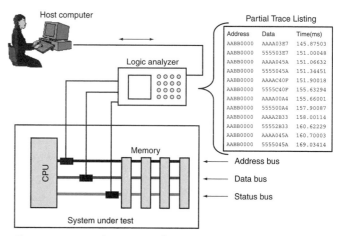

Figure 15.7 Software performance measurement made using a logic analyzer to record the function entry and exit point.

Function	Entry/Exit(msec)	Time difference
03E7	145.87503 / 151.00048	5.12545
045A	151.06632 / 151.34451	0.27819
C40F	151.90018 / 155.63294	3.73276
00A4	155.66001 / 157.90087	2.24086
2B33	158.00114 / 160.62229	2.62115
045A	160.70003 / 169.03414	8.33411

Referring to the table, notice how the function labeled 045A has two different execution times, 0.27819 and 8.33411, respectively. This may seem strange but it actually quite common. For example, a recursive function may have different execution times as well as functions which call math library routines. However, it might also indicate that the function is being interrupted and that the time window for this function may vary dramatically depending upon the current state of the system and I/O activity.

The key here is that the measurement is almost as unobtrusive as you can get. The overhead of a single write to noncached memory should not distort the measurement too severely. Also, notice the logic analyzer is connected to another host computer. Presumably this host computer was the one that was used to do the initial source code instrumentation. Thus, it should have access to the symbol table and link map. Therefore, it could present the results by actually providing the function's names rather than a identifier code.

Thus, if were to run the system under test for a long enough span of time we could continue to gather data like that shown in Figure 15.7 and then do some simple statistical analyses to determine min, max and average execution times for the functions.

What other types of performance data would this type of measurement allow us to obtain? Some measurements are summarized below:

1. Real-time trace: Recording the function entry and exit points provides a history of the execution path taken by the program as it runs in real-time. Rather than single-stepping, or running to a breakpoint, this debugging technique does not stop the execution flow of the program.
2. Coverage testing: This test keeps track of the portions of the program that were executed and portions that were not executed. This is valuable for locating regions of dead code and additional validation tests that should be performed.
3. Memory leaks: Placing tags at every place where memory is dynamically allocated and deallocated can determine if the system has a memory leakage or fragmentation problem.
4. Branch analysis: By instrumenting program branches these tests can determine if there are any paths through the code that are not traceable or have not been thoroughly tested. This test is one of the required tests for any code that is deemed to be *mission critical* and must be certified by a government regulatory agency before it can be deployed in a real product.

While a logic analyzer provides a very low-intrusion testing environment, all computer systems can't be measured in this way. As previously discussed, if an operating system is available, then the tag generation process and recording can be accomplished as another O/S task. Of course, this is obviously more intrusive, but may be a reasonable solution for certain situations.

At this point, you might be tempted to suggest, "Why bother with the tags? If the logic analyzer can record everything happening on the system busses, why not just record everything?" This is a good point and it would work just fine for noncached processors. However, as soon as you have a processor with on-chip caches, bus activity ceases to be a good indicator of processor activity. That's why tags work so well.

While logic analyzers work quite well for these kinds of measurements, they do have a limitation because they must stop collecting data and upload the contents of their trace memory in batches.

This means that low duty cycle events, such as interrupt service routines, may not be captured. There are commercially available products, such as CodeTest® from Metrowerks®[9] that solves this problem by able to continuously collect tags, compress them, and send them to the host without stopping. Figure 15.8 is a picture of the CodeTest system and Figure 15.9 shows the data from a performance measurement.

Figure 15.8: CodeTest software performance analyzer for real-time systems. Courtesy of Metrowerks, Inc.

Designing for Performance

One of the most important reasons that a software student should study computer architecture is to understand the strengths and limitations of the machine and the environment that their software will be running in. Without a reasonable insight into the operational characteristics of the machine, it would be very easy to write inefficient code. Worse yet, it would be very easy to mistake inefficient code for limitations in the hardware platform itself. This could lead to a decision to redesign the hardware in order to increase the system performance to the desired level, even though a simple re-write of some critical functions may be all that is necessary.

Row	Function Name	# of Calls	Minimum(us)	Maximum(us)	Average(us)	Cumulative(us)	% Total Time
0	addMemEntryToList	13,367	17.0	194.0	22.0	293,680.1	2.63%
1	removeMemEntryFromList	13,364	8.4	16.1	10.3	137,142.4	1.23%
2	clearMemListEntry	13,364	4.2	167.7	8.6	114,767.6	1.03%
3	deallocateMemListEntry	14,143	1.1	29.8	13.5	191,355.4	1.71%
4	registerAllocation	13,367	9.8	176.8	15.7	209,507.3	1.87%
5	deregisterAllocation	13,364	9.8	17.5	12.1	162,305.0	1.45%
6	getMemoryPtr	10,982	4.4	6.6	4.7	51,105.5	0.46%
7	getDataType	2,419	4.4	4.6	4.5	10,944.5	0.10%
8	getMemoryAllocator	12,195	2.3	10.3	4.6	58,699.7	0.51%
9	allocator1	1,243	35.3	59.8	45.6	56,674.7	0.51%
10	allocator2	1,291	38.9	210.6	50.7	65,443.7	0.59%
11	deallocator1	1,243	45.1	76.6	54.4	67,647.1	0.61%
12	allocator4	1,291	45.8	76.8	56.1	72,451.7	0.65%
13	allocator3	1,224	159.9	209.1	183.4	224,503.0	2.01%
14	deallocator4	1,213	102.5	380.6	203.8	247,201.7	2.21%
15	deallocator3	1,224	46.8	76.7	57.6	70,446.0	0.63%
16	deallocator6	1,214	42.9	77.7	56.9	69,094.6	0.62%
17	allocator5	1,206	49.5	73.1	60.3	72,674.3	0.65%
18	allocator6	1,213	81.8	5,115.0	2,401.1	2,912,556.6	26.05%
19	deallocator5	1,206	64.1	239.3	79.6	95,944.3	0.86%
20	deallocator6	1,213	101.6	6,211.8	2,915.7	3,538,719.1	31.64%

Figure 15.9: CodeTest screen shot showing a software performance measurement. The data is continuously updated while the target system runs in real-time. Courtesy of Metrowerks, Inc.

Here's a story of an actual incident that illustrates this point:

A long time ago in a career far, far away I was the R&D Director for the CodeTest product line. A major Telecomm manufacturer was considering making a major purchase of CodeTest equipment so we sent a team from the factory to demonstrate the product. The customer was about to go into a redesign of a major Telecomm switching system that they sold because they thought that they had reached the limit of the hardware's performance.

Our team visited their site and we installed a CodeTest unit in their hardware. After running their switch for several hours we all examined the data together. Of the hundreds of functions that we looked at, none of the engineers could identify the one function that was using 15% of the CPU's time. After digging through the source code the engineers discovered a debug routine that was added by a student intern. The intern was debugging a portion of the system as his summer project. In order to trace program flow, he created a high priority function that flashed a light on one of the switch's circuit boards. Being an intern, he never bothered to properly identify this function as a temporary debug function and it somehow it got wrapped into the released product code.

After removing the function and rebuilding the files, the customer gained an additional 15% of performance headroom. They were so thrilled with the results that they thanked us profusely and treated us to a nice dinner. Unfortunately, they no longer needed the CodeTest instrument and we lost the sale. The moral of this story is that no one bothered to really examine the performance characteristics of the system. Everyone assumed that their code ran fine and the system as whole performed optimally.

Stewart[10] notes that the number one mistake made by real-time software developers is not knowing the actual execution time of their code. This is not just an academic issue, even if the software that is being developed is for a PC or workstation, getting the most performance from your system is like any other form of engineering. You should endeavor to make the most efficient use of the resources that you have available to you.

Performance issues are most critical in systems that have limited resources, or have real-time performance constraints. In general, this is the realm of most embedded systems so we'll concentrate our focus in this arena.

Ganssle[11] argues that you should never write an interrupt service routine in C or C++ because the execution time will not be predictable. The only way to approach predictable code execution is by writing in assembly language. Or is it? If you are using a processor with an on chip cache, how do you know what the cache hit ratio will be for your code? The ISR could actually take significantly longer to run than the assembly language cycle count might predict.

Hillary and Berger[12] describe a 4-step process to meet the performance goals of a software design effort:

1. Establish a performance budget,
2. Model the system and allocate the budget,
3. Test system modules,
4. Verify the performance of the final design.

The performance budget can be defined as:

$$\text{Performance budget} = \text{sum(operations require under worst case conditions)}$$
$$= [1/ (\text{data rate})] - \text{Operating system overhead} - \text{headroom}$$

The data rate is simply the rate that data is being generated and will need to be processed. From that, you must subtract the overhead of the operating system and finally, leave some room for the code that will invariably need to add as additional features get added-on.

Modeling the system means decomposing the budget into functional blocks that will be required and allocating time for each block. Most engineers don't have a clue about amount of time required for different functions, so they make "guesstimates". Actually, this isn't so bad because at least they are creating a budget. There are lots of ways to refine these guesses without actually writing the finished code and testing it after the fact. The key is to raise an awareness level of the time available versus time needed.

Once software development begins it makes sense to test the execution time at the module level, rather than wait for the integration phase to see if your software's performance meets the requirements specifications. This will give you instant feedback about how the code is doing against budget. Remember, guesses can go either way, too long or too short, so you might have more time than you think (although Murphy's Law will usually guarantee that this is a very low probability event).

The last step is to verify the final design. This means performing accurate measurements of the system performance using some of the methods that we've already discussed. Having this data will enable you to sign off on the software requirements documents and also provide you with valuable data for later projects.

Best Practices

Let's conclude this chapter with some best practices. There are hundreds of them, far too many for us to cover here. However, let's get a flavor for some performance issues and some do's and don'ts.

1. Develop a requirements document and specifications before you start to write code. Follow an accepted software development process. Contrary to what most students think, code hacking is not an admired trait in a professional programmer. If possible, involve yourself in the system's architectural design decision before they are finalized. If there is no other reason to study computer architecture, this is the one. Bad partitioning decisions at the beginning of project usually lead to pressure on the software team to fix the mess at the back end of the project.

2. Use good programming practices. The same rules of software design apply whether you are coding for a PC or for an embedded controller. Have a good understanding of the general principles of algorithm design. For example, don't use $O(n^2)$ algorithms if you have a large dataset. No matter how good the hardware, inefficient algorithms can stop the fastest processor.

3. Study the compiler that you'll be using and understand how to take fullest advantage of it. Most industrial quality compilers are extremely complicated programs and are usually not documented in a way that mere mortals could comprehend. So, most engineers keep on using the compiler in the way that they've used it in the past, without regard for what kind of incremental performance benefits they might gain by exploring some of the available optimization options. This is especially true if the compiler itself is architected for a particular CPU architecture. For example, there was a version of the GNU®[12] compiler for the Intel i960 processor family that could generate performance profile data from an executing program and then use that data on subsequent compile-execute cycles to improve the performance of the code.

4. Understand the execution limits of your code. For example, Ganssle[14] recommends that in order to decide how much memory to allocate for the stack, you should fill the stack region with an identifiable memory pattern, such as 0xAAAA or 0x5555. Then run your program for enough time to convince yourself that it has been thoroughly exercised. Now, look at *the high water mark* for the stack region by seeing where your bit pattern was overwritten. Then add a safety factor and that is your stack space. Of course, this implies that your code will be deterministic with respect to the stack. One of the biggest *don'ts* in high-reliability software design is to use recursive functions. Each time a recursive function calls itself,

it creates a stack frame that continues to build the stack. Unless you absolutely know the worst-case recursive function call sequence, don't use them. Recursive functions are elegant, but they are also dangerous in systems with strictly defined resources. Also, they have a significant overhead in the function call and return code, so performance suffers.

5. Use assembly language when absolute control is needed. You know how to program in assembly language, so don't be afraid to go in and do some handcrafting. All compilers have mechanisms for including assembly code in your C or C++ programs. Use the language that meets the required performance objectives.

6. Be very careful of dynamic memory allocation when you are designing for any embedded system, or other system with a high-reliability requirement. Even without a designed in memory leak, such as forgetting to free allocated memory, or a bad pointer bug, memory can become fragmented if the memory handler code is not well-matched to your application.

7. Do not ignore all of the exception vectors offered by your processor. Error handlers are important pieces of code that help to keep your system alive. If you don't take advantage of them, or just use them to vector to a general system reset, you'll never be able to track down why the system crashes once every four years on February 29th.

8. Make certain that you and the hardware designers agree on which Endian model you are using.

9. Be judicious in your use of global variables. At the risk of incurring the wrath of Computer Scientists I won't say, "Don't use global variables" because global variables provide a very efficient mechanism for passing parameters. However, be aware that there are dangerous side effects associated with using globals. For example, Simon[15] illustrates the problem associated with memory buffers, such as global variables in his discussion of the shared-data problem. If a global variable is used to hold shared data then a bug could be introduced if one task attempts to read the data while another task is simultaneously writing it. System architecture can affect this situation because the size of the global variable and the size of the external memory could create a problem in one system and not be a problem in another system. For example, suppose a 32-bit value is being used as a global variable. If the memory is 32-bits wide, then it takes one memory write to change the value of the variable. Two tasks can access the variable without a problem. However, if the memory is 16 bits wide, then two successive data writes are required to update the variable. If the second task interrupts the first task after the first memory access but before the second access, it will read corrupted data.

10. Use the right tools to do the job. Most software developers would attempt to debug a program without a good debugger. Don't be afraid to use an oscilloscope or logic analyzer just because they are "Hardware Designer's Tools."

Summary of Chapter 15

Chapter 15 covered:

- How various hardware and software factors will impact the actual performance of a computer system.
- How performance is measured.
- Why performance does not always mean "as fast as possible."
- Methods used to meet performance requirements.

Chapter 15: *Endnotes*

[1] Linley Gwennap, *A numbers game at AMD, Electronic Engineering Times,* October 15, 2001.

[2] http://www.microarch.org/micro35/keynote/JRattner.pdf (Justin Rattner is an Intel Fellow at Intel Labs.).

[3] Arnold S. Berger, *Embedded System Design,* ISBN 1-57820-073-3, CMP Books, Lawrence, KS, 2002, p. 9.

[4] http://www.spec.org/cpu2000/docs/readme1st.html#Q1.

[5] R.P. Weicker, *Dhrystone: A Synthetic Systems Programming Benchmark, Communications of the ACM,* Vol. 27, No. 10, October, 1984, pp. 1013–1030.

[6] Daniel Mann and Paul Cobb, *Why Dhrystone Leaves You High and Dry, EDN,* May 1998.

[7] http://www.eembc.hotdesk.com/about%20eembc.html

[8] Jackie Brenner and Markus Levy, *Code Efficiency and Compiler Directed Feedback, Dr. Dobb's Journal,* #355, December 2003, p. 59.

[9] www.metrowerks.com.

[10] Dave Stewart, *The Twenty-five Most Common Mistakes with Real-Time Software Development,"* a paper presented at the Embedded Systems Conference, San Jose, CA, September 2000.

[11] Jack Ganssle, *The Art of Designing Embedded Systems,* ISBN 0-7506-9869-1, *Newnes,* Newnes, Boston, MA, p. 91.

[12] Nat Hillary and Arnold Berger, *Guaranteeing the Performance of Real-Time Systems, Real Time Computing,* October, 2001, p. 79.

[13] www.gnu.org.

[14] Jack Ganssle, *op cit,* p. 61.

[15] David E. Simon, *An Embedded Software Primer,* ISBN 0-201-61569-X, Addison-Wesley, Reading, MA, 1999, p. 97.

Exercises for Chapter 15

1. People who enjoy playing video games on their PC's will often add impressive liquid cooling systems to remove heat from the CPU. Why?

2. Why will adding more memory to your PC often have more of an impact on performance then replacing the current CPU with a faster one?

3. Assume that you are trying to compare the relative performance of two computers. Computer #1 has a clock frequency of 100 MHz. Computer #2 has a clock frequency of 250 MHz. Computer #1 executes all of its instructions in its instruction set in 1 clock cycle. On average, computer #2 executes 40% of its instruction set in one clock cycle and the rest of its instruction set in two clock cycles. How long will it take each computer to run a benchmark program consisting of 1000 instructions in a row, followed by a loop of 100 instructions that executes 200 times.

    Note: You may assume that for computer #2 the instructions in the benchmark are randomly distributed in a way that matches the overall performance of the computer as stated above.

4. Discuss three ways that, for a given instruction set architecture, processor performance may be improved.

5. Suppose that, on average, computer #1 requires 2.0 cycles per instruction, and uses a 1GHz clock frequency. Computer #2 averages 1.2 cycles per instruction and has a 500 MHz clock. Which computer has the better relative performance? Express your answer as a percentage.

6. Suppose that you are trying to evaluate two different compilers. In order to do this you take one of the standard benchmarks and separately compile it to assembly language using each compiler. The instruction set architecture of this particular processor is such that the assembly language instructions may be grouped into 4 categories according to the number of CPU clock cycles required to execute each instruction in the category. This is shown in the table below:

Instruction category	CPU cycles required to execute
Category A	2
Category B	3
Category C	4
Category D	6

You look at the assembly language output of each of the compilers and determine the relative distribution of each instruction category produced by the compiler.

Compiler A compiled the program to 1000 assembly language instructions and produced the distribution of instructions shown below:

Instruction category	% of instructions in each category
Category A	40
Category B	10
Category C	30
Category D	20

Compiler B compiled the program to 1200 assembly language instructions and produced the distribution of instructions shown below:

Instruction category	% of instructions in each category
Category A	60
Category B	20
Category C	10
Category D	10

Which compiler would you expect to give better performance with this benchmark program? Why? Be as specific as possible.

7. Suppose that you are considering the design trade-offs for an extremely cost-sensitive product. In order to reduce the hardware cost you consider using a version of the processor with an 8-bit wide external data bus instead of the version with the 16-bit wide data bus. Both versions of the processor run at the same clock frequency and are fully 32-bits wide internally. What type of performance difference would you expect to see if you were trying to sum 2, 32-bit wide memory variables with the sum also stored in memory.

8. Why would a compiler try to optimize a program by maximizing the size and number of basic blocks in the code? Recall that a basic block is a section of code with one entry point, one exit point and no internal loops.

CHAPTER 16

Future Trends and Reconfigurable Hardware

● ●

Objectives

When you are finished this lesson, you will be able to describe:
▶ *How programmable logic is implemented;*
▶ *The basic elements of the ABEL programming language;*
▶ *What is reconfigurable hardware and how it is implemented;*
▶ *The basic architecture of a field programmable gate array;*
▶ *The architecture of reconfigurable computing machines;*
▶ *Some future trends in molecular computing;*
▶ *Future trends in clockless computing.*

● ●

Introduction

We've come a long way since Chapter 1, and this chapter is a convenient place to stop and take a forward look to where all of this seems to be going. That is not to say that we'll make a leap to the Starship Enterprise's on-board computer (although that would be a fun thing to do), but rather, let's look a where the trends that are in place today seem to be leading us. Along the way, we'll look at a topic that is emerging in importance but we've just not had a convenient place to discuss it until now.

The focus of this text has been to view the hardware as it is relevant to a software developer. One of the trends that has been going on for several years and continues to grow is the blurring of the lines between what is hardware and what is software. Clearly, software is the driving code for the hardware state machine. But what if the hardware itself was programmable, just like a software algorithm? Can you imagine a computing engine that had no personality at all until the software is loaded? In other words, the distinction between 68K, x86 or ARM would not exist at all until you load a new program. Part of the program actually configures the hardware to the desired architecture. Science fiction you say? Not all. Read on.

Reconfigurable Hardware

From a historical perspective, configurable hardware arrived in the 1970's with the arrival of a digital integrated circuit called a *PAL*, or *programmable array logic*. A PAL was designed to contain a collection of general purpose gates and flip-flops, organized in a manner that would allow a designer to easily create simple to moderately complex sum-of-products logic circuits or state machines.

The gate shown in Figure 16.1 is a non-inverting buffer gate. A gate that doesn't provide a logic function may seem strange, but sometimes the purity of logic must yield to the realities of the electronic properties of digital circuits and we need a circuit with noninverting logic from input to output. What is of interest to us in this particular example is the configuration of the output circuitry of the gate. This type of circuit configuration is called either *open collector* or *open drain*, depending upon the type of integrated circuit technology that is being used. For our purposes, open collector is the more generic term, so we'll use it exclusively.

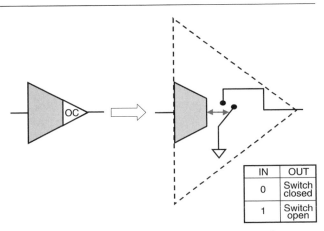

IN	OUT
0	Switch closed
1	Switch open

Figure 16.1: Simplified schematic diagram of a non-inverting buffer gate with an open collector output configuration.

An open-collector output is very similar to a tri-state output, but with some differences. Recall that a tri-state output is able to isolate the logic function of the device (1 or 0) from the output pin of the device. An open collector device works in a similar manner, but its purpose is not to isolate the input logic from the output pin. Rather, the purpose is to enable multiple outputs to be tied together in order to implement logic functions such as AND and OR. When we implement an AND function by tying together the open collector gate outputs we call the circuit a *wired-AND* output.

In order to understand how the circuit works imagine that you are looking into the output pin of the gate in Figure 16.1. If the gate input is at logic zero, then you would see that the open collector output "switch" is closed, connecting the output to ground, or logic level 0. If the input is 1, the output switch is opened, so there is no connection to anything. The switch is in the high impedance state, just like a tri-state gate. Thus, the open collector gate has two states for the output, 0 or high impedance.

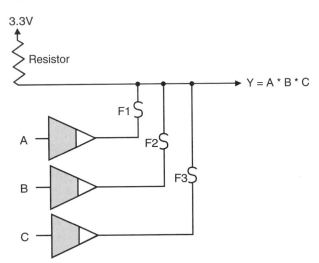

Figure 16.2 illustrates a 3-input wired AND function. For the moment please ignore the circuit elements labeled F1, F2 and F3. These are fuses that can be permanently "blown". We'll discuss their purpose in a moment. Since all three gates are open collector devices, either their outputs are

Figure 16.2: 3-input wired AND logic function. The circuit symbols labeled F1, F2 and F3 are fuses which may be intentionally destroyed, leaving the gate output permanently removed from the logic equation.

connected to ground (logic 0) or in the Hi-Z state. If all three inputs A, B and C are logic 1, then all three outputs are Hi-Z. This means that none of the outputs are connected to the common wire. However, the resistor ties the wire to the system power supply, so you would see the wire as a logic level 1. The difference is that the logic 1 is being supplied by the power supply through the resistor, rather than the output of a gate. We refer to the resistor as a *pull-up resistor* because it is pulling the voltage on the wire "up" to the voltage of the power supply.

If input A, B or C is at logic 0, then the wire is connected to ground. The resistor is needed to prevent the power supply from being directly connected to ground, causing a short circuit, with lots of interesting circuit pyrotechnics and interesting odors. The key is that the resistor limits the current that can flow from the power supply to ground to a safe value and also provides us with a reference point to measure the logic level. Also, it doesn't matter how many of the gates inputs are at logic 0, the effect is still to connect the entire wire to ground, thus forcing the output to a logic 0.

The fuses, F1, F2 and F3 add another dimension to the circuit. If there was a way to vaporize the wire that makes up the fuse, say with a large current pulse, then that particular open collector gate would be entirely removed from the circuit. If we blow fuse F3, then the circuit is a 2-input and gate, consisting of inputs A and B and output Y. Of course, once we decide to blow a fuse we can't put it back the way it was. We call such a device *one-time programmable*, or an OTP device.

Figure 16.3 shows how we can extend the concept of the wired AND function to create a general purpose device that is able to implement an arbitrary sum of products logic function in four inputs and two outputs within a single device. The circles in the figure represent programmable *cross-point switches*. Each switch could be an OTP fuse, as in Figure 16.2, or an electronic switch, such as the output switch of a tri-state gate. With electronic switches we would also need some kind of additional configuration memory to store the state of each cross-point switch, or some other means to reprogram the switch when desired.

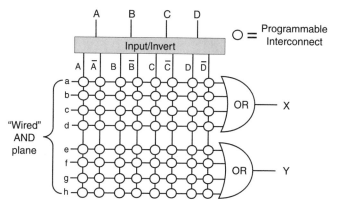

Figure 16.3: Simplified schematic diagram of a portion of a programmable array logic, or PAL device.

Notice how each input is converted to the input and its complement by the input/invert box. Each of the inputs and complements then connect to each of the horizontal wires in the array. The horizontal wires implement the wired "AND' function for the set of the vertical wires. By selectively programming the cross-point switches of the AND plane, each OR gate output can be any single level sum of products term.

Figure 16.4 shows the pin outline diagram for two industry standard PAL devices, the 16R4 and 16L8. Referring to the 16L8 we see that the part has 10 dedicated inputs (pins 1 through 9 and pin 11), 6 pins that may be configured to be either inputs or outputs (pins 12 through 18), and two pins that are strictly for outputs (12 and 19). The other device in Figure 16.4, the 16R4 is similar to the 16L8 but it includes 4 'D' type flip-flop devices that facilitate the design of simple state machines.

Even with relatively simple devices such as these, the number of interconnection points which must be programmed quickly numbers in the hundreds or thousands. In order to bring the complexity under control, the *Data I/O Corporation®*[1], a manufacturer of programming tools for programmable devices, invented one of the earliest hardware description languages, called ABEL, which is an acronym for *Advanced Boolean Equation Language*. Today ABEL is an industry-standard hardware description language. The rights to ABEL are now owned by the Xilinx Corporation, a San Jose, California-based manufacturer of programmable hardware devices and programming tools.

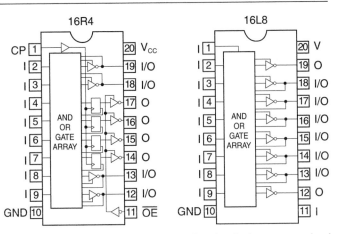

Figure 16.4: Pin-out diagrams for the industry standard 16L8 and 16R4 PALs.

ABEL is a simpler language than Verilog or VHDL, however it is still capable of defining fairly complex circuit configurations. Here's a partial example of an ABEL source file. The process is as follows:

1. Create an ASCII based source file, *source.abl*
2. Compiler the source file to create a programming map called *source.jed*.
3. The source.jed file is downloaded to a device programmer, such as the type built by Data I/O, and the device is programmed by "blowing" (removing) the appropriate fuses for a OTP part, or turning on the appropriate cross-point switches if the device is reprogrammable.

Some of the appropriate declarations are shown below. The keywords are shown in bold.

```
module      AND-OR;

title

Designer   Arnold Berger
Revision   2
Company    University of Washington-Bothell
Part Number      U52




declarations
```

```
" Inputs

AND-OR        Device           PAL20V10 ;

a1    pin    1        ;
b1    pin    2        ;
b2    pin    3        ;
b3    pin    4        ;
b4    pin    5        ;
c1    pin    6        ;
c2    pin    7        ;
c3    pin    8        ;
c4    pin    9        ;
a4    pin    19       ;
a3    pin    18       ;
a2    pin    17       ;

"Outputs

zc    pin    11    istype    'com,buffer'      ;
yc    pin    12    istype    'com,buffer'      ;
yb    pin    13    istype    'com,buffer'      ;
zb    pin    14    istype    'com,buffer'      ;
za    pin    15    istype    'com,buffer'      ;
ya    pin    16    istype    'com,buffer'      ;

Equations

ya    =     a1 # a2 # a3 ;
!za   =     a1 # a2 # a3 ;

yb    =     b1 # b2 # b3 # b4 ;
!zb   =     b1 # b2 # b3 # b4 ;

yc    =     c1 # c2 # c3 # c4 ;
!zc   =     c1 # c2 # c3 # c4 ;

Test_vectors

( [ a1,a2,a3,a4,b1,b2,b3,b4,c1,c2,c3,c4]) -> [ya,za,yb,zb,yc,zc] ;

[1,1,1,x,1,1,1,1,1,1,1,1]  ->  [1,0,1,0,1,0] ;
[0,1,1,x,0,1,1,1,0,1,1,1]  ->    [1,0,1,0,1,0] ;
[1,0,1,x,1,0,1,1,1,0,1,1]  ->  [1,0,1,0,1,0] ;
[1,1,0,x,1,1,0,1,1,1,0,1]  ->  [1,0,1,0,1,0] ;
[1,1,1,x,1,1,1,0,1,1,1,0]  ->  [1,0,1,0,1,0] ;
[0,0,0,x,0,0,0,0,0,0,0,0]  ->  [0,1,0,1,0,1] ;

end ;
```

Much of the structure of the ABEL file should be clear to you. However, there are certain portions of the source file that require some elaboration:

- The *device* keyword is used to associate a physical part with the source module. In this case, our AND-OR circuit will be mapped to a 20V10 PAL device.
- The *pin* keyword defines which local term is mapped to which physical input or output pin.
- The *istype* keyword is used to assign an unambiguous circuit function to a pin. In this case the output pins are declared to be both *com*binatorial and noninverting (*buffer*)
- In the *equations* section, the '*#*' symbol is used to represent the logical OR function. Even though the device is called an AND-OR module, the AND part is actually AND in the negative logic sense. DeMorgan's theorem provides the bridge to using OR operators for negative logic AND function.
- The *test_vectors* keyword allows us to provide a reference test for representative states of the inputs and the corresponding outputs. This is used by the compiler and programmer to verify the logical equations and programming results.

PLDs were soon surpassed by *complex programmable logic devices, or CPLDs*. CPLDs offered higher gate counts and increased functionality. Both families of devices are still very popular in programmable hardware applications. The key point here is that an industry standard hardware description language (ABEL) has provides hardware developers the same design environment (or nearly the same) as the software developers.

The next step in the evolutionary process was the introduction of the *field programmable gate array, or FPGA*. The FPGA was introduced as a prototyping tool for engineers involved in the design of custom integrated circuits, such as ASICs. ASIC designers faced a daunting task. The stakes are very high when you design an ASIC. Unlike software, hardware is very unforgiving when it comes to bugs. Once the hardware manufacturing process is turned on, hundreds of thousands of dollars and months of time become committed to fabricating the part. So, hardware designers spend a great deal of time running simulations of their design in software. In short, they construct reams of test vectors, like the ones shown above in the ABEL source file, and use them to validate the design as much as possible before committing the design to fabrication. In fact, hardware designers spend as much time testing their hardware design before transferring the design to fabrication as they spend actually doing the design itself.

Worse still, the entire development team, must often wait until very late in the design cycle before they actually have working hardware available to them. The FPGA was created to offer a solution to this problem of prototyping hardware containing custom ASIC devices, but in the process, it created an entirely new field of computer architecture called *reconfigurable computing*. Before we discuss reconfigurable computing, we need to look at the FPGA in more detail. In Figure 16.3 we introduced the concept of the cross-point switch. The technology of the individual switches can vary. They may include:

- *Fusible links:* The switches are all initially 'closed'. Programming is achieved by blowing the fuses in order to disconnect unwanted connections.
- *Anti-fusible links:* The links are initially opened. Blowing the fuse cases the switching element to be permanently turned on, closing the switch and connecting the two crossbar conductors.

- Electrically Programmable: The cross-point switch is reprogrammable device which, when programmed with a current pulse, retains its state until it is reprogrammed. This technology is similar to the FLASH memory devices we use in our digital cameras, MP3 players and BIOS ROMs in your computers.
- RAM-based: Each cross-point switch is connected to bit in a RAM memory array. The state of the bit in the array determines whether the corresponding switch is on or off. Obviously, RAM-based devices can be rapidly reprogrammed by re-writing the bits in the configuration memory.

The FPGA introduced a second new concept, the *look-up table*. Unlike the PLD and CPLD devices which implemented logic using the wired AND architecture and combined the AND terms with discrete OR gates, a look-up table is a small RAM memory element that may be programmed to provide any combinatorial logical equation. In effect, the look-up table is a truth table directly implemented in silicon. So, rather than use the truth table as a starting point for creating the gate level, or HDL implementation of a logical equation, the look-up table simply takes the truth table as a RAM memory and directly implements the logical function.

Figure 16.5 illustrates the concept of the look-up table. Thus, the FPGA could readily be implemented using arrays of look-up tables, usually presented as a 5-input, two-output table combined with registers in the form of D-flip flops and cross-point switch networks to route the interconnections between the logical and storage resources. Also, clock signals would also need to be routed to provide synchronization to the circuit designs.

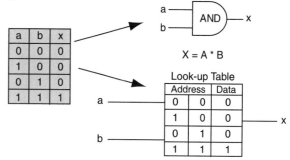

Figure 16.5: Logical equation implemented as a gate or a look-up table.

In Figure 16.6 we see the complete architecture of the FPGA. What is remarkable about this architecture is that it is entirely RAM-based. All of the routing and combinatorial logic is realized by programming memory tables within the FPGA. The entire personality of the device may be changed as easily as reprogramming the device.

The introduction of the FPGA had a profound effect upon the way that ASICs were designed. Even though FPGAs were significantly more expensive than a custom ASIC device; could not be

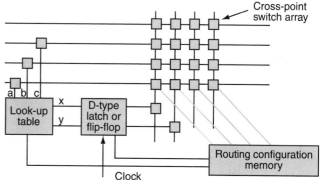

Figure 16.6: Schematic diagram of a portion of a Field Programmable Gate Array.

clocked as fast, and had a much lower gate capacity, the ability to build working prototype hardware that ran nearly as fast as the finished product greatly reduced the inherent risks of ASIC design and also, greatly reduced the time required to develop an ASIC-based product.

Today, companies such as Xilinx®[2] and Actel®[3] offer FPGAs with several million equivalent gates in the system. For example, the Xilinx XC3S5000 contains 5 million equivalent gates and has 784 user I/Os in a package with a total of 1156 pins. Other versions of the FPGA contain on-chip RAM and multiplier units.

As the FPGA gained in popularity software support tools also grew around them. ABEL, Verilog and VHDL will compile to configuration maps for commercially available FPGAs. Microprocessors, such as the ARM, MIPS and PowerPC families have been ported so that they may be directly inserted into a commercial FPGA.[4]

While the FPGA as a stand-alone device and prototyping tool was gaining popularity, researchers and commercial start-ups were constructing reconfigurable digital platforms made up of arrays of hundreds or thousands of interconnected FPGAs. These large systems were targeted at companies with deep pockets, such as Intel and AMD, who were in the business of building complex microprocessors. With the high stakes and high risks associated with bringing a new microprocessor design to market, systems that allowed the designers to load a simulation of their design into a hardware accelerator and run it in a reasonable fraction of real time were very important tools.

While it is possible to simulate a microprocessor completely in software by executing the Verilog or VHDL design file in an interpreted mode, the simulation might only run at 10s or 100s of simulated clock cycles per second. Imagine trying to boot your operating system on a computer running at 5 KHz. Two competing companies in the United States, Quickturn and PiE built and sold large reconfigurable *hardware accelerators*. The term hardware accelerator refers to the intended use of these machines. Rather than attempt to simulate a complex digital design, such as a microprocessor, in software, the design could be loaded into the hardware accelerator and executed at speeds approaching 1 MHz.

Quickturn and PiE later merged and then was sold again to Cadence Design Systems® of San Jose, CA. Cadence is the world's largest supplier of electronic design automation (EDA) tools.

When these hardware accelerators were first introduced they cost approximately $1.00 per equivalent gate. One of the first major commercial applications of the Quickturn system was when Intel simulated an entire Pentium processor by connecting several of the hardware accelerators into one large system. They were able to boot the system to the DOS prompt in tens of minutes[5].

While the commercial sector was focusing their efforts on hardware accelerators, other researchers were experimenting with possible applications for reconfigurable computing machines. One group at the Hewlett-Packard Company Laboratories that I was part of was building a custom reconfigurable computing machine for both research and hardware acceleration purposes[6]. What was different about the HP machine was the novel approach taken to address the problem of routing a design so that it is properly partitioned among hundreds of interconnected FPGAs.

One of the biggest problems that all FPGAs must deal with is that of maximizing the utilization of the on-chip resources, which can be limited by the availability of routing resources. It is not uncommon for only 50% of the available logic resources to be able to be routed within an FPGA. Very complex routing algorithms are needed to work through the details. It was not uncommon for a route of a single FPGA to take several hours and the routing of a complex design into a reconfigurable computing engine to take several weeks, even with the routing algorithm distributed among a cluster of UNIX workstations.

There were a number of reasons for this situation but the primary one was that the commercially available FPGAs were not designed to be interconnected in arrays, but rather were designed to be stand-alone ASIC prototyping tools. They were rich in logic resources but were limited in their routing resources. The reason for this is easy to see. In a large array of FPGAs signals may need to be routed through certain FPGAs just to get to other ones. Thus, routing resources are being consumed even though no on-chip logical resources were used.

The HP team took a different approach. From the start they made sure that Rent's Rule was always obeyed. Rent's Rule was first articulated by Richard Rent at IBM in the 1960s. He wrote-up his results in an internal memo and the results were never published, although subsequent researchers[7] verified its validity.

Rent was investigating the relationship between the number of components on a printed circuit board, the number of interconnections between them, and the number of inputs and outputs the were required to support the circuitry. Rent was able to express the relationship in a mathematical equation.

$$N = k * G^e$$

where:

    N = number of I/O signals into an arbitrary partition, or grouping of circuit devices.
    k = average number of I/O signals per device
    G = number of devices within the partition.
    e = an exponent in the range of $0.5 <= e <= 0.7$

Figure 16.7 illustrates Rent's Rule for a simple case of several gates.

Here we see that there are 6 I/O pins into the partition. Each gate has 3 I/O pins and there are 4 gates within the partition. Using a value of e = 0.5, Rent's Rule correctly models the partition. The value of the exponent can vary over a small range because different types of circuits, such as very regular memory arrays, lead to different results then random logic arrays.

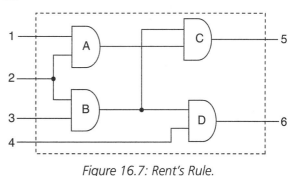

Figure 16.7: Rent's Rule.

Rent's Rule can be used to predict the amount of routing (I/O) that would be necessary to support any arbitrary grouping. In order to design a reconfigurable computing machine which followed Rent's Rule, HP designed a custom FPGA called the *Plasma* chip. Plasma is an acronym for "Programmable Logic And Switch Matrix"[8]. The layout of the Plasma chip is shown schematically in Figure 16.8.

Each PLASMA chip consisted of 16 groupings of 16 Programmable Atomic Logic Elements, or *PALEs*. A grouping of 16 PALEs, with its own set of crossbar interconnects was called a *Hextant*.

Each PALE consisted of a 6-input, 2-output look-up table, followed by a D-type storage register. Each hextant fed a large central crossbar matrix.

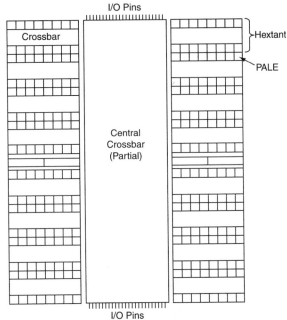

The central crossbar is only one quarter populated. This means that only ¼ of the main signal wires attach to each signal wire from a given hextant. This was done for space considerations. As it is, the central crossbar contained 400 vertical lines and 1600 lines from the hextants. As always, Rent's Rule was observed to insure that no partition of circuits would create an unroutable situation. A photograph of the PLASMA chip is shown in Figure 16.9. What is striking in the photo is the relative size of the routing resources relative to the logic resources. This serves to emphasize the importance of Rent's Rule in providing a context model for any generalized reconfigurable computer. In earlier chapters we discussed how busses were used to carry information through the machine. Here we see how the size of the bus structure must grow in order to direct the arbitrary routing of the data within the machine.

Figure 16.8: Schematic representation of the PLASMA chip.

The reconfigurable computing machine that the HP team created was named *Teramac*. Teramac was designed to perform 10^{12} gate operations per second (one million gates switching at 1 MHz), hence the prefix 'Tera'. Since it was reconfigurable, it was called a Multiple Architecture Computer (mac).

Teramac consisted of 1728 Plasma chips arranged in a complex network. By following Rent's Rule, any PALE in any one of the Plasma chips could be connected to a PALE in any other of the chips.

One other aspect of Teramac was unique, and perhaps is more relevant to this chapter on the future of computing. Teramac was defect tolerant. There

Figure 16.9: Photograph of the PLASMA chip.

could be defects in the Plasma chips or the interconnections between them and the machine could still function. In a sense, the defect tolerance design architecture is similar to that of your hard disk drive. Very few hard disks are perfect. Each surface or platter may contain minor defects that would render one or more of the magnetic bits unreadable. Rather than throw out the entire disk, the defective regions are mapped during the low level formatting of the drive. A special region of the disk stores mapping information so that bad regions are not used.

Today, microprocessor yields typically measure less than 25%. That means that 75% of the chips on the wafer are defective. The reason is that modern microprocessors must be perfect in order to work. The Teramac architecture had no such restriction. During the testing process, special characterization software is run on the machine. As the results come back, regions where the test circuits fail to run properly are mapped out of the design database. The circuits are tested using a pseudo-random number generator circuit. The circuit consists of a 32-stage shift register made up of D-Type flip-flops. Each stage of the shift register feeds into a 3-input XOR gate. The output of the XOR gate feeds the input of the next stage. The remaining two inputs of each XOR gate is randomly connected to the Q output of the other 31 stages.

Using this circuit, long strings of pseudo-random numbers can be generated. If a failure is detected in the circuit, then the random number string will dramatically diverge from the expected result. This type of testing is called *signature analysis* and is also used for detecting open circuits and short circuits in printed circuit boards. In order to detect a defect resources are grouped into the sequences of pseudo-random number generators and tested. If a defect is detected, each element of the failed circuit is independently tested again as part of an orthogonal test circuit. The defective resource exists at the intersection of two failed tests. This is shown in Figure 16.10.

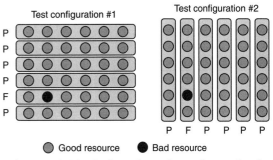

Figure 16.10: Defect detection scheme in the Teramac system.

Within the Plasma chip itself, only 7% of the chip area is considered critical. This means that one or more defects in this area would render the chip unusable. This means that there could be large numbers of other defects in the Plasma chips or the associated interconnections between them, and the system would still function. However, testing the system for defects takes over a week to test just one of the 16 circuit boards in the machine.

Now, we're finally getting to the reason for all of the previous discussion. On numerous occasions we've looked at the parallelism between the hardware solution and software solution to algorithms. Hardware could easily accelerate the solution to a specific algorithm by factors of a thousand times or more. Consider the graphics processor chip on your video card. Imagine how the shoot'em-up game would play if the CPU was calculating all of the image transformations by itself.

Hardware accelerates the algorithmic solution because it can dedicate specific resources in parallel. Typically, as many hardware resources are used as are needed by the algorithm. If you need to process a 1024 bit image vector, then use a 1024 bit register set.

In fact, assuming that the appropriate algorithm could be designed and loaded into Teramac, it has been estimated[10] that a machine like Teramac, running at 1 MHz, could perform DNA string matching algorithms 1000 times faster than the fastest general-purpose supercomputer. Another area of application of machines like Teramac is in the field of data encryption and security. The ability of future computers to perform massively parallel calculations is necessary to both break codes and to create unbreakable codes. For example, today's fastest computers would require billions of years to decode a 300-digit encryption key.[12] Of course, researchers today are also working on ways to couple large numbers of conventional computers into parallel networks. The SETI project (Search for Extraterrestrial Intelligence) is one such example of this. You can volunteer to "loan" SETI your PC when it is idling and the project will utilize it to analyze radio telescope data.

Now, imagine that sometime in the future, our computers consist of large FPGA-like arrays of uncommitted resources. A future compiler compiles both our object code and simultaneously, the optimal computer architecture necessary to implement the solution. If the design of the hardware is such that the configuration memory could be quickly loaded with a block memory move instruction, then it would be possible to reconfigure the hardware to its optimal configuration at every function call.

While this may seem farfetched, companies are already building the early versions of this architecture. Triscend Corporation[11], Mountain View, CA builds configurable systems-on-chip platforms. These products consist of uncommitted logic and routing resources surrounding an industry-standard microprocessor core. The configuration memory is mapped to the memory space of the microprocessor, so that it can reconfigure the peripheral hardware as necessary to implement I/O devices, such as Ethernet controllers, timer, ports, etc. Also, the uncommitted gates can be configured as algorithmic accelerators, like floating point units, graphics processors a like. The Triscend A7 family consists of an ARM7TDMI processor core surrounded by a sea of gates.

Molecular Computing

One of the most amazing constants of the computer industry has been the validity of Moore's Law. Since Intel founder Gordon Moore first proposed it in 1965[14], Moore's Law has been amazingly accurate at predicting the exponential rise in the density of integrated circuits. Advances in integrated circuit process technology have led the charge. The original Intel 4004 processor, introduced in 1971 had 2,250 transistors and a Pentium 4 processor, introduced in 2000, has 42,000,000.[15]

Already device physicists are predicting that Moore's Law may begin to break down in the near future. This is not to say that computers will stop growing in complexity and processing power, but it is a statement of the ability of semiconductor devices to continue to shrink indefinitely. Some of these limitations are economic, and some have to do with the Laws of Physics.

Economic factors could easily force a change in direction. Each time that the integrated circuit manufacturing process takes a step forward, the capital costs of the equipment rises as well. You could say that Moore's Law also holds for the manufacturing costs as well. Today, we are nearly at the end of our ability to use light waves to sensitive the circuit masks. As the dimensions of the integrated circuits continue to shrink, higher energy radiation sources, such as synchrotrons, will be necessary to produce the kind of "light" needed to expose the silicon wafers.

The Laws of Physics are even more demanding. As the dimensions continue to shrink, quantum mechanical effects will become more significant. We already make use of quantum effects to program our FLASH memory devices. By applying a voltage pulse to the FLASH memory cell, quantum mechanical effects allow us to "inject" charge carriers through an insulating barrier of silicon dioxide. When the pulse is removed, the charge is trapped on the other side of the barrier. As we continue to shrink the dimensions, less and less voltage will be necessary to lower the barrier, and the probability that the charge can escape will increase to the point that the circuit is no longer reliable.

Another effect of the shrinkage process is called *electromigration.* Here the conductor paths become so small that the flow of electrons can actually cause impurities in the metal conductors to be swept along with the current. This "electron wind" will move the impurities to regions where they can aggregate and eventually cause the circuit to fail.

Certainly IC Process Engineers and Device Physicists will continue to address these fabrication problems and we'll continue to build even more dense circuits for years to come. However, it is clear that we can't keep shrinking the dimensions of the circuit elements until they are the size of electrons and subatomic particles. Or can we?

What Teramac has shown in a very convincing way is that massively parallel, reconfigurable computing machines can be built using a strategy of "less than perfect" will still work. Let's take a very brief look at this technology.

Molecular electronics is the premise that it is possible to build arrays of individual molecules that can mimic the behavior of today's logic and other electronic circuit components. Various teams are working on methods to store charge within molecules and affect the electrical conductivity of the molecule, much like a MOS transistor behaves today. Other groups have produced rudimentary logic functions (an AND gate) using organic molecules.

The possible commercial viability of this research is hard to assess (at least for me) at this time. What is clear is that the dimensions of these "devices" are at least 1000 times smaller than the smallest integrated circuit of today. Of course, a roomful of very expensive apparatus is also needed to make the molecular switch "switch", but that's just because we are looking at a technology in its infancy. It is reasonable to recall that the original Eniac computer had 17,000 vacuum tubes, filled a room at the University of Pennsylvania, operated at a clock rate of 20 KHz and cause the lights in Philadelphia to dim when it was turned on.

There are many huge hurdles that must be overcome in order to realize molecular computers. For example, what is a wire? How do you interconnect the molecules so that electrical current can easily flow between them? Another problem, how do you detect the minute signals in these circuits? A roomful of powerful analytical tools that are used today won't fit in the dorm rooms with the same ease of today's laptops.

Thus, the basic problem reduces to finding molecular equivalents of these basic circuit elements:[16]

- Switching device: The molecular equivalent of a transistor.
- Memory cell.
- Interconnect technology.
- Signal amplification.

Local Clocks

The last future that I want to discuss is the concept of local clocks. Before we look at this phenomenon, we should spend some time looking into the problem that we are trying to solve. First, let's try to scope the problem. At this writing (August, 2004) the fastest microprocessor clock frequencies are approximately 3.5 GHz. Predictions are that we will easily be at 5 GHz in the next year or so, and that 10 GHz is not far behind.

A 5 GHz clock rate corresponds to a clock period of 200 picoseconds (ps). Since the speed of light is roughly 12 inches per nanosecond in free space and 6 inches per nanosecond through a wire, this means that in 200 ps, light can travel about 1.2 inches. A modern microprocessor is about ¾ of an inch on a side, so this means that 62% of the clock period will be wasted just getting the clock signal from one edge of the chip to the other. Since our microprocessor is a fully synchronous machine, this is a very serious problem. We call this problem *clock skew*. Clock skew is simply the difference in time between corresponding portions of the clock (phase difference) because of the problems associated with simultaneously distributing the clock to all portions of the chip. In Teramac, clock skew was a major design issue that had to be factored into all elements of the machine design. Also, the original Cray supercomputer controlled clock skew by adjusting the lengths of the coaxial cables carrying the clock to various circuit boards in the machine.

Another potential problem is that all transistors don't switch in exactly the same way. There can be slight differences in the switching characteristics of the clock circuitry at various portions of the chip. Measurements have shown these differences in switching characteristics to be as large as about 180 ps[17]. Thus, as the chips get bigger and faster, our ability to keep the clock uniformly distributed across the chip becomes more problematic.

Today, most clock distribution networks are hierarchical. Figure 16.11 shows a typical clock distribution network. The circuit block labeled *phase-locked loop* represents the method used in modern computers to multiply the internal clock frequency to a higher value than the external clock input. For example, if your external clock frequency is 200 MHz, a multiplier value that you might set in the BIOS, or is locked into the chip, could be a factor of 11. Thus, the internal clock frequency is 2200 MHz, or 2.2 GHz. As you can see, simple variations in IC process parameters could lead to clock skew problems as the clock is distributed to all of the synchronous circuitry on the chip.

Recall that the modern processor is a pipeline-driven device with different

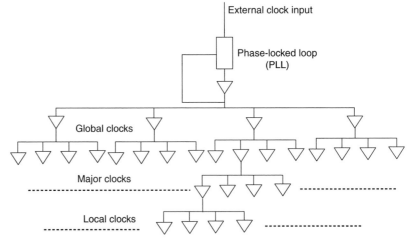

Figure 16.11: Synchronous clock distribution network.

combinatorial logic circuits functioning within the various stages of the pipe. All of the stages are driven from the same synchronous clock, as shown in Figure 16.12. Here we can see the reason why

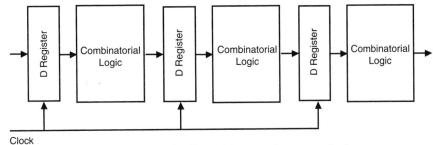

Figure 16.12: Pipeline with a synchronous clock.

limiting clock skew is so critical. Each stage of the pipeline must complete its work before the clock arrives to latch the result into the next stage of the pipeline. The combinatorial logic within each pipeline stage depends upon the time budget it has to complete its work before the next clock edge comes along. Skewing of the clock edges means that some pipeline stages will be clocked sooner than others, destroying the synchronicity of the pipeline.

Now, let's modify the architecture slightly to allow each combinatorial block to execute at its own pace. Figure 16.13 shows a schematic diagram of an asynchronously clocked pipeline.

The system clock is used to drive local clock controllers for each stage of the

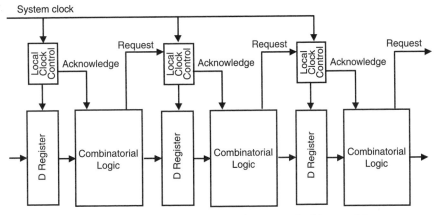

Figure 16.13: Pipeline with an asynchronous clocking architecture.

pipeline. However, each pipeline stage is autonomous, and its local clock is not synchronized with the clock of either the previous stage or the next stage of the pipeline.

When the combinatorial logic of a particular stage has completed its work, the stage logic outputs a request to the local clock controller to latch the result to into the D register that feeds the next stage. When the data is latched into the input register for the next stage, the local clock controller issues an acknowledge signal to the next stage, indicating that valid data is now available to work with. The net effect is that we've created a pipeline with *handshake control* between the stages. Each stage must request a data transfer and the latch mechanism responds with an acknowledgement of the transfer to the next stage.

The drawback of this scheme is that because the local clocks are not synchronized, the handshake may miss a clock edge and the data may have to wait another for clock cycle before the transfer to the next stage may occur. Since each stage is waiting for the previous to complete, this delay in the pipe could easily propagate back and stall the pipe. However, the advantages of such a scheme could far outweigh the disadvantages when we are asking our processors to run at clock speeds in excess of 10 GHz. Given that we may still be able to build digital logic circuits capable of running at such high clock rates, local clocking of the system is probably the only solution.

This raises an interesting question, "Why use clocks at all?" Can we build a completely asynchronous (clockless) computer. According to Marculescu *et al*[17] fully asynchronous designs are probably still a ways away. The computer-aided design (CAD) tools used for design and verification of modern processors still have not reached a level of sophistication that would allow them to deal with a fully asynchronous design. Also, there's the problem of inertia. We just don't design computers this way. However, the local clock remains a viable compromise to the problem of clock skew.

Several start-up companies have already formed to exploit the idea of a fully asynchronous microprocessor design. Fulcrum Microsystems[18] grew out of work done at Caltech. Figure 16.14 illustrates one of the potential advantages to asynchronous processors.

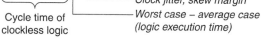

Figure 16.14: Advantage of clockless logic over traditionally clocked logic. Courtesy of Fulcrum Microsystems.

With an asynchronous system, the data in the pipeline flows through at its own rate. Additional circuitry is needed to prevent the runaway condition that clocks and registers are used to prevent in traditional clocked microprocessor systems.

This concept is similar to the use of local clocks, but in this case, additional logic is necessary to detect when a stage has completed its work so that the next stage in the pipeline may be enabled. This is shown in Figure 16.15.

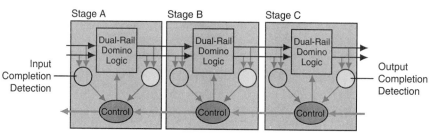

Figure 16.15: Clockless pipeline. Courtesy of Fulcrum Microsystems.

Summary of Chapter 16

In Chapter 16, we covered:

- The architecture of programmable logic devices
- The architecture of field programmable gate arrays
- The development of reconfigurable computing machines based upon arrays of field programmable gate arrays
- Future trends in molecular computing, local clocks and clockless computers.

Chapter 16: *Endnotes*

[1] http://www.datio.com.

[2] http://www.xilinx.com.

[3] http://www.actel.com.

[4] http://www.xilinx.com/company/press/kits/v2pro/backgrounder.pdf.

[5] "Inside Intel: It's Moving at Double-Time to Head Off Competitors," Business Week, June 1, 1992.

[6] Greg Snider, Philip Kuekes, W. Bruce Culbertson, Richard J. Carter, Arnold S. Berger, Rick Amerson, *The Teramac Configurable Computer Engine*, Proceedings of the 5th International Workshop on Field-Programmable Logic and Applications, edited by Will Moore and Wayne Luk, Oxford, UK, September 1995, p. 44.

[7] B.S. Landman and R.L. Russo, *IEEE Trans. Comp., C20,* 1469, 1971.

[8] Rick Anderson, Richard J. Carter, W. Bruce Culbertson, Philip Kuekes, Greg Snider, Lyle Albertson: *Plasma: An FPGA for Million Gate Systems.* FPGA '96. Proceedings of the 1996 Fourth International Symposium on Field Programmable Gate Arrays, February 11-13, 1996, Monterey, CA, USA. ACM, 1996, pp. 10–16.

[9] B. Culbertson, R. Amerson, R. Carter, P. Kuekes, G. Snider, *The Teramac Custom Computer: Extending the limits with defect tolerance*, IEEE International Symposium on Defect and Fault Tolerance in VLSI Systems, November 1996.

[10] Barry Shakleford, HP Labs, Private Communication.

[11] http://www.triscend.com.

[12] Daniel Tynan, "Silicon is Slow," Popular Science, June, 2002, p. 25.

[13] http://setiathome.ssl.berkeley.edu/.

[14] Gordon E. Moore, *Cramming More Components onto Integrated Circuits*, *Electronics*, Volume 28, Number 8, April 19, 1965.

[15] http://www.intel.com/research/silicon/mooreslaw.htm.

[16] Mark A. Reed and James M. Tour, *Computing with Molecules*, *Scientific American*, June, 2000, p. 89.

[17] Diana Marculescu, Dave Albonesi, Alper Buyuktosunoglu, *Tutorial: Partially Asynchronous Microprocessors*, *Micro-35*, Istanbul, Turkey, Nov. 18, 2002.

[18] http://www.fulcrummicro.com.

Exercises for Chapter 16

1. Consider the circuit for a portion of a PLD as shown below. Indicate a fuse that is "blown" by a solid black interconnect and a connection as an open white circle. Make a copy of the diagram and "program" the device by filling in the interconnect circles of the fuses that you want to blow. Program the logical equation:

$$X = (A \oplus B) + C * \overline{D}$$

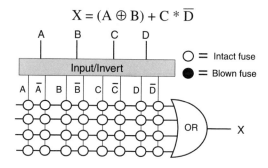

2. Does the circuit shown below obey Rent's Rule?

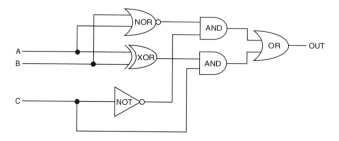

3. Circuits similar to the circuit shown below, consisting of 16–32 stages, are used to detect defective interconnects or defective logic elements in defect tolerant computing machines. Why is this circuit particularly a particularly good choice for such a task?

4. Suppose that you want to design a synchronous CPU with a 10 GHz clock rate. The worst case propagation delay through the logic gates is 28 picoseconds. No stage of the pipeline has more than three levels of logic circuitry. You also need to maintain a safety margin of 10 picoseconds to allow for manufacturing uncertainties, device set-up times, and differences between the switching characteristics of the devices in the circuitry. Approximately what is largest difference in the length of the clock paths that this design can tolerate?

Chapter 1: Solutions for Odd-Numbered Problems

1. Moore's Law states that the number of transistors on an integrated circuit die doubles approximately every 18 months. Since the number of transistors that circuit designers can place on a single die is constantly going up, this means that the complexity of the type of computers and memories that they use is also going up. Also, since the numbers of transistors is increasing, the size of the transistors is decreasing, so transistors are being packed more closely and the distance that the electrical signals have to travel goes down. This means that circuits can run faster.

   Thus, there are two effects going on. Computers can achieve higher performance in areas such as bus bandwidth and complexity because we can take advantage of the number of circuits we can place on a single die. Also, these complex designs can run faster. Finally, complex circuit designs allow even more complex software applications to run because we have memories with higher speed and capacity to implement the algorithms.

3. An advantage of an abstraction layer concept is that you can hide the details and differences of the lower level details so that programs at the upper level need only be written once and will be able to run on a wide range of different machines. A disadvantage is that you may lose efficiency as calls to the lower level functions must progress through the different layer and be translated at each step.

5. On average, semiconductor memory is 34,286 times faster than the hard drive.

7. Convert the following hexadecimal numbers to decimal:

   (i) 0xFE57 = 65,111
   (j) 0xA3011 = 667,665
   (k) 0xDE01 = 56,833
   (l) 0x3AB2 = 15026

9. 545 microfeet per second or 545×10^{-6} feet per second.

[Solutions to the even-numbered problems are available through the instructor's resource website at http://www.elsevier.com/0750678860.]

1. The AND circuit becomes an OR circuit and the OR circuit becomes an AND circuit.

3.

Part a

a	b	c	F
0	0	0	0
1	0	0	1
0	1	0	1
1	1	0	0
0	0	1	0
1	0	1	0
0	1	1	1
1	1	1	0

Part b

a	b	c	F
0	0	0	0
1	0	0	0
0	1	0	0
1	1	0	1
0	0	1	1
1	0	1	0
0	1	1	1
1	1	1	1

5. The truth table is shown on the right.

a	b	c	d	X
0	0	0	0	1
1	0	0	0	0
0	1	0	0	0
1	1	0	0	1
0	0	1	0	0
1	0	1	0	1
0	1	1	0	1
1	1	1	0	0
0	0	0	1	0
1	0	0	1	1
0	1	0	1	1
1	1	0	1	0
0	0	1	1	1
1	0	1	1	0
0	1	1	1	0
1	1	1	1	1

7. The circuit is shown below:

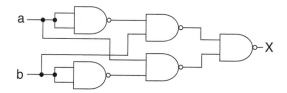

Chapter 3: Solutions for Odd-Numbered Problems

1. The truth table and K-maps are shown below:

A	B	Cin	SUM	Cout
0	0	0	0	0
1	0	0	1	0
0	1	0	1	0
1	1	0	0	1
0	0	1	1	0
1	0	1	0	1
0	1	1	0	1
1	1	1	1	1

	Karnaugh Map for SUM					Karnaugh Map for Cout			
	*A*B	*AB	AB	A*B		*A*B	*AB	AB	A*B
Cin	1		1		Cin		1	1	1
*Cin		1		1	*Cin			1	

$$SUM = \overline{A} * \overline{B} * Cin + A * B * Cin + \overline{A} * B * \overline{Cin} + A * \overline{B} * \overline{Cin}$$

$$SUM = Cin * (\overline{A} * \overline{B} + A * B) + \overline{Cin} * (\overline{A} * B + A * \overline{B})$$

We can simplify the second term by realizing that $\overline{A} * B + A * \overline{B}$ is just the equation for the Exclusive OR (XOR) gate. Also, the first term, $\overline{A} * \overline{B} + A * B$ is just the complement of the exclusive OR function. Thus, there are two nested XOR terms.

$$SUM = Cin \oplus [A \oplus B]$$

We can Use the Karnaugh map to simplify the logic for Cout. There are three loops:

$$Cout = B * Cin + A * Cin + A * B$$

Following is the logic circuitry for SUM and Cout.

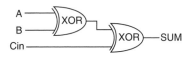

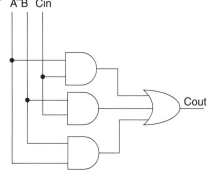

3. Assume that at T = 0 the logic level changes from 0 to 1, as shown, above. We can see that as the change propagates through each gate an additional 10 ns delay is introduced. When the signal gets to point A, 50 ns later, it puts the opposite polarity signal on the first gate and the sequence starts over again in the opposite direction. At T = 100 ns the situation is the same as T = 0, but 100 ns have elapsed. Thus, the circuit oscillates with a period of 100 ns. Therefore, the frequency at point A is 10 MHz.

The waveform seen at point A is:

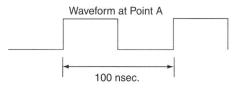

Waveform at Point A

100 nsec.

5. The truth tables and K-maps are shown below:

The simplified equations are:

$$X = A * \overline{B} * \overline{D} + C * \overline{D}$$

$$Y = C * \overline{D} + B * \overline{D} = \overline{D} * (C + B)$$

$$Z = D$$

Truth Table

A	B	C	D	X	Y	Z
0	0	0	0	0	0	0
1	0	0	0	1	0	0
0	1	0	0	0	1	0
1	1	0	0	0	1	0
0	0	1	0	1	1	0
1	0	1	0	1	1	0
0	1	1	0	1	1	0
1	1	1	0	1	1	0
0	0	0	1	0	0	1
1	0	0	1	0	0	1
0	1	0	1	0	0	1
1	1	0	1	0	0	1
0	0	1	1	0	0	1
1	0	1	1	0	0	1
0	1	1	1	0	0	1
1	1	1	1	0	0	1

K-Map for X

	$\overline{A}\,\overline{B}$	$A\overline{B}$	AB	$\overline{A}B$
$\overline{C}\,\overline{D}$		1		
$\overline{C}D$	1	1	1	1
CD				
$C\overline{D}$				

K-Map for X

	$\overline{A}\,\overline{B}$	$A\overline{B}$	AB	$\overline{A}B$
$\overline{C}\,\overline{D}$			1	1
$\overline{C}D$	1		1	1
CD				
$C\overline{D}$				

K-Map for X

	$\overline{A}\,\overline{B}$	$A\overline{B}$	AB	$\overline{A}B$
$\overline{C}\,\overline{D}$				
$\overline{C}D$				
CD	1	1	1	1
$C\overline{D}$	1	1	1	1

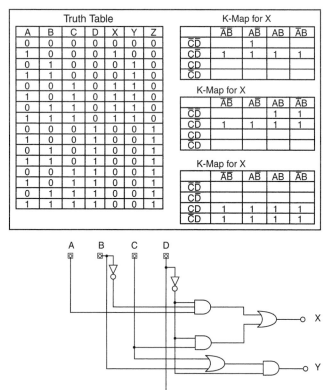

7. Let's walk through the logic of the solution. The pump motor logic is designed so that if the temperature is too low, the pump would not automatically start the pump motor and the heater. Another possible interpretation is that a low temperature would automatically start the pump motor and the heater. The circuitry for the pump shows both options for the solution.

    a. Pump motor: The pump motor is on (f = 1) when the timer (B) is on OR the manual switch (F) is on AND the key switch (E) is on. Note in the alternative solution the temperature being low can also turn on the pump, so we've added a term to account for that case.

    b. Heater: The heater should go on (h = 1) when the temperature sensor (A) indicates that the temperature of the water is below the set temperature on the control panel. We also have the practical consideration that the heater shouldn't be turned on unless the pump is also operating. This could be dangerous if the water isn't flowing while it is being heated. The solution is shown in the circuit diagram for the heater, h.

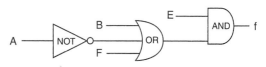

Solution

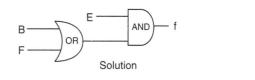

Alternative Solution

Thus, in the above circuit there are three AND conditions for the heater to be turned on.

    1. The key switch (E) must be enabled,

    2. The pump must be on (B + F),

    3. The temperature is low ($\overline{A}$).

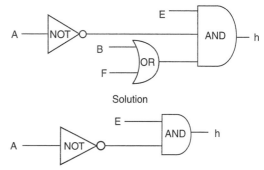

Solution

The alternative solution leads to a simpler arrangement. Only the key switch AND low temperature are required to turn on the heater. We don't have to worry about the pump because $\overline{A}$ also turns it on.

Alternative Solution

    c. Blower: The air blower (g) is pretty simple. The key switch must be on (E = 1) AND the blower switch must be on (D = 1) to turn on the soothing bubbles after a hard day of solving homework problem sets. The solution is shown, right:

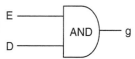

9. The circuit is shown below:

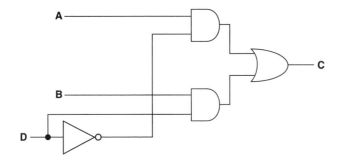

Chapter 4: Solutions for Odd-Numbered Problems

1. Following is the state machine diagram:

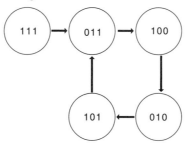

3. The table is shown below:

Clock Pulse	Qa	Qb
Before clock pulse	*0*	*0*
After clock pulse 1	*1*	*0*
After clock pulse 2	*1*	*1*
After clock pulse 3	*0*	*1*
After clock pulse 4	*0*	*0*

5. The table is shown below. The pattern repeats itself after six clock pulses.

	BEFORE PULSE				AFTER PULSE			
	A	B	C	D	A	B	C	D
1	0	0	0	0	0	0	1	1
2	0	0	1	1	1	0	1	0
3	1	0	1	0	0	1	1	0
4	0	1	1	0	1	0	0	0
5	1	0	0	0	0	1	1	1
6	0	1	1	1	0	0	0	0
7	0	0	0	0	0	0	1	1
8	0	0	1	1	1	0	1	0

7. The synchronous counting circuit is shown below:

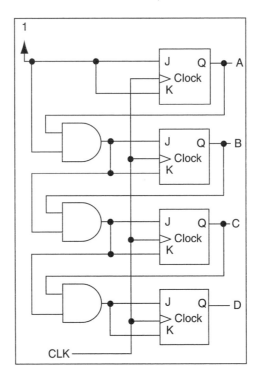

Chapter 5: Solutions for Odd-Numbered Problems

1. The truth table is shown below. The state diagram is shown to the right:

A in	B in	Z	A out	B out
0	0	0	0	1
1	0	0	0	0
0	1	0	1	1
1	1	0	1	0
0	0	1	1	0
1	0	1	0	1
0	1	1	1	1
1	1	1	0	0

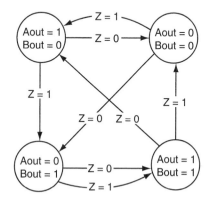

3. The solution is shown below:

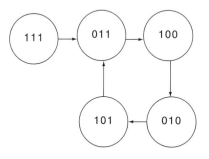

447

5. We have four states, S0 through S3, so we need two variables, X and Y, to provide the outputs to the register and to provide two inputs to the truth table. Thus, we can make the following assertions:

$S0 \rightarrow X = 0, Y = 0$
$S1 \rightarrow X = 1, Y = 0$
$S2 \rightarrow X = 0, Y = 1$
$S4 \rightarrow X = 1, Y = 1$

Let's first analyze the system in words. Once we do that, we can begin to fill in the truth table. Suppose that the system is in state S0 and no money is deposited. It just stays there, so we can describe that with the table entry shown below:

a	b	x	y	X	Y	z
0	0	0	0	0	0	0

Now, assume that we're in state S0 ($S0 \rightarrow X = 0, Y = 0$). The possibilities are:

1. No coin is deposited, stay in S0.
2. A dime is deposited (a = 0, b = 1) transition to state S1.
3. A quarter is deposited (a = 1, b = 0) transition to state S3.

We can express this condition as follows:

a	b	x	y	X	Y	z
0	0	0	0	0	0	0
0	1	0	0	1	0	0
1	0	0	0	1	1	0

Now, assume that we're in state S1 ($S1 \rightarrow X = 1, Y = 0$). The possibilities are:

1. No coin is deposited, it stays in S1.
2. A dime is deposited, it transitions to S2.
3. A quarter is deposited, it returns to S0 and dispenses the merchandise.

We can show this as the following conditions:

a	b	x	y	X	Y	z
0	0	1	0	1	0	0
0	1	1	0	0	1	0
1	0	1	0	0	0	1

Now, assume that we're in state S2 ($S2 \rightarrow X = 0, Y = 1$). The possibilities are:

1. No coin is deposited, it stays in S2.
2. A dime is deposited, it transitions to S0 and dispenses merchandise.
3. A quarter is deposited, it returns to S0 and dispenses the merchandise.

We can show this as the following conditions:

a	b	x	y	X	Y	z
0	0	0	1	0	1	0
0	1	0	1	0	0	1
1	0	0	1	0	0	1

Now, assume that we're in state S3 (S3 → X = 1, Y = 1). The possibilities are:

1. No coin is deposited, it stays in S3.
2. A dime is deposited, it transitions to S0 and dispenses merchandise.
3. A quarter is deposited, it returns to S0 and dispenses the merchandise.

We can show this as the following conditions:

a	b	x	y	X	Y	Z
0	0	1	1	1	1	0
0	1	1	1	0	0	1
1	0	1	1	0	0	1

That covers all the possibilities. Let's now fill in the truth table with what we know:

	a	b	x	y	X	Y	Z
S0	0	0	0	0	0	0	0
S0	0	1	0	0	1	0	0
S0	1	0	0	0	1	1	0
S0	1	1	0	0	X	X	X
S1	0	0	1	0	1	0	0
S1	0	1	1	0	0	1	0
S1	1	0	1	0	0	0	1
S1	1	1	1	0	X	X	X
S2	0	0	0	1	0	1	0
S2	0	1	0	1	0	0	1
S2	1	0	0	1	0	0	1
S2	1	1	0	1	X	X	X
S3	0	0	1	1	1	1	0
S3	0	1	1	1	0	0	1
S3	1	0	1	1	0	0	1
S3	1	1	1	1	X	X	X

The X's indicate "don't care conditions." They'll never occur in real operation, so we'll save them to see if they help us to simplify the K-map of the circuit.

The K-map of the state variable X, is shown below:

	$\bar{a}*\bar{b}$	$\bar{a}*b$	$a*b$	$a*\bar{b}$
$\bar{x}*\bar{y}$		1	1	1
$\bar{x}*y$				
$x*y$	1			
$x*\bar{y}$	1			

I added the term in gray (a * b * $\bar{x}$ * $\bar{y}$) because it simplifies the equation by a bit.

$$X = \bar{a} * \bar{b} * x + b * \bar{x} * \bar{y} + a * \bar{x} * \bar{y}$$

The K-map of the state variable Y, is shown next:

	$\overline{a}*\overline{b}$	$\overline{a}*b$	$a*b$	$a*\overline{b}$
$\overline{x}*\overline{y}$			1	1
$\overline{x}*y$	1			
$x*y$	1			
$x*\overline{y}$		1	1	

$$Y = a * \overline{x} * \overline{y} + \overline{a} * \overline{b} * y + b * x * \overline{y}$$

Finally, the K-map for the output variable Z, is shown below:

	$\overline{a}*\overline{b}$	$\overline{a}*b$	$a*b$	$a*\overline{b}$
$\overline{x}*\overline{y}$				
$\overline{x}*y$		1	1	1
$x*y$		1	1	1
$x*\overline{y}$			1	1

This gives us three loops:

$$Z = b * y + a * y + a * x * \overline{y}$$

The gate diagram is shown below:

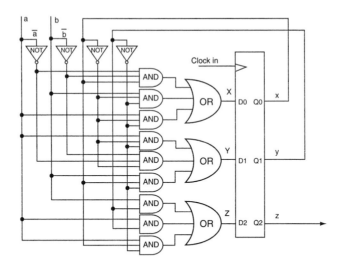

7. After the RESET all of the outputs are zero. This guarantees that the machine starts from a known state. The state of the system after each clock pulse is shown in the table, below:

Clock	RESET	1	2	3	4	5	6	7	8	9	10	11	12	13	14
Output	0000	1100	1011	1000	1001	0001	0100	1110	0010	0101	0110	0111	1111	1010	0000

Thus, after 14 clock pulses the states begin to repeat.

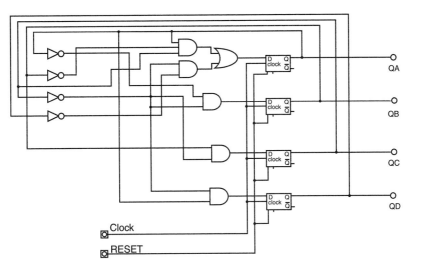

Chapter 6: Solutions for Odd-Numbered Problems

1. The gate design is shown below. Note that this is more a problem of conversion from negative to positive logic then anything else.

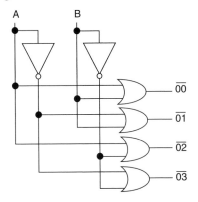

3a. Since the memory width is 32 bits, we need 4 memory chips to form 1 32-bit page. We have a total of 2^{26} address bits. Each page is 512K, which requires 2^{19} address lines per page. Thus, $2^{26} - 2^{19} = 2^7$, so we have 128 pages of memory. Since each page requires four devices, we need a total of *512 memory chips.*

3b.

Page number	Starting Address(hex)	Ending Address (hex)
0	*0000000*	*007FFFF*
1	*0080000*	*00FFFFF*
2	*0100000*	*017FFFF*

5a. Direct memory access: A method of improving the efficiency of data transfers between a peripheral device and the computer's memory. The DMA process allows a peripheral device to take control of the memory bus while the processor idles and the peripheral handles the data transfer directly to memory, bypassing the processor.

5b. Tri-state logic: A circuit design that adds an additional output control to memory or other devices that enables their outputs to be tied together on busses. When the tri-state logic turns off the output of the device, the output presents a high impedance to the bus. In other words, it is as if it isn't connected to the bus, thus enabling another output to drive the bus.

5c. Address bus, data bus, status bus: These are the three main busses of the processor. The address bus presents the address of the next memory operation to the memory system. It is unidirectional, that is all signals are outputs from the processor to memory. The data bus is bidirectional. Data flows into the processor and out to memory on the same bus signals. The status bus is heterogeneous. Some signals are input only, some are output only and others are bidirectional. The status bus carries all of the housekeeping signals of the processor.

7a. The circuit for the memory decoder is shown, right:

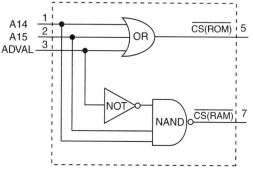

U6 MEMORY DECODING CIRCUIT "CHIP"

7b. The net list is shown below:

Net name					
addr0	U1-36	U2-1	U3-1	U4-1	U5-1
addr1	U1-35	U2-2	U3-2	U4-2	U5-2
addr2	U1-34	U2-3	U3-3	U4-3	U5-3
addr3	U1-33	U2-4	U3-4	U4-4	U5-4
addr4	U1-32	U2-5	U3-5	U4-5	U5-5
addr5	U1-31	U2-6	U3-6	U4-6	U5-6
addr6	U1-30	U2-7	U3-7	U4-7	U5-7
addr7	U1-29	U2-8	U3-8	U4-8	U5-8
addr8	U1-28	U2-9	U3-9	U4-9	U5-9
addr9	U1-27	U2-10	U3-10	U4-10	U5-10
addr10	U1-26	U2-11	U3-11	U4-11	U5-11
addr11	U1-25	U2-12	U3-12	U4-12	U5-12
addr12	U1-24	U2-13	U3-13	U4-13	U5-13
addr13	U1-23	U2-14	U3-14	U4-14	U5-14
addr14	U1-22	U6-1			
addr15	U1-21	U6-3			
data0	U1-1	U2-15	U4-15		
data1	U1-2	U2-16	U4-16		
data2	U1-3	U2-17	U4-17		
data3	U1-4	U2-18	U4-18		
data4	U1-5	U2-19	U4-19		
data5	U1-6	U2-20	U4-20		
data6	U1-7	U2-21	U4-21		
data7	U1-8	U2-22	U4-22		
data8	U1-9	U3-15	U5-15		
data9	U1-13	U3-16	U5-16		

(continued)

Net name				
data10	U1-14	U3-17	U5-17	
data11	U1-15	U3-18	U5-18	
data12	U1-16	U3-19	U5-19	
data13	U1-17	U3-20	U5-20	
data14	U1-18	U3-21	U5-21	
data15	U1-19	U3-22	U5-22	
$\overline{ADVAL}$	U1-12	U6-3		
$\overline{WR}$	U1-10	U2-23	U3-23	
$\overline{RD}$	U1-11	U2-24	U3-24	U4-24	U5-24
$\overline{CSROM}$	U6-5	U4-25	U5-25	
$\overline{CSRAM}$	U6-7	U2-25	U3-25	

Chapter 7: Solutions for Odd-Numbered Problems

1. The sizes do not have to match. There are plenty of examples exhibiting internal memory busses that are either smaller or larger then the external busses. For example, the modern PC has a 64-bit wide external memory bus but the current Athlon and Pentium processors are 32-bits wide internally.

   In cost-sensitive systems, the external bus may be narrower than the internal data bus in order to allow designers to build a more economical system. However, narrow memory width means more memory fetches must occur and overall performance will go down.

3a. MOVE.W $1000,A3: You must use MOVEA.W to load an address register

3b. ADD.B   D0,#$A369: The destination operand can't be a literal.

3c. ORI.W   #$55AA007C,D4: The size of the operation, in this case a word, must match the size of the literal source operand.

3d. MOVEA.L D6,A8: There is no address register A8

3e. MOVE.L $1200F7,D3: This is a nonaligned access violation. For a word or long word access the source or destination must be on an even word address boundary.

5. $AA.

7a. MOVE.L	D0,D7	This is legal
7b. MOVE.B	D2, #$4A	Illegal: Can't store a value to a literal
7c. MOVEA.B	D3,A4	Illegal: Can't store an address value as a byte
7d. MOVE.W	A6,D8	Illegal: D8 is not a valid register
7e. AND.L	$4000,$55AA	Illegal: A data register must, at least, be the source or destination operand of the AND operation

9. The logical operation of an XOR instruction is to bit-wise do the "EXCLUSIVE OR" of the bits. Thus, any bit pairs that are both 1 will give a zero result, any bit pair that is a 1 and a 0 will give a 1 result. The FFFF word has the effect of causing the AAAA word to become complemented to 5555. Adding 1 to it makes it 5556.

$$FFFF\ XOR\ AAAA = 5555 + 1 = 5556$$

11. <004000> = $4515

Chapter 8: Solutions for Odd-Numbered Problems

1.

```
*************************************************************
*
* Subroutine timer:
*
* This subroutine counts down from a number between 1 and 9
* passed into it in register D0.B.
* The count-down rate is one digit every two seconds, using a
* 500 millisecond hardware timer located at address $00001002
* The seven-segment display is located at address $00001000
*
* No error checking is done in the routine
*
* All registers used are save upon exit.
*************************************************************
disp0        EQU     $3F            *Bit patterns for the display
disp1        EQU     $06
disp2        EQU     $5B
disp3        EQU     $4F
disp4        EQU     $56
disp5        EQU     $6D
disp6        EQU     $7D
disp7        EQU     $07
disp8        EQU     $7F
disp9        EQU     $67
trigger      EQU     $10            *This starts the timer
time_out     EQU     $01            *This test for done
display      EQU     $00001000      *Memory location of the display
delay        EQU     $00001002      *Memory location of the timer hardware
* Code begins here
timer        MOVEM.L     A0/A1/D1/D2/D3,-(SP)  *Save the registers on
entry
  LEA               patterns,A0             *A0 points to patterns to
display
             MOVEA.L     #display,A1              *A1 points to the display
```

```
                MOVEA.L      #delay,A2        *A2 points to the time delay
circuit
                CLR.L        D1               *D1 is the index register
                MOVE.B       D0,D1            *Get index value
loop1           MOVE.B       00(A0,D1),(A1)   *Send pattern to the display
                CMPI.B       #00,D1           *Is D1 = 0 yet?
                BEQ          return           *Yes, go home
                MOVE.B       #4,D2            *Count down timer set-up

loop2           MOVE.B       #trigger,(A2)    *Start timer
loop3           MOVE.B       (A2),D3          *Get status
                ANDI.B       #time_out,D3     *Isolate DB0
                BEQ          loop3            *Keep waiting
                SUBQ.B #1,D2                  *Decrement D2
                BNE          loop2            *Go back
                SUBQ         #1,D1            *Point to next pattern
                BRA          loop1
return          MOVEM.L      (SP)+,D3/D2/D1/A1/A0  *Restore the registers
                RTS
patterns        DC.B
        disp0,disp1,disp2,disp3,disp4,disp5,disp6,disp7,disp8,disp9
```

3. **\<D0\> = $0000002A**

5. The ROM is a read-only device. The last instruction, MOVE.W D0,(A2) is doing a write to ROM. This is incorrect.

7.

```
delay       equ       2000           *2 second delay
mask        equ       $8000          *timer status
bits        equ       01             *bit pattern
io_port     equ       $4000          *location of I/O port
timer       equ       $8000          *timer port


            org   $400

start       move.b    #bits,io_port  *load io port
loop        move.w    #delay,timer   *load timer
            move.b    io_port,d0     *get port
            rol.b     #bits,d0       *roll it left
            move.b    d0,io_port     *put it back
            move.w    #delay,timer   *set delay
wait        andi.w    #mask,timer    *check status
            beq       wait           *not zero, keep waiting
            bra       loop           *do it again
            end $400
```

9.

```
************************************************************
*
* This is a program to fill memory with the word pattern $5555
*
************************************************************
          OPT     CRE
fill_st   EQU     $00002000        * Start of block to fill
fill_end  EQU     $000020FF        * Last address to fill
pattern   EQU     $5555            * Fill pattern

start     EQU     $400             * Program begins here

          ORG     start
          LEA     fill_st,A0       * Load starting address
          LEA     fill_end,A1      * Load ending address
          MOVE.W  #pattern,D0      * Load pattern to write
loop      MOVE.W    D0,(A0)+       * Move it and advance pointer
          CMPA.L    A1,A0          * Are we done yet?
          BLE       loop           * No? Go back and repeat
          STOP      #$2700         * Pops us back to the simulator
          END       start
```

Chapter 9: Solutions for Odd-Numbered Problems

1. This is an example of the addressing mode known as "address register indirect with index and displacement". The effective address is the sum of the address value in A0, the index value, D0, and the 2's complement displacement. Since $84 is a negative number, –7C. Thus, the effective address, EA = $2000 + $0400 + (–$7C).

$$EA = \$2384$$

The program is not relocatable for two reasons:
1. There is a jump to an absolute address, *start*
2. An absolute address is loaded into A0. The program could still be relocatable by managing what gets loaded into A0 and D0, but the jump instruction forces it to be absolute.

3. The value in D0 after the highlighted instruction is $0000002A.

5.
```
00000400  067955550000AAAA        ADDI.W   #$5555,$0000AAAA
00000408  06B9AAAA55550000FFFE    ADDI.L   #$AAAA5555,$0000FFFE
00000412  0640AAAA                ADDI.W   #$AAAA,D0
```

7.
```
****************************************************************

*
* CSS 422 HW #4: Relocatable Memory test program
*
****************************************************************


* System equates

pattern1    EQU     $AAAA           * First test pattern
pattern2    EQU     $FFFF           * Second test pattern
pattern3    EQU     $0001           * Third test pattern
st_addr     EQU     $00000400       * Starting address of test
end_addr    EQU     $0009FFF0       * Ending address of the test
stack       EQU     $000C0000       * Location of the stack pointer
word        EQU     2               * Length of a word, in bytes
byte        EQU     1               * One byte long, NO MAGIC NUMBERS!
bit         EQU     1               * Shifting by bits
exit_pgm    EQU     $2700           * Simulator exit code
data        EQU     $500            * Data storage region
start       EQU     $400            * Program starts here
```

```
new_ad      EQU     $000A0000       * Relocated program runs here
pr_cmd      EQU     00              * Command to print message


* Main Program

            OPT     CRE             * Turn on cross references
            ORG     start           * Program begins here
            LEA     stack,SP        * Initialize the stack pointer
            LEA     relo,A0         * Starting address pointer
            LEA     last_addr,A1    * End pointer
            LEA     new_ad,A3       * Destination
relo_lp     MOVE.W  (A0)+,(A3)+     * Move a word
            CMPA.L  A0,A1           * Have we moved enough?
            BPL     relo_lp
            JMP     new_ad
relo        LEA     test_patt(PC),A3 * A3 points to the test pattern to use
            LEA     bad_cnt(PC),A4  * A4 points to bad memory counter
            LEA     bad_addr(PC),A5 * A5 points to the bad addr location
            LEA     data_read(PC),A6 * A6 points to data storage
            CLR.B   (A4)            * Clear bad address count
            MOVE.W  (A3)+,D0        * Get current pattern, point to next
one
            BSR     do_test         * Run first test
            NOT.W   D0              * Complement bits for next test
            BSR     do_test         * Run second test
            MOVE.W  (A3)+,D0        * Get next pattern
            BSR     do_test         * Run third test
            NOT.W   D0              * Complement bits for fourth test
            MOVE.W  (A3),D0         * Get last pattern
shift1      BSR     do_test         * Run shift test

            ROL.W   #bit,D0         * Shift bits
            BCC     shift1          * Done yet? No go back
            MOVE.W  -(A3),D0        * Get test pattern 3 again
            NOT.W   D0              * Complement test pattern 3
shift2      BSR     do_test         * Run the test
            ROL.W   #bit,D0         * Shift the bits
            BCS     shift2          * Done yet? If not go back
message     MOVE.B  #pr_cmd,D0      * Load command to print banner
            LEA     string(PC),A1   * Point to message
            MOVE.W  str_len(PC),D1
            TRAP    #15             *Do it!


done        STOP    #exit_pgm       * Quit back to simulator
```

```
***********************************************************************

* Subroutine: do_test
*
* Performs the actual memory test. Fills
* the memory with the test pattern of interest.
* Registers used: D1,A0,A1,A2
* Return values: None
* Registers saved: None
* Input parameters:
* D0.W = test pattern
* A4.L = Points to memory location to save the count of bad addresses
* A5.L = Points to memory location to save the last bad address found
* A6.L = Points to memory location to save the data_read back and data
* written
*
* Assumptions: Saves all registers used internally

***********************************************************************


do_test        MOVEM.L   A0-A2/D1,-(SP) * Save registers
               LEA       st_addr,A0      * A0 points to start address
               LEA       end_addr,A1     * A1 points to last address
               MOVE.L    A0,A2           * Fill A2 will point to memory
fill_loop      MOVE.W    D0,(A2)+        * Fill and increment pointer
               CMPA.L    A1,A2           * Are we done?
               BLE       fill_loop
               MOVE.L    A0,A2           * Reset pointer
test_loop      MOVE.W    (A2),D1         * Read value back from memory
          CMP.W      D0,D1              * Are they the same?
               BEQ       addr_ok         * OK, check next location
not_ok         MOVE.L    A0,(A5)         * Save the address of the bad loca-
tion
               ADDQ.W    #byte,(A4)      * Increment the counter
               MOVE.W    D1,(A6)+        * Save the data read back
               MOVE.W    D0,(A6)         * Save the data written
               SUBQ.L    #word,A6        * Restore A6 as a pointer
addr_ok     ADDQ.L     #word,A2          * A2 points to next memory location
               CMPA.L    A1,A2           * Have we hit the last address yet?
               BLE       test_loop       * No, keep testing
               MOVEM.L   (SP)+,D1/A0-A2 * Restore registers
               RTS                       * Go back
* Data Space
```

```
test_patt DC.W        pattern1,pattern2,pattern3 * Memory test patterns
bad_cnt   DS.W 1                     * Keep track of # of bad addresses
bad_addr  DS.L 1                     * Store last bad address found here
data_read DS.W 1                     * What did I read back?
data_wrt  DS.W 1                     * What did I write?
string    DC.B 'End of test'         * Exit message
str_len   DC.W str_len-string
last_addr DS.W 1

          END         start
```

Chapter 10: Solutions for Odd-Numbered Problems

1. <0C0020h> = 15C7h. The word is aligned.

3. 0F57Ch

5.
```
        MOV CX,4
        MOV BX,10

loop1:
        inc BX
        dec CX
        jnz loop1
```

7.
```
<AX> = 0AF3DH
```

9.
```
        MOV   AX,8200H    ;Get segment value
        MOV   DX,AX       ;Load segment register
        MOV   SI,0000     ;Load source index register
        MOV   DI,0200H    ;Load destination index register
        MOV   CX,1000     ;Load counter
loader:
        MOV   AL,[SI]     ;Get byte
        MOV   [DI],AL     ;Store byte
        INC   SI          ;advance pointers
        INC   DI
        DEC   CX
        JNZ   loader
```

Chapter 11: Solutions for Odd-Numbered Problems

1. The 68K has two operational modes, *user and supervisor*. The ARM architecture allows for 7 operational modes. User mode is the lowest privilege level.. The other modes are: System, Supervisor, Abort, Fast Interrupt Request, Interrupt Request and Undefined.

3. The biggest difference is that, with the exception of registers, r13-r15, all registers are completely general-purpose. Any register may be used as part of an arithmetic operation or as an address pointer. This is in sharp contrast to the distinction that the 68K architecture makes between the address registers, A0-A6 and the data registers, D0-D7.

5. ```
   MOV     r4,#&100
   ORR     r4,r4,#3
   ```

7. `<r11> = &0013E94C`

9. If the Z flag = 0, then the value in register r1, &DEF02340, is incremented by 4 to &DEF02344 and that value is used as an address pointer to retrieve the 16-bit data object stored in that memory location. The 16-bit value is then loaded into general-purpose register r4. If the Z flag = 1, then the instruction is not executed.
   ```

Chapter 12: Solutions for Odd-Numbered Problems

1.

```
************************************************************************
*    Subroutine: xmitStr
*    Purpose: Transmits a string of characters to the UART
*    serial port.
*    Input register list:
*       A6- Pointer to the data string to be sent.
*    Return register list:
*       A6- Pointer to the character after the string terminating
*    character.
*    Register usage: All registers used by xmitStr will be saved and
*    restored upon exit
*
*    Assumptions:
*      - There is at least one character to transmit.
*      - String is terminated by $FF.
*
************************************************************************


* Data definitions

eom        EQU         $FF             *End of message character
status     EQU         $2001           *Status register
xmit       EQU         $2000           *Transmit data register
tbmt_mask  EQU         $01             *Isolate transmit buffer

* Subroutine starts here

xmitStr    MOVEM.L     D0/D1/A0/A1,-(SP) *Save the registers
           LEA.L       xmit,A0          *A0 points to transmitter
           LEA.L       status,A1        *A1 points to the status reg.
xmit_loop  MOVE.B      (A1),D1          *Get status
           ANDI.B      #tbmt_mask,D1    *Isolate bit
```

```
         BEQ          xmit_loop          *Still busy, keep waiting
         MOVE.B       (A6)+,D0           *Get byte
         CMPI.B       #eom,D0            *Last byte?
         BEQ          quit               *Yes, we're done
         MOVE.B       D0,(A0)            *Ship it
         BRA          xmit_loop          *Go back

 quit    MOVEM.L      (SP)+,D0/D1/A0/A1 *Restore the registers
         RTS
```

3. The successive approximation always takes the same number of clock cycles to digitize the unknown signal. Since it is 16 bits of resolution, it takes 16 clock cycles. Since this is a 1 MHz clock rate, it takes 16 microseconds to do the digitization.

   The single ramp A/D must count up to the unknown voltage. Therefore, we need to determine how many counts it takes to get to 1.5001 volts. However, we can easily see that the range of a 16-bit converter is 0 through 65,535 ($0000 to $FFFF in hexadecimal). Thus, the minimum voltage increment of the A/D converter is 0.0001 volts. Thus, it takes 15,001 clock cycles or 15,001 microseconds to digitize the unknown voltage.

5a. An 11-bit, 2's complement number can represent a number in the range of −1028 to +1027, so each change of 1 digital value corresponds to 0.010 volts. Anything smaller might not be detectable.

5b. Since we know that each digital code increment represents 0.01 volts, we know that +5.11 volts would be represented as 511 (5.11 volts / 0.01 volts/count = 511 counts). In binary, +511 would be 00011111111, so the 2's complement negative value (−5.11) would be 11100000001.

5c. 8.96 volts would correspond to a digital value of 896, or 01110000000. In order to properly represent this as a 16-bit number we need to add the appropriate number of leading zeros. Thus, the result is 0000 0011 1000 0000 or 0x0380.

5d. In order to digitize an 11-bit value using successive approximation, which is the hardware analogy of a binary search algorithm, we would need LOG_2 2^{11} or 11 samples.

5e. Since we take a sample on every rising edge of the clock and we need 11 samples, we need 11 rising edges. The clock frequency is 1 MHz, so the clock period is 1 microsecond. Thus, it takes 11 microseconds to digitize the analog signal.

7a. 25 microseconds = 40 KHz frequency. In order to collect 4 samples per cycle, the maximum frequency of the unknown waveform must be no greater than 10 KHz.

7b. 14 bit conversion = 1 part in 16,384. 10V/16,384 = $.000\overline{6}V$

7c. In one millisecond it droops 1 volt. in 25 microseconds it droops. (25/1000) * 1 = .025 volts. Since this is significantly greater than the 0.0006 resolution of the converter, the S/H would introduce an unacceptably large error. Thus, it can't be used.

9. Solution:
    A. f - Initialize hardware
    B. c - Confidence check
    C. e - Select channel
    D. g - S/H
    E. d - Digitize
    F. a - Wait
    G. b - Get data

    Alternative solution:
    A. c - Confidence check
    B. f - Initialize hardware
    C. e - Select channel
    D. g - S/H
    E. d - Digitize
    F. a - Wait
    G. b - Get data

Chapter 13: Solutions for Odd-Numbered Problems

1. In this particular example, Segment A would be result in better pipeline efficiency. The reason is that each of the instructions is independent of the others; there is no dependencies between them. In Segment B, each instruction must complete before the next instruction has enough information to complete. Thus, the MOVE.W D1,D0 must put the result in D0 before the ADD.W instruction can begin to operate. Likewise, MULU, can't begin until the result of the ADD operation in the previous instruction has completed. Thus, in a pipelined operation, the instructions must each complete before the next one can finish.

3. a. No, because it involves a memory to memory transfer.
    b. Yes, the addition occurs between two registers.
    c. Yes, the move is a *store* operation that transfers a register to memory.
    d. No, The AND operation takes place between an immediate value and memory.
    e. Yes, the operation is an immediate *load* operation that transfers data from memory to a register.

5. There are several RISC characteristics illustrated here. The most important RISC characteristic is that the ADD operation could only take place between data stored in the general purpose registers. Also, there was no effective addressing mode that allowed the memory address to be directly specified, the memory addresses had to be loaded as literals into the registers and then the registers were used as memory pointers. Thus, we see only two addressing modes used.

7a. Since the pipeline has seven stages, and each stage requires 2 clock cycles, then it takes 14 clock cycles for the first instruction to move down the pipeline. Since each clock cycle takes 10 nanoseconds, the total time for the first instruction is 140 nanoseconds.

7b. If we assume no stalls, after the first instruction is retired, the next 9 instructions would follow at intervals of 2 clock cycles, so we would have 9 times 20 nanoseconds, or 180 nanoseconds for the basic block to completely execute. However, the pipeline will stall twice for 4 clock cycles, this adds another 80 nanoseconds ($2 \times 4 \times 10$), so the total elapsed time is:

$$ET = 140 \text{ ns} + 180 \text{ ns} + 80 \text{ ns} = 400 \text{ ns}$$

Chapter 14: Solutions for Odd-Numbered Problems

1. a. The memory hierarchy is often represented as a pyramid with the CPU at the top. It illustrates the point that the fastest memory, but least amount of memory, is closest to the CPU and that as we get further from the CPU the amount of memory goes up, but the speed goes down. Thus, there is a reciprocal relation between the access speed of memory and the size of the memory. Also, the cost per bit goes down as you get further from the top.

   b. Spatial locality refers to the fact that instructions and data tend to be grouped together. Instructions are located in sequence and data tends to stored in clusters. For caches, this means that a cache can be much smaller than main memory but still be efficient in terms of the probability that if instructions or data are already in the cache, then it is likely that successive instructions or data will be there as well.
   Temporal locality refers to the fact that if an instruction or data was recently accessed, it is likely to be accessed soon, again. Thus, if something is in the cache and has been recently accessed, it is likely that it will be accessed again, thus improving the efficiency of the cache.

   c. With caches, we want to maximize the hit rate and minimize the miss penalty. One way to minimize the miss penalty is to refill a portion of the cache in a burst, rather than one word at a time whenever there is a cache miss. Modern SDRAM memory is designed to refill the on-chip cache in a burst of data reads, thus minimizing the penalty or reloading.

   d. A write through cache will always write the data into the cache and to main memory at the same time, thus avoiding the problem of data differences between the cache and main memory, but sacrificing some performance. The write back cache will hold the data written only to the cache and then write it to main memory when the bus is available. Performance is improved but runs the risk of memory being corrupted.

3. Spatial locality can be demonstrated in three ways.

   a. The compiled instructions occupy a very small region of memory, only 32 bytes in length. Thus, we may assume that they are located close to each other.

   b. Since the variables in the array DataStream are being accessed by de-referencing the pointer variable, DataStream + an offset value, count, the individual elements of the array must be located adjacent to each other in memory.

   c. The variables, count and maxcount are local variables to the function, main(). As such the compiler has created a stack frame on the system stack just large enough to hold two integer values, so they must also be located near each other.

Temporal locality can be demonstrated as follows:

a. Since the main part of the program is a *for loop*, the instructions in the loop are executed 11 times in a row.

b. The variables count and maxcount are repeatedly accessed because count is being incremented and compared with maxcount each time through the loop.

c. The pointer variable DataStream is repeatedly being de-referenced to place the values of count squared into successive memory locations.

5. a. Main memory has an address range of 00000...FFFFF, or 2^{20} discreet addresses. This is approximately 1 Mbytes of addresses. If each refill line has 64 bytes, or 2^6, then the number of refill lines = $2^{20} / 2^6 = 2^{14}$

$$16,384 \text{ refill lines in main memory}$$

b. The cache memory is 4096 bytes in size. Using the same method as in A, above, the number of refill lines in the cache = $2^{12} / 2^6 = 2^6$

$$64 \text{ refill lines in the cache memory}$$

c. Since this is a direct mapped cache there must be the same number of rows of refill lines in the cache memory as there are in the main memory. Therefore, the number of rows of refill lines multiplied by the number of columns of refill lines = 16,384

$$\text{Number of columns of refill lines} = 16,384/64 = 2^{14} / 2^6 = 2^8 = 256$$

$$256 \text{ columns of refill lines in the main memory}$$

d. Since there are 256 columns, the TAG memory must contain 8 bits in order to be able to address any one of the 256 columns. Thus, tag memory requires 8 bits.

e. See the below diagram:

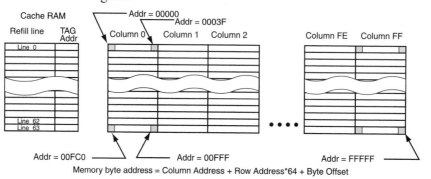

7. Effective execution time = *hit rate * hit time + miss rate * miss penalty*

Effective execution time = $.98*10 + .02 * 100 * 10 = 9.8 + 20 = 29.9$ *nsec.*

9. When the processor is initialized at start-up, all TLB entries are invalid. The validity bit is needed to know when a valid entry has been placed in the TLB or if it is just garbage.

Chapter 15: Solutions for Odd-Numbered Problems

1. Video gamers are notorious for overclocking their CPUs to gain the last ounce of performance from the machine. However, overclocking generates more heat, which slows down the internal processes, and also causes the processor to run closer to its design limits. Liquid cooling is more efficient at removing heat, so the CPU can run cooler with a higher heat load on the CPU.

3. For computer #1:
   1. Each instruction executes in 1 clock cycle, or 1/100MHz = 10 nanoseconds.
   2. It must execute a total of 1000 + 200 * 100 = 21000 instructions
   3. Total execution time = 21000 × 10 nanoseconds = 2.1×10^4 times 10×10^{-9}
      $$= 21 \times 10^{-5} = 0.210 \times 10^{-6} \text{ or } 210 \text{ microseconds.}$$

   For computer #2:
   1. It must execute the same 21000 instructions, but some take twice as long as others. Therefore 40% of the 21000 instructions take 1 clock cycle and 60% of the 21000 instructions take 2 clock cycles.
   2. At 250 MHz, 1 clock cycle takes 4 nanoseconds and 2 clock cycles take 8 nanoseconds.
   3. Therefore, the total execution time is $0.4 \times 21000 \times 4 \times 10^{-9} + 0.6 \times 21000 \times 8 \times 10^{-9}$
      $$= (8.4 \times 10^3) \times (4 \times 10^{-9}) + (12.2 \times 10^3) \times (8 \times 10^{-9})$$
      $$= (33.6 \times 10^{-6}) + (97.6 \times 10^{-6}) = 131.2 \times 10^{-6} = 131.2 \text{ microseconds.}$$

5. Cycles per instruction x seconds per clock cycle = seconds per instruction
   This is the measure we want:
   Computer #1 requires 2 cycles per instruction and each clock cycle takes 1 ns (1/1GHz).
   Therefore, *computer #1 requires 2 nanoseconds to execute 1 instruction.*
   Computer #2 requires 1.2 cycles per instruction and each clock cycle takes 2 ns (1/500MHz).
   Therefore, *computer #2 requires 2.4 nanoseconds to execute 1 instruction.*
   Thus, performance = 2.4/2.0 = 1.2. Or computer #1 has 20% better performance.

7. Analyzing this problem requires that we consider the number of accesses required for both the instructions and the actual add operation. Let's use 68000 assembly language for this example. Here's a representative code snippet:

```
MOVE.L    var1,D0        *6 bytes long
ADD.L     var2,D0        *6 bytes long
MOVE.L    D0,var3        *6 bytes long
```

Thus, the add operation requires 18 bytes to be read from, or written to, memory. The 8-bit wide bus would require 18 memory accesses and the 16-bit wide bus would require 9 accesses, so in this case, the additional 9 accesses would have to be accounted for.

Chapter 16: Solutions for Odd-Numbered Problems

1. The fuse map is shown below:

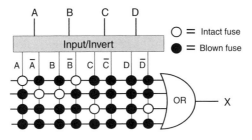

3. A circuit such as this, with a large number of stages, can generate a long sequence of pseudo-random numbers. If there are defective elements, the sequence of numbers will quickly diverge from the expected sequence if the circuitry was perfect. In a sense, this is the hardware analog of a good hashing function. Thus, any imperfection quickly generates a result that is very different from the standard.

About the Author

Arnold S. Berger is a Senior Lecturer in the Computing and Software Systems Department at the University of Washington-Bothell. He received his BS and PhD degrees from Cornell University. Prior to joining UWB, Dr. Berger was an R&D Director at Applied Microsystems Corporation, a manufacturer of specialized hardware and software tools for embedded systems developers. Prior to coming to the Pacific Northwest 5 years ago, he was the Embedded Tools Marketing Manager at Advanced Micro Devices and an R&D Project Manager at Hewlett-Packard's Logic Systems Division in Colorado Springs, Colorado.

Dr. Berger has published over 40 papers on embedded systems development methods and the tools needed to design them. He holds three patents in the area of embedded systems design tools and embedded systems simulation. He is the author of Embedded Systems Design: An Introduction to Processes, Tools and Techniques. During the two-year period prior to the Y2K date changeover, Dr. Berger consulted for the electric power industry on testing and remediation of their embedded systems.

When not teaching or consulting, Arnie is an avid cyclist, electronic hobbyist, and woodworker. His stable of three bicycles collectively log more mileage in a year than does his car.

Index

ELSEVIER SCIENCE DVD-ROM LICENSE AGREEMENT

PLEASE READ THE FOLLOWING AGREEMENT CAREFULLY BEFORE USING THIS DVD-ROM PRODUCT. THIS DVD-ROM PRODUCT IS LICENSED UNDER THE TERMS CONTAINED IN THIS DVD-ROM LICENSE AGREEMENT ("Agreement"). BY USING THIS DVD-ROM PRODUCT, YOU, AN INDIVIDUAL OR ENTITY INCLUDING EMPLOYEES, AGENTS AND REPRESENTATIVES ("You" or "Your"), ACKNOWLEDGE THAT YOU HAVE READ THIS AGREEMENT, THAT YOU UNDERSTAND IT, AND THAT YOU AGREE TO BE BOUND BY THE TERMS AND CONDITIONS OF THIS AGREEMENT. ELSEVIER SCIENCE INC. ("Elsevier Science") EXPRESSLY DOES NOT AGREE TO LICENSE THIS DVD-ROM PRODUCT TO YOU UNLESS YOU ASSENT TO THIS AGREEMENT. IF YOU DO NOT AGREE WITH ANY OF THE FOLLOWING TERMS, YOU MAY, WITHIN THIRTY (30) DAYS AFTER YOUR RECEIPT OF THIS DVD-ROM PRODUCT RETURN THE UNUSED DVD-ROM PRODUCT AND ALL ACCOMPANYING DOCUMENTATION TO ELSEVIER SCIENCE FOR A FULL REFUND.

DEFINITIONS

As used in this Agreement, these terms shall have the following meanings:

"Proprietary Material" means the valuable and proprietary information content of this DVD-ROM Product including all indexes and graphic materials and software used to access, index, search and retrieve the information content from this DVD-ROM Product developed or licensed by Elsevier Science and/or its affiliates, suppliers and licensors.

"DVD-ROM Product" means the copy of the Proprietary Material and any other material delivered on DVD-ROM and any other human-readable or machine-readable materials enclosed with this Agreement, including without limitation documentation relating to the same.

OWNERSHIP

This DVD-ROM Product has been supplied by and is proprietary to Elsevier Science and/or its affiliates, suppliers and licensors. The copyright in the DVD-ROM Product belongs to Elsevier Science and/or its affiliates, suppliers and licensors and is protected by the national and state copyright, trademark, trade secret and other intellectual property laws of the United States and international treaty provisions, including without limitation the Universal Copyright Convention and the Berne Copyright Convention. You have no ownership rights in this DVD-ROM Product. Except as expressly set forth herein, no part of this DVD-ROM Product, including without limitation the Proprietary Material, may be modified, copied or distributed in hardcopy or machine-readable form without prior written consent from Elsevier Science. All rights not expressly granted to You herein are expressly reserved. Any other use of this DVD-ROM Product by any person or entity is strictly prohibited and a violation of this Agreement.

SCOPE OF RIGHTS LICENSED (PERMITTED USES)

Elsevier Science is granting to You a limited, non-exclusive, non-transferable license to use this DVD-ROM Product in accordance with the terms of this Agreement. You may use or provide access to this DVD-ROM Product on a single computer or terminal physically located at Your premises and in a secure network or move this DVD-ROM Product to and use it on another single computer or terminal at the same location for personal use only, but under no circumstances may You use or provide access to any part or parts of this DVD-ROM Product on more than one computer or terminal simultaneously.

You shall not (a) copy, download, or otherwise reproduce the DVD-ROM Product in any medium, including, without limitation, online transmissions, local area networks, wide area networks, intranets, extranets and the Internet, or in any way, in whole or in part, except that You may print or download limited portions of the Proprietary Material that are the results of discrete searches; (b) alter, modify, or adapt the DVD-ROM Product, including but not limited to decompiling, disassembling, reverse engineering, or creating derivative works, without the prior written approval of Elsevier Science; (c) sell, license or otherwise distribute to third parties the DVD-ROM Product or any part or parts thereof; or (d) alter, remove, obscure or obstruct the display of any copyright, trademark or other proprietary notice on or in the DVD-ROM Product or on any printout or download of portions of the Proprietary Materials.

RESTRICTIONS ON TRANSFER

This License is personal to You, and neither Your rights hereunder nor the tangible embodiments of this DVD-ROM Product, including without limitation the Proprietary Material, may be sold, assigned, transferred or sub-licensed to any other person, including without limitation by operation of law, without the prior written consent of Elsevier Science. Any purported sale, assignment, transfer or sublicense without the prior written consent of Elsevier Science will be void and will automatically terminate the License granted hereunder.

TERMS

This Agreement will remain in effect until terminated pursuant to the terms of this Agreement. You may terminate this Agreement at any time by removing from Your system and destroying the DVD-ROM Product. Unauthorized copying of the DVD-ROM Product, including without limitation, the Proprietary Material and documentation, or otherwise failing to comply with the terms and conditions of this Agreement shall result in automatic termination of this license and will make available to Elsevier Science legal remedies. Upon termination of this Agreement, the license granted herein will terminate and You must immediately destroy the DVD-ROM Product and accompanying documentation. All provisions relating to proprietary rights shall survive termination of this Agreement.

LIMITED WARRANTY AND LIMITATION OF LIABILITY

NEITHER ELSEVIER SCIENCE NOR ITS LICENSORS REPRESENT OR WARRANT THAT THE INFORMATION CONTAINED IN THE PROPRIETARY MATERIALS IS COMPLETE OR FREE FROM ERROR, AND NEITHER ASSUMES, AND BOTH EXPRESSLY DISCLAIM, ANY LIABILITY TO ANY PERSON FOR ANY LOSS OR DAMAGE CAUSED BY ERRORS OR OMISSIONS IN THE PROPRIETARY MATERIAL, WHETHER SUCH ERRORS OR OMISSIONS RESULT FROM NEGLIGENCE, ACCIDENT, OR ANY OTHER CAUSE. IN ADDITION, NEITHER ELSEVIER SCIENCE NOR ITS LICENSORS MAKE ANY REPRESENTATIONS OR WARRANTIES, EITHER EXPRESS OR IMPLIED, REGARDING THE PERFORMANCE OF YOUR NETWORK OR COMPUTER SYSTEM WHEN USED IN CONJUNCTION WITH THE DVD-ROM PRODUCT.

If this DVD-ROM Product is defective, Elsevier Science will replace it at no charge if the defective DVD-ROM Product is returned to Elsevier Science within sixty (60) days (or the greatest period allowable by applicable law) from the date of shipment.

Elsevier Science warrants that the software embodied in this DVD-ROM Product will perform in substantial compliance with the documentation supplied in this DVD-ROM Product. If You report significant defect in performance in writing to Elsevier Science, and Elsevier Science is not able to correct same within sixty (60) days after its receipt of Your notification, You may return this DVD-ROM Product, including all copies and documentation, to Elsevier Science and Elsevier Science will refund Your money.

YOU UNDERSTAND THAT, EXCEPT FOR THE 60-DAY LIMITED WARRANTY RECITED ABOVE, ELSEVIER SCIENCE, ITS AFFILI-ATES, LICENSORS, SUPPLIERS AND AGENTS, MAKE NO WARRANTIES, EXPRESSED OR IMPLIED, WITH RESPECT TO THE DVD-ROM PRODUCT, INCLUDING, WITHOUT LIMITATION THE PROPRIETARY MATERIAL, AN SPECIFICALLY DISCLAIM ANY WARRANTY OF MERCHANTABILITY OR FITNESS FOR A PARTICULAR PURPOSE.

If the information provided on this DVD-ROM contains medical or health sciences information, it is intended for professional use within the medical field. Information about medical treatment or drug dosages is intended strictly for professional use, and because of rapid advances in the medical sciences, independent verification f diagnosis and drug dosages should be made.

IN NO EVENT WILL ELSEVIER SCIENCE, ITS AFFILIATES, LICENSORS, SUPPLIERS OR AGENTS, BE LIABLE TO YOU FOR ANY DAMAGES, INCLUDING, WITHOUT LIMITATION, ANY LOST PROFITS, LOST SAVINGS OR OTHER INCIDENTAL OR CON-SEQUENTIAL DAMAGES, ARISING OUT OF YOUR USE OR INABILITY TO USE THE DVD-ROM PRODUCT REGARDLESS OF WHETHER SUCH DAMAGES ARE FORESEEABLE OR WHETHER SUCH DAMAGES ARE DEEMED TO RESULT FROM THE FAILURE OR INADEQUACY OF ANY EXCLUSIVE OR OTHER REMEDY.

U.S. GOVERNMENT RESTRICTED RIGHTS

The DVD-ROM Product and documentation are provided with restricted rights. Use, duplication or disclosure by the U.S. Government is subject to restrictions as set forth in subparagraphs (a) through (d) of the Commercial Computer Restricted Rights clause at FAR 52.22719 or in subparagraph (c)(1)(ii) of the Rights in Technical Data and Computer Software clause at DFARS 252.2277013, or at 252.2117015, as applicable. Contractor/Manufacturer is Elsevier Science Inc., 655 Avenue of the Americas, New York, NY 10010-5107 USA.

GOVERNING LAW

This Agreement shall be governed by the laws of the State of New York, USA. In any dispute arising out of this Agreement, you and Elsevier Science each consent to the exclusive personal jurisdiction and venue in the state and federal courts within New York County, New York, USA.